T0276549

Multimedia Signals: Image, Audio and Video Processing

Multimedia Signals: Image, Audio and Video Processing

Editor: Anna Sanders

New York

Published by NY Research Press
118-35 Queens Blvd., Suite 400,
Forest Hills, NY 11375, USA
www.nyresearchpress.com

Multimedia Signals: Image, Audio and Video Processing
Edited by Anna Sanders

International Standard Book Number: 978-1-63238-529-1 (Hardback)

Cataloging-in-publication Data

Multimedia signals : image, audio and video processing / edited by Anna Sanders.
p. cm.
Includes bibliographical references and index.
ISBN 978-1-63238-529-1
1. Multimedia systems. 2. Signal processing--Digital techniques. 3. Digital communications. I. Sanders, Anna.
QA76.575 .M85 2017
006.7--dc23

Printed in the United States of America.

Contents

Preface

Multimedia signal processing uses computational techniques to read digital data. This book on multimedia signals explains the processes that are involved in audio and video processing, data conversion and data compression. Signals are arranged in a sequence related to time, space or frequency. Algorithms that correlate to simple digital signal processes are discussed in detail. This book is a valuable compilation of topics ranging from the basic to the most complex advancements in the field of multimedia signal processing. The aim of this text is to present researches that have transformed this discipline and aided the advancement of this field. It will be helpful to students, experts, researchers and professionals engaged in the fields of digital electronics, computer engineering and telecommunication theory.

This book has been an outcome of determined endeavour from a group of educationists in the field. The primary objective was to involve a broad spectrum of professionals from diverse cultural background involved in the field for developing new researches. The book not only targets students but also scholars pursuing higher research for further enhancement of the theoretical and practical applications of the subject.

It was an honour to edit such a profound book and also a challenging task to compile and examine all the relevant data for accuracy and originality. I wish to acknowledge the efforts of the contributors for submitting such brilliant and diverse chapters in the field and for endlessly working for the completion of the book. Last, but not the least; I thank my family for being a constant source of support in all my research endeavours.

Editor

1

Detecting fingering of overblown flute sound using sparse feature learning

Yoonchang Han[1] and Kyogu Lee[1,2*]

Abstract

In woodwind instruments such as a flute, producing a higher-pitched tone than a standard tone by increasing the blowing pressure is called overblowing, and this allows several distinct fingerings for the same notes. This article presents a method that attempts to learn acoustic features that are more appropriate than conventional features such as mel-frequency cepstral coefficients (MFCCs) in detecting the fingering from a flute sound using unsupervised feature learning. To do so, we first extract a spectrogram from the audio and convert it to a mel scale. Then, we concatenate four consecutive mel-spectrogram frames to include short temporal information and use it as a front end for the sparse filtering algorithm. The learned feature is then max-pooled, resulting in a final feature vector for the classifier that has extra robustness. We demonstrate the advantages of the proposed method in a twofold manner: we first visualize and analyze the differences in the learned features between the tones generated by standard and overblown fingerings. We then perform a quantitative evaluation through classification tasks on six selected pitches with up to five different fingerings that include a variety of octave-related and non-octave-related fingerings. The results confirm that the learned features using the proposed method significantly outperform the conventional MFCCs and the residual noise spectrum in every experimental condition for the classification tasks.

Keywords: Musical instrument, Flute overblowing detection, Unsupervised feature learning, Feed-forward neural networks, Sparse filtering, Timbre analysis, Music information retrieval

1 Introduction

Tone production on a flute involves the speed of an air jet across the embouchure and the fingering used [1]. Every pitch on the flute has its own standard fingering with an appropriate blowing pressure. However, it is possible to generate a tone that is higher in frequency than the usual tone by increasing the blowing pressure [2]; this is referred to as "overblowing" or "harmonic" by flutists. This implies that it is also possible to generate the same-pitched sound with different fingerings. For instance, a C6 tone can be produced using a C4 or C5 fingering as well as a C6 fingering. However, each fingering produces a sound with a slightly different timbre.

Although standard fingering generates a brighter and more stable tone than an overblown sound, alternative fingerings are frequently used to minimize mistakes in note transition or to add a special color to the tone, especially in contemporary repertoire [3]. On the other hand, novice players unintentionally produce overblown sounds because the fingering rules of a flute are not always consistent. In particular, most of the octave-4 and octave-5 fingerings are identical except for C, C#, D, and D#, and thus, many beginner flute players use octave-4 fingerings for octave-5 notes.

Since the timbral differences among the same-pitched notes with distinct fingerings are not obvious, music transcription systems typically focus only on detecting the onset and pitch of a note, discarding the fingering information. However, it is valuable to figure out which fingering is used to generate a specific note, particularly for musical training purposes. Handcrafted audio features such as mel-frequency cepstral coefficients (MFCCs) are commonly used for musical timbre analysis, but there is an increasing interest in learning features from data in an unsupervised manner.

This paper proposes a method that attempts to estimate the fingering of a flute player from an audio signal

* Correspondence: kglee@snu.ac.kr
[1]Music and Audio Research Group (MARG), Graduate School of Convergence Science and Technology, Seoul National University, 599 Gwanak-ro, Seoul, Republic of Korea
[2]Advanced Institutes of Convergence Technology (AICT), Suwon, Republic of Korea

based on sparse filtering (SF), which is an unsupervised feature-learning algorithm [4]. In our previous work, we demonstrated that the learned features could be used to detect the mistakes of beginner flute players [5]. However, the performance of the learned features was nearly identical to that of the MFCCs, which are among the most widely used handmade features for timbre analysis. Furthermore, the evaluation was limited in that only octave-related fingerings were considered for binary classification, and the same flute was used to create all of the sound samples, thus failing to generalize for all types of overblowing.

In this paper, we extend our previous work with a more complete evaluation by developing a fingering dataset that contains flute sound samples from various materials and makers. A major motivation for this work is to determine whether the performance of learned features shows a consistent increase in performance over conventional features under various types of real-world situations, because learned sparse features have been shown to outperform handmade features such as MFCCs for various music information tasks such as genre classification and automatic annotation [6, 7].

To avoid classification errors caused by the blowing skill of the performer, we recorded sound samples from players with a minimum of 3 years' experience. In addition, we expanded the dataset to include more target pitches up to octave 6 and experimented with up to five different fingerings for each selected note. In terms of our algorithm, we changed the input for feature learning from linear-scale spectrogram and MFCCs to a mel-scale spectrogram. We also added max-pooling to make the system more robust.

The remainder of the paper is organized as follows. We present related works in Section 2 and describe the spectral characteristics of an overblown tone in Section 3. In Section 4, we present the overall structure of the proposed system, including the preprocessing, feature-learning, max-pooling, and classification steps. In the following section, we explain the evaluation procedure, including the process of dataset construction, in detail. In Section 6, we present the results of the experiments and discussions, followed by the conclusions and directions for future work in Section 7.

The term "flute" is widely used across cultures and can refer to one of the various types of woodwind instruments. In this paper, "flute" means a modern open-hole Boehm flute, which is the most common member of the flute family and a regular member of symphony orchestras in the West [8].

2 Related work

A study by Kereliuk et al. [9] and Verfaille et al. [10] attempted to detect overblown flute fingering using the residual noise spectrum with principal component analysis (PCA) and linear discriminant analysis (LDA). This approach uses energy measurements of multiples of the fundamental frequency submultiples $F0/l$ where ($l = \{2,3,4,5,6\}$) in the first octave and a half (i.e., between F0 and 1.5F0). This is where, in their observation, the most noticeable differences appear [10]. The spectrum energy is measured using a Hann window centered on the region of interest. In addition, the researchers use a comb-summed energy measure, which simply sums the energies of the same harmonic comb, to reduce dimensionality in the experiment. In addition, the researchers recorded a flute sound from a microphone attached to the flute head joint, and only the attack segments of the notes were used in the experiment. Their proposed method allows for a detection error below 1.3 % for notes with two and three possible fingerings. However, this error dramatically increased up to 14 % for four and five possible fingerings. Although this system is specifically designed for detecting overblown flute sounds by measuring energy in the region of interest, from the experimental results, it is possible to say that the system does not capture all existing spectral differences between the sounds from each fingering. We plan to solve this limitation by replacing hand-designed features with sparse feature learning.

It is reported that using learned features has matched or outperformed handmade features in a range of different modalities [4]. Recently, there have been an increasing number of attempts to apply various feature-learning methods to the music information retrieval field. Henaff et al. applied a sparse coding algorithm to a single frame of a constant-Q transform spectrogram for musical genre classification [11], and Schülter et al. applied restricted Boltzmann machines (RBMs) to similarity-based music classification [12]. In addition, Nam et al. applied sparse RBMs to music annotation and piano transcription [13, 14], and Lee et al. used convolutional deep-belief networks for music genre and artist classification [7].

As shown in the works listed above, feature learning shows promising results in many audio and music information retrieval fields when appropriate input data, learning algorithms, and pre/postprocessing methods are used.

3 Spectral characteristics of overblown tone

Timbre is often described by musicians with adjectives such as full, rich, dull, mellow, and round [15] because it is associated with the perceived feeling of the sound derived from various spectral characteristics. A timbral difference between the sound generated by standard fingering and overblown fingering is actually very small; thus, it is difficult for non-musicians to spot the difference. As shown in Fig. 1, spectrograms of a D5 tone with proper D5 fingering and D4 fingering look quite alike. However, flute experts can easily distinguish the difference. Usually, a standard tone is described as "clearer," and an overblown flute sound is slightly more "airy" than the original tone. The sound of a regular flute tone is highly sine-wave-like, and only multiples of F0 are visibly strong, while the other part of the spectrum has very low energy. The "airy" timbre of the overblown tone is mainly caused by the spectral energy existing at multiples of other than F0. It is difficult, but still possible, to observe from Fig. 1 that the spectrum of the overblown sound has strong peaks at multiples of F0 (587 Hz) as well as minor energy around multiples of F0 of the original fingering (293 Hz) between F0 and 2F0, and 2F0 and 3F0. Interestingly, this tendency is not clear between 3F0 and 4F0 but appears again between 4F0 and 5F0 for a D5 tone. As shown above, it is certain that an airy timbre is caused by residual noise at multiples of other than F0, but it is difficult to set a concrete handmade rule for the spectral characteristics to distinguish every fingering.

4 System architecture

We perform flute-fingering detection using the system architecture presented in Fig. 2. The details of each block in the diagram, including the parameter values, are explained in this section.

4.1 Preprocessing

Although sparse feature learning effectively learns the feature from the data, appropriate preprocessing is a prerequisite to obtaining a good feature. In addition, we need to decrease the dimensionality as much as possible without losing important parts of the information. We perform several steps for preprocessing, as described below.

4.1.1 Downsampling

In the first preprocessing step, the input audio is downsampled to 22,050 Hz from the original 44,100-Hz sampling rate. Since now the Nyquist frequency of the input signal is 11,025 Hz, we can get rid of unnecessary noisy information above this frequency while retaining enough numbers of harmonics for the highest note of the flute. This experiment covers frequencies up to the seventh harmonic of G6.

4.1.2 Time-frequency representation

We compute a discrete Fourier transform (DFT) for the spectrogram of the input audio with a 25-ms window and a 10-ms hop size. Then, its linear frequency scale is converted to a mel scale, and the spectral magnitude is compressed by a natural logarithm. We chose 128 for the number of mel-frequency bins, following Hamel's [16] and Nam's work [14]. By using a moderate number of mel-frequency bins instead of a linear-scale DFT result, it is possible to reduce the input dimension

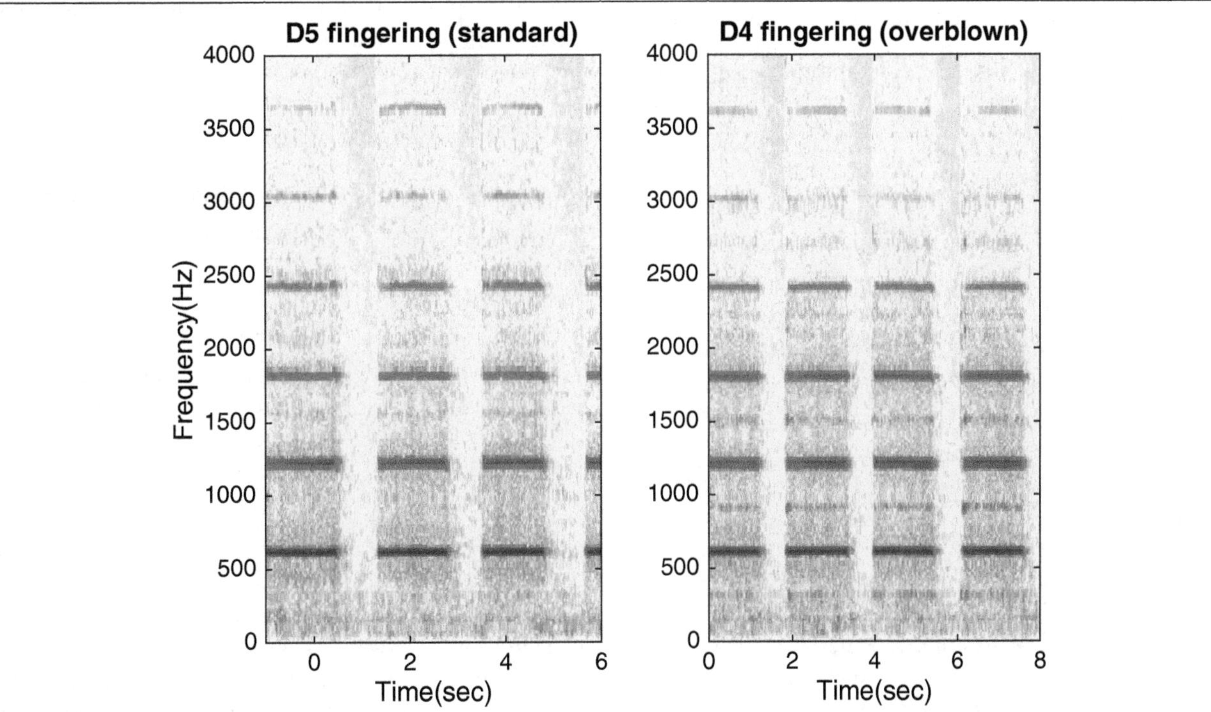

Fig. 1 Log spectrogram example of standard and overblown tone. Standard tone is D5 tone played with regular D5 fingering, and overblown tone is D5 tone played with D4 fingering with a sharper and stronger air jet

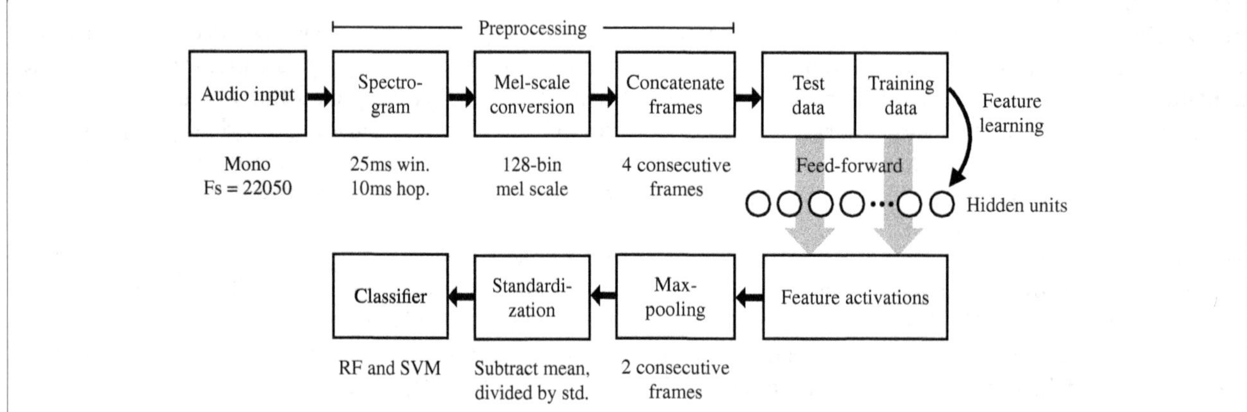

Fig. 2 Overall schematic of proposed flute fingering detection system. The system takes an audio waveform as an input, and four consecutive frames of a 128-bin mel-scale spectrogram are concatenated to learn timbral features. Note that features are learned only from training data. Obtained feature activations are max-pooled and standardized prior to training the classifier

significantly while preserving the audio content effectively enough. This is very helpful for decreasing the overall computational complexity of the system because the DFT of an audio signal generates a high-dimensional vector, and the next frame-concatenation step multiplies the number of feature dimensions.

4.1.3 Frame concatenation

As a final stage of preprocessing, we concatenate several consecutive frames as a single input for feature learning. This process can be seen as learning temporal information within the size of a concatenation for feature learning and to make the system more robust than when we use a single frame. We concatenate four frames as a single-input example. Thus, the time resolution of the data in this step is now 70 ms according to the window and hop size mentioned above.

4.2 Feature learning and representation

We learn sparse features from the preprocessed data described above. This section explains the sparse filtering algorithm, activation function, and details about how we generated the sparse feature representation.

4.2.1 Unsupervised feature learning

Recent efforts in machine learning have sought to define a method to automatically learn higher-level representations of input data [17]. Such an approach is extremely valuable, particularly when designing features by hand is challenging, and the timbre is an example where it is difficult to determine a distinctive difference in a sound spectrum by observation. Many choices are available for unsupervised feature learning, such as restricted Boltzmann machines [18], autoencoder [19], and sparse coding [20]. These approaches have been successfully applied to a variety of modalities, but they require extensive tuning of the hyperparameters [4].

From these approaches, we chose to employ sparse filtering as a feature-learning algorithm in the proposed system because it has only one hyperparameter (the number of features to learn) and converges more quickly than other algorithms, especially when the number of input data and the feature dimension are large. This characteristic is suitable for our task because it can be easily implemented in a real-time score-following system for mobile applications without fine-parameter tuning for each device and environment. Furthermore, a short-time Fourier analysis of audio wave inherently generates a significant number of input data.

4.2.2 Sparse filtering algorithm

Sparse filtering first normalizes each feature to be uniformly active in total by dividing each feature by its ℓ_2-norm throughout all examples as follows:

$$f_j' = f_j / \|f_j\|_2 \tag{1}$$

where f_j represents the jth feature value. In a similar manner, it then normalizes each example by dividing each example by its ℓ_2-norm across all features as follows:

$$\tilde{f}^{(i)} = f'(i) / \|f'(i)\|_2 \tag{2}$$

where $f^{(i)}$ represents the ith example. By computing Eq. 1 and Eq. 2, now all values lie in the unit-ℓ_2 hypersphere. This feature normalization introduces competition between features. For example, if one component of $f^{(i)}$ is increased, other components will decrease because of the normalization. Finally, the normalized features are optimized for sparseness using an ℓ_1 penalty, and it can be written as

$$\text{minimize} \sum_{i=1}^{N} \left\| \tilde{f}^{(i)} \right\|_1 = \sum_{i=1}^{N} \left\| \frac{f'(i)}{\|f'(i)\|_2} \right\|_1 \qquad (3)$$

for a dataset of N examples. Consequently, each example is represented by a small number of active sparse units, and each feature is active only for a small number of examples at the same time. A more detailed description of sparse filtering can be found in [4].

As an activation function, we used t soft-absolute function shown below:

$$f_j^{(i)} = \sqrt{\epsilon + \left(w_j^T x^{(i)}\right)^2} \approx \left|w_j^T x^{(i)}\right| \qquad (4)$$

where w is the weight matrix, x is the visible node (i.e., input data), and we set $= 10^{-8}$. An off-the-shelf L-BFGS [21] package is used to optimize the sparse filtering objective until convergence.

4.2.3 Learning strategy and feature representation

We obtained a weight matrix from training data such that the feature activation of the test data can be simply obtained by a feed-forward process. As mentioned above, sparse filtering only has one parameter to tune: the number of features to learn. To determine the effect of hidden unit size on the overall detection accuracy, we used 39, 512, and 1024 for the number of hidden units. The top 10 most active feature bases for six different notes (C5, D5, C6, C#6, E6, and G6) are shown in Fig. 3. It is possible to observe that harmonic partials of the mel spectrum are distributed into bases.

4.3 Max-pooling

Using short-term features provides high time resolution but is prone to local short-term errors. Max-pooling is the process that takes the largest value in the region of interest such that only the most responsive units are retained. We apply max-pooling over the time domain to every two non-overlapping consecutive frames. This temporal maximum value-selection process is included to add an extra robustness to the system and to reduce the computational complexity of the next step, a classifier.

It is important to note that max-pooling is a nonlinear downsampling; hence, the time resolution becomes half of the original resolution when it is pooled every two frames. In the proposed system, the original time resolution of the spectrogram is 25 ms, and it becomes 70 ms after we concatenate four frames with a hop size of 10 ms. Since we are pooling over two frames, the time resolution after max-pooling is 140 ms. Increasing the size of the max-pooling region might increase the detection accuracy further, but this is an appropriate resolution for flute transcription because a typical sixteenth note with 100 beats per minute (BPM) is 150 ms, which is still greater than our frame size.

4.4 Classification

There are a variety of choices available for classifier. The support vector machine (SVM) and random forest (RF) are both highly popular classifiers across various applications. The underlying idea behind the SVM is to calculate a maximal margin hyperplane that performs a binary classification of the data [22]. By contrast, RF is an algorithm that uses a combination of decision trees that have a randomly selected subset of variables [23].

The performances of the SVM and RF are a highly arguable topic, and there is significant variability in their problems and metrics [24]. We use both SVM and RF as classifiers to compare the performance. We first standardize the values by subtracting the mean and divide them by the standard deviation prior to feeding the data into the classifier. Thus, the data have their mean at zero and standard deviation at one. We use a radial basis function (RBF) kernel for SVM, and the number of trees for RF is set as 500.

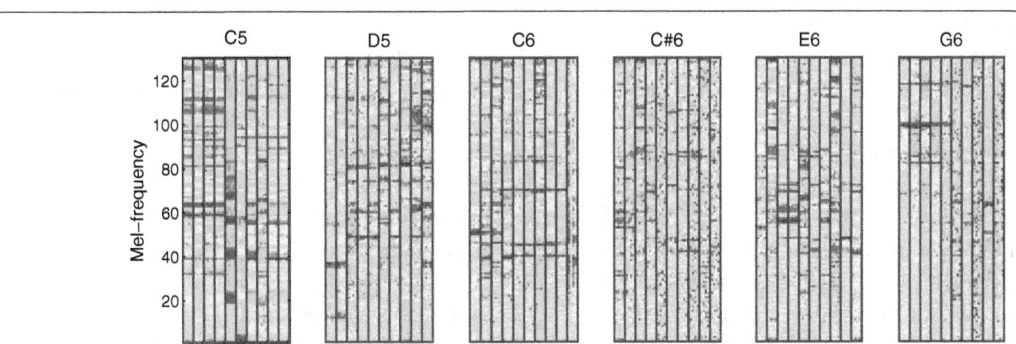

Fig. 3 Top 10 most active feature bases of each note. It is possible to observe that harmonic partials are distributed into bases. The bases are learned from sparse filtering with 39 hidden units

Table 1 Selected fingering set

Played note	Fingerings
C5	C4, C5
D5	D4, D5
C6	C4, C6, F4
C#6	C#4, C#6, F#4
E6	A4, C4, E4, E6
G6	C4, C5, E♭4, G4, G6

Played note is a pitch of the note, and each pitch is played with different fingerings by altering only the blowing pressure. The evaluation is composed of six independent classification problems, and the number of classes is the number of fingerings

5 Evaluation

5.1 Data specification

We created a dataset that resembles the one in a previous study by Verfaille et al. [10]. This includes diverse overblown fingering cases such as octave-related to non-octave-related fingerings and "few keys changing" to "many keys changing" positions, as illustrated in Table 1. We recorded flute sounds from two expert performers with more than 20 years of experience and two intermediate performers with 3 years of experience. Each performer played six pitches with all possible fingerings as they appeared in Table 1. The average length of each tone was 16 s, excluding noises and silences. The length of the audio recorded from each performer was 305 s on average, and the total audio length was 1218 s. For flute tones, performers sustained the target tone for each fingering without vibrato, special articulations, or melodic phrases. We extracted 122 K frames of spectrogram in total using a 25-ms window and 10-ms hop size before frame concatenation and the max-pooling step.

The flutes used for the recording were modern Boehm flutes with B foot joints made by multiple flute makers and of different materials. Two intermediate players used silver heads with nickel body joints (Yamaha). One expert used a sterling silver flute (Powell), and another expert used a rose gold flute (Brannen-Cooper). Note that only the first nickel body flute had a "split E" mechanism, which facilitates the production of E6. Every flute used in the experiment was an open-hole flute without any cylindrical acrylic plugs.

Flutes made of several different materials are used for the experiment because not every flutist uses a golden flute, and the general timbre of the flute changes according to the material used in its construction. For instance, the flute made of gold is often described as having a fuller, richer, and more liquid timbre, while the silver flute is more delicate and shrill in the loud and high tones [25].

Although we used the same fingering set as [10], our data is recorded without attaching a microphone to the flute head joint. To simulate a real-world situation, audio samples were recorded using the built-in microphone of a Samsung NT900 laptop rather than by using high-end studio microphones. Flute sounds from four different performers were all recorded in different locations. Three performers played their flutes in their apartment living rooms, and one performer played the flute in a small office. The dataset is available to download from (http://marg.snu.ac.kr/DFFD), which includes the audio files with annotations.

5.2 Experiments

A stratified eightfold cross validation was used to evaluate the performance of the proposed method. As shown in Table 1, our experiment is composed of six independent classification problems, and the number of classes for each note is the number of fingerings. First, we collected flute tones from every performer and partitioned the data into eightfold that were proportionally representative of each class. Then, onefold was used for a test; the remaining folds were used for training. Note that this procedure was performed frame-wise, and the order of the frames was random. In addition, each fold includes the audio recorded from flutes of various materials and in various recording environments. The mean and standard deviation of the detection error probability P_f (i.e., inverse accuracy) was calculated eight times via repeated cross validation for each fingering of each selected pitch.

We also evaluated MFCCs to compare them with our proposed feature representation. To make a fair comparison, we used the same experiment settings, such as identical frame and hop sizes, concatenating four frames as a single example and standardizing to have a zero mean and unit variance. Thirteen-dimensional MFCCs were used with delta and double delta for a total of 39 dimensions. As mentioned previously, we evaluated the performance of sparse features with 39 hidden units for a fair comparison with MFCCs, and 512 and 1024 units are also evaluated in order to determine the effects of increased hidden unit size on the overall classification accuracy.

6 Results and discussion

The detection accuracy of the proposed flute-fingering detection method and the effect of parameters, classifiers, and max-pooling are presented in this section. We compare the performance of sparse feature representation with existing approaches such as MFCCs and the PCA/LDA method.

6.1 Comparison with MFCCs

In general, the proposed algorithm outperformed MFCCs for every fingering set in the same 39-dimensional setting as indicated in Table 2. The worst detection error rate of

Table 2 Eightfold cross-validation result of the fingering detection system

		SVM		Random forest	
Note	Feature (dim.)	$P_f \pm \sigma^2_{Pf}$	$P_f \pm \sigma^2_{Pf}$ (Mp)	$P_f \pm \sigma^2_{Pf}$	$P_f \pm \sigma^2_{Pf}$ (Mp)
C5	MFCCs (39)	9.87 ± 1.57	8.07 ± 1.00	14.91 ± 1.91	12.20 ± 1.05
	SF (39)	0.55 ± 0.47	0.20 ± 0.27	0.68 ± 0.52	0.13 ± 0.24
	SF (1024)	0.26 ± 0.24	0.00 ± 0.00	0.36 ± 0.24	0.07 ± 0.19
D5	MFCCs (39)	7.43 ± 1.30	8.21 ± 0.82	10.16 ± 1.59	10.20 ± 2.42
	SF (39)	2.20 ± 0.78	1.41 ± 1.12	1.48 ± 0.54	0.40 ± 0.51
	SF (1024)	1.67 ± 0.72	1.18 ± 0.68	1.00 ± 0.36	0.39 ± 0.38
C6	MFCCs (39)	8.47 ± 0.86	6.91 ± 1.48	11.78 ± 1.59	8.91 ± 0.82
	SF (39)	2.08 ± 0.88	1.50 ± 0.80	1.30 ± 0.50	0.67 ± 0.40
	SF (1024)	0.76 ± 0.42	0.13 ± 0.17	0.56 ± 0.34	0.04 ± 0.12
C#6	MFCCs (39)	7.01 ± 0.62	5.47 ± 0.90	9.31 ± 1.01	6.02 ± 0.98
	SF (39)	1.39 ± 0.72	1.07 ± 0.82	1.26 ± 0.52	0.56 ± 0.33
	SF (1024)	0.31 ± 0.29	0.08 ± 0.15	0.29 ± 0.27	0.04 ± 0.11
E6	MFCCs (39)	6.36 ± 0.78	4.56 ± 0.90	7.43 ± 0.96	4.80 ± 1.27
	SF (39)	2.35 ± 0.79	1.74 ± 0.76	2.03 ± 0.46	1.31 ± 0.51
	SF (1024)	1.06 ± 0.48	1.11 ± 0.30	1.02 ± 0.37	0.34 ± 0.24
G6	MFCCs (39)	13.80 ± 1.14	12.77 ± 1.28	13.76 ± 1.12	10.63 ± 0.96
	SF (39)	5.71 ± 1.30	4.49 ± 0.76	4.62 ± 1.26	3.22 ± 0.58
	SF (1024)	1.53 ± 0.53	1.06 ± 0.54	1.55 ± 0.36	1.06 ± 0.44

Mean and standard deviations of detection error ($P_f \pm \sigma^2_{Pf}$) probabilities (%) using MFCCs and sparse filtering (SF) for 39 and 1024 units. SVM and random forest were used for the classifications, and probabilities are given for features without and with max-pooling (Mp)

the MFCCs occurred when P_f = 14.91 ± 1.91 % for C5 with RF; the highest error rate of the proposed method occurred when P_f = 5.71 ± 1.30 % for G6 with SVM.

The proposed system does not use PCA for orthogonalization or dimension reduction. However, the first and second highest eigenvalues of PCA are computed from MFCCs and sparse features in order to visualize two-dimensional feature representation, as shown in Fig. 4. In addition, we also present a visualization using the t-SNE algorithm described in [26] in Fig. 4. This algorithm is capable of learning high-order dependencies. From this figure, it is possible to observe that MFCCs overlap significantly for all fingerings, whereas the generated sparse feature, by comparison, overlaps much less.

6.2 Effect of max-pooling

Performing max-pooling on the obtained feature activation visibly improved the performance for both MFCCs and sparse features, as shown in Table 2. It can also be observed from Fig. 4 that adding max-pooling effectively removes a significant amount of feature noise. We perform max-pooling over two consecutive frames. Increasing the pooling size might increase the performance further, but we decided not to increase the pooling size in order to keep the time resolution smaller than the length of a sixteenth note at 100 BPM (150 ms).

6.3 Effect of the number of hidden units

Although we used 39 dimensions for a fair comparison with the MFCCs, we evaluated higher values for the number of hidden units to determine their effects. Using additional hidden units means that more bases are used to describe the input signal. This effectively decreased the detection error rate. The classification performance nearly approached saturation at 256 units with a slight improvement up to 1024 units, as shown in Fig. 5.

6.4 Effect of classifier

It is interesting to note that RF demonstrated generally matched or better performance for sparse features, and SVM exhibited superior performance for MFCCs, as shown in Fig. 6. For example, RF showed a lower error rate for D5, C6, C#6, E6, and G6 for SF, while SVM performed better for C5, D5, C6, C#6, and E6 for MFCCs. However, the performance gap between RF and SVM for MFCCs becomes smaller as the number of possible fingerings increases, while RF constantly returns a better performance for sparse features. We can conclude from this result that RF is more suitable for classifying sparse features, especially for multiclass tasks, and that using SVM is marginally better for MFCCs when there exist only a few possible fingerings.

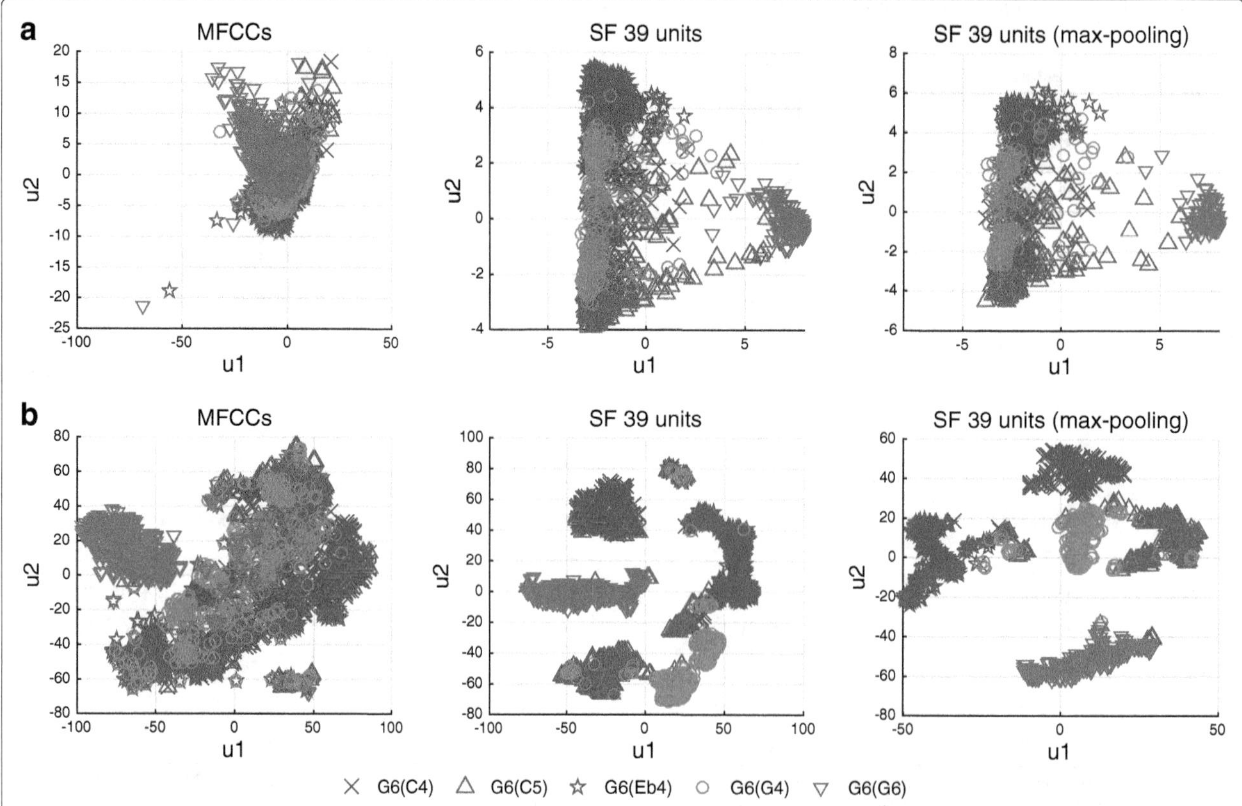

Fig. 4 MFCCs and sparse features with and without max-pooling projected onto a factorial map. The factorial map is built from the first and second highest eigenvalues of PCA and t-SNE (u_1, u_2) for visualization purposes. **a** PCA factorial map of five fingerings (C4, C5, E♭4, G4, G6) producing a G6 tone. **b** t-SNE factorial map of the same tones. It is possible to observe that sparse features are "cleaner" features for flute fingerings than the MFCCs. In addition, max-pooling removes a considerable amount of noise from the feature

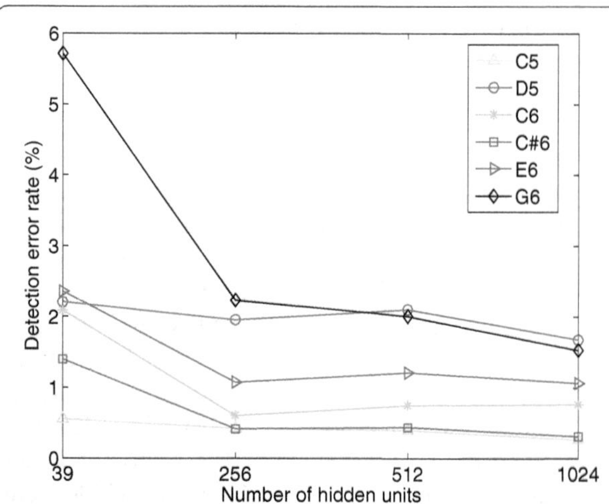

Fig. 5 Effect of number of hidden units. It is possible to observe that the detection error rate decreases as the number of hidden units increases. This nearly approaches saturation at 256 units, with minor improvement up to 1024 units

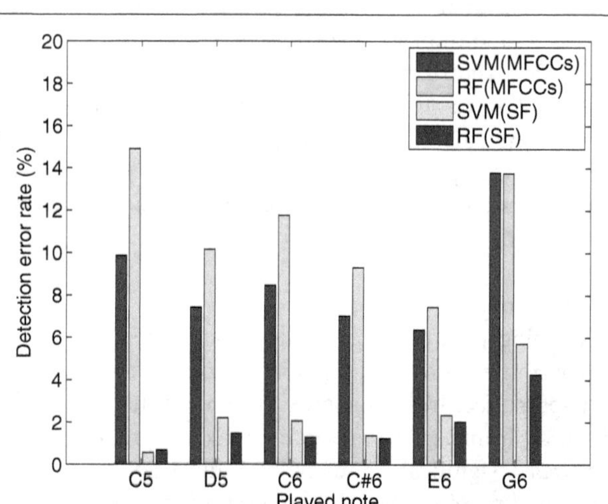

Fig. 6 Effect of classifier. In general, SVM and RF showed better performance for MFCCs and SF, respectively. However, using RF returned an improved result for both features as the number of possible fingerings increase

6.5 Comparison with PCA/LDA method

A direct comparison of the PCA/LDA method from Verfaille et al. [10] with the proposed method would be inappropriate because Verfaille recorded sounds from a microphone attached to the flute head joint and used the mean energy measured from the initial 100 ms of the attack segment only. In our proposed system, we used entire notes rather than attack segments because attack parts do not exist when the notes are played with a slur during the actual flute performance. Further, to simulate a real-world smart-device application, we used a laptop's built-in microphone at a distance to record, rather than attaching an extra microphone to the flute. However, it is meaningful to observe that the error probabilities of the proposed method were less than 5.71 % for 39 sparse feature units and less than 1.11 % for 1024 units with max-pooling applied to four- and five-fingering configurations. The error rates of the PCA/LDA method were below 1.3 % for two and three fingerings; however, these rates dramatically increased to 13.3 % for four and five fingerings in the eightfold cross validation.

This performance improvement was a consequence of the spectral differences between signals successfully captured by sparse feature learning, as visualized in Fig. 7. From the magnitude response plot of C#6 in Fig. 7, it can be observed that the eighth basis is activated for C#4 fingering; however, the eighth basis was not activated for C#6 and F#4. We can see that this eighth basis effectively captures the spectral characteristics of the C#4 fingering between the first and third octave and is particularly clear around 1348 and 1905 Hz, which are annotated with blue circles. Similarly, 1855 Hz of the 32nd basis, annotated with a red circle, is significantly activated for C#4 and F#4 fingering; however, it is not significantly activated for C#6 fingering. This occurs because the sounds from C#4 and F#4 fingerings have strong energy in this region with a peak at 1905 and 1806 Hz, respectively, whereas C#6 has no significant energy in this region.

The handmade feature of [10] uses energy measurements of the multiples of the fundamental frequency submultiples between the first octave and a half (i.e., F0/l where $l = \{2,3,4,5,6\}$ between F0 and 1.5F0) because this is where different acoustical characteristics appear the most significant in their observation, as mentioned above. The proposed system achieved improved performance by learning spectral characteristics with sparse features and describing the sound spectrum with activations of learned bases rather than by restricting a region of interest and measuring energies at certain points.

7 Conclusions

We designed a flute-fingering detection system based on the sparse feature learning method. The results obtained in this study indicate that the learned sparse features delivered improved performance compared with other conventional features for flute-fingering detection, especially as the number of possible fingerings increased. The performance gap between the MFCCs and sparse features for flute-fingering detection was not significant in our previous study [5]; however, the more complicated task with a larger dataset described in this paper confirmed that the learned sparse features were able to capture the

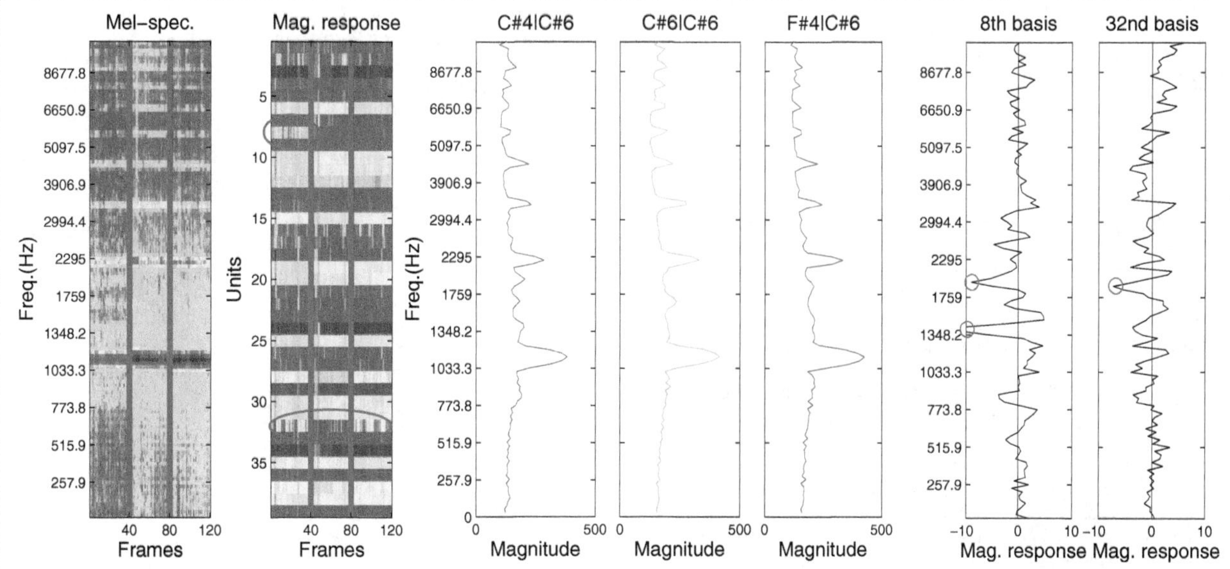

Fig. 7 Mel spectrogram, feature activations, mean mel spectrum, and selected basis of all possible C#6 fingerings. Mel-spectrogram and feature activations of C#4, C#6, and F#4 fingerings are separated by red vertical lines. Unique feature activations for each fingering and important spectral peaks of basis are annotated with circles

spectral differences among the same-pitched notes with distinct fingerings and outperformed the conventional features such as MFCCs that focus on the spectral envelope.

The proposed method achieved a detection error rate of less than 5.71 % for all cases, while the error rate of the existing PCA/LDA method dramatically increased to 13.3 % for four and five fingerings. Increasing the number of units and adding max-pooling to the generated feature further improved the performance and achieved error rates of up to 1.18 % for all cases. In addition, a comparison of classifier performance between SVM and RF showed that RF is generally more suitable for sparse features and is more robust against the number of classes.

The proposed system is well suited for potential use as a flute-fingering detection application for flute education. The window/hop size for the spectrogram and the max-pooling size for the feature activation were determined while considering real-time flute transcriptions. This could be used in multiple recording environments with mobile devices because the system does not require excessive computational power or extensive parameter tuning related to the recording environment. In addition, this framework can be easily applied to other instruments or timbre analysis tasks with minor changes because it does not use the deterministic rule but learns the differences from the input signal.

Competing interests
The authors declare that they have no competing interests.

Acknowledgements
This work was supported partly by the Ministry of Science, ICT (Information and Communication Technologies) and Future Planning by the Korean Government (NRF-2014K2A2A6049835), and partly by a National Research Foundation of Korea (NRF) grant funded by the Korean government (MSIP) (No. 2014R1A2A2A04002619).

References

1. JW Coltman, Sounding mechanism of the flute and organ pipe. J. Acoust. Soc. Am. **44**, 983–992 (1968)
2. J Wolfe, J Smith, J Tann, NH Fletcher, Acoustic impedances of classical and modern flutes. J. Sound Vib. **243**, 127–144 (2001)
3. JR Smith, N Henrich, J Wolfe, The acoustic impedance of the Boehm flute: standard and some non-standard fingerings. Proc. Inst. Acoust. **19**, 315–330 (1997)
4. J Ngiam, PW Koh, Z Chen, S Bhaskar, AY Ng, Sparse filtering, in *Proceedings of the Advances in Neural Information Processing Systems 18 (NIPS'11)*, 2011, pp. 1125–1133
5. Y Han, K Lee, Hierarchical approach to detect common mistakes of beginner flute players, in *Proceedings of the 15th International Society for Music Information Retrieval Conference (ISMIR'14)*, 2014, pp. 77–82
6. P Hamel, D Eck, Learning features from music audio with deep belief networks, in *Proceedings of the 11th International Society for Music Information Retrieval Conference (ISMIR'10)*, 2010, pp. 339–344
7. H Lee, P Pham, Y Largman, AY Ng, *Unsupervised feature learning for audio classification using convolutional deep belief networks*. Adv. Neural Inf. Process Syst, 2009, pp. 1096–1104
8. SJ Maclagan, *A dictionary for the modern flutist* (Scarecrow Press, Plymouth, 2009), p. 19
9. C Kereliuk, B Scherrer, V Verfaille, P Depalle, MM Wanderley, Indirect acquisition of fingerings of harmonic notes on the flute, in *Proceedings of the International Computer Music Conference (ICMC'07)*, 2007, pp. 263–266
10. V Verfaille, P Depalle, MM Wanderley, Detecting overblown flute fingerings from the residual noise spectrum. J. Acoust. Soc. Am. **127**, 534–541 (2010)
11. H Henaff, K Jarrett, K Kavukcuoglu, Y LeCun, Unsupervised learning of sparse features for scalable audio classification, in *Proceedings of the 12th International Society for Music Information Retrieval Conference (ISMIR'11)*, 2011, pp. 77–82
12. J Schülter, C Osendorfer, Music similarity estimation with the mean-covariance restricted Boltzmann machine, in *Proceedings of the 10th International Conference on Machine Learning and Applications (ICMLA'11)*, 2011, pp. 118–123
13. J Nam, J Ngiam, H Lee, M Slaney, A classification-based polyphonic piano transcription approach using learned feature representations, in *Proceedings of the 12th International Society for Music Information Retrieval Conference (ISMIR'11)*, 2011, pp. 175–180
14. J Nam, J Herrera, M Slaney, JO Smith, Learning sparse feature representations for music annotation and retrieval, in *Proceedings of the 13th International Society for Music Information Retrieval Conference (ISMIR'12)*, 2012, pp. 565–570
15. KW Berger, Some factors in the recognition of timbre. J. Acoust. Soc. Am. **36.10**, 1888–1891 (1964)
16. P Hamel, S Lemieux, Y Bengio, D Eck, Temporal pooling and multiscale learning for automatic annotation and ranking of music audio, in *Proceedings of the 12th International Society for Music Information Retrieval Conference (ISMIR'11)*, 2011, pp. 729–734
17. A Coates, B Carpenter, C Case, S Satheesh, B Suresh, T Wang, DJ Wu, AY Ng, Text detection and character recognition in scene images with unsupervised feature learning, in *Proceedings of the International Conference on Document Analysis and Recognition (ICDAR'11)*, 2011, pp. 440–445
18. GE Hinton, S Osindero, YW Teh, A fast learning algorithm for deep belief nets. Neural Comput. **18**, 1527–1554 (2006)
19. P Vincent, H Larochelle, Y Bengio, PA Manzagol, Extracting composing robust features with denoising autoencoders, in *Proceedings of the 25th International Conference on Machine Learning (ICML'08)*, 2008, pp. 1096–1103
20. BA Olshausen, DJ Field, Sparse coding with an overcomplete basis set: a strategy employed by V1? Vis. Res. **37**, 3311–3325 (1997)
21. M Schmidt. minFunc. http://www.cs.ubc.ca/~schmidtm/Software/minFunc.html, 2005.
22. A Statnikov, L Wang, CF Aliferis, A comprehensive comparison of random forests and support vector machines for microarray-based cancer classification. BMC Bioinf. **9.1**, 319 (2008)
23. L Breiman, Random forests. Mach. Learn. **45.1**, 5–32 (2001)
24. R Caruana, A Niculescu-Mizil, An empirical comparison of supervised learning algorithms, in *Proceedings of the 23rd International Conference on Machine Learning (ICML'06)*, 2006, pp. 161–168
25. DC Miller, The influence of the material of wind-instruments on the tone quality. Science **29**, 161–171 (1909)
26. LJP van der Maaten, GE Hinton, Visualizing high dimensional data using t-SNE. J. Mach. Learn. Res. **9**(Nov), 2579–2605 (2008)

2

iSargam: music notation representation for Indian Carnatic music

Stanly Mammen[1*], Ilango Krishnamurthi[1], A. Jalaja Varma[2] and G. Sujatha[3]

Abstract

Indian classical music, including its two varieties, Carnatic and Hindustani music, has a rich music tradition and enjoys a wide audience from various parts of the world. The Carnatic music which is more popular in South India still continues to be uninfluenced by other music traditions and is one of the purest forms of Indian music. Like other music traditions, Carnatic music also has developed its musicography, out of which, a notation system called Sargam is most commonly practiced. This paper deals with development of a music representation or encoding system for the Sargam notation scheme which enables easy music notation storage, publishing, and retrieval using computers. This work follows a novel idea of developing a Unicode-based encoding logic and allows storage and easy retrieval of music notation files in a computer. As opposed to many existing music representation systems for western music notation, iSargam is the only music notation encoding system developed for Indian music notation.

Keywords: Music representation, Music encoding, Music processing, Music information retrieval, Indian music, Computer music

1 Introduction

Music information representation systems, in general, encompass musical information contained in music notation or recordings. Since the whole purpose of music representation is computer analysis, retrieval, or synthesis, most of the representation systems focused on representation of musical notation, particularly, the Common Western Notation (CWN), because of its effectiveness in representing discrete elements of music. Thus, many systems like ENP [1], LilyPond [2], Humdrum [3], Guido [4], and Cadenza [5] used ASCII-based textual descriptions of western music notations. Similarly, music representations for non-CWN systems like Gregorian chant, Django for tabulature, GOODFEE for Braille notation etc. also developed [1]. An example of a Gregorian chant is given in Fig. 1 [6]. Another recent trend was the development of Web-viewable notation applications such as Scorch by Sibelius software and ScoreSVG [7].

The textual descriptions of music notations can be further classified as record-based, command-based, symbolic codes, and LISP-based. While systems like DARMS [8], Guido [4] etc. used symbolic codes, systems like CMN [9] used command-based representations. A few examples of LISP-based representations are MUZACS [10], Rhythm-Editor of Patch Work [11], and CMN [9]. Also, to take advantage of XML features like structuring and portability, many popular XML-based notations like MusicXML [12], MEI [13] [14], and WEDELMUSIC [15] emerged.

2 Background and history

Indian classical music is one of the oldest music traditions in the world, and it enjoys the next position to western music in its popularity. The music tradition of India can be divided into two large traditions, namely, Hindustani and Carnatic music. The Hindustani classical music which is predominant in the northern part of the Indian subcontinent, originates from the ancient Vedic, Persian, and many folk traditions. While the Carnatic classical music, uninfluenced by non-Indian music traditions, is the purest form of Indian music and is prevalent in the southern parts of the Indian subcontinent. It is generally homophonic in nature with emphasis on vocal music. If performed on an instrument, it assumes a singing style.

Carnatic music is usually performed by an ensemble of musicians consisting of a principal performer, usually a

* Correspondence: stanly.mammen@gmail.com
[1]Department of Computer Science and Engineering, Sri Krishna College of Engineering and Technology, Coimbatore, Tamil Nadu, India
Full list of author information is available at the end of the article

Fig. 1 Example of a Gregorian chant, an excerpt from "Antiphonale Monasticum" by Solesmes, licensed under Creative Commons [36]

vocalist, accompanied by a rhythm instrument, melodic instrument, and a monophonic drone instrument.

The tradition of using music notations for singing was practiced from Vedic times. But there was no uniformity in the notation system used from time to time. However, all the notations developed so far can be classified as script as opposed to staff notation used for western music. Different systems of notation were prevalent in each period of history in both Carnatic and Hindustani traditions. Also, many notation systems existed in parallel.

In regard to Carnatic music, there were two popular notation systems, viz, the old notation and modern notation. The old notation, as the name implies, can be found in many ancient books, including the "Sangita Sampradaya Pradishini" [16] written by the late Subbarama Diksitar, a popular music theorist. Currently, a new music notation system, namely, modern notation or Sargam notation is widely used for notating Carnatic music in various books, reviews, and in academia. The Sargam notation system uses a subset of symbols used in old notation. A Carnatic music composition notated using Sargam notation is given in Fig. 2.

The main objective of this work is to develop a unified representation system for storing Indian Carnatic music notations in computer files. The system aims at encoding music notation symbols and other associated information like title, composer name etc., while ignoring the layout specific details. Most of the encoding systems developed so far are intended at encoding western music staff notation or a derivate of it. This work aims at developing a machine readable music notation system for Carnatic music which can support playback, printing, retrieval, and searching within the composition.

JATISWARAM Raga – Sankarabharana Tala – Rupaka Mela 29.			{s r g m p d n s {ṡ n d p m g r s
O	l_4	o	l_4
Pallavi			
Ṡ;	; Ṙ Ṡ N	D P	; M G M
P ;	; Ṡ D P	M G	R g m P d n (Ṡ ; ;)
Anupallavi			
P ;	; dp M p m	G m g	R g r S s n
S ;	; sn S g r	G m g	M p m P d n (Ṡ ; ;)
Charanam 1			
M ;	; p m g s r g	M ;	; p d p p m g
M ;	; p m ṡ N d p	M ;	; g r g s r g
M g s	r s M g . s p p	M g - d	p p m g m p d n (Ṡ ; ;)

Fig. 2 An excerpt from "Jatiswaram," notated in Sargam

Since Unicode is currently supported by most text editors and web browsers, the iSargam file can be read directly and displayed without the need for any additional code similar to the ASCII-based text files. Additionally, they can employ plug-ins to provide different styles and layouts for printing purposes. Since most of the current programming languages support Unicode, application programmers and music analysts can easily build algorithms for the music information retrieval or computer-based analysis of the iSargam music databases.

The next section gives details about Sargam notation system, and the section following provides details of related works done elsewhere. We then describe the iSargam encoding system, explaining its approach and encoding algorithm. For increasing readability of western readers, we give comparisons with western music concepts wherever applicable. We also present some example encoding compositions as proof of our approach.

3 The Sargam notation system

Sargam notation is a music notation language for Carnatic music. Each notation starts with specification of raga, tala, and mela. Sometimes, the notes used in ascending scale and descending scale, known as arohana and avarohana, are explicitly defined in the start of the composition. This is followed by the time signature and actual music notation following it as illustrated in Fig. 2. In this section, we attempt to describe the terminologies used and the notation scheme.

3.1 Raga

Raga [17] is one of the most distinguished features of Carnatic music. The raga can be defined by a melodic scheme characterized by a definite scale or notes, order or sequence in which the notes can be used, melodic features, pauses and stresses, and tonal graces. Some ragas define the same set of music notes (swara) but are still differentiated by some other features like the order of appearance of swara, melodic punctuation, accent, intonation, and melodic phrases. The raga in Carnatic music is analogous to key signatures in western music.

A related field is the specification of arohana and avarohana. The arohana (meaning "ascending") follows from the raga specification and explicitly lists the set of allowed swara syllables when the music is following an ascending flow. Similarly, the avarohana (meaning "descending") specifies the name of the parent raga.

3.2 Tala

The term "tala" [17] refers to the rhythm system which controls and establishes the music. There are hundreds of defined rhythm styles (talas) in Indian music. The name of the tala used in the notated composition is given above the notation as in Fig. 2.

Now, we attempt to describe the tala system explaining its elements and symbols. Each rhythm pattern or tala is derived as a combination of six basic elements called "angas," each measured with a unit known as "aksharakala." The name, notation, and duration measurement of the six basic elements are given in Table 1. This is similar to various note types like crotchets and semiquavers in western music notation. Also, each element (anga) has a defined reckoning mode. The reckoning mode specifies the actual delivery of the rhythm.

The rhythm pattern is repeated in a cyclic manner throughout the music and hence is known as "avarta," which means repetition. The basic rhythm used is notated at the beginning of the notation, and notes are grouped according to it as seen in Fig. 2. The grouping method used is similar to grouping notes according to time signature as in western music notation but with bars of different measure.

3.3 Mela

Mela or Melakartas [17] are parent ragas from which the other ragas evolved. There are twenty-two of them. A distinguished feature of the mela or parent ragas is that they contain all the seven notes in order. The melakartas have a numbering scheme and are identified by the number. The parent raga of the composition is so indicated by an integer number as illustrated in Fig. 2.

3.4 Notation

The notation style is primarily script-based where the note symbols are placed on a straight line. Suitable signs and symbols are also used to indicate various other musical features. A detailed review of the Sargam transliteration scheme can be found in [18].

Table 1 Basic elements of Indian Carnatic music rhythms (tala)

Element	Symbol	Duration in aksharakala units
Anudrutam	˘	1
Drutam	o	2
Laghu	$\vert_3$, $\vert_4$, $\vert_5$,$\vert_6$ $\vert_7$, $\vert_9$	3,4,5,6,7,9 respectively
Guru	8	8
Plutam	8̊	12
Kakapadam	+	16

As shown in Fig. 2, the notation part starts with tala symbols called "anga" which group the other music notes according to time duration. The symbols used in notation can be classified as music note (swara), gamaka symbols, and other symbols. The following section details the concept and notation of musical note, gamaka, and other symbols.

3.4.1 Musical note

A swara or music note usually denotes the note name indicating the pitch, duration, octave, and whether it is played with expressions (called gamaka) or not. The notes are named differently according to its pitch as shadja (sa), rishabha (ri), gandhara (ga), madhyama (ma), panchama (pa), dhaivata (dha), and nishada (ni) and is abbreviated as given in brackets. The current style of written forms is with vowels removed and expressed with single letters as shadja (S or s), rishabha (R or r), gandhara (G or g), madhyama (M or m), panchama (P or p), dhaivata (D or d), and nishada (N or n).

3.4.2 Octave or sthayi

In Indian music, there are five referred octaves, with a middle octave and two upper and two lower octaves. The first upper octave is denoted by adding a dot below the music note and the second upper octave denoted by adding two dots below the music note as shown in Fig. 3c. Similarly, the immediate lower octave is denoted by adding a dot above the music note and the next lower octave is denoted by adding two dots above the music note as shown in Fig. 3b.

3.4.3 Duration

The duration of each music note is measured in terms of aksharakala, which is the unit of measurement in the Indian tala (rhythm) system and is analogous to "beats" in western music. To denote the duration of a music note, uppercase or lowercase letters with or without comma, semicolon, or with underline or over line are used. A swara letter in lowercase indicates one aksharakala duration, and an upper case swara letter indicates two aksharakala duration. A comma placed near a music note increases its duration by one aksharakala and a

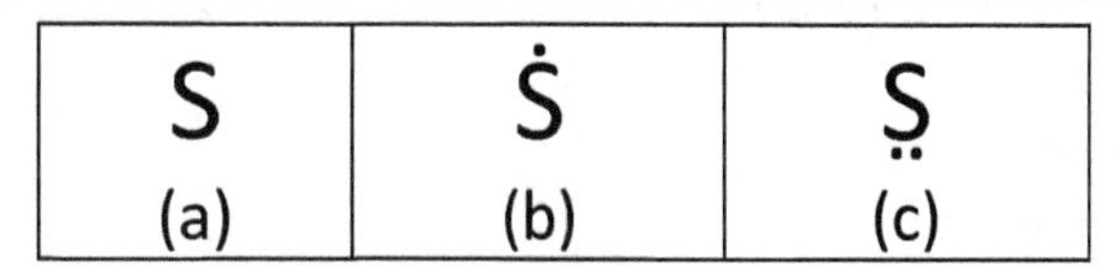

Fig. 3 A musical note shadja with **a** mid octave, **b** lower octave, and **c** two octaves higher

semicolon by two. Similarly, a single horizontal line over the swara reduces the swara duration to its half and double over or under line reduces it to its quarter.

The duration of a rest note is indicated using the necessary number of semicolon or comma symbols placed inside simple parenthesis, e.g., (,;).

3.4.4 Note variety

In Carnatic music, each music note (swara) can represent more than one pitch value, usually two, according to the raga followed by the composition. Because of this characteristic, they are called "note varieties" as each swara note provides many colors to choose from.

Generally, there are no special signs or symbols to represent the variety of the note. This information is implicitly associated with the raga of the song. However, some subscripted numerals with Swara symbols are rarely used to denote sharp and flat varieties of the notes which comprise the twelve "swarasthanas" [18] [19]. For example, r_1 denotes the musical note "Sudhhari" or "Komalari" and r_2 denotes "Chatusrutiri" or "Tivrari" [18] [19].

3.4.5 Additional symbols

There are a few special symbols used in the notation scheme for denoting some musical features like articulation, ornaments etc. All of them are notated using symbols attached with the swara letter. This includes symbols for an ascending or descending glide, foreign note, stressed note, repeat symbols, and gamaka mark. A complete list of such symbols and the symbol used is given in Table 2.

Rarely, some notes which are not part of the raga specification are used, and such notes are represented by an asterisk mark over the swara symbol. The repeat symbol, usually found at the end of an avarta (measure) denotes that the portion of music should be repeated. A stressed note, denoted by letter "w" over the swara, is similar to staccato in functionality, as used in western music. The gamaka mark, represented by a tilde symbol over the swara symbol symbolizes ornamentation which is of utmost importance to Indian music. A music phrase, unlike in western music, represents a set of musical notes which has to be sung together in one breath duration and is symbolized by hyphens at the start and end of the phrase.

Table 2 List of additional symbols used and their notation

Name of symbol	Symbol
Ascending/descending glide	/ and \
Foreign note	*
Repeat avarta	(r)
Stressed note	W
Gamaka	~
Musical phrase	- -

3.5 Notation arrangement

The music notes with adjoint symbols are written on a straight line similar to tonic solfa notation in western music. The music notes are then grouped according to the rhythm structure (tala) of the composition, which is similar to the grouping of notes with time signatures in western music notation. Here, we explain the grouping mechanism in comparison with grouping in western notation for easy understanding of readers.

In western notation, notes are grouped according to the indicated time measure to form equi-measured bars as demonstrated in Fig. 4. In Carnatic music, the grouping or structuring of music notes is done according to the rhythm pattern (tala). The tala (rhythm) specification consists of a set of basic elements (angas), each with a specific duration. The basic rhythm pattern repeats over the entire composition. The music notes are grouped in such a way that the total duration of the music notes is equal to the corresponding anga duration as illustrated in Fig. 4. This shows that Indian music follows variable measured bars as opposed to equi-measured bars in western music. Once the music notes are grouped for one cycle of the rhythm pattern, it is called "avarta." Thus, the grouping of music notes is done until the end of the composition. This is illustrated in Fig. 5.

4 Related works

Music representation systems encompass musical information in any of the three levels: sound, music notation, or data for analysis [20]. Music notations are generally an encoding of abstract representations of music. They contain instructions for performance and representation of sound. Several structured representation like the hierarchical music structures representation [8], "Music Structures" [21], "TTrees" [22], hierarchical representation of scores [23], musical events [20], musical tones [23], abstract datatype representation in [24], generative approaches like in grammars [24], Petri Nets [25], Markov Models [26], and object-oriented approaches like SmOKe [27], Aspect Music [20], and graph-based approaches [28], for representing musical data.

Music representation systems can be classified as audio signal representations, resulting from the recording of sound sources or from direct electronic synthesis, and symbolic representations which represent discrete musical events such as notes, rhythm etc. [28]. *The proposed system is a symbolic representation, and is content-aware and can relate musical events to formalized concepts of Carnatic music theory.*

The musical representation systems can also be classified according to the encoding system (format) used for

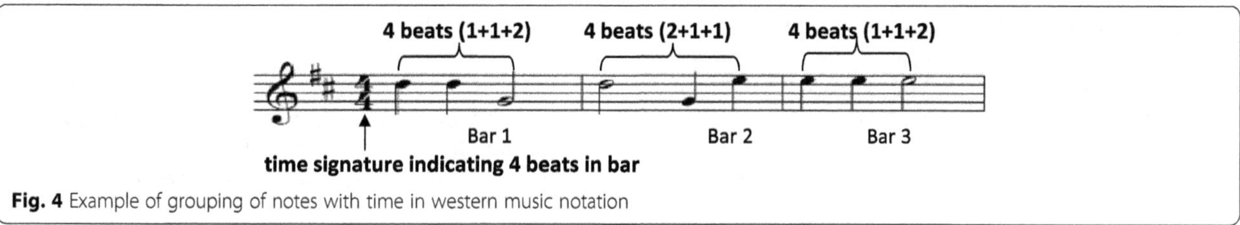

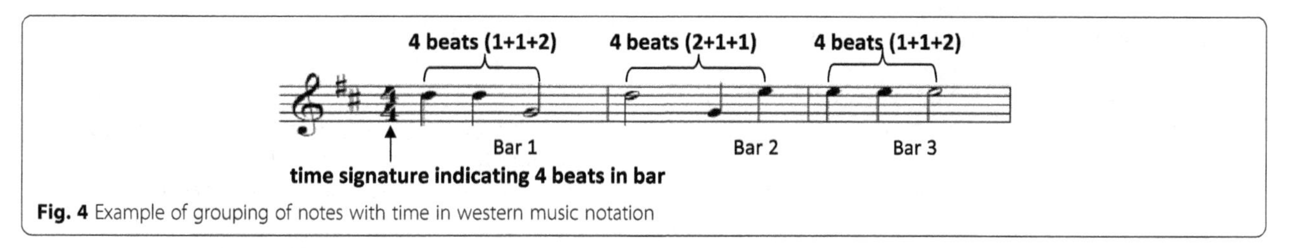

Fig. 4 Example of grouping of notes with time in western music notation

storing the information. Thus, it can be classified into binary, ASCII-based, XML-based, and proprietary formats. The popular binary formats are MP3, WAV etc. The ASCII formats like Cadenza [5], DARMS [29], Guido [4], ENP [30], LilyPond [11], and Humdrum [3] encode musical score information using ASCII-based text. They can be further classified as record-based, command-based, codes, and LISP-based. The XML formats like MIDI XML [31], MusicXML [32], MEI [13] etc. provide hierarchical XML representation of musical information. Also, many popular score-writing programs like Rhapsody and Sibelius use proprietary formats. *We propose to use Unicode standards to encode the Carnatic music notation called "Sargam" (or modern notation), and it can be considered as the first Unicode-based music representation system.*

Even though most of these music representation systems were evolved around western music tradition/notation, there were a few attempts to extend its applicability to other regional music traditions like Korean [24], Greek [33], **Bhat [34] etc. *Unlike these extensions of western music, the proposed system is a unique approach to representation of South Indian Carnatic music based on Indian music theory.*

5 The iSargam language

As opposed to many ASCII-based representations like DARMS [15], Guido [4] etc., iSargam is formed as a Unicode-based music notation representation language. That means we use various Unicode symbols to represent musical entities in Carnatic music.

In this section, we describe the iSargam representation system by explaining its approach, encoding logic, and algorithm.

5.1 Terminologies

Before describing the encoding logic, we would like to illustrate a few concepts which we developed as part of the encoding logic, viz, singleton/grouped entity, and music constituent.

5.1.1 Singleton/grouped entity

The musical symbols used in Sargam notation are classified as singleton or grouped entity according to whether they have meaning or sense in single form or they make sense only when they combine with another musical entity. For example, anumandra, the octave specification symbol makes sense only when it is joined with a musical note (Swara). Singleton musical entities are always found independent in the notation and have semantics of their own. An example of singleton entities are tala marking symbols, anga/avarta mark etc. This classification among music symbols is required due to difference in encoding single and group entities, where group entity symbols can be encoded together only and not individually.

5.1.2 Music constituent

Here, we attempt to define the term "music constituent" which is a concept developed as part of the encoding logic. A music constituent is the most basic unit of music. Here, it consists of a pitch symbol (swara), a sthayi (octave), and duration information. It may be noted that a rest note does not have pitch and octave but has duration. Thus, the basic constituent elements are swara and rest. So, the general syntax of a pitched music constituent can be defined as

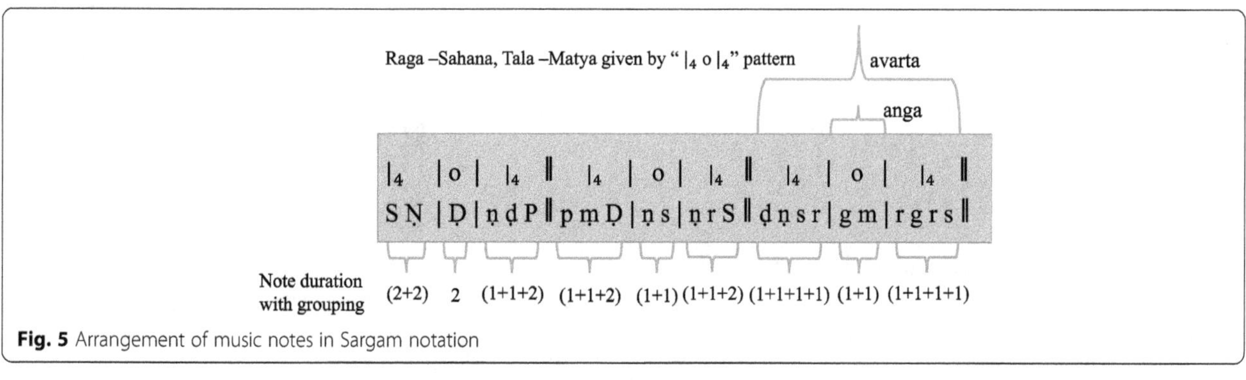

Fig. 5 Arrangement of music notes in Sargam notation

$$< \text{swara-syllable} > [< \text{octave} >][< \text{duration} >] \quad (1)$$

or in the case of an unpitched (rest) note defined or expressed as

$$([< \text{duration-symbol-comma} > | < \text{duaration-symbol-semicolon} >]*) \quad (2)$$

It may be noted that the swara syllable alone is a complete musical constituent since it already contains octave and duration information.

The basic element can be grouped according to some rhythmic pattern or it can be further augmented with additional symbols or other music notes, forming various grouped entities. In our approach, the former is called rhythmic group and the latter is called notational grouping. This latter is again classified into intra notational and inter notational group entities. Inter notational groupings are always associated with music constituents.

Intra notational grouping occurs when the music constituent is further augmented by adding parameters which apply in a single note level. This is denoted by adding extra signs or symbols to the base syllable. The musical entities in this category are stressed note symbol (w), gamaka mark (~), foreign note symbol (*), violin marks, upward stroke of the bow (v), and downward stroke of the bow (^). In case of inter notational grouping, multiple musical notes are grouped together, mostly to give a musical expression such as a musical phrase and ascending or descending glides.

5.2 Encoding logic

Having defined the basic terminologies, now we attempt to present our encoding logic. Initially, our system maps every Sargam notation symbol to a Unicode symbol. The chosen Unicode character resembles the Sargam notation symbol used. Each symbol is also assigned a priority number.

The encoding logic depends on whether the music symbol is a singleton or grouped entity. So, for encoding of a notated composition, we take every symbol and check if it can be further split into different characters as illustrated in Fig. 6. This is done by checking the baseline and upperline of the character. An atomic symbol is a singleton entity and it is directly mapped to its representation Unicode symbol.

If the character is a grouped entity, then it might be a Swara symbol indicating a pitched note or it is a rest note, where both are music constituents. We process the constituent symbols together with a priority queue [35]. The priority queue orders them according to the preassigned symbol priority and thus produces unambiguous encoding. A pseudo code for the proposed encoding scheme is given in Table 3.

The iSargam system chooses the unique numbers carefully so as to make sure that the corresponding Unicode character almost fully resembles the actual music notation in appearance, even in the case of the grouping or joining of music notation symbols. The advantage here is that Unicode symbols appear discrete in encoding, which favors easy identification of music entities for music processing, but in appearance, it appears joined, resembling the original notation. Also, it may be noted that in such a representation, a combined notation can be easily split to its constituent basic music entities.

We use Unicode full width forms for standalone music elements like swara syllable or duration, and Unicode combining diacritical marks for adjunct symbols like octave, stress, foreign note indication, duration symbols which symbolize duration less than one unit, etc. More specifically, all the intra notational symbols and violin marks are represented by combining diacritical marks. Additionally, we use Unicode full width symbols for representing symbols in the rhythmic group like anga, avarta symbols, which are analogous to measure and bar markings for western music, and other rhythm-specific elements like laghu, plutum etc., which mark the number of beats within a measure.

Having said encoding logic, we now attempt to illustrate our encoding approach for each music entity, illustrated with an example, as given in Table 4.

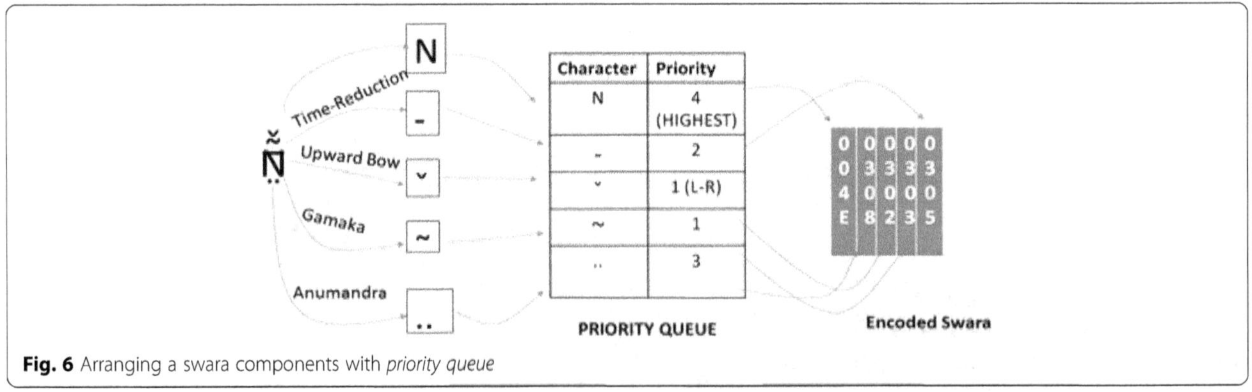

Fig. 6 Arranging a swara components with *priority queue*

Table 3 Abstract encoding logic

```
Function iSargamEncoding(Document doc)
For i=1 to doc.length
  X=readInputChar(i)
If stand-alone- item(X)
  write(Unicode(X))
Else If grouped-item(X)
  If MusicConstitutent(X)
    If pitchedSwara(X)
      Components= split(symbol)
      For j=0 to length(Components)
        pushToPriorityQueue(Components[j]))
      next j
      For j=0 to length(Components)
        y=popFromPriorityQueue(Components[j]))
        write(Unicode(y));
      next j
    else If unpitchedNote(X)
      rest=readFullRestSymbol(i);
      i=i+length(rest);
      For j=0 to length(rest)
        write(Unicode(rest[j]))
      next j
      End If
      Else
       write(Unicode(X));
      End If
    End If
Next i
End Function
```

The encoded file consists of various sections, viz, the header, rhythm markup, and actual composition, explained as follows.

5.2.1 Header section

The header section accompanies every music notation. It mainly consists of two sections, viz, a compulsory part and an optional part. The compulsory part is known as the music description part, and it specifies the most important elements for interpreting the notation. These most important fields are raga name and tala name. The optional part consists of the fields composition title, composer, arohana/avarohana, and mela. The header elements are considered as keywords which are case-insensitive and are separated by a colon character. These values are case-sensitive.

5.2.2 Tala (rhythm) markup section

The tala section marks the rhythm pattern of the composition. Usually, a rhythm pattern is defined as a combination of its basic elements called angas, as described in the previous section. For encoding, each anga symbol is assigned a Unicode-based identifier and is separated from the others by using the vertical line symbol (U +007C). The avarta is marked by a double pipeline symbol (U+01C1) at the beginning and the end as shown in Table 4.

5.2.3 Music notation section

This section contains the notation of the actual composition. It consists of music constituents along with required signs and symbols with notational and rhythmic grouping. The encoding of the actual music notation is illustrated in the following subsections.

5.2.3.1 Encoding of rhythm grouping The music notes are grouped according to the tala specification, splitting them into many anga and avarta as described in the previous section. The angas are separated by vertical line symbol (U+007C), and avarta is marked by double pipeline symbol (U+01C1) at the beginning and the end.

The general form of avarta can be given by

$$\| < \text{music-note} > \ldots \mid\ < \text{music-note} > \ldots \mid \ldots (\mathrm{r}) \| \tag{3}$$

Sometimes, a repeat symbol is inserted in front of the avarta end to denote repetition of an avarta. The symbol used is "(r)"and the encoding is encoded by the Unicode symbol parenthesized Latin small letter R (U+24AD).

Table 4 iSargam encoded file

```
<!—Header--!>

Raga:sankarabharana

Tala:Matya

Mela:28

Title:JANYA RAGA LAKSHANA GITA

Composer:PAIDALA GURUMURTI SASTRI

Araohana:s,r,g,m,p,m,d,n,s

Avarohana:s,n,d,p,m,G,m,R,g,r,s

<!—Tala specification--!>

‖|4|°| |4‖

<!—Composition--!>

‖S N. |D.|n.d.P.‖ p.m.D.|n.s |n.rS‖

‖d.n.sr|gm|rgrs ‖ n.sgr|n.s |D.d.n‖

‖d.p.m.d.|D.|n.srs‖Mmd|D|Nns.‖‖d.n.sr|gm|rgrs‖n.sgr|n.s |D.d.n‖

‖d.p.m.d.|D.|n.srs‖Mmd|D|Nns.‖
```

5.2.3.2 Encoding of music note The encoding strategy followed for a pitched music note and an unpitched note is different. The encoding logic for an unpitched note is straightforward like a singleton entity. But a pitched note is regarded as a grouped entity and we use priority-based encoding for this set as illustrated in Fig. 6.

Unlike the encoding strategy used for singleton entities, encoding for grouped entities is done together and not as individual elements. It can be observed that any pitched note is an extension of a pitched music constituent. In case of a pitched music constituent, the swara symbol is followed by octave and duration symbols, as mentioned in the previous section. Additionally, musical notes can contain additional characters which augment the basic music note like stressed note symbol, foreign note symbol etc. The musical note may then be part of another group in case of occurrence of musical phrase or glide expressions. It might look straightforward to assign symbol priorities in the same order. But this does not work due to a difference in the type of notation symbols used. So, iSargam develops a new encoding logic for notating a music note, which is explained here. In this context, we would like to redefine the general syntax of music notes given in the previous section to enable easy encoding.

A musical note can be defined as a music constituent optionally followed by other additional qualifiers which may denote some kind of ornamentation.

$$< \text{music-note} >=< \text{music-constituent} > [\text{additional-symbols}][\text{glide} < \text{musical-note} >] \quad (4)$$

$$< \text{music-constituent} >=< \text{swara-syllable} > [< \text{octave} >][< \text{duration-1} >] \quad (5)$$

The iSargam system modifies this general syntax to construct the syntax for our encoding, to take advantage of certain properties of Unicode combining diacritics, as given below. Also, we assign priorities for these symbols to ensure proper ordering in the encoded script, which avoids ambiguous representations.

$$< \text{swara-syllable} > [< \text{octave} >][< \text{duration-1} >]\\ [\text{additional-symbols}] \tag{6}$$

or

$$< \text{swara-syllable} > [< \text{octave} >][\text{additional-symbols}]\\ [< \text{duration-2} >] \tag{7}$$

The <duration-1> symbol consists of duration symbols which have duration of less than one aksharakala unit, and we use Unicode diacritics symbols to represent them. <additional-symbols> are also represented by Unicode diacritics symbols. The <duration-2> consists of duration which have duration greater than one aksharakala unit, and we use basic Latin Unicode symbols comma and semicolon to represent them.

Now, we introduce symbol priority which defines the encoding order within music note symbols as illustrated in Fig. 6. The swara syllable letters are from the English alphabet and are assigned with the maximum priority. We use Unicode Latin letters to represent them. All the octave symbols are adjoint symbols with the swara letter, and they form the next priority level. They are represented by Unicode diacritic symbols. Next, we deal with the duration elements by splitting it into two groups, viz, duration-1 and duration-2. The first set consists of duration symbols which are adjoint symbols with the swara character. So, their position is placed close to the swara letter and is assigned a priority value of three. The second set of duration symbols consists of standalone symbols which are encoded with Unicode full width forms. The symbols used are comma and semicolon, and they are assigned only the next priority after the additional marks, which is a priority value of five. The term <additional-symbols> represent the set of musical entities which augment the basic music note, which are, a foreign note marked by an asterisk, a stressed note marked by wavy line, a gamaka mark symbolized by a tiled symbol, and violin marks upward and downward circumflex accent. All these additional symbols are to be placed above the swara syllable and so we assign them a priority value of four and we use Unicodes of combining diacritical marks to encode them. The full encoding algorithm is given in Table 5. Here, we use Unicode character "combining x above" for foreign note, "combining inverted double arch" symbol for stressed note, "combining tilde" symbol for indicating gamaka, and combining carons and combining circumflex accent for violin marks.

5.2.3.3 Inter notational grouping The inter notation grouping occurs within an avarta. For providing certain musical effects, more than one music note is grouped

Table 5 iSargam encoding algorithm

```
Algorithm Convert-MUSIC-NOTATION-TO-REPRESENTATION. Given a vector V
containing N elements, this algorithm converts music notation to its representation

1. [Initialize variables and vector]
   X← ' '
   OUTPUT← ' '
2. [CHECK FOR START OF NOTATION]
   If V[i]="|" then
     OUTPUT ← Unicode(V[i]))
   i ← i + 1
   else
   report ERROR.
   End If
3. Repeat thru Step 4-5 for i=0 to LENGTH(vector)
4. If isSwara(V[i]) then
   i ← i+1
    Q ← SplitToChars[V[i]]
   for j=0 to LENGTH(Q) do
   Select case (Q[j])
   Case ['S' or 'R' or 'G' or 'P' or 'D' or 'N' or 's' or 'r' or 'g' or 'p' or 'd' or 'n'] :
     PRIORITY_VALUE ← 1
   ENQUEUE(PRIORITY_QUEUE,Unicode(Q[j]), PRIORITY_VALUE)
    Case ['*' or 'ʷ'or '˜' or '' or '^' or 'ˇ':
     PRIORITY_VALUE ← 4
   ENQUEUE(PRIORITY_QUEUE,Unicode(Q[j]), PRIORITY_VALUE)
    Case ['.' or '..' or '' or '¨'] :
    PRIORITY_VALUE ← 2
   ENQUEUE(PRIORITY_QUEUE,Unicode(Q[j]), PRIORITY_VALUE)
   OUTPUT ← Unicode(Q[j]))
    Case [',' or ';' ] :
   PRIORITY ← 5
   ENQUEUE(PRIORITY_QUEUE, Unicode(Q[j]), PRIORITY_VALUE)
   Case ['‾' or '=' or '≡']
   PRIORITY ← 3
     Return
     End Select Case
    OUTPUT=OUTPUT+DEQUEUE_IN_PRIORITY_ORDER(PRIORITY_QUEUE)
   OUTPUT=OUTPUT+Unicode(V[i])
   Goto Step 4.
   For k=0 to PRIORITY_QUEUE.LENGTH
   OUTPUT=OUTPUT+DEQUEUE_IN_PRIORITY_ORDER(PRIORITY_QUEUE)
   End If
5. Case '-' : OUTPUT=OUTPUT+Unicode('-')
   i ← i + 1
Goto Step 4.
6. Case '(' : temp ← Unicode(V[i])
     i ← i + 1
     While V[i] ≠ ')'
       If V[i] = ')' or V[i] = ',' or V[i] = ';' then
         temp ← temp + Unicode(V[i])
          i ← i + 1
       End If
     Next
7. Case '/' or '\' : OUTPUT=OUTPUT+Unicode(V[i])
   i ← i + 1
    If isSwara(V[i]) then
   Goto Step 4
   Else
     Error
   End If
8. Case '|' : OUTPUT=OUTPUT+Unicode('|')
   i ← i + 1
   If isSwara(V[i]) then
   Goto Step 4
   Else
     Error
   End If
9. Case 'I' : OUTPUT=OUTPUT+Unicode('I')
    Return
```

Table 6 Actual display in Unicode-supported text editor

<!—Header--!>

Raga:sankarabharana

Tala:Matya

Mela:28

Title:JANYA RAGA LAKSHANA GITA

Composer:PAIDALA GURUMURTI SASTRI

Araohana:s,r,g,m,p,m,d,n,s

Avarohana:s,n,d,p,m,G,m,R,g,r,s

<!—Tala specification--!>

‖|₄|°| |₄‖

<!—Composition--!>

‖S Ṅ|Ḋ|ṅ.đṖ‖ṗṁḊ|ṅs |ṅrS‖

‖đṅsr|gm|rgrs ‖ ṅsgr|ṅs |Ḋđn‖

‖đṗṁđ|Ḋ|ṅsrs‖Mmd|D|Nnṡ‖

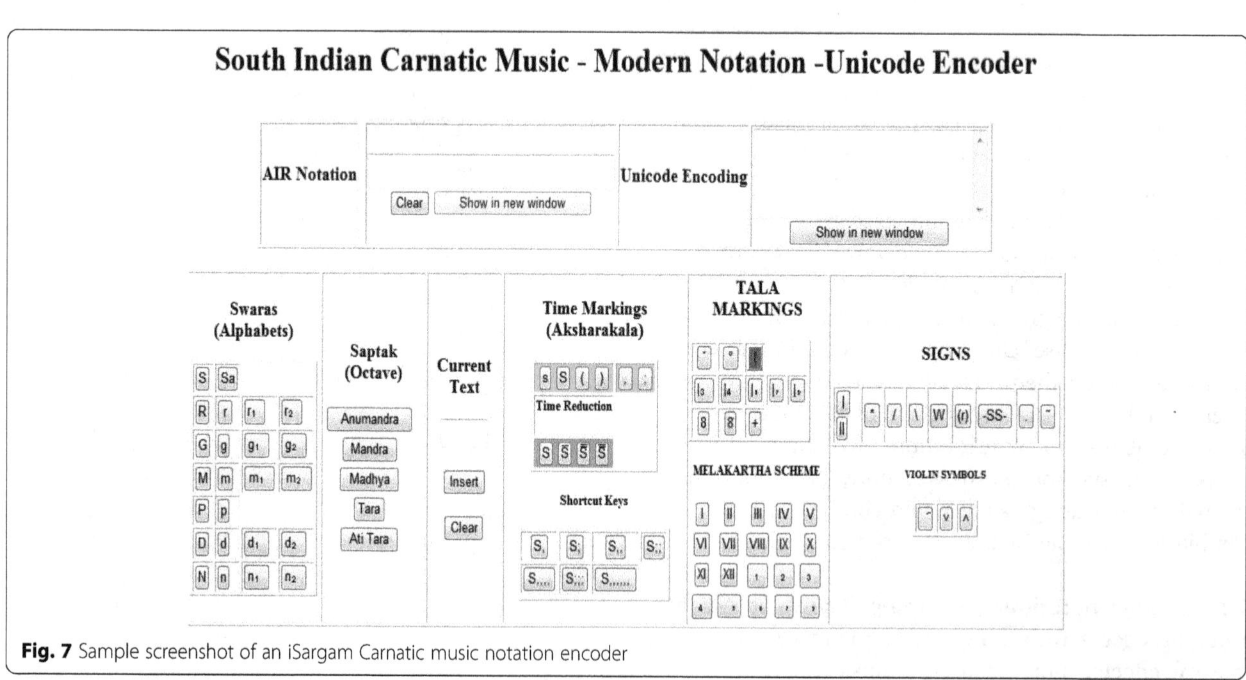

Fig. 7 Sample screenshot of an iSargam Carnatic music notation encoder

together. In the case of Carnatic music, there are two occurrences of such inter notation grouping, viz, the musical phrase and the ascending/descending glides. In Carnatic music, the meaning of music phrase takes on a different meaning than the one used in western music. In Carnatic music, the intended meaning of music phrase is that the given set of notes has to be sung or played within one breath time in the case of vocal or in a single bow in the case of violin. In Carnatic notation scheme, the musical notes belonging to a phrase expression are placed within two hyphens. The general syntax is as given below, and the encoding places Unicode of hyphen-minus (U+002D) symbol before and after the first and last music note belonging to the expression.

$$- < \text{music-note} >< \text{music-note} > [< \text{music-note} > *]- \tag{8}$$

In the case of the glide marks, it can be treated as an operation between two musical notes. The ascending glide places a glyph character between the music notes, while the descending glide places forward slash character between the notes.

$$< \text{music-note} > [/\text{or}] < \text{music-note} > \tag{9}$$

So, the encoding also follows a simple strategy of placing the Unicode of the backward or forward slash between the two musical note elements. Even though they are grouped symbols, the encoding strategy used encodes them in a straightforward manner by just inserting the Unicode symbol solidus (U+002F) for ascending glide and reverse solidus (U+OO5C) for descending glide.

5.3 Encoding algorithm

Now, according to our prioritized encoding algorithm, we process note by note, encode them by splitting each musical note into basic music elements, represent them with an assigned unique id taken from the Unicode set, and arrange them in priority order. This procedure is illustrated in Table 5. A sample view of output viewed with a Unicode-based text editor is given in Table 6.

6 Implementation

So far, implementation covers the development of an editor for Sargam notation. Similar to the virtual keypad present in a calculator application program, we have designed an internal keypad which displays a visual keypad on the computer screen with buttons corresponding to the symbols used in the Sargam notation. By pressing the appropriate buttons, a user can enter the desired music notations. On hitting each button, the corresponding entry can be viewed in the editor. The editor uses the iSargam encoding system to store the musical information entered by the user. A screenshot of the encoder application is given in Fig. 7.

7 Conclusions

The proposed work is an encoding system for Carnatic music notation. The limitation of this work is that it only stores musical information in a retrievable form. A possible extension of this work is to integrate search and retrieval mechanisms upon the encoded form.

Most of the popular music retrieval systems employ search mechanisms with note patterns fed into the system as a set of note or swara symbols using the basic ASCII character set. Similarly, most of the query-by-humming-based music information retrieval systems internally convert the user-hummed query into music notations before the actual comparison process. Since iSargam is a Unicode-based encoding system, an extension of the existing MIR systems to support iSargam files or an application of the MIR approaches to the iSargam musical database will only require an integration of a simple ASCII to Unicode conversion module. The same approach can be used to apply the musical analysis mechanisms to the iSargam files or database.

Another suggested improvement is to integrate the work with a knowledgebase developed for Indian music and to incorporate a playback facility with the editor.

Competing interests

The authors declare that they have no competing interests.

Acknowledgements

This work has been carried out as part of a sponsored project entitled "Representation, Retrieval and Analysis Mechanisms for South Indian Carnatic Music using Computers" with the sponsorship of Ministry of Culture, Government of India, for a period of 2 years. The authors acknowledge the financial support made the agency to carry out the proposed work.

Author details

[1]Department of Computer Science and Engineering, Sri Krishna College of Engineering and Technology, Coimbatore, Tamil Nadu, India. [2]School of Drama and Fine Arts, Dr. John Matthai Centre, University of Calicut, Thrissur, Kerala, India. [3]Department of Music, Government Women's college, Trivandrum, Kerala, India.

References

1. M Balaban, *Musical structures: interleaving the temporal and hierarchical aspects to music*. (In Understanding Music with AI: Perspectives in Music Cognition, MIT Press, Cambridge, 1992), pp. 110–138
2. N Han-Wen, N Jan, *LILYPOND, A System for Automated Music Engraving* (Paper presented at the XIV Colloquium on Musical Informatics, Firenze, 2003)
3. D Huron, Music information processing using the Humdrum Toolkit: concepts, examples, and lessons. Comp. Music J. MIT Press **26**(2), 11–26 (2002)
4. HH Hoos, KA Hamel, K Renz, J Kilian, *Representing score-level music using the GUIDO music notation format*. (Computing in Musicology, MIT Press, Cambridge, 2001), p. 12
5. HS Field-Richards, Cadenza: a music description language. Comput. Music. J. **17**(4), 60–72 (1993)
6. Solesmes, Antiphonale Monasticum, vol 1, (Paraclete Press, 2005), pp. 542–543
7. G Bays, *ScoreSVG: a new software framework for capturing the semantic meaning and graphical representation of musical Scores Using Java2D, XML,*

and SVG (Master's Thesis, College of Arts and Sciences, Georgia State University, 2005)
8. MM Erin, R Jenny, Saffran, music and language: a developmental comparison. Music. Percept. **21**(3), 289–311 (2004)
9. Center for Computer Research in Music and Acoustics, Common Music Notation, (Stanford University, 2010), https://ccrma.stanford.edu/software/cmn/cmn/cmn.html. Accessed on 12 April 2015
10. W Kornfeld, Machine Tongues VII: LISP. Comput. Music. J. **4**(22), 6–12 (1980)
11. M Laurson, *Dissertation, Sibelius Academy*, 1996
12. M Good, G Actor, *Using MusicXML for file interchange, In Proceedings Third International Conference on WEB Delivering of Music, Leeds, UK, September 15–17, 2003* (IEEE Press, Los Alamitos, 2003), p. 153
13. P Roland, *Design Patterns in XML Music Representation* (Paper presented at the Fourth International Conference on Music Information Retrieval, University of Johns Hopkins, Baltimore, 2003)
14. R Perry, The music encoding initiative, in *Proceedings First International Conference on Musical Application using XML*, ed. by H Goffredo, L Maurizio, 2002
15. P Bellini, P Nesi, *WEDELMUSIC Format: An XML Music Notation Format for Emerging Applications, First International Conference on WEB Delivering of Music* (IEEE Computer Society, Washington, 2001), pp. 79–87
16. PP Narayanaswami, VS Jayaraman (eds.), *Sangita Sampradaya Pradarsini English Web Edition*, 2006. http://ibiblio.org/guruguha/ssp.htm. Accessed on 15 June 2014
17. P Sambamurthy, *A Practical Course in Karnatik Music*, 1st edn. (The Indian Music Publishing House, India, 1963), pp. 23–56
18. L Issac, Theory of Indian music. (Shyam Printers, 1967), pp. 18–24
19. S Bhagyalekshmy, *Ragas in Carnatic Music, (CBH Publications*, 2003)
20. P Hill, S Holland, R Laney, An introduction to aspect oriented music representation. Comput. Music. J. **31**(4), 47–58 (2007)
21. A Smaill, G Wiggins, M Harris, Hierarchical music representation for composition and analysis, computers and the humanities. Kluwer Acad. Publishers **27**(1), 7–17 (1993)
22. X Serra, *The Musical Communication Chain and Its Modeling, Mathematics and Music, 243–255*, 2002
23. M Besson, AD Friederici, Language and music: a comparative view. Music. Percept. **16**(1), 1–9 (1998)
24. JH Lee, JS Downie, A Renear, Representing Korean traditional music notation in XML, in Proceedings of the Third International Conference on Music Information Retrieval. (IRCAM Centre Pompidou, Paris, 2002)
25. AD Patel, Music and the brain: three links to language. *Oxford Handbook of Music Psychology*. (2008). doi:10.1093/oxfordhb/9780199298457.013.0019
26. F Ramus, M Nesport, J Mehler, Correlates of lingustic rhythm in the speech signal. Cognition **73**, 265–292 (1999)
27. J Kippen, B Bel, Modelling music with grammars: formal language representation in the Bol Processor, Computer Representations and Models in Music. (Academic Press Limited, 1992), pp. 207–232
28. F Lerdahl, R Jackendoff, *A generative theory of tonal music* (MIT Press, Cambridge, 1985), pp. 130–162
29. R F Erickson, The DARMS project: A status report, Computing and the Humanities. Springer **9**(6), 291–298 (1975)
30. M Kuuskankare, M Laurson, Expressive notation package. Comput. Music. J. **30**(4), 67–79 (2006)
31. MIDI Manufactures Association. Making music with MIDI, https://www.midi.org/. Accessed on 25 March 2015
32. Michael Good, MusicXML. http://www.musicxml.com/. Accessed on 1 April 2015
33. D Politis, D Margounakis, S Lazaropoulos, Leontios, Papaleontiou, George Botsaris, Konstantinos Vandikas, Emulation of ancient Greek music using sound synthesis and historical notation. Comput. Music. J. **32**(4), 48–63 (2008)
34. P Chordia, *A system for the analysis and representation of Bandishes and Gats using Humdrum syntax, in Proceedings of the 2007 Frontiers of Research in Speech and Music Conference* (Mysore, India, 2007)
35. MA Weiss, *Data Structures and Algorithm Analysis in C++, Pearson Education*, 1993
36. Creative Commons. http://creativecommons.org/licenses/by-nc-sa/2.0/fr/legalcode. Accessed on 28 Dec 2015

3

Screen reflections impact on HDR video tone mapping for mobile devices: an evaluation study

Miguel Melo[1,2*], Maximino Bessa[1,2], Luís Barbosa[1,2], Kurt Debattista[3] and Alan Chalmers[3,4]

Abstract

This paper presents an evaluation of high-dynamic-range (HDR) video tone mapping on a small screen device (SSD) under reflections. Reflections are common on mobile devices as these devices are predominantly used on the go. With this evaluation, we study the impact of reflections on the screen and how different HDR video tone mapping operators (TMOs) perform under reflective conditions as well as understand if there is a need to develop a new or hybrid TMO that can deal with reflections better. Two well-known HDR video TMOs were evaluated in order to test their performance with and without on-screen reflections. Ninety participants were asked to rank the TMOs for a number of tone-mapped HDR video sequences on an SSD against a reference HDR display. The results show that the greater the area exposed to reflections, the larger the negative impact on a TMO's perceptual accuracy. The results also show that under observed conditions, when reflections are present, the hybrid TMOs do not perform better than the standard TMOs.

Keywords: High-dynamic-range video, Tone mapping operator evaluation, Mobile devices, Screen reflections

1 Introduction

Current imaging techniques, also known as standard dynamic range (SDR) or low dynamic range (LDR), are not capable of representing all the real-world color gamut and contrast in a way that matches the human visual system (HVS)'s dynamic range. To overcome this limitation, high-dynamic-range (HDR) imaging was developed. Ensuring HDR is maintained along the entire imaging pipeline from capture to display allows the full range of captured scene data to be used in a number of applications, including security, broadcasting in difficult lighting conditions, etc.

When an HDR display is available, it is possible to deliver HDR content in a relatively straightforward manner [1]; however, the majority of displays currently available are still LDR. This is particularly true to mobile devices where there is, as yet, no HDR display. It is thus necessary to map any content's dynamic range to match that of the targeted display. The dynamic range reduction can be achieved by employing tone mapping operators (TMOs). These take into account scene characteristics and/or the HVS properties in order to provide the best viewing experience on the LDR display from the available HDR data. A large variety of TMOs have been proposed, with only a few dedicated to HDR video and none designed specifically for HDR video on small screen devices (SSDs). TMOs for SSDs may need to take into account their portability which can result in situations in which there is a sudden exposure to widely differing luminance levels and reflections that could impact on the viewing experience.

Mobile devices have become widespread, and their penetration rate is reaching nearly 100 %, that is, one mobile device per person in the world [2]. Furthermore, mobile devices are already being widely used to consume multimedia, and indeed, it is estimated that around 51 % of the traffic on mobile devices is now video [3]. In fact, a recent report [4] showed that the online video requests are more and more made by mobile devices and, while in 2002 only 6 % of all online video was requested by mobile devices, in 2014, the number has increased to 30 % and it is projected to reach over 50 % by the end of 2015.

*Correspondence: emekapa@sapo.pt
[1]Universidade de Trás-os-Montes e Alto Douro, Quinta de Prados, 5000-801 Vila Real, Portugal
[2]INESC-TEC, Rua Dr. Roberto Frias, 4200-465 Porto, Portugal
Full list of author information is available at the end of the article

With this growing popularity, it becomes important to consider possible challenges posed by mobile device displays compared to the traditional desktop devices. These include limitations in terms of dynamic range, viewing angle, and distance as well as size. In addition, to help ensure an optimal viewing experience, it is important to take into account external factors such as luminance levels and screen reflections. Furthermore, the current diversity of screens for mobile devices may require the re-targeting of content to play properly on any delivery screen. This can be a major challenge [5].

While previous work has addressed comparisons of diverse TMOs on mobile devices, none of this has addressed the impact of reflections on the screen when viewing HDR video on SSDs. As an SSD is very likely to be subjected to reflections, this paper presents a novel investigation on the visualization of HDR video on mobile devices under conditions where the display is exposed to reflections in order to understand their impact on the visualization quality. As reflections are more likely to happen outdoors, this work was concerned with simulating an outdoor environment with bright luminance levels. The insights gained from this study should suggest if there are advantages in developing a hybrid TMO to account for reflections and minimize any negative impact on the viewing experience. One possible future application arising from this study is the creation of automated methods that detect reflections on a screen and applies the best TMO according to the usage scenario. The evaluation of the HDR video tone mappers in this paper is carried out with three different scenarios:

- With a reflection across the entire screen
- With no reflections
- With a reflection on half of the screen (this can be either the left or right side)

The division of the screen was chosen so we would be capable of defining precise areas of the display where we could employ a specific TMO. This study considered one of the larger size SSDs, an iPad 4 which has a 9.7-in. screen. This is representative of current SSDs including mobile phones, whose screen size is increasing.

For the experiments, six HDR videos were used (Fig. 1). Two TMOs were considered: the model of visual adaptation [6] and the display adaptive technique [7], both successful TMOs in a number of previous experiments, for example, [8].

Fig. 1 Frames from each of the six HDR videos that were used for the evaluations

2 Related work

A wide variety of TMOs have been proposed to date. These have all been developed with different concerns or goals in mind. TMOs can be based on simple mathematical operations such as exponential or logarithmic functions performing linear transformations as well as be more elaborate and inspired by, for example, features of the HVS [9]. They can be broadly divided on two categories: global and local. As the name suggest, global TMOs process the image as a whole, applying the same computation to every pixel while local TMOs process the image pixel by pixel taking into account the adjacent pixels. Standard global and local approaches work well with images but when the target is video, there is another important feature to account for: temporal coherence. A new category of TMOs has thus emerged: time-dependent TMOs.

Due to the wide variety of TMOs proposed and the advantages that HDR can bring, it has been important to perform evaluations in order to understand and identify which are the best TMOs for certain scenarios. Despite the efforts on TMO evaluation for HDR video, few papers have explored TMO evaluation for mobile devices and none has addressed the problem of reflections on a screen as often happens when using mobile devices, especially outdoors. The next section presents a brief survey of previous methodologies to evaluate TMOs.

2.1 TMO evaluation

As several TMOs have been proposed over the years, several TMO evaluation studies were also conducted adopting two main methodologies: error metrics and psychophysical experiments.

The error metrics methodology consists of objective measures based on theoretical models using computers to compare images/videos. This methodology can be based on simple approaches such as measuring and comparing individually pixel values as well as something more complex as, for example, simulating the HVS and identifying the differences between the original content and the produced content based on visual perceptible differences. One popular example of such a technique is the HDR-VDP (visual difference predictor) [10] and HDR-VDP2 [11].

Psychophysical experiments, on the other hand, are based on studies conducted with human participants and therefore subjective. One key aspect of the experiments is to guarantee the preservation of the variables over the experiments in order to accomplish a well-controlled scenario where the participants take the tests and perform their judgment over the contents being evaluated. Another key aspect is the proper randomization of the variables to be evaluated between participants in order to avoid bias. The evaluations can be made with or without a reference. For HDR, the reference is typically the real-world scene or an image or video shown on an HDR display.

One of the first TMO psychophysical studies conducted was by Drago et al. that evaluated seven different TMOs applied to four different scenes with 11 participants on a pairwise comparison without reference [12]. The first evaluation study made using an HDR display as reference was conducted by Ledda et al. that evaluated six TMOs applied to 24 images that were evaluated by 18 participants [13]. An example of experiments using real-world scenes as reference are the ones conducted by Čadik [14] that evaluated 14 TMOs in three different scenes.

More recently, evaluations started addressing HDR video, for example, the work by Eilertsen et al. [15] that evaluated 11 TMOs on a set of HDR videos that were both camera captured and computer generated. A total of 36 participants took part in the experiments where they were asked to make pairwise comparisons between the tone-mapped footage without reference. Regarding mobile devices, only a few studies have been done. The paper by Urbano et al. was the first one that was aimed specifically at SSDs [16]. The study evaluated TMOs on different sized displays (17" for conventional sized and 2.8" for small sized) through pairwise comparison against the real-world scene and concluded that for mobile devices, content that offered stronger detail reproduction, more saturated colors, and overall brighter image appearance was preferred.

Akyüz et al. also evaluated both TMOs and exposure fusion algorithms on SSDs [17]. The study was divided in two pairwise comparison experiments where some scenes were evaluated with a reference and some without a reference. The displays used were 24" for the conventional sized display and 3" for the SSD. The participants were asked to evaluate color, contrast, and detail.

Other work which considered HDR video tone mapping for mobile devices was conducted by Melo et al. [18]. The evaluation was performed using a 37" LCD display or a 9.7" display with an HDR display as reference. The results demonstrated that there was a statistically significant difference between the choice of TMOs between the SSD and the large screen display although the TMOs accuracy order remained the same across the two displays. Further work by the same authors investigated the impact on visualization of HDR video on mobile devices under different lighting levels. Three scenarios were considered: dark, dim, and bright lighting levels. The study showed that under dark and dim environments, the TMOs' accuracy ranking obtained was different than that for the bright lighting level environments. The paper concluded that participants gave more importance to contrast and naturalness over details and color saturation in bright environments [8].

2.2 Video tone mapping operators

HDR video tone mapping is a growing field and a number of successful video TMOs have been proposed and successful in previous evaluations using mobile devices ([8, 15]) such as the time-dependent visual adaptation TMO [6], the display adaptive TMO [7], the visual adaptation for realistic images TMO [19], or the temporal coherence tone mapping method [20]. Other examples of well-known TMOs that were designed to address HDR video are the encoding of HDR with a model of human cones [21], the temporally coherent local tone-mapping of HDR Video [22], or the real-time noise-aware tone mapping [23].

The time-dependent visual adaptation TMO [6] exploits the fact that the HVS does not adapt instantly to big changes in luminance intensities. In this method, the appearance is modified in order to match the viewers' visual responses so they can perceive the scene as they would in reality. It uses a global adaption model based on Hunt's static model of color vision and uses the retina response signals for rods for calculating the luminance information and the response vector for color information. In addition, temporal coherency is added. The method is not computationally complex and thus is suitable to use in real-time applications.

Regarding the time-dependent visual adaptation TMO [7], it is a TMO that is capable of adapting to the display features. This TMO offers a set of default, ready to use profiles that are pre-configured, and it is also possible to configure parameters individually such as display reflectivity, the peak luminance, or black level of the display. This TMO uses an HVS model in order to minimize the visible contrast distortions taking into account the characteristics of the given display. The TMO ensures temporal coherency through the limitation of the temporal variations above 0.5 Hz as this is the peak sensitivity of the HVS for temporal changes.

The TMO proposed by [19] is based on a model of visual adaptation from psychophysical experiments. It considers key aspects of the HVS such as visibility, visual acuity, and color appearance. For modeling photopic and scotopic vision, this operator uses TVI functions. In order to achieve the mesotopic range, the authors use a linear combination of both the photopic and scotopic ranges.

The method proposed by [20] follows a different approach as it post-processes the HDR content as it is not capable of doing real-time processing. This is because initially, the method analyzes the whole video sequence in order to preserve the temporal stability of the video sequence. It is used in combination with static TMOs, and the focus was on optimizing it to be used with the Reinhard's photographic tone reproduction method. This operator first processes each frame of the video individually with a static TMO, and then, it considers the luminance of each frame taking into account the features of the whole HDR video sequence. The encoding of HDR with a model of human cones TMO developd by [21] is based on the dynamical response characteristics of primate cones and deals with the temporal coherency by employing temporal filters to handle noise through the absorption of the retinal illuminance by visual pigment. This is achieved by using two low-pass filters where the first is responsible for reducing the dynamic range of the content in order to fit the displays' dynamic range by applying a combination of dynamic non-linearities. The second low-pass lter is based on a non-linear differential equation that reduces noise and automatically adapts to the prevailing scene luminance.

The real-time noise-aware TMO developed by [25] offers a video tone-mapping process that controls the visibility of noise as well as it is capable of adapting itself to the display and viewing conditions and minimizes contrast distortions. Authors describe their method based on three main parts: edge-stopping spatial filter, local tone curves, and noise-aware control over image details. The first is responsible for transforming the input into a log domain, a base layer that describes luminance variances over time and a detail layer that describes the local features. Then, the local tone curves block compresses the dynamic range of the base layer using a set of local tone-curves that are distributed spatially through the scene. Each tonecurve is responsible for mapping the luminance range of the input into the range of luminance that is afforded by the target display. Regarding the noise-aware control over image details block, it gets as input a base layer, a detail layer and the tonecurves in order and allows users to preserve or enhance the local contrast and details of the scene and the visibility noise is controlled based on the noise characteristics on the input layers. The final step consists in applying an inverse model in order to transform the colometric values into pixels.

The temporally coherent local tone-mapping of HDR Video proposed by [23] was designed having as concern the temporal artifacts and the limited local contrast reproduction capability common to TMOs in general. In order to avoid these problems, the authors worked on a temporal domain extension of the common spatial base-detail layer decomposition. The pipeline of this TMO can be divided into 3 main steps: a spatiotemporal filtering that is performed on adjacent frames and uses optical ow estimates to warp each frame's temporal neighbourhood and avoid artifacts; a temporal filtering that reduces temporal artifacts by penalizing ow vectors with high gradients; and the nal tone mapping step that is capable of maintaining the average value of brightness over time as well as an high local contrast.

3 Experimental setup

This section describes the experiments undertaken. In the experiments, the participants evaluated six HDR video sequences with four TMOs under three different scenarios (with a reflection across the entire screen, with no reflections, and with reflections on half of the screen).

3.1 Method

An experimental framework that makes use of randomization of the videos and TMO combinations was used in order to minimize selection bias. A rank-based evaluation was carried out across the TMO methods over six HDR videos. Since the goal was to evaluate HDR video tone mapping for mobile devices, a method was needed that allowed us to define reflections on the screen with precision under controlled conditions. The solution adopted was to point a photographic softbox directly towards an area of the screen as such a device allows the distribution of light in a well-defined area. The vertical division point from the part of the screen that was under reflections and the other part of the screen that was not under reflections was the middle of the screen. To ensure the reflections were the same for all relevant experiments, we had some markers (that were invisible to participants) that indicated that the reflection was being applied correctly. To avoid bias in the scenarios where there was reflection, the scenario was always randomized between participants so that the reflection could be over the left or right half of the display. Figure 2 shows the mobile device with the reflection applied to half of the screen.

The experiments were conducted in a room in which all the environmental variables could be controlled. There are three independent variables: the set of TMOs used, the reflections on the display, and the scene groups. The scene groups and the reflections on the display were

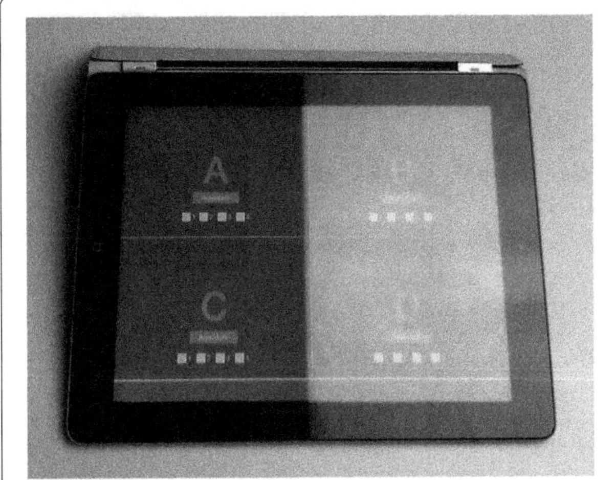

Fig. 2 Mobile device with half of the screen under reflection

in-between participant independent variables; the TMOs was a within-participant variable.

The overall results were analyzed using a 2 (scene groups) × 3 (reflections on display) × 4 (TMOs) mixed factorial ANOVA. The main effects calculated across all videos were of the group (each viewed three videos), TMO (the four TMOs used), and the scenario (with reflections, without reflections, and with reflections on half of the screen). Regarding each evaluated scenario, the data gathered is relative to 30 participants and was analyzed using a 2 (scene groups) × 4 (TMOs) mixed factorial ANOVA.

The set of TMOs used were the display adaptive TMO, the time-dependent visual adaptation for fast realistic image display TMO, and two hybrid approaches of both. The reflections on the screen variable consisted of three conditions: reflections across the whole screen (scenario 1), no reflections on the screen (scenario 2), and reflections on half of the screen (scenario 3). The HDR videos were tone mapped in four different ways: using Pat across the whole frame; using only Man across the whole of the frame and dividing the video vertically in half and applying Man to one side and Pat to the other (referred from now on as ManPat); and dividing the video vertically in half and applying Pat on one side and Man (referred from now on as PatMan) on the other. The scene groups variable consisted of two groups, the participants were divided, where half of them evaluated the first three scenes and the other half the second three scenes.

To avoid bias, one concern was to ensure that the participant was at a correct distance from the HDR display. The participant was placed at a distance of approximately 1.8 m since the suggested viewing distance for high-definition displays is three times the height of the display (which was approximately 60 cm) according to the International Telecommunication Union Recommendation BT.500-13 [24]. A further concern was the luminance adaption of the human eye. To avoid maladaptation, we thus gave the participant some time before beginning the experiment to adapt himself/herself to the experiment scenario. Another concern was the auto-brightness feature of mobile devices that increases display brightness in brighter scenarios. To address this, one has set the devices' brightness to the maximum level ensuring that the mobile device was having its best performance on the given conditions.

Bias is always an important factor in subjective studies since this type of study is based on participants' answers that could be influenced by many factors. It is important, therefore, to pay as much attention as possible to account for possible disturbances. In the case of mobile devices, visual angle and viewing distance are important variables. The mobile device was thus placed on a stand (Fig. 3), and we have been careful to place each participant at approximately the same position so the viewing angle and viewing distance were approximately the same during the experiments. It was not possible to use a chin rest to ensure a stable position of the participant but between each video the position of the participant was checked and he/she was instructed to move his/her head only in a vertical axis in order to look at the HDR display and at the mobile device to evaluate the TMO accuracy. Reflection is a complex issue and since there is no previous work in the field, we considered only a larger size SSD. Future work will consider the effect of reflection on mobile devices with different screen sizes.

3.2 Materials

The generic setup for all the experimental scenarios is presented in Fig. 4. This work arises from previous work that evaluated a set of TMOs for mobile devices under different ambient lighting levels [8]. As the main goal is to evaluate the impact of reflections on TMO performance rather than evaluating which of the existent TMOs is the best for the depicted scenarios, the evaluations

Fig. 3 Position of the mobile device during the experiments

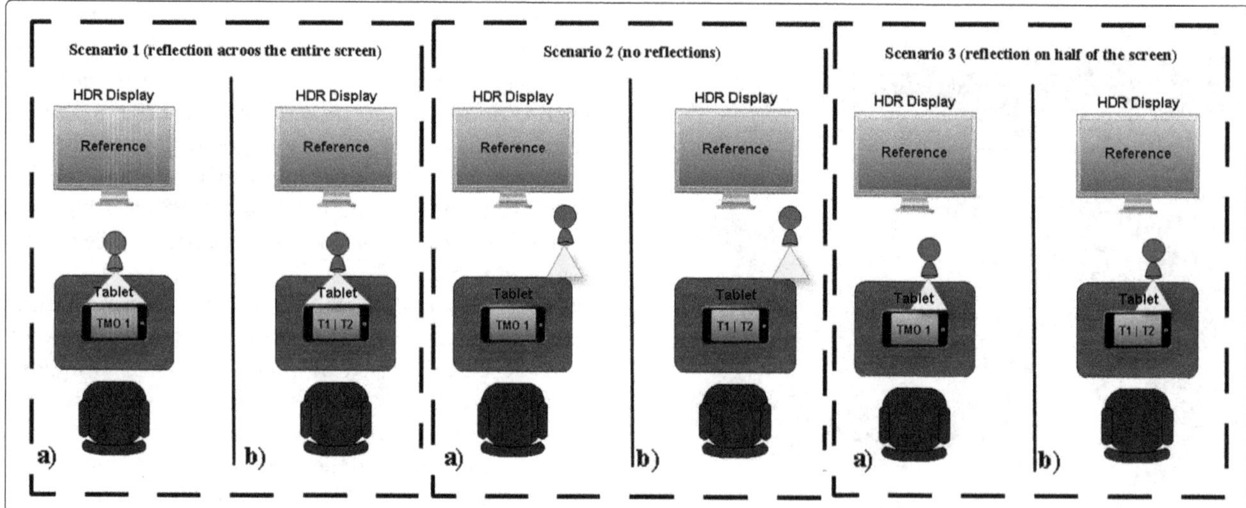

Fig. 4 Experimental setup scheme for the three scenarios. **a** A scenario where the same TMO is applied to the whole image and **b** where two TMOs are applied to the same image

considered two TMOs that performed well on different ambient lighting conditions. Therefore, we selected the best ranked TMO for dark and medium scenarios as well as the best ranked TMO for the bright scenario: the display adaptive TMO ([7]) (that will be referred in this paper as Man) and the time-dependent visual adaptation TMO ([6]) (that will be referred to as Pat). This choice was made since the two TMOs can give insights on reflection impact due to their differences such as, for example, Pat intends to be a visual system simulator that includes temporal coherence by simulating the HVS, while Man is considered a best subjective quality TMO that was designed to produce the best looking images based on subjective studies. In general, looking at the tone-mapped footage produced for the experiments, Pat seems to produce more natural and higher contrast images while Man deliver more detailed and saturated images.

The "join" between the two TMOs was attenuated by a fade between the TMOs so there was no visible division. To achieve it, one filled 60 % of right side with one TMO and 60 % of the left side with the other TMO. Where both TMOs are present, there was applied a sort of gradient with the alpha being that at the middle of the join, there was 50 % of each. An example of processed frames is shown in Fig. 5. The photographic softbox was always placed so it was not between the participant and the reference and in the experimental scenario with no reflections the photographic softbox was also on to ensure the same lighting level across all the scenarios.

For the experiments, the HDR reference display was a properly calibrated 47" SIM2 display and the tablet used was an iPad 4 from Apple. The technical specifications of the displays are presented in Table 1.

Six different HDR videos were considered, labeled "CGRoom","Explosion", "Jaguar", "Kalabsha", "Medical" and "Morgan Lovers". Table 2 shows the features of these videos where the average dynamic range (avg. DR) is expressed log units and the length in seconds. The measurement of the avg. DR was obtained by disregarding the top 1 % and bottom 1 % of the values in each frame and averaging them; this was performed to avoid possible error introduced by noise in the frames.

The experimental room had a controlled luminance level of 1450 cd/m^2 which is equivalent to the average local outdoor luminance level recorded at the time of the experiments and corresponds to a typical partially cloudy day. This luminance level was obtained by using a strong ceiling illumination as well as four photographic spotlights.

Specific experimental software was devised for ranking the TMOs (Fig. 6). For each video, the participants were able to see the tone-mapped content on the iPad and the corresponding HDR reference on the HDR display simultaneously. They could watch the videos as many times they wanted before ranking the TMOs according to how well the tone-mapped video matched the HDR reference. After ranking the four combinations for a video sequence, a button appeared that allowed participants to proceed to the next video. When all videos were evaluated, a message appeared to inform the participant that he/she had finished.

The optimal TMO settings were determined by a group of three experts (who each had at least 2 years of experience in HDR imaging). The global version of Pat was used since the local version does not support the time-dependent effects. The configurable settings for this TMO are the adaptation levels for cones, and the adaptation

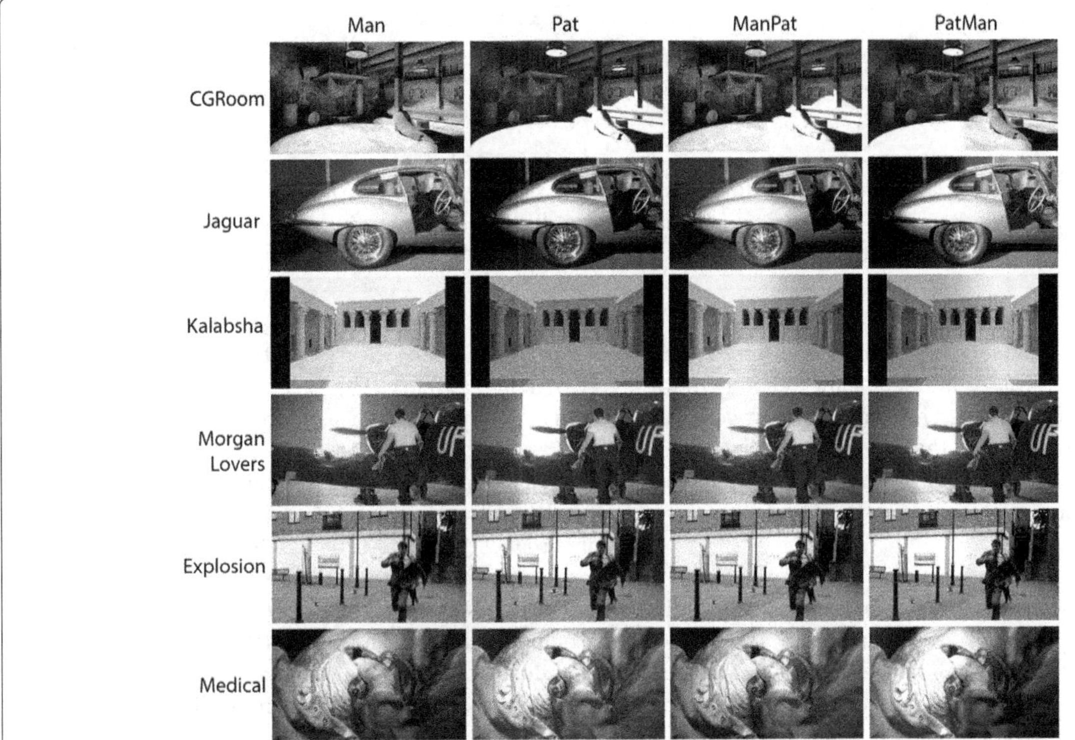

Fig. 5 Example of processed frames used in the experiments

levels for rods that were calculated for each case using the average luminance. For Man, when under reflection-affected areas, the following settings were applied: gamma correction $\gamma = 2.2$, maximum display luminance $L_{\max} = 200$, the black levels of the display $L_{\text{black}} = 0.8$, reflectivity of the display $k = 0.01$, and the ambient illumination $E_{\text{amb}} = 400$. When there were no reflections, the following parameters were used: gamma correction $\gamma = 2.2$, maximum display luminance $L_{\max} = 200$, black levels of the display $L_{\text{black}} = 0.8$, the reflectivity of the display $k = 0.01$, and the ambient illumination $E_{\text{amb}} = 1450$.

Table 1 Technical specifications of the displays used in the experiments

	HDR display	Mobile device
Brand	SIM2	Apple
Model	HDR47ES4MB	iPad 4
Size	47"	9.7"'
Resolution	1920×1080	2048×1536
Contrast ratio	> 1,000,000:1	877:1
Max luminance	> 4000 cd/m^2	476 cd/m^2
Min luminance	0 cd/m^2	0.48 cd/m^2
View angle (horizontal)	40°	175°
View angle (vertical)	15°	175°

3.3 Participants

A total of 90 participants, 51 men and 39 women aged between 19 and 28 years, were randomly assigned between the different experimental scenarios and between the six scenes (half of them evaluated a set of three videos—CGRoom, Jaguar, and Kalabsha—and the other half the set containing the remaining three videos—Explosion, Medical, and Morgan Lovers). For each experimental scenario, there were a total of 15 evaluations for each video as it was the minimum participants required to obtain significant results. All the participants reported normal or corrected to normal vision. The grouping of the scenes was random.

3.4 Procedure

The participant stood at approximately 1.8 m from the HDR display at a table on which the tablet was placed. Each experimental scenario (reflection conditions and scene group) was randomly assigned between the participants. Before each experiment, the experimental room was prepared accordingly. The participants had a brief explanation of how they would participate in the experiments. As quality is a key factor, the participants were

Table 2 Features of the HDR videos used

Video	Length (s)	Avg. DR (log units)	Capture	Device max F-stops
CGRoom	7	16.6	CG	20
Jaguar	13	13.4	Canon 1D Mark II	14
Kalabsha	11	18.5	CG	20
Morgan Lovers	15	19.5	Spheron HDRv	20
Explosion	8	12	Canon 5D	12
Medical	4	15.2	Spheron HDRv	20

asked to take into account color, contrast, naturalness, and details as a whole as these parameters are well known for characterizing an image [25]. The evaluation software was shown to the participants so they could familiarize themselves with it. The average time for each experiment was around 10 to 15 min.

4 Results

A large amount of data was collected and analyzed. To facilitate the presentation of the results, they are divided into subsections with a discussion following in the next section.

As mentioned in Section 3.1, the overall results were analyzed using a 2 (scene groups) × 3 (reflections on display) × 4 (TMOs) mixed factorial ANOVA. The main effects calculated across all videos were of the group (each viewed three videos), TMO (the four TMOs used—two standard TMOs and the two hybrid combinations of the standard TMOs used), and the scenario (with reflections, without reflections, and with reflections on half of the screen). Regarding each evaluated scenario, the data gathered is relative to 30 participants and was analyzed using a 2 (scene groups) × 4 (TMOs) mixed factorial ANOVA. In scenario 3 (when there is reflection only on half of the display) and for ManPat and PatMan, the results were treated so the TMO under reflection is the first (Man in ManPat and Pat in PatMan). The results also present Kendall's coefficient of concordance W that serves as an estimate of agreement among participants where $W = 1$ signifies perfect agreement among participants and $W = 0$ completes disagreement. Based on Kendall's coefficient of concordance W, we test the null hypothesis as to whether there is no agreement among the participants at $p < 0.05$. This provides an indication of the agreement among the participants. Table 3 summarizes the obtained results. We highlight that, despite the fact that Kendall's coefficient of concordance does not range high in values (it is between 0.14 and 0.24).

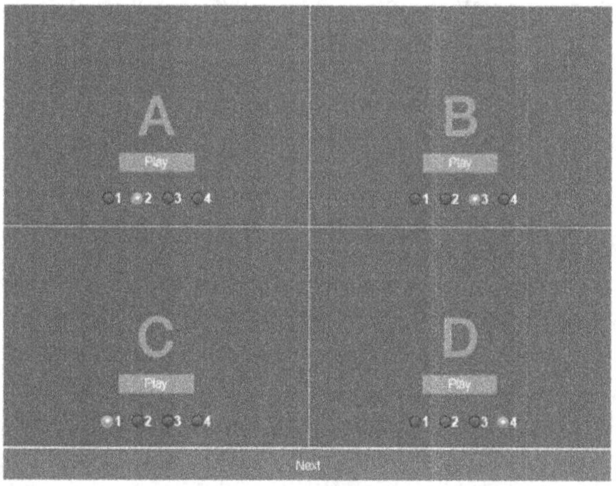

Fig. 6 Software used in the experiments

4.1 Overall results

For the different scenarios and TMOs that were evaluated, the statistically significant difference is given by their main effect that was $F(5.86, 245.95) = 2.13\,(p < 0.05)$ and $F(2.93, 245.95) = 20.98\,(p > 0.05)$, respectively. Regarding the group main effect, the value reported was $F(2.93, 245.95) = 6.59\,(p > 0.05)$. On all cases, sphericity was violated; therefore, Greenhouse-Geisser correction was applied.

These overall results show that differences are statistically significant between the evaluated TMOs as well as between the different scenarios that were considered. The consistency of the rankings was also significant since the computed Kendall's coefficient of concordance was $W = 0.17$. The rank order obtained shows that Pat was the best ranked TMO. ManPat was second grouped with PatMan and Man while PatMan was third grouped with Man. Also, although there were two groups ranking the scenes, there was no significant difference between them which is a good result indicating coherence between choices.

As for the TMOs, looking closely to the groupings, it is noticeable that there is a TMO that clearly outperforms the others. Pat has been consistently ranked as the most accurate TMO, and the difference between this TMO and the remaining ones is statistically significant. The second ranked TMO was ManPat grouped together (meaning that their difference is not statistically significant) with PatMan and Man. There is a third grouping differentiating PatMan and Man from the others.

Table 3 Overall results obtained for each scenario

	Kendall's coefficient of concordance	1st	2nd	3rd	4th
Across all scenarios	0.17	Pat	ManPat	PatMan	Man
Scenario 1 (reflections across the entire screen)	0.14	Pat	ManPat	Man	PatMan
Scenario 2 (no reflections on the screen)	0.24	Pat	ManPat	PatMan	Man
Scenario 3 (reflections on half of the screen)	0.19	Pat	PatMan	ManPat	Man

Colored groupings represent TMOs that were not found to be significantly different using pairwise comparisons to each other, via Bonferroni adjustment, at $p < 0.05$

4.2 Scenario 1—reflections across the whole screen

Scenario 1 was where the participants were less consistent while ranking the TMOs, and therefore, the groupings were more complex, although the obtained results show that there is significant difference between TMOs. Despite the complex groupings, it is possible to see that Pat and ManPat were the top two TMOs, with no significant difference between them, followed by Man and PatMan.

The reported main effect of TMO was $F(2.68, 74.97) = 5.31$, and the Greenhouse-Geisser correction as sphericity was violated (Maulchy's test for sphericity, $p < 0.01$). As before, the main effect of the group was not significant $F(2.68, 74.97)$ ($p > 0.05$). The Kendall's coefficient of concordance was significant at $W = 0.14$. Table 4 shows the results obtained for each scene on this scenario.

4.3 Scenario 2—no reflections on the screen

In scenario 2, Pat was once again ranked as the top TMO and there was a significant difference from the other TMOs which were all grouped together (ManPat, PatMan, and Man was the ranking order). In this scenario, the ranking concordance was the highest compared with the other scenarios. There was also no significant difference between the groups' rankings.

Table 4 Results obtained for each video on Scenario 1 (reflections across whole screen). Colored groupings represent TMOs that were not found to be significantly different using pairwise comparisons to each other, via Bonferroni adjustment, at $p < 0.05$.

	Kendall's Co-efficient of Concordance	1st	2nd	3rd	4th
CGRoom	0.03	Man	Pat	PatMan	ManPat
Jaguar	0.12	Pat	ManPat	Man	PatMan
Kalabsha	0.4	Pat	PatMan	ManPat	Man
Morgan Lovers	0.09	Pat	PatMan	ManPat	Man
Explosion	0.03	ManPat	Man	PatMan	Pat
Medical	0.04	ManPat	Man	PatMan	Pat

Table 5 Results obtained for each video on Scenario 2 (no reflections on the screen). Colored groupings represent TMOs that were not found to be significantly different using pairwise comparisons to each other, via Bonferroni adjustment, at $p < 0.05$.

	Kendall's Co-efficient of Concordance	1st	2nd	3rd	4th
CGRoom	0.05	PatMan	Pat	ManPat	Man
Jaguar	0.07	Pat	Man	PatMan	ManPat
Kalabsha	0.05	Pat	PatMan	Man	ManPat
Morgan Lovers	0.04	Pat	ManPat	PatMan	Man
Explosion	0.19	PatMan	Pat	ManPat	Man
Medical	0.12	ManPat	Pat	PatMan	Man

The results reported for TMO main effect were $F(2.82, 81.23) = 2.55$ (Greenhouse-Geisser correction as sphericity was violated, $p < 0.01$)). The reported group main effect was not significant $F(3, 84) = 3.58$ ($p > 0.05$). The computed Kendall's coefficient of concordance was $W = 0.24$ ($p < 0.05$). For further reference, the results for each scene are shown on Table 5.

4.4 Scenario 3—reflections on half of the screen

The third scenario reported a significant difference between TMOs and, once again, Pat was classified as first. However, in this scenario, it was grouped together with PatMan meaning that there were no significant differences between them. The level of concordance in this scenario was higher than in the scenario in which the screen was entirely under reflection.

The TMOs' main effect was $F(2.91, 81.23) = 7.74$ (Greenhouse-Geisser correction was applied, $p < 0.01$)). As before, there was no significant difference between the groups' rankings as the main effect was $F(2.91, 81.23) = 1.57$ ($p > 0.05$). The Kendall's coefficient of concordance on this scenario was $W = 0.19$ ($p < 0.05$). The results for each scene are shown on Table 6 for completeness.

5 Discussion

The main goals of the experiment were to study the impact of reflections on mobile device displays and to provide some insights on whether the perceptual accuracy of the TMOs changed depending on the different scenarios. With this knowledge, it is possible to understand if there is a need to develop a new or hybrid TMO that can deal with reflections better and which features should be taken

Table 6 Results obtained for each video on Scenario 3 (reflections on half of the screen). Colored groupings represent TMOs that were not found to be significantly different using pairwise comparisons to each other, via Bonferroni adjustment, at $p < 0.05$.

	Kendall's Co-efficient of Concordance	1st	2nd	3rd	4th
CGRoom	0.07	PatMan	Pat	ManPat	Man
Jaguar	0.06	Pat	PatMan	Man	ManPat
Kalabsha	0.12	Pat	PatMan	Man	ManPat
Morgan Lovers	0.16	Pat	Man	PatMan	ManPat
Explosion	0.05	ManPat	Man	PatMan	Pat
Medical	0.04	Pat	PatMan	Man	ManPat

into account when working with TMOs with the purpose of dealing with reflective scenarios.

For the three scenarios, the calculated Kendall's coefficient of concordance was significant for $p < 0.05$ meaning that there was significant agreement between the participants, giving statistical significance to the obtained results. Furthermore, the results showed that there was no significant difference between the two groups that performed the experiment indicating coherence between choices. This gives more strength to the results. A first conclusion that can be extracted from the results is that overall results reported significant differences between the three experimental scenarios and all the scenarios reported significant differences between the TMOs. This indicates that the reflections do indeed have an impact on a TMOs' perceptual accuracy and that their perceptual accuracy can change according to the usage scenario.

Scenario 1 (reflection across the whole display) had a lower Kendall's coefficient of concordance indicating a lower agreement between the participants' rankings. On top of that, the groupings were complex in this case suggesting that it is more difficult to choose which TMO is the best for this condition. This may indicate that when the display is fully exposed to reflections, a TMO's perceptual accuracy can be compromised and the visualization experience negatively affected. Further studies are required to validate this assumption.

Scenario 2 (no reflection) is similar to the experiment described in [8]. As in this paper, Pat was the best ranked TMO. One significant difference between scenario 2 and [8] was that the Kendall's coefficient of concordance reported was $W = 0.24$ for this work compared with $W = 0.89$ in [8]. This can be because in this paper, we selected only the top two TMOs reported in [18] so the choice in quality may not have been that obvious.

Interestingly, in scenario 2, although PatMan and ManPat are a combination of two TMOs on the same frame, they were grouped together with Man and indeed slightly preferred than Man. This result suggests a preference towards the video that includes the most preferred TMO, even if it has not been used for the whole frame. Eye tracking will be used in future work to help confirm whether the participants spent more time looking at the part of the frame computed with the preferred TMO or not.

In scenario 3 (reflection across half of the screen) the participants agreed more on their choices than in scenario 1 (reflection across whole screen) ($W = 0.19$ against $W = 0.14$) which means that reflections have a negative impact on the TMOs' perceptual accuracy and the more the area exposed to reflections the greater this negative impact. The TMOs rank order was Pat, PatMan, ManPat, and Man. An important result is that Pat and PatMan were significantly better than the other two TMOs.

Urbano et al. [16] identified that stronger detail reproduction, more saturated colors, and overall brighter image appearance were preferred. Our findings do not corroborate with those, and we attribute this to visual attention mechanisms as it has also been shown that the mechanisms of visual attention are much more significant for images than for video due to the interframe correlation and shorter viewing time [26]. An additional factor that could have contributed to the differences between Urbano et al.'s [16] work and ours is that that while viewing an image, participants focus more on all the details across the whole image whereas when viewing a video, it is most likely that they devote more attention to regions were motion occurs [27].

Another interesting result is that PatMan and ManPat are always grouped together with Man and that they were slightly preferred than Man. This result indicates a preference towards the video that have been tone mapped with the preferred TMO even if this is not on the entire frame. As in scenario 2, this needs to be confirmed with eye tracking.

6 Conclusions

In this paper, we set out to undertake an evaluation of the impact of reflections on a screen by understanding how different HDR video TMOs perform under reflective conditions. In addition, this study intended to clarify if there is a need to develop a new or hybrid TMO that can deal with reflections better than existing TMOs.

Overall, the results have shown that there are significant differences between scenarios where there is no reflection and scenarios with reflections as well as differences between the TMOs' perceptual accuracy. The results further show that reflections have a negative impact on the visualization experience since there was less consistency and coherence in participants' responses in the scenarios where there were reflections on the display. An important result is that Pat was the top TMO for all the scenarios, i.e., with and without reflections. It outperformed the hybrids, PatMan and ManPat. Furthermore, the hybrids were never significantly better than the second preferred TMO, Man. We can conclude therefore that, at least in the scenarios we studied, there is no need for a hybrid TMO. This might also suggest that, similarly to what happens when comparing TMOs' performance on dark environments against TMOs' performance on bright environments, participants seems to prefer contrast and naturalness over color and details, and therefore, these image features should be carefully addressed when developing TMOs to deal with reflective scenarios.

The paper did show that the scenario 3 results were more coherent and consistent than the results of scenario 1. This demonstrates that the reflections' negative impact in this scenario was less that in scenario 1, suggesting that the more reflections on the display, the more the negative the impact on the viewing experience. The paper did not answer, however, whether having only half of the screen exposed to reflections was more negative for the viewing experience than having the full screen exposed, especially as the visual mechanisms could not adapt properly to the different exposed regions of the screen.

Finally, the results suggest that having two TMOs simultaneously applied to the same frame does not have a negative effect on the perceptual accuracy rankings. Here, the participants demonstrated a preference for the cases where Pat was applied (even only partially). This has raised the question: "Did participants unconsciously rank the videos based on the most accurate region of the frame rather than a whole?" This will need to be investigated further in future work with the help of eye tracking to better understand how and which features participants value more when evaluating videos' perceptual accuracy in a variety of different scenarios. This study was the starting point of a new research question regarding reflection impact on TMO performance which has shown that reflection does indeed have a negative impact. Future work will consider a more extensive set of state-of-the-art TMOs in order to verify if there are TMOs that can minimize these negative effects. As there is a wide spectrum of mobile devices with different screen features, an important variable that will also need to be taken into account is the impact of the absolute reflection index of the device and how can TMOs take advantage of it.

Competing interests

The authors declare that they have no competing interests.

Acknowledgments

We would like to thank to the the participants for taking part in the study and to Elmedin Selmanovic for providing the Jaguar HDR footage.
This work was partially supported by the Portuguese government, through National Foundation for Science and Technology - FCT (Fundação para a Ciência e a Tecnologia) for supporting this PhD through the grant SFRH/BD/76384/2011. I would also like to acknowledge the European Union (COMPETE, QREN and FEDER) for the partial support through the project REC I/EEI-SII 0360/2012 entitled "MASSIVE - Multimodal Acknowledgeable multiSenSorial Immersive Virtual Enviroments". This work was also partially supported by the ICT COST Action IC1005 "HDRi: The digital capture, storage, transmission and display of real-world lighting". Debattista and Chalmers are partially supported by a Royal Society Industrial Fellowship. This work was also partially supported by the ICT COST Action IC1005 "HDRi: The digital capture, storage, transmission and display of real-world lighting".

Author details

[1]Universidade de Trás-os-Montes e Alto Douro, Quinta de Prados, 5000-801 Vila Real, Portugal. [2]INESC-TEC, Rua Dr. Roberto Frias, 4200-465 Porto, Portugal. [3]WMG, University of Warwick, Gibbet Hill Road, CV4 7AL Coventry, UK. [4]goHDR Ltd., Gibbet Hill Road, CV4 7AL Coventry, UK.

References

1. H Seetzen, W Heidrich, W Stuerzlinger, G Ward, L Whitehead, M Trentacoste, A Ghosh, A Vorozcovs, High dynamic range display systems. ACM Trans. Graph. **23**(3), 760–768 (2004). doi:10.1145/1015706.1015797
2. IT Union, The World in 2013: ICT facts and figures (2011). http://www.itu.int/ITU-D/ict/facts/material/ICTFactsFigures2011.pdf. Accessed: 21/11/2013
3. CISCO, Cisco visual networking index: global mobile data traffic forecast update, 2012-2017 (2013). http://www.cisco.com/c/en/us/solutions/collateral/serv\ice-provider/visual-networking-index-vni/white_paper_c11-520862.html. Accessed: 21/11/2013
4. Ooyala, *Global video index q3 2014*. (Report, International Telecommunication Union, Silicon Valley, USA, 2014). http://go.ooyala.com/rs/OOYALA/images/Ooyala-Global-Video-Index-Q3-2014.pdf
5. J Ross, R Simpson, B Simpson, Media richness, interactivity and retargeting to mobile devices: a survey. Int. J. Arts Technol. **4**(4), 442–459 (2011). doi:10.1504/IJART.2011.043443
6. SN Pattanaik, J Tumblin, H Yee, DP Greenberg, in *Proceedings of the 27th Annual Conference on Computer Graphics and Interactive Techniques. SIGGRAPH '00*. Time-dependent visual adaptation for fast realistic image display (ACM Press/Addison-Wesley Publishing Co., New York, NY, USA, 2000), pp. 47–54. doi:10.1145/344779.344810
7. R Mantiuk, S Daly, L Kerofsky, Display adaptive tone mapping. ACM Trans. Graph. **27**(3), 68–16810 (2008). doi:10.1145/1360612.1360667
8. M Melo, M Bessa, K Debattista, A Chalmers, Evaluation of Tone-Mapping Operators for HDR Video Under Different Ambient Luminance Levels. Computer Graphics Forum. **34**(8), 38–49 (2015). doi:10.1111/cgf.12606 http://dx.doi.org/10.1111/cgf.12606
9. F Banterle, A Artusi, K Debattista, A Chalmers, *Advanced High Dynamic Range Imaging: Theory and Practice*, 1st edn. (AK Peters, Ltd (CRC Press), Natick, MA, USA, 2011). http://vcg.isti.cnr.it/Publications/2011/BADC11
10. R Mantiuk, K Myszkowski, H-P Seidel, in *Proceedings of IEEE International Conference on Systems, Man and Cybernetics*. Visible difference predicator for high dynamic range images (IEEE, New Jersey, USA, 2004), pp. 2763–2769
11. R Mantiuk, KJ Kim, AG Rempel, W Heidrich, Hdr-vdp-2: A calibrated visual metric for visibility and quality predictions in all luminance conditions. ACM Trans. Graph. **30**(4), 40–14014 (2011). doi:10.1145/2010324.1964935
12. F Drago, W Martens, K Myszkowski, H-P Seidel, *Perceptual evaluation of tone mapping operators with regard to similarity and preference*. (Research Report MPI-I-2002-4-002, Max-Planck-Institut für Informatik, Germany, 2002)
13. P Ledda, A Chalmers, T Troscianko, H Seetzen, in *ACM SIGGRAPH 2005 Papers. SIGGRAPH '05*. Evaluation of tone mapping operators using a high dynamic range display (ACM, New York, NY, USA, 2005), pp. 640–648. doi:10.1145/1186822.1073242
14. M Čadík, M Wimmer, L Neumann, A Artusi, Evaluation of HDR tone mapping methods using essential perceptual attributes. Comput. Graph. **32**(3), 330–349 (2008)
15. G Eilertsen, J Unger, R Wanat, R Mantiuk, in *ACM SIGGRAPH 2013 Talks*. Survey and evaluation of tone mapping operators for hdr video (ACM, New York, USA, 2013)
16. C Urbano, L Magalhães, J Moura, M Bessa, A Marcos, A Chalmers, Tone mapping operators on small screen devices: an evaluation study. Comput. Graph. Forum. **29**(8), 2469–2478 (2010). doi:10.1111/j.1467-8659.2010.01758.x
17. AO Akyüz, ML Eksert, MS Aydin, An evaluation of image reproduction algorithms for high contrast scenes on large and small screen display devices. Comput. Graph. **37**(7), 885–895 (2013). doi:10.1016/j.cag.2013.07.004
18. M Melo, M Bessa, K Debattista, A Chalmers, Evaluation of HDR video tone mapping for mobile devices. Signal Process. Image Commun. **29**(2), 247–256 (2014). doi:10.1016/j.image.2013.09.010 Special Issue on Advances in High Dynamic Range Video Research
19. JA Ferwerda, SN Pattanaik, P Shirley, DP Greenberg, in *Proceedings of the 23rd Annual Conference on Computer Graphics and Interactive Techniques. SIGGRAPH '96*. A model of visual adaptation for realistic image synthesis (ACM, New York, NY, USA, 1996), pp. 249–258. doi:10.1145/237170.237262
20. R Boitard, K Bouatouch, R Cozot, D Thoreau, A Gruson, in *Applications of Digital Image Processing XXXV. Proc. SPIE*. Temporal coherency for video tone mapping, vol. 8499 (SPIE, San Diego, CA, USA, 2012), pp. 84990–8499010. doi:10.1117/12.929600
21. JH Van Hateren, Encoding of high dynamic range video with a model of human cones. ACM Trans. Graph. **25**(4), 1380–1399 (2006). doi:10.1145/1183287.1183293

22. TO Aydin, N Stefanoski, S Croci, M Gross, A Smolic, Temporally coherent local tone mapping of hdr video. ACM Trans. Graph. **33**(6), 196–119613 (2014). doi:10.1145/2661229.2661268
23. G Eilertsen, RK Mantiuk, J Unger, Real-time noise-aware tone mapping. ACM Trans. Graph. **34**(6) (2015). doi:10.1145/2816795.2818092
24. IT Union, *Methodology for the subjective assessment of the quality of television pictures*. (International Telecommunication Union, Geneve, Switzerland, 2012). https://www.itu.int/dms_pubrec/itu-r/rec/bt/R-REC-BT.500-13-201201-I!!PDF-E.pdf
25. M Čadík, M Wimmer, L Neumann, A Artusi, in *Proceedings of Pacific Graphics 2006 (14th Pacific Conference on Computer Graphics and Applications)*. Image attributes and quality for evaluation of tone mapping operators (National Taiwan University Press, Taipe, Taiwan, 2006), pp. 35–44. http://www.cg.tuwien.ac.at/research/publications/2006/CADIK-2006-IAQ/
26. M Narwaria, MP Da Silva, P Le Callet, R Pepion, in *Signal Processing Conference (EUSIPCO), 2014 Proceedings of the 22nd European*. Single exposure vs tone mapped high dynamic range images: A study based on quality of experience (Springer International Publishing, Switzerland, 2014), pp. 2140–2144
27. Y Zhai, M Shah, in *Proceedings of the 14th Annual ACM International Conference on Multimedia. MULTIMEDIA '06*. Visual attention detection in video sequences using spatiotemporal cues (ACM, New York, NY, USA, 2006), pp. 815–824. doi:10.1145/1180639.1180824, http://doi.acm.org/10.1145/1180639.1180824

4

Grid-based approximation for voice conversion in low resource environments

Hadas Benisty, David Malah* and Koby Crammer

Abstract

The goal of voice conversion is to modify a source speaker⊠ speech to sound as if spoken by a target speaker. Common conversion methods are based on Gaussian mixture modeling (GMM). They aim to statistically model the spectral structure of the source and target signals and require relatively large training sets (typically dozens of sentences) to avoid over-fitting. Moreover, they often lead to muffled synthesized output signals, due to excessive smoothing of the spectral envelopes.

Mobile applications are characterized with low resources in terms of training data, memory footprint, and computational complexity. As technology advances, computational and memory requirements become less limiting; however, the amount of available training data still presents a great challenge, as a typical mobile user is willing to record himself saying just few sentences. In this paper, we propose the grid-based (GB) conversion method for such low resource environments, which is successfully trained using very few sentences (5–10). The GB approach is based on sequential Bayesian tracking, by which the conversion process is expressed as a sequential estimation problem of tracking the target spectrum based on the observed source spectrum. The converted Mel frequency cepstrum coefficient (MFCC) vectors are sequentially evaluated using a weighted sum of the target training vectors used as grid points. The training process includes simple computations of Euclidian distances between the training vectors and is easily performed even in cases of very small training sets.

We use global variance (GV) enhancement to improve the perceived quality of the synthesized signals obtained by the proposed and the GMM-based methods. Using just 10 training sentences, our enhanced GB method leads to converted sentences having closer GV values to those of the target and to lower spectral distances at the same time, compared to enhanced version of the GMM-based conversion method. Furthermore, subjective evaluations show that signals produced by the enhanced GB method are perceived as more similar to the target speaker than the enhanced GMM signals, at the expense of a small degradation in the perceived quality.

Keywords: Bayesian tracking, Global variance (GV), Mel cepstral distortion (MCD), Grid-based approximation, Spectral conversion

1 Introduction

Voice conversion systems aim to modify the perceived identity of a source speaker saying a sentence to that of a given target speaker. This kind of transformation is useful for personalization of text-to-speech (TTS) systems, voice restoration in case of vocal pathology, obtaining a false identity when answering the phone (for safety reasons, for example), and also for entertainment purposes such as online role-playing games.

The identity of a speaker is associated with the spectral envelope of the speech signal, and with its prosody attributes: pitch, duration, and energy. Most voice conversion methods aim to transform the spectral envelope of the source speaker to the spectral envelope of the target speaker. The pitch contour is commonly converted by a linear transformation based on the global mean and standard deviation values of the pitch frequency.

The classical conversion method, based on modeling the spectral structure of the speech signals using Gaussian mixture model (GMM), is the most commonly used method to date. The conversion function is linear, trained using either least squares (LS) [1], or a joint source-

*Correspondence: malah@ee.technion.ac.il
Electrical Engineering Department, Technion–Israel Institute of Technology, Technion City, Haifa, Israel

target GMM training (JGMM) [2]. These linear conversion methods produce over-smoothed spectral envelopes leading to muffled synthesized speech [3, 4]. Several modifications of the GMM-based conversion have been proposed since, among these are as follows: GMM with dynamic frequency warping (DFW) [4], GMM and codebook selection [5], and a combined pitch and spectral envelope GMM-based conversion [6]. Still, these GMM-based conversion methods have been reported to produce muffled output signals, probably due to excessive smoothing of the temporal evolution of the spectral envelope. Recently, a different approach aiming to capture the temporal evolution of the spectral envelope was presented [7]. A GMM is trained using concatenated sequences of the source and target spectral features, and the conversion function is evaluated using maximum likelihood (ML) estimation. To reduce the muffling effect, the global variance (GV) of the spectral features was considered in the trained statistical model. A GV enhancement method called CGMM was also proposed, [8], in the framework of the classical GMM-based conversion, where the GV of the converted features is constrained to match the GV of the features related to the target speaker. These two conversion schemes (with integrated GV enhancement) improve the quality of the converted signals, at the expense of some increase in the spectral distance between the converted and target signals. A real-time implementation for the ML approach have also been proposed [9]. This implementation is based on a low-delay estimation of the conversion parameters [10] using recursive parameter generation and GV enhancement.

In order to estimate a conversion function from a source speaker to a target speaker, voice conversion methods use training sets of both speakers. Most training algorithms require parallel data sets, that is, prerecorded sentences of the source and target speakers saying the same text. In such a setup, evaluation of a conversion function is based on coupled feature vectors—source and target. Alternatively, some methods have been proposed, suggesting training algorithms which avoid the need for pre-alignment altogether. A probabilistic approach presented by Nankaku et al. includes statistical modeling for optimizing the conversion function and the correspondence between source-target segments [11]. Another method which does not require time alignment as a preprocessing stage is the iterative combination of a nearest neighbour search step and a conversion step alignment method (INCA) [12]. This method uses iterative estimation of the alignment (using nearest neighbour search) and conversion estimation (classical GMM conversion). Recently, we proposed a modified version of this method called temporal-context INCA (TC-INCA), using context vectors instead of single spectral vectors, which lead to improved estimation of the alignment and to higher quality and similarity to the target speaker [13]. Although these methods were designed for a non-parallel setup, they can be used in a parallel setup, when aligned data is unavailable.

Even when a parallel training set is available, matching an analysis frame of the source speaker to one of the analysis frames of the target speaker is not straightforward, since the two speakers generally do not pronounce the text at the exact same rate. A time alignment is usually carried out using dynamic time warping (DTW), constrained by starting and ending of speech utterances [14]. These time stamps are commonly obtained by phonetic labeling, representing the beginning and ending of each phoneme. Since the source and target training sentences are not spoken in exactly the same rate, DTW often replicates or omits feature vectors, artificially producing a match. The importance of correct time alignment was recently demonstrated as having a large influence on the quality of the synthesized converted speech [15]. A different approach was suggested by [16], where a statistical model for an eigen-voice was trained using several parallel data sets. The conversion function is trained using the eigen-voice model and speech sentences related to a target speaker (not necessarily parallel to the source data sets).

The GMM-based conversion methods mentioned above, using either parallel or non-parallel data, typically require several dozens of sentences for training, and therefore when applied in a mobile environment impose a long recording session on the user. Even the low delay GMM-based approach suggested by Toda et al. was reported to be trained using 60–250 mixtures and 50 training sentences [9]. Therefore, applying them in a mobile environment would compel the user to a long recording session.

Some approaches for training a conversion function that are not based on GMM have been proposed, among them training using a state-space representation [17], and using exemplar-based sparse-representation [18]. Since these methods are closely related to the proposed GB method, we address them and discuss the differences between them and the GB approach in more details after describing the proposed method in this work (see Section 4). Still, these method are also not suitable for mobile environment since they require several hundreds of parallel training sentences and/or very high computational load during conversion and a substantial memory footprint.

In this paper, we propose a method for spectral conversion based on a grid-based (GB) approximation [19]. We express the spectral conversion process as a sequential Bayesian estimation problem of tracking the target spectrum using observed samples from the source spectrum. We propose models for evaluation of the evidence and likelihood probabilities needed for the GB formulation. Using these approximated probabilities, the algorithm sequentially evaluates the converted spectrum as a weighted sum of the target training vectors. Recently, we presented a similar method using GB approximation which requires phonetic labeling during the test stage [20]. In this paper, we propose a modified version of this method, which does not require any labeling for testing. Additionally, as in TC-INCA [13], we use context vectors instead of single vectors in order to improve the estimation of the likelihood probability. Altogether, we present here a more thorough description of the GB approximation method and its modified application for voice conversion under low resources constraints, followed by an extended analysis and detailed results.

Furthermore, as opposed to previously proposed methods that use parallel and time-aligned training sets, the GB conversion approach does not require a one-to-one correspondence between the source and target training vectors. The training process uses parallel sentences but is based on soft correspondence between the source and target vectors, obtained by phonetic labeling of the training sentences without frame alignment, thus eliminating the need for DTW.

Unlike other GMM-based methods that use statistical modeling of the spatial structure of the source and target spectra, the GB method is data-driven, so it is easily trained using merely 5–10 sentences. Its training stage involves simple computations based on the Euclidean distance between the training vectors.

Objective evaluations show that the GB conversion method proposed here leads to GV values that are closer to the GV values of the target speaker than the classical GMM conversion method and to lowest (or very close to it) spectral distance to the target spectra, at the same time. To further improve the quality, we applied a GV enhancement post-processing block. We recently proposed this GV enhancement approach and examined its effect on signals converted by a classical GMM conversion method [21]. In this paper, we present an overall scheme, enhanced GB (En-GB), consisting of GB conversion, followed by GV enhancement. We used objective measures and performed extensive subjective evaluations to compare our proposed En-GB scheme to joint GMM (JGMM) [2], also followed by the same GV enhancement block (En-JGMM) and to a GMM-based conversion, trained with a GV constraint (CGMM) [8]. Objectively, En-GB leads to better performance than En-JGMM and CGMM in terms of both spectral distance and GV, using 10 sentences. Listening tests show that in terms of similarity to the target, En-GB outperforms the other examined methods. In terms of quality, CGMM was rated as best, where En-GB was rated as comparable to En-GMM.

This paper is organized as follows. In Section 2, a brief description of GB approximation is presented. The GB conversion method is described in Section 3. The difference between the GB approach and some related methods is discussed in Section 4. Experimental results, demonstrating the performance of the proposed En-GB scheme compared to En-GMM-based methods, are presented in Section 5. Conclusions and further research suggestions are given in Section 6.

2 Grid-based formulation

A brief formulation of sequential estimation using Bayesian tracking is presented in Section 2.1. In many practical cases, applying this formulation yields a high computational load, which is sometimes unfeasible. The GB method provides a discrete approximation for Bayesian tracking with much less computational complexity, as described in Section 2.2.

2.1 Bayesian tracking

Let $\mathbf{y}_t$ denote a hidden state vector that follows a first-order Markov dynamics as

$$\mathbf{y}_t = f_t\left(\mathbf{y}_{t-1}, \mathbf{u}_t\right), \tag{1}$$

where f_t is a function (not necessarily linear) of $\mathbf{y}_{t-1}$ and of an i.i.d. noise sequence $\mathbf{u}_t$. The observed signal, $\mathbf{x}_t$, depends on the hidden state and on an i.i.d. measurement noise,$\mathbf{v}_t$:

$$\mathbf{x}_t = h_t\left(\mathbf{y}_t, \mathbf{v}_t\right), \tag{2}$$

where $h_t(\cdot)$ may also be non-linear.

The Bayesian optimal estimate for the state vector $\mathbf{y}_t$ in terms of minimizing the mean square error, given t vectors sequentially sampled from the observed process—$\mathbf{x}_{1:t} \triangleq \{\mathbf{x}_1, \ldots, \mathbf{x}_t\}$, is obtained by[1]

$$\hat{\mathbf{y}}_t = E\left[\mathbf{y}_t|\mathbf{x}_{1:t}\right] = \int p\left(\mathbf{y}_t|\mathbf{x}_{1:t}\right)\mathbf{y}_t d\mathbf{y}_t. \tag{3}$$

The posterior probability $p\left(\mathbf{y}_t|\mathbf{x}_{1:t}\right)$ can be obtained recursively in two stages:

1. Prediction: obtain the prior probability

$$p\left(\mathbf{y}_t|\mathbf{x}_{1:t-1}\right) = \int p\left(\mathbf{y}_t|\mathbf{y}_{t-1}\right)p\left(\mathbf{y}_{t-1}|\mathbf{x}_{1:t-1}\right)d\mathbf{y}_{t-1}. \tag{4}$$

2. Update: use the current observation $\mathbf{x}_t$ to update the posterior probability

$$p\left(\mathbf{y}_t|\mathbf{x}_{1:t}\right) = \frac{p\left(\mathbf{x}_t|\mathbf{y}_t\right)p\left(\mathbf{y}_t|\mathbf{x}_{1:t-1}\right)}{p\left(\mathbf{x}_t|\mathbf{x}_{1:t-1}\right)}, \tag{5}$$

where

$$p\left(\mathbf{x}_t|\mathbf{x}_{1:t-1}\right) = \int p\left(\mathbf{x}_t|\mathbf{y}_t\right)p\left(\mathbf{y}_t|\mathbf{x}_{1:t-1}\right)d\mathbf{y}_t. \tag{6}$$

This recursion is initialized by setting the prior probability to be equal to the initial probability of the state vector: $p(\mathbf{y}_0|\mathbf{x}_0) = p(\mathbf{y}_0)$, where $p(\mathbf{y}_0)$ is assumed to be known (in practice, mostly taken as a uniform distribution). The likelihood function $p(\mathbf{x}_t|\mathbf{y}_t)$ that appears in Eq. (5) is determined according to the measurement model (Eq. (2)) and the statistics of the measurement noise $\mathbf{v}_t$.

When the noise signals $\mathbf{u}_t$ and $\mathbf{v}_t$ are Gaussian, and the functions $f_t(\cdot)$ and $h_t(\cdot)$ are linear and time invariant (meaning that $f_t(\cdot) \equiv f(\cdot)$ and $h_t(\cdot) \equiv h(\cdot)$), this recursion can be computed analytically, leading to Kalman filtering [22]. Yet, in most practical cases where these conditions are not sustained, this derivation is hard and often performed using approximation methods such as GB approximation or particle filtering [19]. These methods sequentially evaluate the posterior probability as a discrete weighted sum using a given set of samples in case of GB or a randomly drawn set in case of particle filtering.

In this paper, we express the spectral conversion process as a sequential estimation problem tracking the target spectrum, using observed samples from the source spectrum. We propose models for the evidence and likelihood probabilities needed for the GB formulation. Using these approximated probabilities the algorithm sequentially evaluates the converted spectrum as a weighted sum of the target training vectors. It is well known that the performance of particle filtering crucially depends on successful statistical modeling of the state-space temporal evolution. The performance of GB, on the other hand, depends on dense modeling of the state space by a set of predetermined grid points. Nevertheless, in the following sections, we show that 5–10 training sentences alone, which still result in several thousands of spectral feature vectors, are sufficient for training a GB conversion. Our subjective evaluations show that the GB conversion is found to be better or comparable, at least, to the classical GMM conversion method, when trained by this small set.

2.2 Grid-based approximation

The main principle of GB approximation is to provide a Bayesian sequential estimation framework while avoiding the integral computations in Eqs. (4) and (6) by using a discrete evaluation of the posterior probability.

Let $\left\{\mathbf{y}_t^k\right\}_{k=1}^{N_y}$ be a set of predetermined grid points taken from the state space $\left\{\mathbf{y}_t\right\}$. We divide the state space into cells, so that each cell has a grid point $\mathbf{y}_t^k$ as its center. Thus, the posterior probability can be approximated by[2]

$$p(\mathbf{y}_t|\mathbf{x}_{1:t}) \approx \sum_{k=1}^{N_y} w_{t|t}^k \delta\left(\mathbf{y}_t - \mathbf{y}_t^k\right), \tag{7}$$

where the posterior weights $\left\{w_{t|t}^k\right\}_{k=1}^{N_y}$ denote the conditional probabilities

$$w_{t|t}^k = p\left(\mathbf{y}_t = \mathbf{y}_t^k|\mathbf{x}_{1:t}\right). \tag{8}$$

Using this discrete approximation, the prior probability is also approximated as a discrete sum

$$p(\mathbf{y}_t|\mathbf{x}_{1:t-1}) \approx \sum_{k=1}^{N_y} w_{t|t-1}^k \delta\left(\mathbf{y}_t - \mathbf{y}_t^k\right). \tag{9}$$

The prior weights can be estimated sequentially [19]

$$w_{t|t-1}^k \approx \sum_{l=1}^{N_y} w_{t-1|t-1}^l p\left(\mathbf{y}_t^k|\mathbf{y}_{t-1}^l\right), \tag{10}$$

where $p\left(\mathbf{y}_t^k|\mathbf{y}_{t-1}^l\right)$, called the *evidence probability*, is derived from the state space dynamics (Eq. (1)). The posterior weights $\left\{w_{t|t}^k\right\}_{k=1}^{N_y}$ are evaluated by the following:

$$w_{t|t}^k \approx \frac{w_{t|t-1}^k p\left(\mathbf{x}_t|\mathbf{y}_t^k\right)}{\sum_{l=1}^{N_y} w_{t|t-1}^l p\left(\mathbf{x}_t|\mathbf{y}_t^l\right)}, \tag{11}$$

where, as stated above, the likelihood probability $p\left(\mathbf{x}_t|\mathbf{y}_t^k\right)$ is derived from the measurement model (Eq. (2)).

Finally, the hidden state vector $\mathbf{y}_t$ is approximated using the posterior weights

$$\hat{\mathbf{y}}_t = E\left[\mathbf{y}_t|\mathbf{x}_{1:t}\right] \approx \sum_{k=1}^{N_y} w_{t|t}^k \mathbf{y}_t^k. \tag{12}$$

Note that Eqs. (10), (11), and (12) are discrete evaluations of Eqs. (4)–(3), correspondingly. It is known [19] that the estimated terms in Eq. (7) and in Eq. (12) are biased for any finite N_y. Still, as more grid points are taken, the bias gets smaller and the approximation improves, since the state space is more densely represented.

The sequential estimation process is initialized using the initial probability of the state vector $p\left(\mathbf{y}_0^k\right)$, which as stated above, is assumed to be known

$$w_{0|0}^k = p\left(\mathbf{y}_0^k\right). \tag{13}$$

Table 1 summarizes the main stages of sequential Bayesian estimation using GB approximation.

3 Voice conversion using grid-based approximation

We now use the GB approximation method described above as a framework for spectral voice conversion. We express the conversion as a sequential estimation problem, where the observed process is the source spectrum, and the tracked state space is the target spectrum. We

Table 1 Bayesian estimation using grid-based approximation

Input: a sequence of states sampled from the observed process $\mathbf{x}_{1:T}$

Initialization: set the initial weights, $\left\{w^k_{0|0}\right\}_{k=1}^{N_y}$, using Eq. (13)

Main iteration: for $t = 1, \ldots T$, perform the following steps:

1. Evaluate the prior weights, $\left\{w^k_{t|t-1}\right\}_{k=1}^{N_y}$, using Eq. (10).
2. Evaluate the posterior weights, $\left\{w^k_{t|t}\right\}_{k=1}^{N_y}$, using Eq. (11).
3. Evaluate the hidden state, $\hat{\mathbf{y}}_t$, using Eq. (12).

Output: a sequence of the estimated hidden states $\hat{\mathbf{y}}_{1:T}$

propose models for both likelihood and evidence densities, required for the sequential estimation process, as described in Eqs. (10)–(12).

The GB conversion method proposed here uses a parallel training set but does not require time alignment between the source and target training vectors since it is trained using soft correspondence between them, rather than matched pairs. The training and conversion stages of the proposed GB conversion method are presented below in Sections 3.1 and 3.2, respectively.

3.1 Training stage

The training process described here includes pre-computation of the evidence and discrete likelihood probabilities. These probabilities are evaluated using all available training data. Note the difference from our previously presented GB method, where these probabilities were evaluated separately for each phoneme [20]. The source and target training sentences are assumed to be parallel and phonetically labeled. The spectral features of the two speakers are extracted from the voiced frames, but, as stated above, no time alignment is performed. Instead, a matching process of the source and target utterances is performed as follows. Each sequence of frames related to a certain phoneme at the source is matched to its corresponding sequence at the target, according to the phonetic labeling. When matching frames extracted from recordings of the word 'father', for example, the sequence of frames related to the phoneme 'f' at the source is matched to the sequence of frames related to the phoneme 'f', taken from the target's recording of this word. The same is done for 'a', 'th', etc. Note that although matched sequences mostly have different lengths, our training process does not require using an alignment procedure such as DTW, unlike GMM-based methods do. Based on the matched sequences, we model the *discrete likelihood probability* used in Eq. (11), as follows:

$$p\left(\mathbf{x}_t = \mathbf{x}^m | \mathbf{y}_t = \mathbf{y}^k\right) \propto \begin{cases} 1 & \mathbf{x}^m, \mathbf{y}^k \text{ belong to the same phonetic sequence} \\ 0 & \text{otherwise,} \end{cases} \tag{14}$$

where $\{\mathbf{x}^m\}_{m=1}^{N_x}$ and $\{\mathbf{y}^k\}_{k=1}^{N_y}$ are source and target training vectors, respectively. We normalize the obtained discrete likelihood probability so that

$$\sum_{m=1}^{N_x} p\left(\mathbf{x}_t = \mathbf{x}^m | \mathbf{y}_t = \mathbf{y}^k\right) = 1, \quad \forall k = 1, \ldots, N_y. \tag{15}$$

The discrete likelihood probability defines a relaxed correspondence between the source and target training vectors, as opposed to a one-to-one match defined in other parallel methods, for which $p\left(\mathbf{x}_t = \mathbf{x}^m | \mathbf{y}_t = \mathbf{y}^k\right) = \delta_{m,k}$.

The evidence probability, as mentioned before, expresses the transition probability from state $\mathbf{y}^l$ to state $\mathbf{y}^k$. In natural speech, spectral feature vectors related to consecutive time frames are typically similar, but not identical. Motivated by this behaviour, we model the transition probability as having the same value for all the states inside a ball, centered at $\mathbf{y}^k$ with a radius R_y. The probability of transitions to farther states, however, is taken as a simple Gaussian distribution, centered at $\mathbf{y}^k$. Altogether, we model the *discrete evidence probability*, used in Eq. (10), as follows:

$$p\left(\mathbf{y}_t = \mathbf{y}^k | \mathbf{y}_{t-1} = \mathbf{y}^l\right) = \frac{e^{-\frac{M_{k,l}^2}{2}}}{\sum_{k=1}^{N_y} e^{-\frac{M_{k,l}^2}{2}}}, \tag{16}$$

where the exponential term in Eq. (16) is the maximum between the Mel cepstral distortion (MCD) of the two states $\mathbf{y}^l$ and $\mathbf{y}^k$ normalized by a parameter R_y, and 1

$$M_{k,l} = \max\left(\frac{\mathrm{MCD}\left(\mathbf{y}^k, \mathbf{y}^l\right)}{R_y}, 1\right), \tag{17}$$

$$\mathrm{MCD}\left(\mathbf{y}^k, \mathbf{y}^l\right) = \frac{10\sqrt{2}}{\ln 10}\sqrt{\sum_{p=1}^{P}\left(y^k(p) - y^l(p)\right)^2}, \tag{18}$$

where $y^p(p)$ and $y^l(p)$ are the pth elements of $\mathbf{y}^k$ and $\mathbf{y}^l$, respectively. An alternative approach would be to take the exponential term, defined in Eq. (17), as a normalized distance. For example, $M_{k,l} = \mathrm{MCD}\left(\mathbf{y}^k, \mathbf{y}^l\right)/R_y$, where R_y is a parameter selected by the user. However, in case of a sparse training set, the most substantial probability would be for staying in the same state. Since the training set is fixed, the likelihood and evidence densities are in fact time invariant.

3.2 Conversion stage

The likelihood probability modeled above in Eq. (14) is defined only for a discrete set consisting of the source training vector. In this section, we extend Eq. (14) to model any input vector $\mathbf{x}_t \in \mathbb{R}^P$, as required by the GB formulation.

In our previous work dealing with GB conversion [20], we modeled the continuous likelihood probability

$p\left(\mathbf{x}_t|\mathbf{y}_t=\mathbf{y}^k\right)$ as a sum of the discrete likelihood probabilities $p\left(\mathbf{x}^m|\mathbf{y}_t=\mathbf{y}^k\right)$, $m=1,\ldots,N_x$, (defined in Eqs. (14) and (15)), each weighted by a Gaussian kernel, centered at $\mathbf{x}^m$

$$p\left(\mathbf{x}_t|\mathbf{y}_t=\mathbf{y}^k\right)=\frac{\sum_{m=1}^{N_x}p\left(\mathbf{x}^m|\mathbf{y}_t=\mathbf{y}^k\right)e^{-\mathrm{MCD}^2(\mathbf{x}_t,\mathbf{x}^m)/2R_x^2}}{\sum_{k=1}^{N_y}\sum_{m=1}^{N_x}p\left(\mathbf{x}^m|\mathbf{y}_t=\mathbf{y}^k\right)e^{-\mathrm{MCD}^2(\mathbf{x}_t,\mathbf{x}^m)/2R_x^2}}, \tag{19}$$

where R_x is a parameter determined by the user. The Gaussian term $e^{-\mathrm{MCD}^2(\mathbf{x}_t,\mathbf{x}^m)/2R_x^2}$ can be viewed as an interpolation factor from the discrete space represented by the source training vectors to the continuous space of the test source vectors.

Denote $\mathbf{X}_t=\left(\mathbf{x}_{t-\tau/2},\ldots,\mathbf{x}_t,\ldots,\mathbf{x}_{t+\tau/2}\right)$ as context test vector—a sequence of test source vectors. Also, denote $\{\mathbf{X}_t^m\}_{m=1}^{N_x}$ as training context vectors similarly obtained from the source training set. Previously, in [13], we have shown that Euclidian distance between context vectors leads to improved spectral matching compared with Euclidian distance between single vectors. Although that was shown for matching spectral segments of two different speakers, it is certainly beneficial for matching spectral segments taken from the same speaker. Therefore, we substitute the MCD term in the Gaussian kernel in Eq. (19) with the mean MCD between context vectors, i.e.,

$$p\left(\mathbf{x}_t|\mathbf{y}_t=\mathbf{y}^k\right)=\frac{\sum_{m=1}^{N_x}p\left(\mathbf{x}^m|\mathbf{y}_t=\mathbf{y}^k\right)e^{-\overline{MCD}^2(\mathbf{X}_t,\mathbf{X}_t^m)/2R_x^2}}{\sum_{k=1}^{N_y}\sum_{m=1}^{N_x}p\left(\mathbf{x}^m|\mathbf{y}_t=\mathbf{y}^k\right)e^{-\overline{MCD}^2(\mathbf{X}_t,\mathbf{X}_t^m)/2R_x^2}}$$

$$\overline{\mathrm{MCD}}^2\left(\mathbf{X}_t,\mathbf{X}_t^m\right)=\frac{1}{\tau}\sum_{\nu=-\tau/2}^{\tau/2}\mathrm{MCD}\left(\mathbf{x}_{t+\nu},\mathbf{x}_{t+\nu}^m\right). \tag{20}$$

Define $w_{t|t}^k$ as the posterior weights corresponding to the training vectors $\left\{\mathbf{y}^k\right\}_{k=1}^{N_y}$

$$w_{t|t}^k\triangleq p\left(\mathbf{y}_t|\mathbf{x}_{1:t}\right). \tag{21}$$

During conversion, the posterior weights are sequentially evaluated, using the corresponding evidence and likelihood probabilities defined in Eqs. (16) and (20), according to Eqs. (10) and (11). The posterior weights are used to obtain the converted outcome as a discrete Bayesian approximation (as defined in Eq. (12))

$$\mathcal{F}\{\mathbf{x}_t\}=E\left[\mathbf{y}_t|\mathbf{x}_{1:t}\right]\approx\sum_{k=1}^{N_y}w_{t|t}^k\mathbf{y}_t^k. \tag{22}$$

Due to the sequential update of the posterior weights, the converted spectral outputs evolve smoothly in time, within each phonetic segment, also during transitions between phonemes. Figure 1 demonstrates the obtained

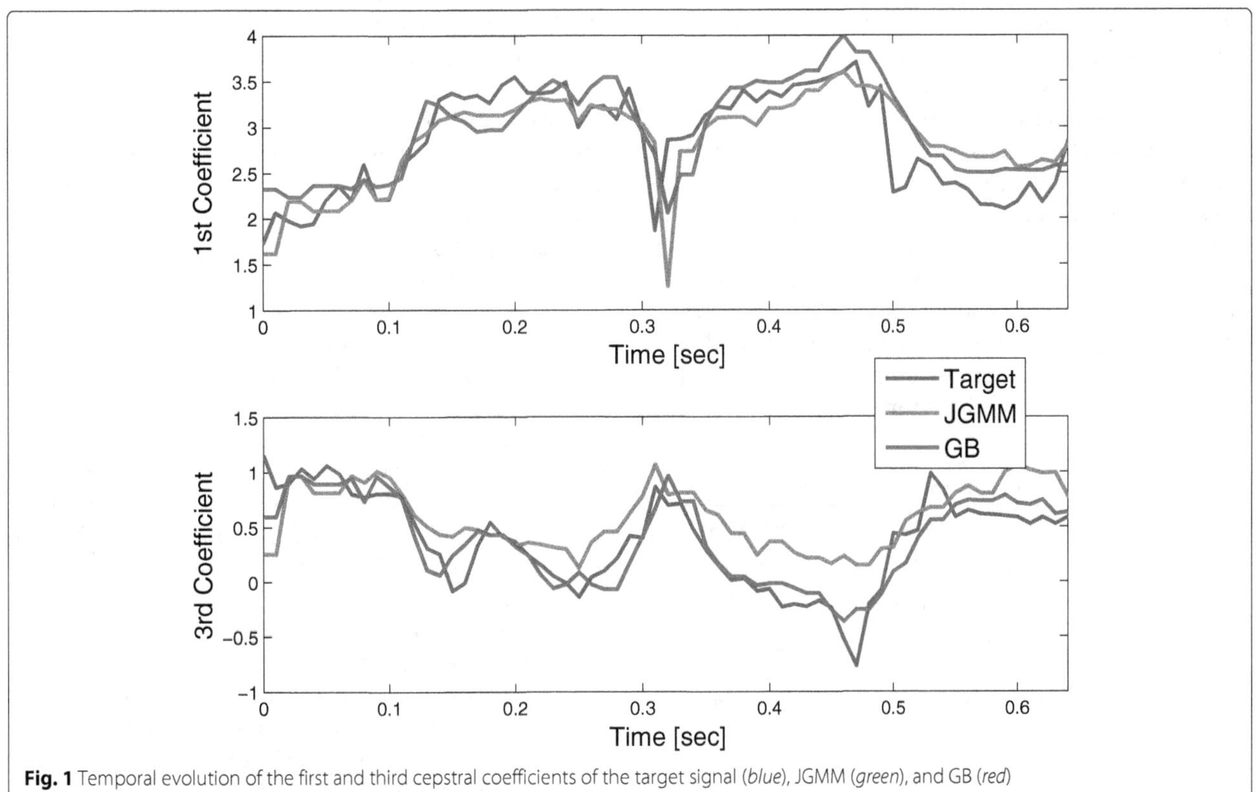

Fig. 1 Temporal evolution of the first and third cepstral coefficients of the target signal (*blue*), JGMM (*green*), and GB (*red*)

time evolution of the first and third Mel frequency cepstrum coefficients (MFCCs) using GB conversion, compared to the classical GMM-based conversion—JGMM [2]. The classical GMM-based conversion is applied frame by frame which may lead to discontinuities. The proposed GB, however, is based on a sequential update leading to a smoother time evolution of the cepstral elements, as seen in Fig. 1.

To conclude, the main stages of converting a sequence of source vectors are summarized in Table 2.

4 Discussion

The GB approach uses a state space representation of the source and target spectra to obtain a converted spectra as a weighted sum of the target training vectors. In this section, we address two related methods: (1) a method based on state space representation [17] and (2) an exemplar-based approach [18], where the converted spectra is evaluated as a weighted sum of the target training vectors. We discuss here the similarities and differences between these methods and our proposed approach.

In [17], a state space approach for representing speech spectra as an observed process generated from an underling sequence of a hidden Markov process has been proposed. The source and target speech are both modeled using this state space representation. The state space parameters are divided into two parts: a common part related to the uttered speech (assuming a parallel training set) and a differentia part related to the difference between the speakers. These parts are evaluated during training time using an iterative algorithm known as expectation maximization (EM) [23]. During the test, the common parameters related to the test utterance are evaluated using EM and then used, along with the trained differentia part to obtain the converted spectra. Both training and conversion stages include iterative training (EM). Conversion results reported by the authors were obtained using several hundreds of parallel training sentences. Although our method and Xu et al.'s method [17] both use state space for representing the temporal evolution of the speech spectra, in our method, the source and the target spectra are linked through a state space dynamics, while in Xu et al.'s approach, the parallel source and target spectra are each modeled as the observed signals of a shared underlined unobserved Markov process.

Table 2 Voice conversion using GB approximation

Input: a sequence of feature vectors related to the source speaker $\mathbf{x}_{1:T}$
Initialization: set the initial weights, $\left\{w_{0\|0}^{k}\right\}_{k=1}^{N_y}$.
Main iteration: for $t = 1, \ldots T$, perform the following steps:
1. Evaluate the prior weights, $\left\{w_{t\|t-1}^{k}\right\}_{k=1}^{N_y}$, using Eqs. (10) and (16).
2. Evaluate the posterior weights, $\left\{w_{t\|t}^{k}\right\}_{k=1}^{N_y}$, using Eqs. (11) and (14).
3. Evaluate $\tilde{\mathbf{y}}_t = \mathcal{F}\{\mathbf{x}_t\}$, using Eq. (22).
Output: a sequence of converted vectors $\tilde{\mathbf{y}}_{1:T}$

An exemplar-based sparse representation approach for voice conversion has been proposed in [18]. Each speech signal is modeled as a linear combination of basis vectors (the training vectors), where the weighting matrix is called an activation matrix. The main assumption used in this method is that the speaker's identity is modeled by the basis vectors, where the information regarding the uttered text lies entirely in the activation matrix. Therefore, given a test source signal, its activation matrix is evaluated and then multiplied by the target training set, used as the target basis vectors, to obtain the converted spectra. Therefore, this method does not require any training, but its testing stage includes high computational load and a substantial memory footprint. As the exemplar-based method, our proposed GB method also uses a linear combination of the target training vectors. Besides the obvious differences in the models used by the two methods, there are two major differences: (1) We use sequential evaluation of the weights to ensure smooth temporal evolution while in the exemplar-based method, the activation matrix is evaluated as a batch. (2) We use scalar weights while the exemplar-based method uses weighting vectors (the activation matrix).

5 Experimental results

5.1 Experiments setup

In our experiments, we used speech sentences of four US English speakers taken from the CMU ARCTIC database [24]: two males (bdl, rms) and two females (clb, slt). Two different sizes of training sets 5 and 10 parallel sentences were used to demonstrate the performance of the examined methods as a function of training set size. The testing set consisted of 50 additional parallel sentences. All sentences were sampled at 16 kHz and were phonetically labeled.

Analysis and synthesis were both carried out using an available vocoder [25]. This vocoder uses a two-band harmonic/noise parametrization, separated by a maximal voicing frequency for representing each spectral envelope [26]. Twenty-five MFCCs were extracted from the harmonic parameters [27]: the zeroth coefficients, related to the energy, were not converted. The other 24 coefficients were used as spectral feature vectors during training and conversion.

The spectral features of unvoiced frames were not converted but simply copied to the converted sentence, since they do not capture much of the speaker's individuality [28] and their conversion often leads to quality degradation [29]. The maximal voicing frequency was also not converted but re-estimated from the converted parameters by the vocoder. The sequences of the training data set used for GB conversion were matched (without alignment), as described in Section 3.1. The training set used for the other examined methods, and the testing set, were each time aligned using a DTW algorithm based on phonetic labeling [14].

Pitch was converted by a simple linear function using the mean and standard deviation values of the source and target speakers,

$$\hat{f}_0^{(y),t} = \mu^{(y)} + \left(\sigma^{(y)}/\sigma^{(x)}\right)\left(f_0^{(x),t} - \mu^{(x)}\right), \tag{23}$$

where $f_0^{(x),t}$ and $\hat{f}_0^{(y),t}$ are the pitch values of the source and converted signals at the tth frame, respectively. The parameters $\mu^{(x)}$ and $\mu^{(y)}$ are the mean pitch values, and $\sigma^{(x)}$ and $\sigma^{(y)}$ are the standard deviations of the source and target pitch values, respectively. In this case, the mean and standard deviation of the converted pitch contour match the mean and standard deviation of the pitch values of the target speaker.

5.2 Objective evaluations

We evaluated the performance of the examined conversion methods by two objective measures: normalized distortion (ND) and normalized GV (NGV), as defined below.

To obtain a fair comparison between different source-target pairs, we normalized the mean spectral distortion between the converted and target signals by the mean spectral distortion between the source and target signals [30]

$$\text{ND}\left(\tilde{\mathbf{Y}}_{1:T}, \mathbf{Y}_{1:T}\right) = \frac{\sum_{t=1}^{T} \text{MCD}\left(\tilde{\mathbf{y}}_t, \mathbf{y}_t\right)}{\sum_{t=1}^{T} \text{MCD}\left(\mathbf{x}_t, \mathbf{y}_t\right)}, \tag{24}$$

where MCD is the distance between two cepstral vectors (defined in Section 3, Eq. (18)) and $\tilde{\mathbf{Y}}_{1:T} \triangleq (\tilde{\mathbf{y}}_1, \tilde{\mathbf{y}}_2, \ldots, \tilde{\mathbf{y}}_T)^{\top}$, $\mathbf{Y}_{1:T} \triangleq (\mathbf{y}_1, \mathbf{y}_2, \ldots, \mathbf{y}_T)^{\top}$, and $\mathbf{X}_{1:T} \triangleq (\mathbf{x}_1, \mathbf{x}_2, \ldots, \mathbf{x}_T)^{\top}$ are time-aligned sequences of cepstral vectors, related to the converted, target, and source utterances, respectively.

The GV of the pth elements of a sequence, $\tilde{\mathbf{Y}}_{1:T}$, representing a converted speech utterance, is as follows:

$$\sigma^2_{\tilde{\mathbf{Y}}_{1:T}}(p) = \frac{1}{T}\sum_{t=1}^{T}\left(\tilde{y}_t(p) - \frac{1}{T}\sum_{\tau=1}^{T}\tilde{y}_\tau(p)\right)^2, \tag{25}$$

In this paper, we use a normalized global variance (NGV) to measure the variability of a sequence of converted vectors

$$\text{NGV}\left\{\tilde{\mathbf{Y}}_{1:T}\right\} \triangleq \frac{1}{P}\sum_{p=1}^{P}\frac{\sigma^2_{\tilde{\mathbf{Y}}_{1:T}}(p)}{\sigma^2_{\mathbf{Y}}(p)}, \tag{26}$$

where $\sigma^2_{\mathbf{Y}}(p)$ is the empirical GV of the pth elements of the target speaker, obtained from the target training vectors

$$\sigma^2_{\mathbf{Y}}(p) = \frac{1}{N_y}\sum_{k=1}^{N_y}\left(y^k(p) - \frac{1}{N_y}\sum_{n=1}^{N_y}y^n(p)\right)^2. \tag{27}$$

Note that the target GV defined in Eq. (27) is evaluated by averaging over the entire training corpus. This evaluation of GV is different from the one proposed in [7] for spectral conversion and GV enhancement, where the GV of each utterance of the target is modeled as a random variable drawn from a Gaussian distribution.

The desired values for these measures are ND $\rightarrow$ 0 and NGV $\rightarrow$ 1, indicating that the converted outcome is close to the target signal in terms of spectral similarity and global variance.

The examined GMM-based methods (JGMM and CGMM) were trained using diagonal covariance matrices and 1–4 Gaussian mixtures, due to the low amount of training data.

We begin with a short examination of the influence of each of the three parameters of the proposed GB method (R_x, R_y, and τ) on its performance. Figure 2 presents the ND vs. NGV values obtained for the proposed GB method using $R_x \in [0.3, 2]$, $R_y \in [1, 4]$, and $\tau = 1$, trained by 10 sentences, for a male-to-male conversion. As the parameter R_x gets higher, more grid points are considered in the weighted sum, so that ND decreases, but the NGV also decreases. Since the evidence probability is solely determined by the training set (see Eq. (16)), we also examined the performance of the GB method using a data-driven value for R_y, specifically, the median of the MCD between all training vector pairs related to the target speaker. These values vary between 2 and 3 dB when using different source-target pairs and data set sizes. As depicted in Fig. 2, the median leads to the best ND-NGV values so all results presented from now on were obtained using this value for R_y.

Figure 3 presents the ND vs. NGV values obtained for the proposed GB method using $R_x \in [0.3, 2]$, $\tau = (0, 1, 2)$, trained by 10 sentences, for a male-to-male conversion. Using $\tau = 1$ leads to higher NGV values than using $\tau = 0$, with a slight increase in the ND. However, increasing τ further leads to the same NGV values with a minor decrease in the ND.

Table 3 summarizes the ND and NGV values achieved by JGMM [2] and the proposed GB conversion method, for all four gender conversions: male-to-male (M2M), male-to-female (M2F), female-to-male (F2M), and female-to-female (F2F), using 5 and 10 training sentences. The number of mixtures for JGMM and parameters for the GB (R_x and τ) were selected for each method and training set so that a minimal ND was attained, while keeping the NGV as high as possible. As mentioned

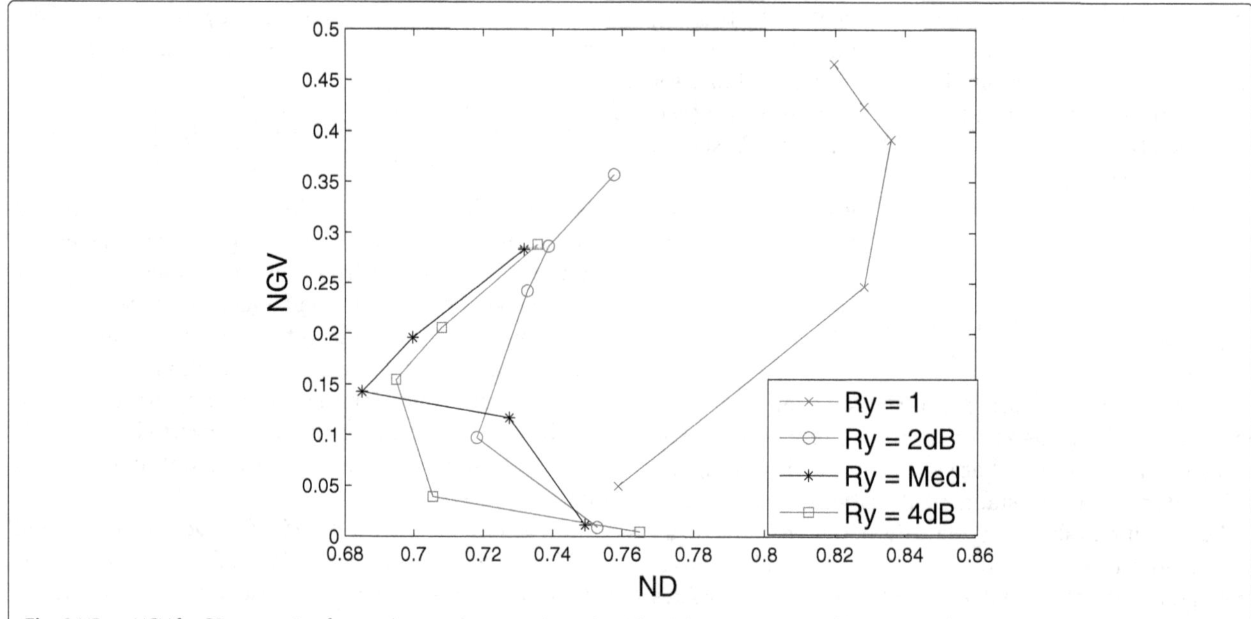

Fig. 2 ND vs. NGV for GB conversion for a male-to-male conversion using 10 training sentences and $R_x \in [0.3, 2]$ dB, $\tau = 1$. $R_y = 1$ dB (*blue x*), $R_y = 2$ dB (*red circle*), $R_y =$ median (*black asterisk*), and $R_y = 3$ dB (*magenta square*)

above, R_y was taken as the median. The proposed GB leads to higher NGV values in all the cases. For five training sentences, JGMM leads to lower ND values (except for F2M), however, using 10 training sentences, the proposed GB achieves lower or very similar ND values. Still, both methods lead to very low NGV values and consequently, muffled sounding synthesized signals.

To further improve the quality of the synthesized speech, we applied the post-processing method for GV enhancement [21]. This method maximizes the GV of an input sequence, under a spectral distortion constraint. The GV of each enhanced sequence is increased up to the level where the MCD between the converted sequence and its enhanced version reaches a preset threshold value,

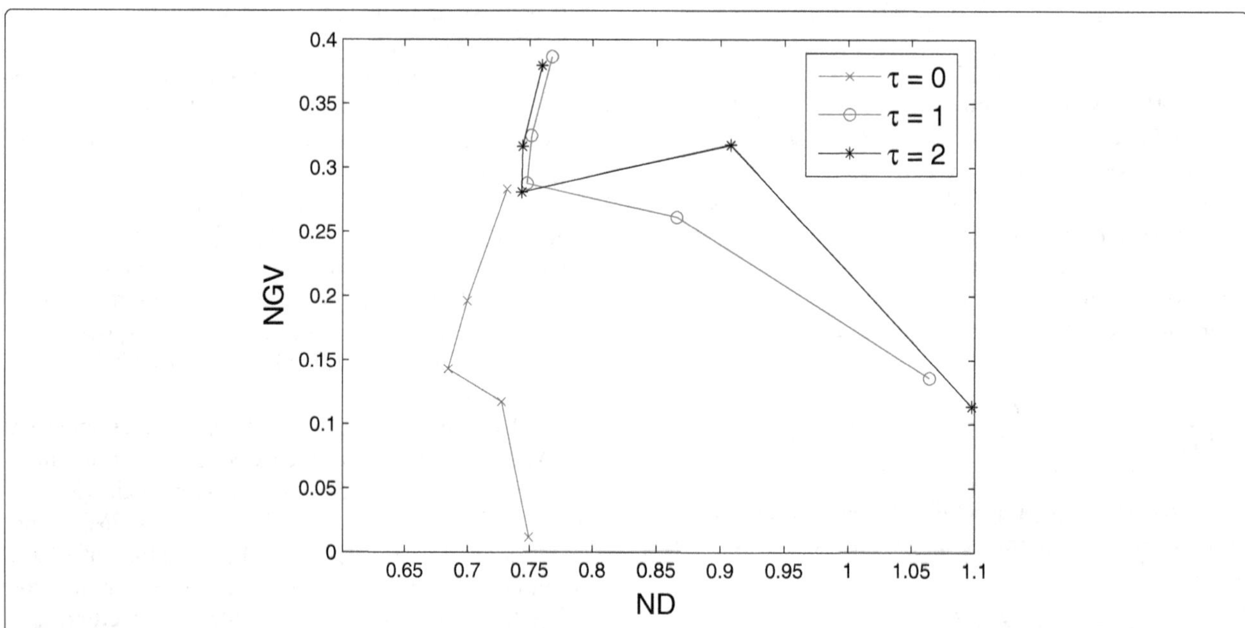

Fig. 3 ND vs. NGV for GB conversion for a male-to-male conversion using 10 training sentences and $R_x \in [0.3, 2]$, $R_y =$ median. $\tau = 0$ (*blue x*), $\tau = 1$ (*red circle*), and $\tau = 2$ (*black asterisk*)

Table 3 Objective performance: ND and NGV values using 5 and 10 training sentences, for all four gender conversions

		5 Training sentences		10 Training sentences	
		ND	NGV	ND	NGV
M2M	JGMM	0.72	0.15	0.71	0.13
	GB	0.73	0.25	0.69	0.14
M2F	JGMM	0.7	0.15	0.7	0.12
	GB	0.71	0.21	0.69	0.19
F2M	JGMM	0.74	0.14	0.71	0.13
	GB	0.71	0.34	0.71	0.42
F2F	JGMM	0.8	0.22	0.8	0.18
	GB	0.88	0.34	0.81	0.31

denoted as θ_{MCD}. We recently showed [21] that this method leads to significant improvement in the perceived quality of signals converted by the classical GMM method [1]. In this work, we applied this GV enhancement method to both JGMM and to our proposed GB conversion outcomes. We also examined the performance of CGMM, which considers GV enhancement at training.

Table 4 summarizes the ND and NGV values achieved by the examined conversion methods, for all four gender conversions using 5 and 10 training sentences.

Again, the GB conversion, followed by GV enhancement with $\theta_{\text{MCD}} = 2$ dB (En-GB) leads to the highest NGV values. Using 5 training sentences, JGMM leads to the lowest ND values, while En-GB comes in second (except for F2F). Using 10 training sentences, En-GB produces the lowest ND and at the same time, the highest NGV, for M2M and M2F conversion. For F2M and F2F conversion, En-GB leads to the highest NGV with very similar ND values to JGMM, which are the lowest.

Table 4 Objective performance: ND and NGV values using 5 and 10 training sentences, for all four gender conversions with GV enhancement ($\theta = 2$ dB)

		5 Training sentences		10 Training sentences	
		ND	NGV	ND	NGV
M2M	JGMM	0.76	0.6	0.74	0.55
	CGMM	0.83	0.46	0.82	0.45
	GB	0.79	0.8	0.73	0.6
M2F	JGMM	0.74	0.57	0.74	0.54
	CGMM	0.83	0.45	0.84	0.46
	GB	0.76	0.73	0.73	0.68
F2M	JGMM	0.77	0.63	0.75	0.69
	CGMM	0.86	0.62	0.85	0.61
	GB	0.76	0.95	0.77	1.1
F2F	JGMM	0.86	0.79	0.85	0.65
	CGMM	0.91	0.63	0.89	0.6
	GB	0.95	1	0.87	0.98

To conclude the objective examination, in terms of NGV, the proposed EN-GB conversion scheme outperforms all the examined methods. In terms of ND, JGMM leads to lower ND values using 5 training sentences. Using 10 training sentences, En-GB leads to the lowest (or very similar to the lowest) ND values.

In the next section, we present subjective evaluation results comparing the proposed En-GB conversion scheme to the classical GMM-based conversion method (with enhancement) and to CGMM, in terms of perceived quality and similarity to the target speaker.

5.3 Subjective evaluations

Listening tests were carried out to subjectively assess the performance of the examined methods (all trained by 10 sentences). In every test, 10 different sentences were examined by 11 listeners (voice samples are available online [31]). The group of listeners included 20–30-year-old, non-experts men and women. The same four speakers (two males and two females) that were used for the objective evaluations were used for the subjective evaluations. The number of mixtures for the GMM-based methods and parameters for the GB conversion (R_x and τ) were set so minimal spectral distortion would be attained while keeping the NGV as high as possible. We used informal listening tests to select the threshold value for GV enhancement from $\theta_{\text{MCD}} = 0.5, 1, 2, 4$ dB. The best perceived quality was obtained with $\theta_{\text{MCD}} = 2$ dB, for both JGMM and GB. All four gender conversions were performed using the same parameters values as described above.

We conducted subjective quality evaluations in a format similar to multi-stimulus test with hidden reference and anchor (MUSHRA) [32]. The listeners were presented with four test signals: (a) a hidden reference—the target speaker, (b) enhanced JGMM, (c) CGMM, and (d) En-GB. The test signals were randomly ordered, and the listeners were not informed about the hidden reference signals being included in the test set. During evaluation, the listeners were asked to compare the test signals to the reference signal (the target speaker) and rate their quality between 0 and 100, where at least one of the test signals (the hidden reference) must be rated 100. As expected, all the listeners rated the hidden reference as 100. The mean scores of the examined methods for M2M, M2F, F2M, and F2F conversions and also their scores averaged over all four conversions are presented in Figs. 4 and 5, respectively. All subjective results are presented with their 95 % confidence intervals. We evaluated the individuality performance using, again, a similar format to MUSHRA, as conducted by Godony et al. [33].

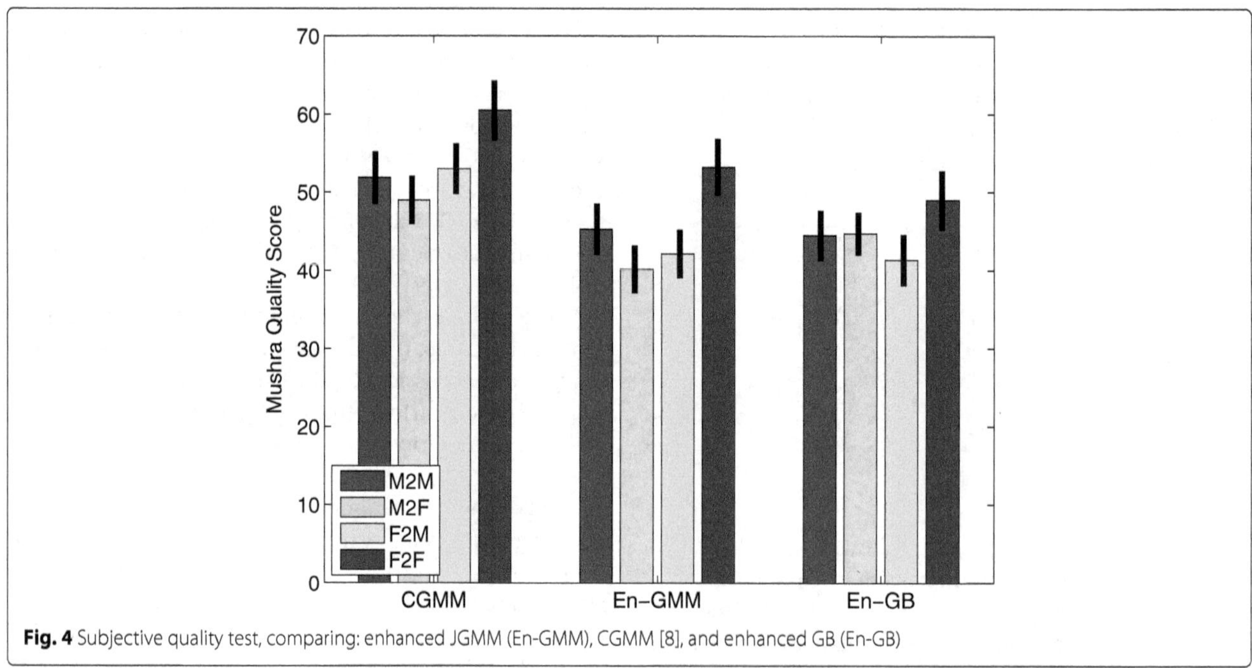

Fig. 4 Subjective quality test, comparing: enhanced JGMM (En-GMM), CGMM [8], and enhanced GB (En-GB)

The listeners were presented with the same test signals (including the hidden reference) and were asked to rate their similarity to the reference signal, in terms of the speaker's identity, while ignoring their perceived quality. The mean individuality scores of the examined methods for M2M, M2F, F2M, and F2F conversions and also their scores, averaged over all four conversions, are presented in Figs. 6 and 7, respectively.

Except for F2F, the proposed EN-GB was rated as most similar to the target speaker (Fig. 6). In terms of perceived quality, CGMM was rated as having the best quality, while EN-JGMM and EN-GB were rated as comparable (Fig. 4). All in all, considering all four gender conversion, the proposed EN-GB was marked as most similar to the target speaker, while CGMM was marked as having the best quality.

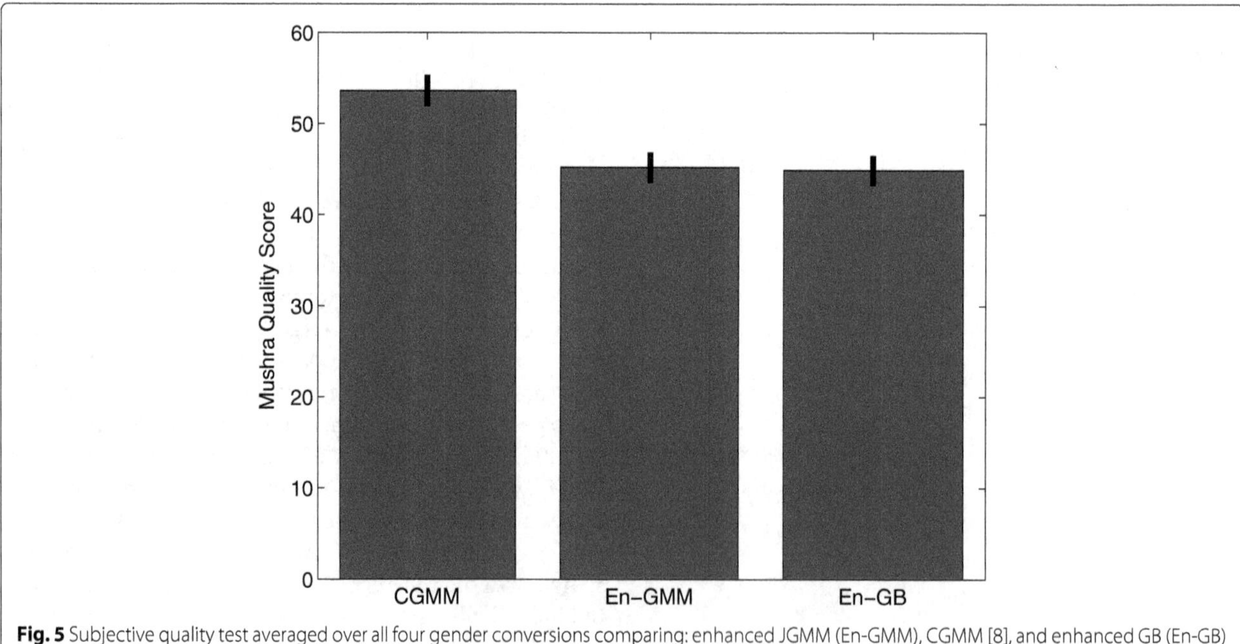

Fig. 5 Subjective quality test averaged over all four gender conversions comparing: enhanced JGMM (En-GMM), CGMM [8], and enhanced GB (En-GB)

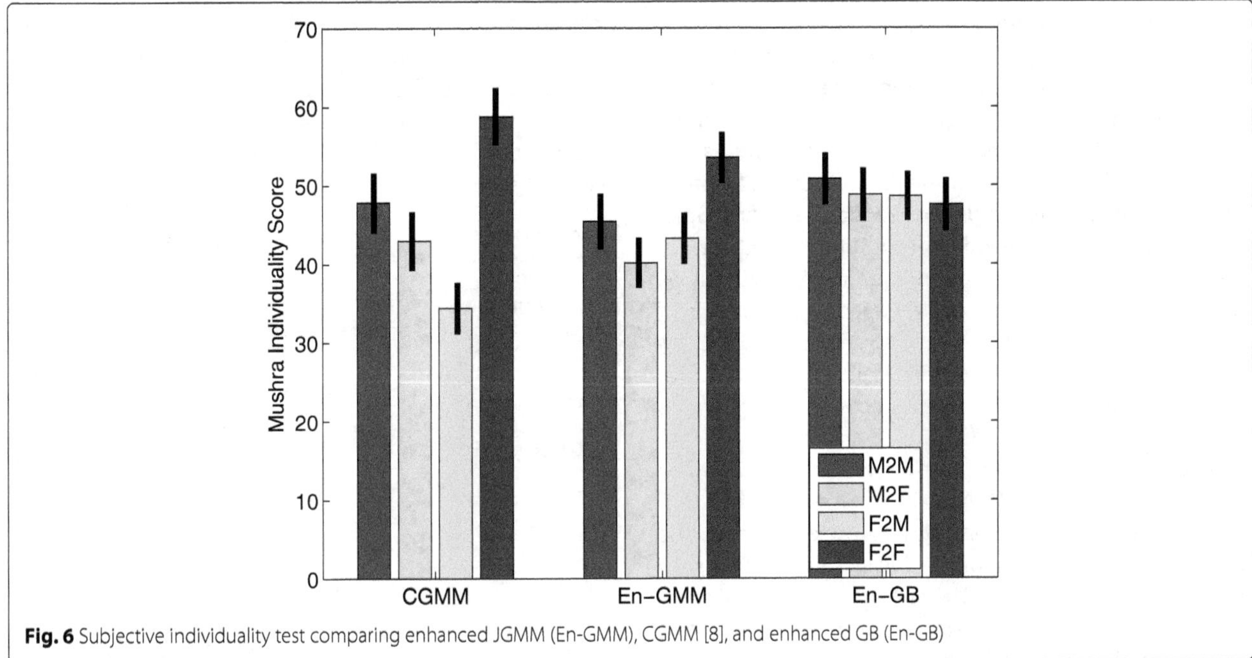

Fig. 6 Subjective individuality test comparing enhanced JGMM (En-GMM), CGMM [8], and enhanced GB (En-GB)

6 Conclusions

Applying voice conversion in low resource environments, such as mobile applications, presents an engineering challenge. While digital processors and memory units become more advanced and less restricting, the amount of available training data remains limited, since most mobile users are not willing to invest much time and effort in recording their own voices. We propose here a GB voice conversion method suitable for such low resource environments. It is based on our recent paper, which presents a GB framework for voice conversion. The modified GB method presented in this paper is successfully trained using very few sentences (5–10) and does not require phonetic labeling of the test signals.

The GB conversion method is based on sequential Bayesian tracking, using a GB formulation. The target

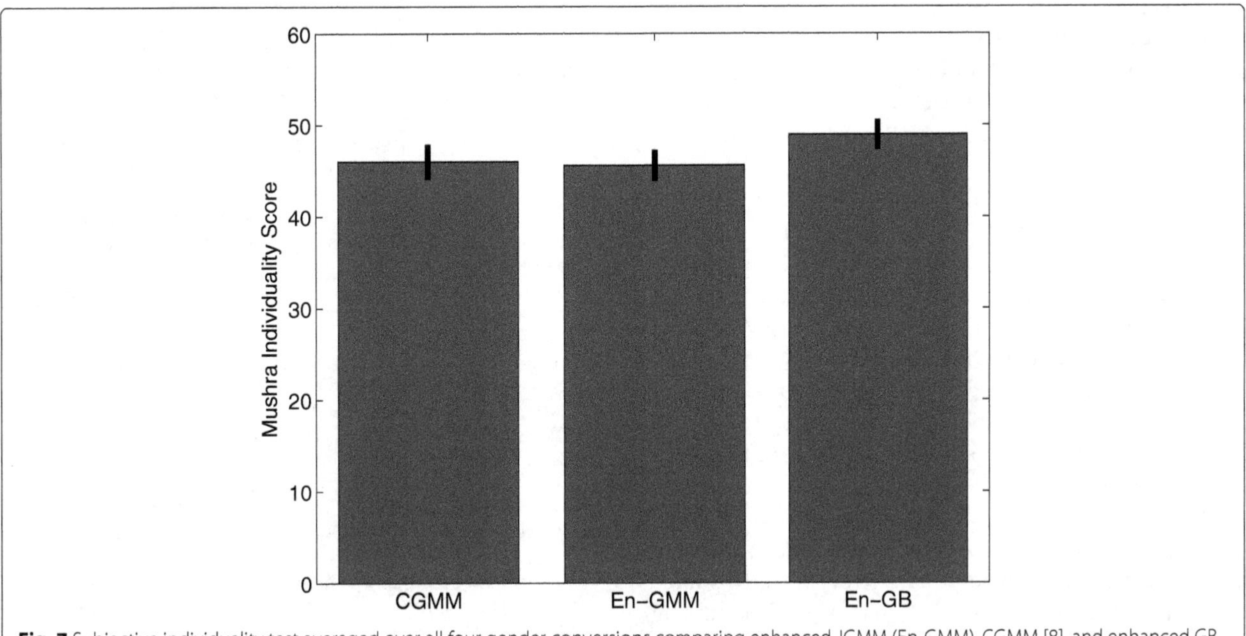

Fig. 7 Subjective individuality test averaged over all four gender conversions comparing enhanced JGMM (En-GMM), CGMM [8], and enhanced GB (En-GB)

spectral evolution is modeled as a hidden Markov process, tracked by using the source spectrum, modeled as the observed process. The training stage is very simple and based on Euclidean distances between the training vectors, and it is successfully performed using very small training sets. Additionally, although GB is trained using a parallel set, time alignment is not needed. During training, the evidence and likelihood probabilities needed for the GB formulation are approximated as discrete densities. During conversion, the converted spectrum is obtained as a weighted sum of the training target vectors, used as grid points. The weights are sequentially evaluated so that a smooth temporal evolution of the converted spectra is produced.

We used a small set of just 10 sentences for training both the classical GMM-based conversion function and our GB method. According to our experiments, the GB conversion method achieves lower spectral distances between the converted and target spectra and GV values which are closer to the target speaker's values than the classical GMM-based conversion. To further improve the quality of the synthesized speech, we increased the variability of the converted vectors by applying GV enhancement as a post-processing block. We compared the proposed En-GB scheme to CGMM and to classical GMM-based conversions, with GV enhancement, using listening tests. This comparison showed that En-GB is the best in terms of similarity to the target speaker and comparable to the enhanced GMM conversion, in terms of quality.

The proposed GB conversion, as most other methods, simply replaces the spectral envelopes extracted from the source signal with the converted outcome. As a result, the synthesized output has the same speaking rate as the source speaker. Further improvement can be obtained by modifying the duration of each converted utterance to match, on average, its corresponding value for the target speaker.

Spectral distortion and GV are commonly used as objective measures since they provide a simple and fully automated way for evaluating conversion systems. These objective measures may express significant trends and phenomena, but as shown here, they do not always agree with subjective evaluation results.

Further research is needed to design alternative measures for objective evaluation of conversion systems, with better correspondence to subjective results. In the mean time, subjective listening tests are imperative to properly evaluate and compare conversion methods.

The proposed GB conversion method, as presented here, is based on soft correspondence between the source and target vectors, obtained by using a parallel training set. Further research is needed to evaluate this correspondence for a non-parallel setup.

Endnotes

[1] In general, any arbitrary integrable function of the state vector $\mathbf{y}_t$ can be evaluated [19].

[2] If the state space is indeed discrete and finite, and the grid points consist of all its states, this evaluation becomes exact.

Competing interests

The authors declare they have no competing interests.

Authors' contributions

This paper is part of the doctoral research work of HB under the supervision of the other two authors.

Acknowledgements

The authors would like to thank Slava Shechtman, and the speech research group headed by Ron Hoory, at the IBM Research Labs, Haifa, Israel, for fruitful discussions.

References

1. OYC Stylianou, E Moulines, Continuous probabilistic transform for voice conversion. IEEE Trans. Speech Audio Proc. **6**(2), 131–142 (1998)
2. A Kain, M Macon, in *Proc. ICASSP*. Spectral voice conversion for text-to-speech synthesis (IEEE, Seattle, Washington, USA, 1998), pp. 285–288
3. T Toda, AW Black, K Tokuda, in *Proc. ICASSP*. Spectral conversion based on maximum likelihood estimation considering global variance of converted parameter (IEEE, Philadelphia, Pennsylvania, USA, 2005), pp. 9–12
4. T Toda, H Saruwatari, K Shikano, in *Proc. ICASSP*. Voice conversion algorithm based on Gaussian mixture model with dynamic frequency warping of STRAIGHT spectrum (IEEE, Orlando, Florida, USA, 2001), pp. 841–844
5. A Kain, MW Macon, in *Proc. ICASSP*. Design and evaluation of a voice conversion algorithm based on spectral envelope mapping and residual prediction (IEEE, Salt Lake City, Utah, USA, 2001), pp. 813–816
6. T En-Najjary, O Rosec, T Chonavel, in *Proc. Interspeech ICSLP*. A voice conversion method based on joint pitch and spectral envelope transformation, (Jeju Island, Korea, 2004), pp. 1225–1225
7. T Toda, AW Black, K Tokuda, Voice conversion based on maximum-likelihood estimation of spectral parameter trajectory. IEEE Trans. Audio Speech Lang. Proc. **15**(8), 2222–2235 (2007)
8. H Benisty, D Malah, in *Proc. Interspeech*. Voice conversion using GMM with enhanced global variance (ISCA, Florence, Italy, 2011), pp. 669–672
9. T Toda, T Muramatsu, H Banno, in *INTERSPEECH*. Implementation of computationally efficient real-time voice conversion (ISCA, Portland, Oregon, U.S, 2012). Citeseer
10. T Muramatsu, Y Ohtani, T Toda, H Saruwatari, K Shikano, in *Interspeech*. Low-delay voice conversion based on maximum likelihood estimation of spectral parameter trajectory (ISCA, 2008), pp. 1076–1079
11. Y Nankaku, K Nakamura, T Toda, K Tokuda, in *Proc. Interspeech*. Spectral conversion based on statistical models including time-sequence matching (ISCA, 2007), pp. 333–338
12. D Erro, A Moreno, A Bonafonte, Inca algorithm for training voice conversion systems from nonparallel corpora. Audio Speech Lang. Process. IEEE Trans. **18**(5), 944–953 (2010)
13. H Benisty, D Malah, K Crammer, in *Proc. ICASSP*. Non-parallel voice conversion using joint optimization of alignment by temporal context and spectral distortion (IEEE, Florence, Italy, 2014), pp. 7909–7913
14. D Erro, A Moreno, A Bonafonte, Voice conversion based on weighted frequency warping. IEEE Trans. Audio Speech Lang. Proc. **18**(5), 922–931 (2010)
15. E Helander, J Schwarz, SHJ Nurminen, M Gabbouj, in *Proc. Interspeech*. On the impact of alignment on voice conversion performance (ISCA, Brisbane, Australia, 2008), pp. 1453–1456
16. T Toda, Y Ohtani, K Shikano, in *Proc. ICSLP*. Eigenvoice conversion based on Gaussian mixture model, (2006), pp. 2446–2449

17. N Xu, Z Yang, L Zhang, W Zhu, J Bao, Voice conversion based on state-space model for modelling spectral trajectory. Electron. Lett. **45**(14), 763–764 (2009)
18. Z Wu, T Virtanen, ES Chng, H Li, Exemplar-based sparse representation with residual compensation for voice conversion. Audio Speech Lang. Process. IEEE/ACM Trans. **22**(10), 1506–1521 (2014)
19. MS Arulampalam, S Maskell, N Gordon, T Clapp, A tutorial on particle filters for online nonlinear/non-Gaussian Bayesian tracking. IEEE Trans. Signal Proc. **50**(2), 174–188 (2002)
20. H Benisty, D Malah, K Crammer, in *Electrical & Electronics Engineers in Israel (IEEEI), 2014 IEEE 28th Convention Of.* Sequential voice conversion using grid-based approximation (IEEE, 2014), pp. 1–5
21. H Benisty, D Malah, K Crammer, in *Proc. EUSIPCO.* Modular global variance enhancement for voice conversion systems, (2012), pp. 370–374
22. B Anderson, J Moore, *Optimal Filtering.* (Prentice-Hall, Englewood Cliffs, NJ, 1979)
23. A Dempster, N Laird, D Rubin, Maximum likelihood from incomplete data via the EM algorithm. J. R. Stat. Soc. B. **39**, 1–38 (1977)
24. J Kominek, AW Black, *CMU ARCTIC Databases for Speech Synthesis*, (2003)
25. Aholab Coder. http://aholab.ehu.es/ahocoder/. Accessed Jan 2013
26. D Erro, I Sainz, I Hernaez, in *Proc. Interspeech.* Improved HNM-based vocoder for statistical synthesizers, (2011), pp. 1809–1812
27. O Cappe, E Moulines, Regularization techniques for discrete cepstrum estimation. IEEE Signal Process. Lett. **3**(4), 100–102 (1996)
28. H Kuwabara, Y Sagisaka, Acoustic characteristics of speaker individuality: control and conversion. IEEE Trans. Signal Proc. **16**(2), 165–173 (1995)
29. D Erro, A Moreno, A Bonafonte, Inca algorithm for training voice conversion systems from nonparallel corpora. IEEE Trans. Audio Speech Lang. Proc. **18**(5), 944–953 (2010)
30. H Ye, S Young, Quality-enhanced voice morphing using maximum likelihood transformations. IEEE Trans. Audio Apeech Lang. Proc. **14**(4), 1301–1312 (2006)
31. Sound Samples. http://sipl.technion.ac.il/Info/hadas/sound-samples.htm Accessed Mar 2015
32. Multi stimulus test with hidden reference and anchors (MUSHRA) (2003). Technical Report ITU-R BS.1534-1, International Telecommunications Union
33. E Godoy, O Rosec, T Chonavel, Voice conversion using dynamic frequency warping with amplitude scaling, for parallel or nonparallel corpora. IEEE Trans. Audio Speech Lang. Proc. **20**(4), 1313–1323 (2012)

5

Driver head pose estimation using efficient descriptor fusion

Nawal Alioua[1,2*], Aouatif Amine[4], Alexandrina Rogozan[3], Abdelaziz Bensrhair[3] and Mohammed Rziza[2]

Abstract

A great interest is focused on driver assistance systems using the head pose as an indicator of the visual focus of attention and the mental state. In fact, the head pose estimation is a technique allowing to deduce head orientation relatively to a view of camera and could be performed by model-based or appearance-based approaches. Model-based approaches use a face geometrical model usually obtained from facial features, whereas appearance-based techniques use the whole face image characterized by a descriptor and generally consider the pose estimation as a classification problem. Appearance-based methods are faster and more adapted to discrete pose estimation. However, their performance depends strongly on the head descriptor, which should be well chosen in order to reduce the information about identity and lighting contained in the face appearance. In this paper, we propose an appearance-based discrete head pose estimation aiming to determine the driver attention level from monocular visible spectrum images, even if the facial features are not visible. Explicitly, we first propose a novel descriptor resulting from the fusion of four most relevant orientation-based head descriptors, namely the steerable filters, the histogram of oriented gradients (HOG), the Haar features, and an adapted version of speeded up robust feature (SURF) descriptor. Second, in order to derive a compact, relevant, and consistent subset of descriptor's features, a comparative study is conducted on some well-known feature selection algorithms. Finally, the obtained subset is subject to the classification process, performed by the support vector machine (SVM), to learn head pose variations. As we show in experiments with the public database (Pointing'04) as well as with our real-world sequence, our approach describes the head with a high accuracy and provides robust estimation of the head pose, compared to state-of-the-art methods.

Keywords: Driver monitoring, Head pose estimation, Support vector machine, Feature selection

1 Introduction

The increasing number of traffic accidents in the last years becomes a serious problem. The enhancement of traffic safety is a high-priority task for different government agencies over the world such as "National Transportation Safety Administration" (NTSA) in USA and "Observatoire National Interministériel de la Sécurité Routière" (ONISR) in France. In addition, automotive manufactures and researcher laboratories are also contributing to this important mission. Some preventive systems such as alcohol test and speed measurement radar are deployed to reduce the number of traffic accidents, but it is obvious that hypovigilance remains one of the most principal causes. In fact, hypovigilance is responsible for 20–30 % of road deaths and this statistic reaches 40–50 % in particular crash types, such as fatal single vehicle semi-trailer crashes [1]. Moreover, there are no standard rules to measure the driver vigilance level; the unique solution is to observe the signs. The first hypovigilance signs are itchy eyes, neck stiffness, back pain, yawning, difficulty to stabilize speed and to maintain trajectory, frequent position changes, and inattention to environment (road signs, pedestrian). Fatigue, sleep deprivation, soporific drugs, driving more than 2 h without break, and driving in a monotone road are the main causes of hypovigilance. The appropriate reactions when those signs appear are to stop driving immediately and take a break, but unfortunately, the drivers are not aware of their vigilance level and overestimate it. For this purpose, several studies have been conducted to develop intelligent systems for continuously

*Correspondence: nawal.alioua@yahoo.fr
[1]Ibn Zohr University, Morocco, BP 32/S, Agadir, Morocco
[2]LRIT-CNRST 29, Faculty of Sciences, Mohammed V University, Ibn Battouta Avenue, Rabat, Morocco
Full list of author information is available at the end of the article

estimating driver vigilance level and emitting visual and acoustic alarms to avert the driver against abnormal state. The warning signals could also activate the vibration of driver's seat or even a mechanism that stops the car at the roadside.

The literature regroups *three categories* of safety systems distinguished by the type of signals used to determine the driver vigilance level. (i) Studying *physiological signals* consists on measuring signal changes represented by brain waves or heart rate using special sensors such as electroencephalography (EEG), electrocardiography (ECG), and electromyography (EMG) [2]. Only few works are proposed in this category since the process is highly intrusive because of the necessity to connect sensing electrodes to the driver body. (ii) Monitoring *vehicle signals* can reveal abnormal driver actions indirectly, by studying several parameters such as vehicle velocity changes, steering wheel motion, lateral position, or lane changes. Some commercial systems already use these techniques since the signals are significant and their acquisition is quite easy compared to the previous category. Mercedes-Benz proposes in 2009 a commercial system named "Attention Assist" based on sensitive sensors allowing precise monitoring of the steering wheel movements and the steering speed. The system is active at 80–180 km/h and calculates an individual behavioral pattern during the first minutes of each trip. Audible and visual signals are emitted when typical indicators of hypovigilance are detected. The major disadvantages of such system are the limitations caused by the dependence to vehicle type, driver experience, and road conditions. (iii) Approaches based on *physical signals* utilize image processing techniques to measure the driver vigilance level reflected through the driver's face appearance and head/facial feature activity. These techniques are based principally on studying facial features, especially eye state [3–5], head pose [6, 7], or mouth state [8]. According to the study performed in [9], monitoring driver eye closure and head pose are the most relevant indicators of hypovigilance. Different kinds of cameras have been used for such systems: visible spectrum (VS) camera [10], infrared (IR) camera [11], stereo cameras [12], and also the Kinect sensor [13]. The Kinect sensor provides color images, IR images, and 3D information. However, this sensor is not very adapted to the real driving conditions since it is designed for indoor use and it is conceived to be placed in a minimal distance of 1.8 m from the target. The IR camera is adapted when driving at night, but it is not recommended when driving at daylight conditions, since the acquisition will suffer from color distortion. The VS camera is the cheapest one, and it provides robust acquisition even if the light is reduced. However, it is a big challenge to monitor the driver vigilance level using a single VS camera without depth information and IR information.

In our previous work [10], we have proposed a real-time system using a very cheap VS camera to determine driver fatigue and drowsiness by analyzing mouth and eyes, respectively. This system suffers from missed detection when the specific facial features are not visible because of non-frontal head position. The aim of this paper is to develop a head pose estimation approach that reveals rapidly driver inattention from monocular visible spectrum images, without prior facial feature extraction. To construct a robust head pose estimator, we follow an appearance-based head pose estimation architecture instead of a model-based one. These two architectures are detailed in Section 2. In fact, the model-based architecture is incompatible with our problem since it requires facial features to construct the face geometrical model, whereas the appearance-based one uses the whole head structure characterized by an image descriptor. Actually, the performance of the appearance-based estimator depends strongly on the image descriptor, which should be chosen carefully in order to reduce the information about identity and lighting contained in the face appearance. In this work, as detailed in Section 3, we first propose a novel descriptor resulting from the fusion of four most relevant orientation-based head descriptors, namely the steerable filters (SF), the histogram of oriented gradients (HoG), the Haar features, and an adapted version of SURF descriptor. Second, in order to construct a compact, robust, and pertinent subset of the descriptor's features, a comparative study is conducted on some well-known feature selection algorithms. Finally, the obtained subset is subject to the classification process, performed by the support vector machine (SVM), to learn head pose variations. In Section 4, an evaluation of the proposed head pose estimator on the public Pointing'04 database is performed to validate our approach and to compare it with the most representative and the best state-of-the-art methods. After that, we have acquired and annotated a driver video sequence simulating attention and inattention states in order to validate the proposed estimator in a real environment. Finally, we present a conclusion and discussion in Section 5.

2 Related works

2.1 Overview of head pose estimation techniques

In computer vision, head pose estimation can be defined as the ability to deduce head orientation relatively to a view of camera and it can refer to different interpretations [14]. At coarse level, a head is identified by a few discrete poses, but it might be estimated by a continuous angular measurement according to multiple degrees of freedom. The discrete representation is adapted to the applications requiring the knowledge of limited number of pose classes instead of the whole possible pose angles corresponding to the continuous representation. Even if muscular rotation of head influences its orientation, it is often ignored and human head is considered as an incorporeal rigid object. This hypothesis allows to represent head pose using only three degrees of freedom which are *pitch*, *yaw*, and *roll*. Pitch corresponds to up and down motion around the X axis, yaw refers to left and right direction around the Y axis, and roll represents tilting the head towards left and right direction around the Z axis (see Fig. 1). Another hypothesis to be considered when building head pose estimator is the *pose similarity assumption*, which means that different people at the same pose look more similar than the same person at different poses. In literature [14–16], three requirements are established to define an efficient head pose estimator.

(R1) Perform head pose estimation from monocular cheap camera. Potentially, the accuracy can be improved using stereo techniques that need additional equipment cost, computation, and memory requirements.
(R2) Ensure autonomy by avoiding manual initialization or adjustment.
(R3) Guarantee invariance to identity and environment in order to make the system more efficient and robust.

In literature, many techniques have been proposed to estimate head pose for diverse applications including monitoring driver state systems. These techniques can be categorized into two main groups [17], namely model-based techniques and appearance-based techniques.

2.1.1 Model-based head pose estimation

Model-based techniques require specific facial features to estimate head pose. In this category, we can find geometric approaches that determine head pose from the relative locations of facial features such as eye corners, mouth corners, and nose tip. The most recent systems based on facial geometry are proposed in [12, 17, 18]. In [18], the authors propose a method for automatic head pose estimation using three features (the eyes and nose locations) ruled by the concept of golden ratio, whereas the majority of geometric approaches require at least five features. The golden ratio is the proportionality constant adopted by Leonardo Da Vinci in his master-work called The Vitruvian Man.

Flexible models based on fitting non-rigid models to the facial structure of each subject also belong to this category since comparisons at feature level are made rather than comparisons at global appearance level. Flexible models include methods such as active shape models (ASM), active appearance models (AAM), and elastic graph matching (EGM). In [19], authors present a probabilistic framework which do not need user initialization unlike most of flexible models which do not respect the requirement (R2).

Model-based techniques are dependent to the performance of the facial feature localization which is, in addition to high-resolution requirement, the major disadvantage.

2.1.2 Appearance-based head pose estimation

Appearance-based techniques work under the assumption that the 3D face pose and some properties of the 2D facial image are linked by a certain relationship [20]. Appearance template methods [21, 22] define this relationship by matching new head images into discrete head templates using image-based comparison metrics. These

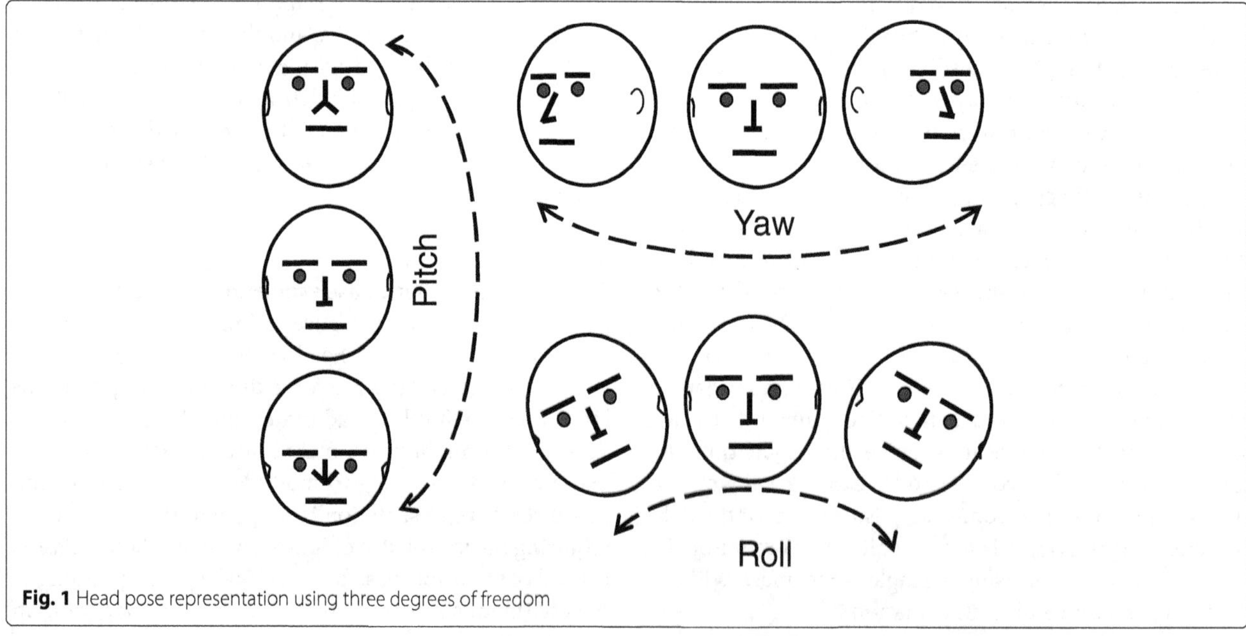

Fig. 1 Head pose representation using three degrees of freedom

methods are the most sensitive to lighting conditions. In our previous work [21], we use a robust descriptor based on steerable filters to construct a reference template for each discrete pose of the training set, and a likelihood parametrized function is learned to match the templates to new entries.

Another way to determine the relationship is the use of classification techniques on a large number of training data in order to learn an efficient separation between pose classes. SVM classifiers are quite used in literature to classify head poses [23–25] since they are adapted to real-time applications. In [23], the authors show that the multi-scale Gaussian derivatives, which are a particular case of steerable filters, combined to SVM give good results. In [26], normalized faces are used to train an auto-associative memory using the Widrow-Hoff correction rule in order to classify head poses. In one of the most recent works [27], the authors consider that object detection and continuous pose estimation are interdependent problems and they jointly formulate them as a structured prediction problem, by learning a single and continuously parameterized object appearance model over the entire pose space. After that, they design a cascaded discrete-continuous inference algorithm to effectively optimize a non-convex objective, by generating a diverse proposal to explore the complicated search space. Then, the model is learned using a structural SVM for joint object localization and continuous state estimation and a new training approach which reduces the processing time. Among the experiments, the authors perform the head pose estimation over the Pointing'04 database without considering the detection task, since they note that the images contain clean backgrounds. Before applying their method, the heads are cropped manually and the HOG descriptor is applied on three scales. Based on the relationship between the symmetry of the face image and the head pose, the authors in [28] propose a face representation method for head yaw estimation which is robust against rotations and illumination variations. First, they extract the multi-scale and multi-orientation Gabor representations of the face image, and then they use covariance descriptors to compute the symmetry between two regions in terms of Gabor representations under the same scale and orientation. Second, they apply a metric learning method named KISS MEtric learning (KISSME) to enhance the discriminative ability and reduce the dimension of the representation. Finally, the nearest centroid (NC) classifier is applied to obtain the final pose.

Regression techniques are also utilized to address head pose estimation problem when the pose angles are ordered, but they are more complex since they need powerful unit process to respect real-time constraints. In this case, the relationship is defined by learning continuous mapping functions between the face image and the pose space [29–31]. In [31], authors extract head feature vector using the robust 3-level HOG pyramid and then the partial least square (PLS) regression is used to determine the coefficients modeling the relationship between the head and its pose. In [24], authors use a dense scale invariant feature transform (SIFT) descriptor to construct feature vector and the random projection (RP) is applied to reduce the vector dimension. Similar to [32], the authors combine classification and regression to obtain an accurate estimation of head pose but this kind of approach is time consuming.

One can also include tracking approaches in the appearance-based techniques since they are based on head appearance to estimate poses in addition to temporal continuity and smooth motion of the heads in the video sequence. Particle filters (PF) [6, 33] are the most used technique to track head poses; in [6], authors propose a hybrid head orientation and position estimation system for driver head tracking based on PF. While tracking techniques can achieve high accuracy, they usually require an initial step such as frontal view or manual initialization which does not respect the requirement (R2).

The major part of the appearance-based techniques presented above is applied on features that verify the pose similarity assumption. In addition, the descriptor must be fast, must be robust to variations of lighting conditions, and should be representative of head orientations in order to respect the requirements (R2) and (R3). Gabor filter [34], steerable filters (SF) [21], SIFT [24], and HOG [33] are the most used descriptors verifying these requirements. Some dimensionality reduction methods can be used to seek a low-dimensional continuous manifold constrained by the pose variations. Principal component analysis (PCA) and linear discriminant analysis (LDA) are the most used dimensionality reduction techniques for head pose estimation [14]. In [35], authors propose to represent each head pose appearance neighborhood by a query point to reduce the size and then apply a piece-wise linear local subspace learning method to map out the global nonlinear structure for head pose estimation.

Each category suffers from some disadvantages. Even if model-based methods are fast and simple, they are sensitive to occlusion and require high-resolution images since the difficulties lie in detecting the specific facial features with high precision and accuracy. Appearance-based approaches are not affected by these limitations, but they are sensitive to information about identity and lighting contained in the face appearance. However, when using a robust and efficient head pose descriptor, the appearance-based techniques become invariant to identity and lighting.

In the following, we expose some head pose estimation techniques for monitoring driver vigilance state.

2.2 Driver head pose estimation

A great interest is focused on driver assistance systems that use driver head pose as a cue to visual focus of attention and mental state [6, 11, 36, 37]. A commercial product called Smart Eye AntiSleep [36, 38] is developed and corresponds to a compact system equipped with one VS camera and two IR flashes designed for automotive applications. AntiSleep measures 3D head position and orientation, gaze direction, and eyelid closures. Authors use a tracking approach and a geometric method as initialization step based on a 3D head model containing the relative distances between specific facial points localized using local Gaussian derivatives [39], SIFT, and Gabor jets [40]. The probability distribution of each point descriptor is learned from a large set of facial training images. Then, an initial head pose is estimated from the positions of the facial features and the generic head model. The detected facial features are then tracked using structure-from-motion algorithms. During tracking, the driver-specific appearance of each generic feature is learned for different views. The obtained information is used to stabilize and speed up tracking. This commercial product is limited to controlled environments and therefore is essentially used for simulation purposes.

The most popular research laboratory working on driver assistance systems is the CVRR Laboratory at the University of California, USA. This team proposes several approaches to monitor driver vigilance [6, 7, 16, 37, 41]. In [6], the problem of estimating driver head pose is addressed using a localized gradient orientation descriptor on 2D video frames acquired by a special camera (sensitive to IR and VS lights) as the input to two support vector regressions (SVRs), one for pitch and the other for yaw. This team has equipped a prototype car with many sensors allowing to look in and look out of a vehicle. Such equipment is too expensive to be widely used in car industry. Unfortunately, we cannot compare with these approaches since their database is not accessible and the systems are not detailed enough to allow reproduction.

The goal of our global work is to propose a system for monitoring driver vigilance level based on low cost equipment. In this paper, we focus our attention on estimating driver head pose respecting the requirements (R1), (R2), and (R3). In Table 1, we summarize the properties of some methods presented above and we precise with the signs "∗" and "+" the approaches that will be used for comparison in Section 4. The sign "∗" is associated to the most used references for benchmarks in literature, and the sign "+" corresponds to the recent works providing the best results. From literature, it is obvious that appearance-based techniques are more adapted to our purpose since they respect the requirement (R3) when the descriptor used to construct the feature vector is chosen carefully. Therefore, we propose an efficient and robust fusion of the most pertinent head pose descriptors and we decide to use the SVM classifier since it is adapted to the real-world applications and it proves its efficiency in literature.

3 Discrete head pose estimation for monitoring driver vigilance level

When analyzing the impact of head orientations on driver inattention, we can observe that the driver is attentive to

Table 1 Overview of the most relevant literature approaches

Reference	Year	Type	Methods	R1	R2	R3
Our(+,0,4)	2015	Cl	Descriptor fusion + SVM	√	√	√
[17](1)	2012	GM	Face symmetry	√	√	×
[12](0)	2012	GM	3D geometry	√	√	√
[18](+,4)	2013	GM	Golden ratio	√	√	×
[19](2,3)	2010	FM	Face model	√	√	√
[21](+,4)	2013	AT	SF + LPF	√	√	√
[31](+,4)	2012	Rg	HOG + PLS Rg	√	√	√
[24](+,4)	2012	Rg	SIFT + RP	√	√	√
[23](+,4)	2013	Cl	Multi-scale SF + SVM	√	√	√
[27](+,4)	2014	Cl	Joint detection and estimation + SVM	√	√	√
[28](+,4)	2014	Cl	Gabor + covariance + learning	√	×	√
[26](∗,4)	2007	Cl	Associative memory	√	√	√
[32](∗,4)	2008	Cl+Rg	SVM + SVR	√	×	√
[35](∗,4)	2007	DR+Cl	LDA + linear learning	√	√	√
[6](0)	2008	Tr	Tracking using particle filters	√	√	√

Best result approaches (*plus sign*), most used references for benchmarks (*asterisk*). Databases: "0": Own; "1": FacePix [48]; "2": BU [49]; "3": MIT [50]; "4": Pointing'04 [34]
GM geometric model, *FM* flexible model, *AT* appearance template, *Cl* classification, *Rg* regression, *Tr* tracking, *DR* dimensionality reduction

the road in frontal position. However, the driver needs to look at the dashboard, the rear-view mirror, and the side-view mirrors which correspond to moving the head to down, up, left, and right positions for a brief time. These positions must be maintained for few seconds; otherwise, they are representative of inattention. We can also conclude that the driver attention is not influenced by the orientation according to the roll angle, which allows us to reduce our degrees of freedom to pitch and yaw angles. According to [32], when one or some head pose labels are considered as a class, the head pose estimation is addressed as a classification problem and if the pose angles are ordered, the problem can be thought as a regression problem. After these observations, we can formulate our problem of estimating head pose to detect driver inattention as the problem of classifying head poses into 3 classes for pitch and 3 classes for yaw presented as follows:

- Pitch: frontal head, up position, down position
- Yaw: frontal head, left profile, right profile

In this work, we study different head pose descriptors able to detect variations in driver head pose and we propose an efficient fusion approach providing a good discrimination of pose variations. Since we address the problem of classifying human heads into discrete poses, we evaluate the ability of these descriptors to represent pose variations by testing their efficiency using the SVM classifier. In the following, we present a brief overview of our global system for monitoring driver vigilance level.

3.1 Global overview of our system for monitoring driver vigilance level

In this subsection, we present an overview of our global system for assessing driver vigilance level, while in the next sections, we focus our attention on studying driver inattention by estimating head pose. The principle of detecting inattention is based on the assumption that driver head is in abnormal position when it is maintained for a certain duration in a non-frontal pose for both pitch and yaw angles. Our system illustrated in Fig. 2 and can be synthesized as follows:

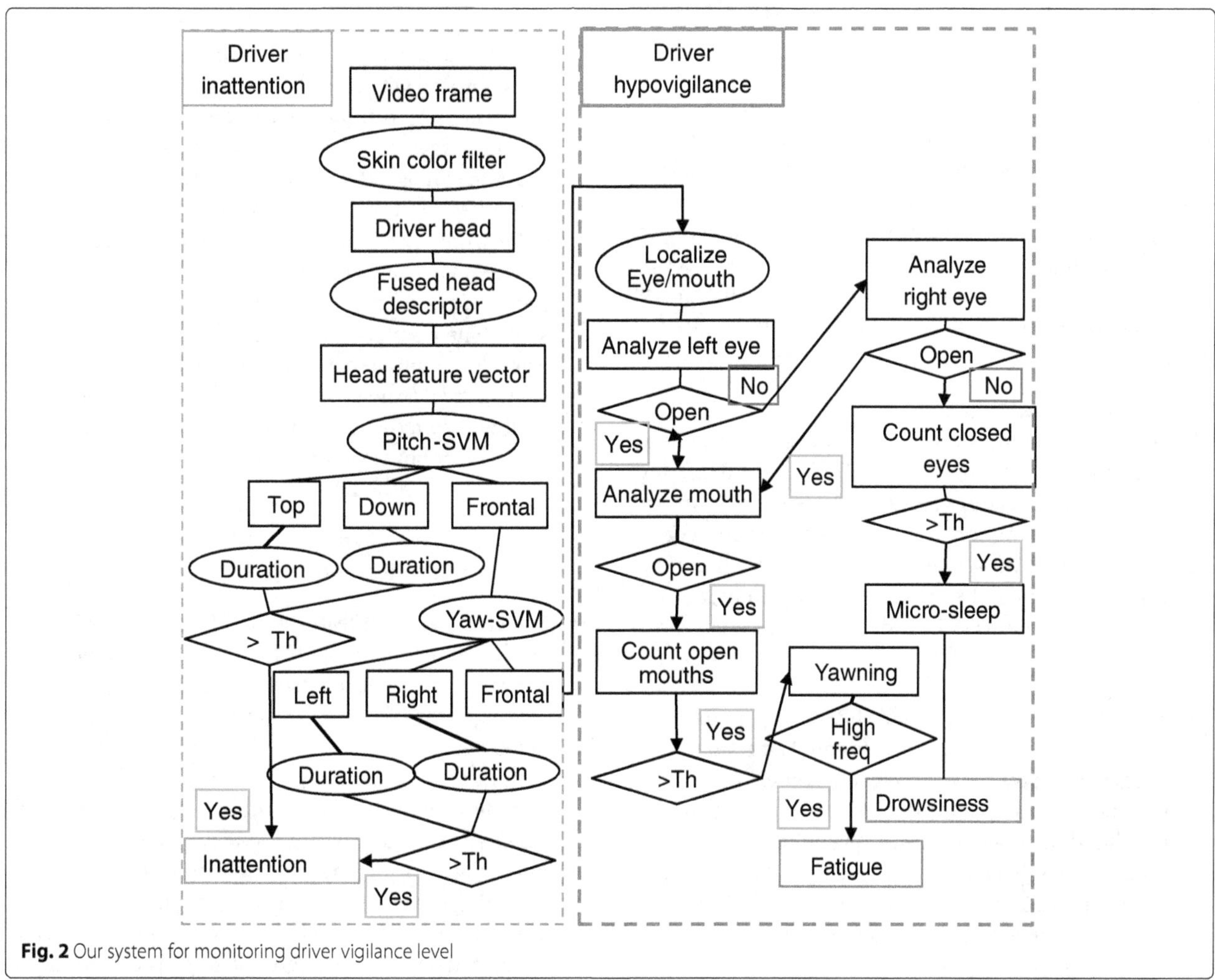

Fig. 2 Our system for monitoring driver vigilance level

Detecting driver inattention:

1. Extract the head from video frame using skin color segmentation.
2. Extract head pose descriptors to obtain a representative feature vector.
3. Apply Pitch-SVM at first since we assume that maintaining head in down position for a certain duration corresponds to the most critical pose and can reveal sleep.
4. If head is in down or up position, observe duration of fixed position and emit inattention warning when it is important.
5. If head is frontal according to pitch, apply Yaw-SVM.
6. If head is in left or right profile, observe duration of fixed position and emit inattention warning when it is important.
7. If head is frontal for both angles, proceed to *hypovigilance detection* (Fig. 2) detailed in [10].

In the following, we focus our attention on studying the most appropriate features to propose a robust fused descriptor representing driver head pose. Moreover, we evaluate the performance of the SVM classifier for estimating head poses.

3.2 Proposed approach for head pose estimation

As mentioned above, we present in this subsection several image descriptors frequently used in literature and judged to be the most representative of head pose variations. Next, we expose feature selection techniques allowing to select the most pertinent attributes among these descriptors. We use the SVM classifier to decide in which class each head image (characterized by its feature vector) is related.

3.2.1 *Head pose descriptors*

We chose to study four descriptors to characterize head pose variations which are SF, HOG, Haar features, and speeded up robust features (SURF). These descriptors are invariant to the common image transformations corresponding to image rotation, scale changes, and illumination variation. Hence, they respect the requirement (R3) allowing them to be used in order to build a head pose estimator.

- *Steerable filters*: The steerable filters [42] are used due to their ability to analyze oriented structures in images. We have proved their robustness in our previous work [21] for estimating head pose using likelihood parametrized function (LPF). Another motivation is given by their capacity to filter an image at any orientation using only a linear combination of its filtered versions obtained by a small set of basis filters. This concept reduces considerably the processing time. We chose a simple SF corresponding to the derivatives of the circularly symmetric Gaussian function $f(x, y) = \exp(-\frac{(x^2+y^2)}{2\sigma^2})$ to describe head poses. In this case, the basis filters are the first derivatives of f according to x and y and correspond to the filters at orientations 0° and 90°, respectively. Hence, a filtered image by an orientation θ can be expressed by $R_1^{\theta} = \cos(\theta)R_1^{0^\circ} + \sin(\theta)R_1^{90^\circ}$, where $R_1^{0^\circ}$ and $R_1^{90^\circ}$ correspond to the image filtered by the two basis filters (see [21] for more details). The performance of the SF depends of the number of filters applied on the image and also the orientation of each filter. We have conducted several experiments, and we find that the following values provide the best result:
    - Number of filers = 2
    - Size of reduced patch image = 15
    - Angular displacement = 50° (i.e., Filter 1 at $\theta = 0°$ and Filter 2 at $\theta = 50°$)
    - SF feature size: 450 ($15 \times 15 \times 2$)
- *Haar features*: The Haar features [43] represent a dense overcomplete representation using wavelets. The two-dimensional Haar decomposition of a square image with n^2 pixels consists of n^2 wavelet coefficients corresponding to a distinct Haar wavelet. The first wavelet is the mean pixel intensity value of the whole image; the rest of the wavelets are computed as the difference in mean intensity values of horizontally, vertically, or diagonally adjacent squares. The contrast variances between the pixel groups are used to determine relative light and dark areas. The Haar coefficient of a particular Haar wavelet is computed as the difference in average pixel value between the image pixels in the black and white regions. From the experiment, we find that the following number of wavelets provides the best estimation of head pose:
    - Number of wavelet = 32
    - Haar feature size : 1024 (32×32)
- *Speeded up robust features*: SURF [44] is a fast and enhanced version of SIFT. It is an algorithm for local, similarity invariant representation and comparison. The algorithm is structured into three steps: detecting interest point, building the descriptor for each interest point, and performing descriptor matching. In our paper, we use an adapted version of SURF since we do not need to perform descriptor matching allowing image comparison. Hence, after obtaining the descriptors of interest points, we sort them according to their orientations. Then, we divide the sorted descriptors in groups before computing

the average of elements of each group. The descriptor size and the number of groups for decomposition are the parameters that influence the SURF performance. We find experimentally that these following values provide the best result of head pose estimation:

  - Descriptor dimension = 64
  - Number of descriptors = 4
  - SURF feature size: 256 (64 × 4)

- *Histogram of oriented gradients*: The basic idea of HOG [45] is that object appearance and shape can be represented by the distribution of local intensity gradients or edge directions, even without precise knowledge of the corresponding gradient or edge positions. This concept can be implemented by splitting the image into small regions (cells) with a defined size adapted to the size and resolution of the object. For each cell, the occurrences of gradient orientation over all the pixels are accumulated in a local histogram. Each orientation histogram divides the gradient angle range into a fixed number of bins. The parameters influencing the HOG are the number of cells per rows and per column in addition to the number of bins. The best performance for head pose estimation using HOG is given by the following configuration:

  - Number of cells per image row = 3
  - Number of cells per image column = 3
  - Number of histogram bins = 10
  - HOG feature size: 90 (3 × 3 × 10)

3.2.2 Feature selection techniques

In our driver head pose estimator, different descriptors are used to extract image features. We choose to extract features as diverse and rich as possible in order to take advantage of their complementarity, but we did not ignore the possibility of redundancy. The aim of the feature selection step is to find a compact, relevant, and consistent set of features for classification task. Feature selection searches through all possible combinations of attributes in the data to find which subset works best for prediction by employing two tasks: search method and attribute evaluator. The search method generates subsets of features and attempts to find an optimal subset while the attribute evaluator determines how good a proposed feature subset is, returning some measures of goodness to the search method. We have evaluated three popular search methods (BestFirst, GreedyStepwise, and Ranker), and we find that the Ranker provides the best results. This can be explained by the individual evaluation of features by the Ranker instead of subset evaluation performed by the two other methods. Therefore, we study three attribute evaluators that can be associated with the Ranker method. The gain ratio (GR) evaluates the worth of an attribute by measuring the gain ratio with respect to the class. The OneR performs evaluation using a simple classification that generates one rule for each predictor in the data and selects the rule with the smallest total error as its "one rule." The evaluation performed by the ReliefF (RF) consists on repeatedly sampling an instance and considering the value of the given attribute for the nearest instance of the same and different class. In Section 4, we will evaluate these feature selection techniques and the best one will be retained to construct the fused feature vector of head pose.

3.2.3 SVM classifier

The SVM is based on structural risk minimization theory [46]. Given a set of training vectors $(x_1, y_1), \ldots, (x_l, y_l)$ composed of observations $x_i \epsilon R^n$ and interpretations $y_i \epsilon \{-1, +1\}$, the binary SVM optimizes a hyperplane to separate positive and negative training samples using their feature vectors. Different kernels could be used to map the classification problem to a higher dimensional feature space. For multiclass problems, the original learning problem must be decomposed into a series of binary learning problems. A standard solution for this problem is the one-against-all approach, which constructs one binary classifier for each class. A faster and more accurate approach for small number of classes is the pairwise classification [47] which is based on transforming the c-class problem into $\frac{c(c-2)}{2}$ binary problems, one for each pair of classes. For our experiments, we used the pairwise classification multiclass SVM with RBF kernel, available in the free software WEKA.

4 Experimental results

Since there is no public database containing various driver head poses, we have acquired video sequences representing a driver in different head poses to perform our experiment. However, this is not enough to prove the robustness of our system which requires to be compared with the state-of-the-art approaches. To guarantee unbiased comparison, we perform experiments on the public Pointing'04 database [34], which is the most used database in literature for head pose estimation [14]. Moreover, this database could represent the driving environment since the distance between the subjects and the camera is comparable to the one between the driver and the dashboard, where the camera is mounted.

4.1 Experiments on public database

4.1.1 The Pointing'04 database

The Pointing'04 database contains head poses labeled according to pitch and yaw angles, and it is composed from 15 sets of near-field images. Each set contains two series of 93 images of the same person at 93 discrete head poses [34]. These ones span both pitch and yaw included in the set {0; ±90; ±60; ±30; ±15} and the interval [−90°; +90°] with a displacement of 15°, respectively. The subjects range in age from 20 to 40 years old, five possessing facial hair and seven wearing glasses. Each subject was photographed against a uniform background, and head pose ground truth was obtained by directional suggestion. In Fig. 3, we show the frontal pose (pitch = yaw = 0°) of the thirty Pointing'04 folds.

We perform several series of experiments on the Pointing'04 database using 80 % as training set (2232 images), 10 % as validation set (279 images), and 10 % as test set (279 images). In the following, the results are given on the test set. We first present the results of the optimization step to fix the best system parameters and also the performance of separate and combined descriptors.

4.1.2 System optimization

In our paper, we deal with the problem of estimating driver head pose according to two degrees of freedom (pitch and yaw angles) in order to identify three classes (cl) for each angle. However, the Pointing'04 database is composed of 9 poses for pitch and 13 poses for yaw. We propose to cluster the poses into three classes for pitch and three classes for yaw to match our problem formulation. For SVMs, we find that the RBF kernel with $\gamma = 0.15$ provides the best classification results. The optimal values of descriptor parameters are presented in the Section 3.2.1 (SF = 450, HOG = 90, SURF = 256, and Haar = 1024).

In Table 2, we show the results of each descriptor evaluated separately in addition to all possible combinations of these descriptors (two, three, and four elements), in terms of accuracy and kappa statistic for both pitch and yaw angles. The accuracy (Acc) is the overall correctness of the model, and it is calculated as the sum of correct classifications divided by the total number of instances, while the kappa statistic (κ) is a chance-corrected measure of agreement between the classifications and the true classes. The highest values of Acc and κ correspond to the best system performance. We also show the processing time in seconds (time) needed to classify one image by pitch-SVM and yaw-SVM. It is obvious that increasing the number of descriptors conduce to increase the processing time of one frame. However, the Haar features are more expensive in terms of computational time because of the large size of their feature vector. From this table, we observe that SF features provide the best result when the descriptors are evaluated separately. The best result of combining two and three descriptors are given respectively by the feature vectors SF, HOG and SF, HOG, SURF. When we combine the four descriptors, the results are less advantageous than those of the best combination of two or three descriptors. This could be explained by an interaction between the attributes of the overall feature vector, which produces contradictions at the decision process performed by the multi-class SVM. This problem could be solved by introducing a feature selection step on the combined descriptor

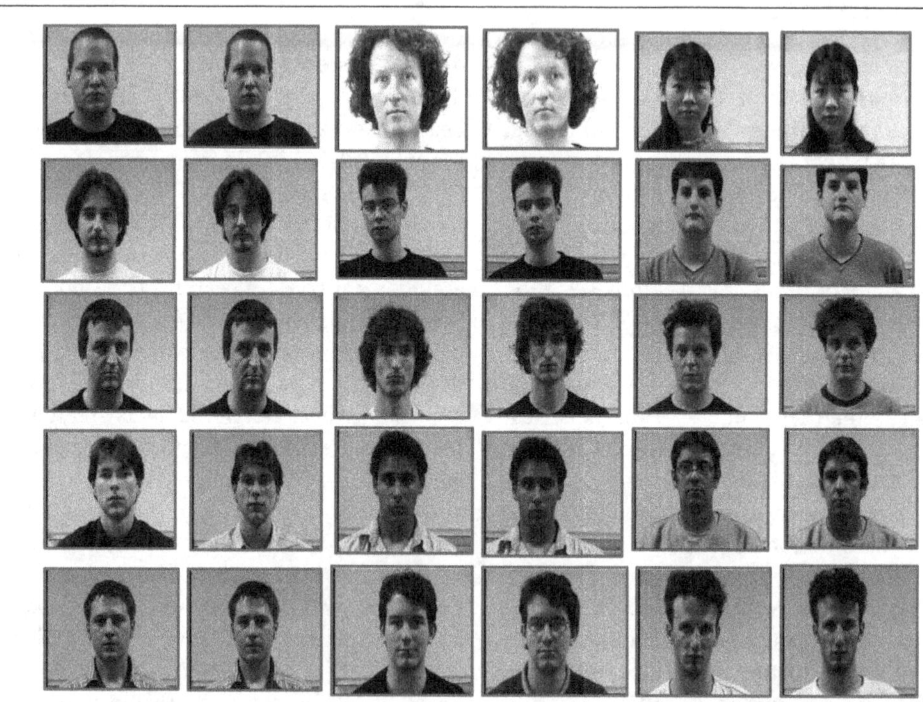

Fig. 3 The frontal pose of the thirty Pointing'04 folds

Table 2 Evaluation of separate descriptors and all possible combinations of them on the test set

	3 classes for pitch-SVM			3 classes for yaw-SVM		
Descriptor	Acc.	κ	Time	Acc.	κ	Time
SF	*87.2*	*0.80*	0.05	*94.3*	*0.91*	0.03
HOG	85.6	0.77	0.01	94	0.90	0.01
SURF	83.8	0.75	0.04	93.7	0.90	0.04
Haar	85.9	0.78	0.2	93	0.89	0.2
(SF, HOG)	*89.3*	*0.83*	0.07	*96.3*	*0.94*	0.06
(SF, SURF)	89.0	0.83	0.13	95.3	0.92	0.12
(SF, Haar)	86.7	0.79	0.5	94.7	0.91	0.47
(HOG, Haar)	88.8	0.82	0.24	94.5	0.91	0.21
(HOG, SURF)	87.2	0.80	0.06	94.9	0.92	0.06
(SURF, Haar)	87.3	0.80	0.35	94.6	0.91	0.32
(SF, HOG, SURF)	*89.1*	*0.83*	0.15	*95.4*	*0.93*	0.11
(SF, HOG, Haar)	85.6	0.77	0.28	95.1	0.92	0.29
(SF, SURF, Haar)	77.9	0.64	0.53	92.3	0.88	0.48
(HOG, SURF, Haar)	87.8	0.81	0.19	94.9	0.92	0.17
(SF, HOG, SURF, Haar)	*87.5*	*0.80*	*0.53*	*94.9*	*0.91*	*0.52*

Italic values in Table 2: Best results obtained by all possible combinations of one, two, three and four descriptors

that allows us to keep the most relevant attributes and reduce the processing time.

4.1.3 Evaluating feature selection techniques

In Table 3, we show the results of evaluating the feature selection techniques presented in Section 3.2.2 using the Ranker as search method, which is equivalent to the evaluation of the performance of three attribute evaluators (Attr. Eval.): GR, OneR, and RF. A first set of tests is conducted on the best combination of three descriptors (SF, HOG, SURF) using the 400 most relevant variables from a total of 796, which corresponds to a reduction of attributes by half. According to Table 3, the best result of these tests is given by the ReliefF algorithm. Hence, in a second set of tests, we apply the ReliefF attribute evaluator on the combination of the four descriptors and we evaluate the impact of varying the number of selected variables on the system performance. We chose to retain 400 relevant variables using ReliefF as the best configuration, since it provides a good compromise between processing time, accuracy, and kappa coefficient. In the last test of this subsection, we show the result of the best configuration when using the k-fold cross validation (CV) process with $k = 10$. The cross-validation reorders the database and divided it into 10 equal parts. Then, for each iteration, one part is used for the test and the other nine parts for learning the classifier. All results are collected and averaged at the end of the cross-validation. From the last line of Table 3, we note that the result obtained by cross-validation (CV) improves the conventional test, which proves that the

Table 3 Performance on the test set of the studied attribute evaluators on the best combination of three and four descriptors using the Ranker search method

		3 classes for pitch-SVM			3 classes for yaw-SVM		
Descriptor	Attr. Eval.	Acc.	κ	Time	Acc.	κ	Time
(SF, HOG, SURF)	(GR,400/796)	87.0	0.79	0.06	94.5	0.91	0.05
(SF, HOG, SURF)	(OneR,400/796)	86.4	0.79	0.05	94.6	0.91	0.05
(SF, HOG, SURF)	(RF,400/796)	90.1	0.84	0.05	95.4	0.93	0.05
(SF, HOG, SURF, Haar)	(RF,600/1820)	90.5	0.85	0.34	96.7	0.94	0.32
(SF, HOG, SURF, Haar)	(RF,400/1820)	90.5	0.85	0.09	96.6	0.94	0.08
(SF, HOG, SURF, Haar)	(RF,200/1820)	88.1	0.80	0.06	94.2	0.91	0.05
(SF, HOG, SURF, Haar)	(RF,400/1820)	91.9	0.87	CV	96.4	0.94	CV

proposed approach allows a good classification of poses even when varying samples.

To visualize which descriptors are more pertinent, we present in Table 4 the total number of each descriptor features (TN), the number of selected features from each descriptor (SN), the rate of the features extracted from each descriptor (FiD), and the participation rate of each descriptor in the fusion (DiF). If we analyze the column (FiD), we can observe that the SF and HOG are the most pertinent descriptors since more than 50 % of their features are selected while less than 10 % of Haar and SURF features are selected. Moreover, the analysis of the column (DiF) shows that the SF features are the most present ones in the final descriptor with more than 65 % of features.

4.1.4 Comparison with existing approaches

The major part of approaches using the Pointing'04 database for evaluation uses its standard representation of poses which corresponds to 9 angles for pitch and 13 angles for yaw. To provide a fair comparison, we increase the number of classes considered by our system in order to respect the standard representation. Therefore, in this experiment, the pitch-SVM and yaw-SVM must classify 9 and 13 head angles, respectively. Moreover, we present the results in terms of angular mean absolute errors (MAE) between the estimated and ground-truth angles for botch pitch and yaw, since all considered approaches for comparison use them. In Table 5, we present the result of our approach compared to the best approaches in literature and also to the most referenced ones (see Table 1). In [26], Gourier et al. measure the human performance for estimating head poses on the Pointing'04 database and find that the angular MAE correspond to 11° for pitch and 11.9° for yaw. From Table 5, we can conclude that our head pose estimator is more precise than the human performance. As can be seen, it provides the best results among all studied approaches.

In the next experiment, we can show the result obtained when using our head pose estimation technique on real video sequence representing driver with various head poses.

Table 4 Number and percentage of selected features from the fused descriptor

		3 classes for pitch-SVM			3 classes for yaw-SVM		
Descriptor	TN	SN	FiD	DiF	SN	FiD	DiF
SF	450	263	58 %	66 %	275	61 %	69 %
HOG	90	62	68 %	16 %	70	78 %	18 %
SURF	256	0	0 %	0 %	11	4 %	2 %
Haar	1024	75	7 %	18 %	44	4 %	11 %

TN the total number of descriptor features, *SN* the number of selected descriptor features after fusion, *FiD* the rate of SN in the descriptor, *DiF* the rate of SN in the fusion

Table 5 Comparison with existing techniques in terms of angular MAE using Pointing'04 database with 9 poses for pitch and 13 poses for yaw

Approach	Year	Pitch	Yaw
Our approach	2015	*4.6°*	*6.1°*
HOG + structural SVM [27]	2014	5.25°	5.91°
Dense SIFT + RP [24]	2012	5.84°	6.05°
Kernel PLS regression [31]	2012	6.61°	6.56°
Gabor + covariance + learning [28]	2014	7.14°	6.24°
Multi-scale SF + SVM [23]	2013	8°	6.9°
SF + LPF [21]	2012	8°	9.37°
Geometric approach (golden ratio) [18]	2013	13.6°	9.6°
Cropped head + SVM + SVR [32]	2008	7.69°	9.23°
Human performance [26]	2007	11°	11.9°
Associative memory [26]	2007	15.9°	10.3°
LDA + linear learning [35]	2007	30.7°	19.1°

Italic values in Table 5: Best angular MAE

4.2 Experiment on driver video sequence

We have acquired a video sequence, as shown in Fig. 4, with a cheap visible spectrum phone camera representing a driver in various head poses and composed from 2636 video frames. Each frame has a resolution of 1280 × 720 pixels.

Fig. 4 Driver acquisition system with cheap phone camera

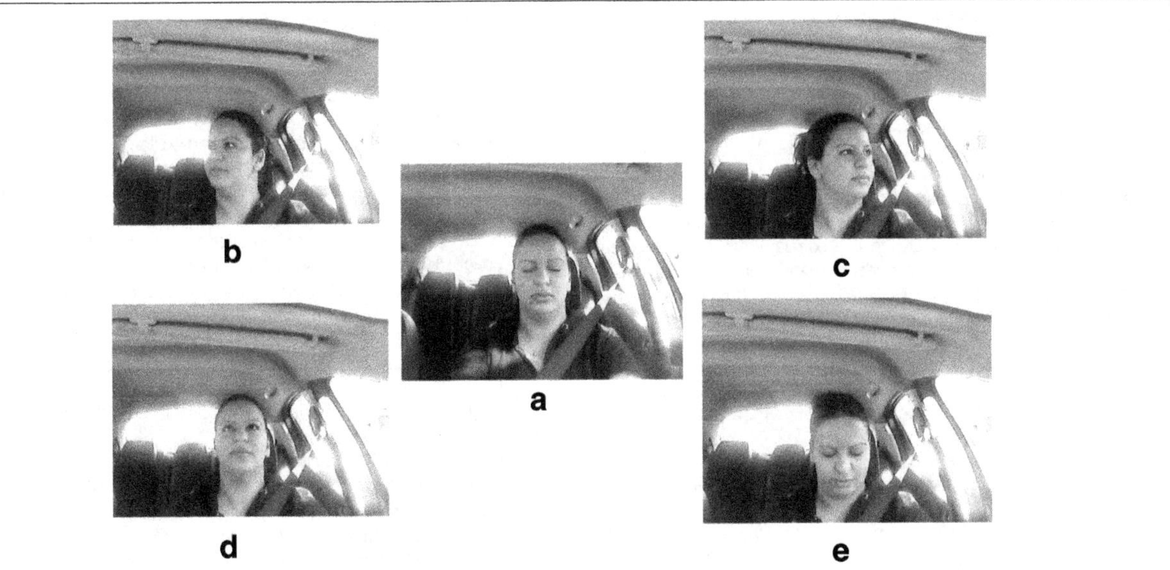

Fig. 5 Examples of driver video frames. **a** Frontal position (pitch and yaw). **b** Left profile (yaw). **c** Right profile (yaw). **d** Up position (pitch); **e** Down position (pitch)

Since we deal with estimating driver head pose, we have annotated our sequence using three classes for pitch and three classes for yaw. Figure 5 represents an example of frames corresponding to each class according to the pitch and yaw angles. The result obtained when applying our head pose estimator on driver video sequence using the best parameters determined in Section 4.1 is given by the second row in Table 6. The first result in this table reports the same experiment applied on Pointing'04 database, previously presented in line 5, Table 3.

Even if our sequence is acquired in real conditions, the results obtained in this experiment are better than the one obtained on Pointing'04 database. This fact might be explained by the inherent problem of annotation caused by the important number of poses in the Pointing'04 database while in our sequence, we annotate 3 poses for each angle.

5 Conclusions

In this paper, we have proposed a head pose estimation approach using a single camera in order to identify driver inattention. Our approach is based on a robust fusion of multiple significant descriptors (SF, HOG, Haar, and SURF) in order to construct an efficient feature vector representing head pose variations. Then, two SVMs are learned to classify the feature vectors according to pitch and yaw angles. Our head pose estimator is not restricted to monitoring driver inattention level and can also be used by diverse applications requiring knowledge of human activity such as human-machine interfaces and game industry. Before applying our estimator, it is important to identify the number of poses that must be estimated for each angle depending on the application requirements. In our paper, we use three classes for both pitch and yaw angles since we deal with the problem of estimating driver head pose to determine its inattention level. Since no public database is available for estimating driver head pose, we perform several experiments on the public database Pointing'04 to validate our approach and compare it with the recent and the most cited state-of-the-art techniques. We have also acquired a video sequence using a cheap visible spectrum camera representing a driver in various attention levels and we find that our head pose estimator can achieve an accuracy of 97.5 % for pitch and 98.2 % for yaw.

As future work, we can improve our global system for monitoring driver vigilance level by adding a gaze estimation approach in order to determine driver focus of attention. Since we use a visible spectrum camera, the acquisition can be perturbed at night and the usage of IR light could be considered to resolve this problem.

Table 6 Results of our head pose estimation on the driver video sequence using 3 classes for pitch and yaw

	3 classes for pitch-SVM			3 classes for yaw-SVM		
Database	Accuracy	Kappa	Time	Accuracy	Kappa	Time
Pointing'04	90.5	0.85	0.09	96.6	0.94	0.08
Our Sequence	97.5	0.96	0.03	98.2	0.98	0.02

Competing interests
The authors declare that they have no competing interests.

Author details
[1]Ibn Zohr University, Morocco, BP 32/S, Agadir, Morocco. [2]LRIT-CNRST 29, Faculty of Sciences, Mohammed V University, Ibn Battouta Avenue, Rabat, Morocco. [3]LITIS, INSA-Rouen, Avenue de l'Université, Saint-Etienne-du-Rouvray, France. [4]LGS, ENSA-Kenitra, Ibn Tofail University, Avenue de l'Université, Kenitra, Morocco.

References

1. State of the Road, A fact sheet of CARRS-Q, Centre of Accident Research and Road Safety-Queensland, Queensland, Australia. Date accessed: March 2013 (2012). http://www.carrsq.qut.edu.au/publications/corporate/hooning_fs.pdf
2. C Berka, DJ Levendowski, MN Lumicao, A Yau, G Davis, VT Zivkovic, RE Olmstead, PD Tremoulet, PL Craven, EEG correlates of task engagement and mental workload in vigilance, learning, and memory tasks. Aviat. Space Environ. Med. **78**, 231–244 (2007)
3. TW Victor, JL Harbluk, JA Engstrom, Sensitivity of eye-movement measures to in-vehicle task difficulty. Transp. Res. Part F: Traffic Psychol. Behav. **8**, 167–190 (2005)
4. N Papanikolopoulos, M Eriksson, Driver fatigue: a vision-based approach to automatic diagnosis. Transpo. Res. Part C: Emerg. Technol. **9**, 399–413 (2001)
5. B Hrishikesh, S Mahajan, A Bhagwat, T Badiger, D Bhutkar, S Dhabe, L Manikrao, Design of DroDeASys (drowsy detection and alarming system). Adv. Comput. Algo. Data Anal. **14**, 75–79 (2009)
6. E Murphy-Chutorian, MM Trivedi, in *Proceeding of the Intelligent Vehicles Symposium (IV)*. Hyhope: Hybrid head orientation and position estimation for vision-based driver head tracking (IEEE, Eindhoven, The Netherlands, 2008), pp. 512–517
7. KS Huang, MM Trivedi, in *Proceedings of the 1st International Workshop on In-Vehicle Cognitive Computer Vision Systems, in Conjunction with the 3rd International Conference on Computer Vision Systems*. Driver head pose and view estimation with single omnidirectional video stream (IAPR, Graz, Austria, 2003), pp. 44–51
8. T Azim, MA Jaffar, M Ramzan, AM Mirza, in *Signal Processing, Image Processing and Pattern Recognition. Communications in Computer and Information Science, vol. 61*. Automatic fatigue detection of drivers through yawning analysis (Springer, Jeju Island, Korea, 2009), pp. 125–132
9. JF May, LC Baldwin, Driver fatigue: The importance of identifying causal factors of fatigue when considering detection and countermeasure technologies. Transpo. Res. Part F: Traffic Psychol. Behav. **12**, 218–224 (2009)
10. N Alioua, A Amine, M Rziza, D Aboutajdine, in *Computer Analysis of Images and Patterns. Lecture Notes in Computer Science, vol. 6855*. Driver's fatigue and drowsiness detection to reduce traffic accidents on road (Springer, Seville, Spain, 2011), pp. 397–404
11. L Bergasa, J Nuevo, M Sotelo, R Barea, E Lopez, Real-time system for monitoring driver vigilance. IEEE Transac. Intell. Transpo. Syst. **7**, 63–77 (2006)
12. S Gurbuz, E Oztop, N Inoue, Model free head pose estimation using stereovision. Pattern Recognit. **45**, 33–42 (2012)
13. L Li, K Werber, CF Calvillo, KcD Dinh, A Guarde, A Konig, in *Online Conference on Soft Computing in Industrial Applications. Advances in Intelligent Systems and Computing, vol. 223*. Multi-sensor soft-computing system for driver drowsiness detection (Springer, 2012)
14. E Murphy-Chutorian, MM Trivedi, Head pose estimation in computer vision: a survey. IEEE Transac. Pattern Anal. Mach. Intell. (PAMI). **31**, 607–626 (2009)
15. X Liu, H Lu, H Luo, in *Proceeding of the 16th International Conference on Image Processing (ICIP)*. A new representation method of head images for head pose estimation (IEEE, Cairo, Egypt, 2009), pp. 3585–3588
16. E Murphy-Chutorian, MM Trivedi, Head pose estimation and augmented reality tracking: an integrated system and evaluation for monitoring driver awareness. IEEE Transac. Intell. Transpo. Syst. **11**, 300–311 (2010)
17. A Dahmane, S Larabi, C Djeraba, IM Bilasco, in *Proceeding of the 21st International Conference on Pattern Recognition (ICPR)*. Learning symmetrical model for head pose estimation (IEEE, Tsukuba, Japan, 2012), pp. 3614–3617
18. G Fadda, G Marcialis, F Roli, L Ghiani, in *International Conference on Image Analysis and Processing (ICIAP). Lecture Notes in Computer Science, vol. 8156*. Exploiting the golden ratio on human faces for head-pose estimation (Springer, Naples, Italy, 2013), pp. 280–289
19. LP Morency, J Whitehill, J Movellan, Monocular head pose estimation using generalized adaptive view-based appearance model. Image Vis. Comput. **28**, 754–761 (2010)
20. B Ma, X Chai, T Wang, A novel feature descriptor based on biologically inspired feature for head pose estimation. Neurocomputing. **115**, 1–10 (2013)
21. N Alioua, A Amine, A Bensrhair, M Rziza, D Aboutajdine, in *Proceedings of the 21st European Signal Processing Conference (EUSIPCO)*. Head pose estimation based on steerable filters and likelihood parametrized function (EURASIP, Marrakech, Morocco, 2013)
22. M Demirkus, B Oreshkin, JJ Clark, T Arbel, in *Proceeding of the 18th International Conference on Image Processing (ICIP)*. Spatial and probabilistic codebook template based head pose estimation from unconstrained environments (IEEE, Brussels, Belgium, 2011), pp. 573–576
23. V Jain, JL Crowley, in *Proceeding of the 18th Scandinavian Conference on Image Analysis*. Head pose estimation using multi-scale gaussian derivatives (Springer, Espoo, Finlande, 2013)
24. HT Ho, R Chellappa, in *Proceeding of the 19th International Conference on Image Processing (ICIP)*. Automatic head pose estimation using randomly projected dense sift descriptors (IEEE, Orlando, Florida, USA, 2012), pp. 153–156
25. R Munoz-Salinas, E Yeguas-Bolivar, A Saffiotti, R Medina-Carnicer, Multi-camera head pose estimation. Mach. Vis. Appl. **23**, 479–490 (2012)
26. N Gourier, J Maisonnasse, D Hall, JL Crowley, in *Multimodal Technologies for Perception of Humans. Lecture Notes in Computer Science, vol. 4122*. Head pose estimation on low resolution images (Springer, Uthampton, UK, 2007), pp. 270–280
27. K He, L Sigal, S Sclaroff, in *European Conference on Computer Vision (ECCV). Lecture Notes in Computer Science, vol. 8692*. Parameterizing object detectors in the continuous pose space (Springer, Zurich, Switzerland, 2014), pp. 450–465
28. B Ma, A Li, X Chai, S Shan, CovGa: a novel descriptor based on symmetry of regions for head pose estimation. Neurocomputing. **143**, 97–108 (2014)
29. Y Ma, Y Konishi, K Kinoshita, S Lao, M Kawade, in *Proceeding of the 18th International Conference on Pattern Recognition (ICPR)*. Sparse Bayesian regression for head pose estimation (IEEE, Hong Kong, China, 2006), pp. 507–510
30. G Fanelli, M Dantone, J Gall, A Fossati, LV Goo, Random forests for real time 3D face analysis. Int. J. Comput. Vision (IJCV). **101**, 437–458 (2013)
31. M Al-Haj, J Gonzalez, LS Davis, in *Proceeding of Computer Society Conference on Computer Vision and Pattern Recognition (CVPR)*. On partial least squares in head pose estimation: how to simultaneously deal with misalignment (IEEE, Providence, USA, 2012), pp. 2602–2609
32. G Guo, Y Fu, CR Dyer, TS Huang, in *Proceeding of the 19th International Conference on Pattern Recognition (ICPR)*. Head pose estimation: classification or regression? (IEEE, Tampa, Florida, USA, 2008), pp. 1–4
33. E Ricci, JM Odobez, in *Proceeding of the 18th International Conference on Image Processing (ICIP)*. Learning large margin likelihoods for realtime head pose tracking (IEEE, Cairo, Egypt, 2009), pp. 2593–2596
34. N Gourier, D Hall, J Crowley, in *Proceedings of Pointing 2004, ICPR, International Workshop on Visual Observation of Deictic Gestures*. Estimating face orientation from robust detection of salient facial structures (IEEE, Cambridge, UK, 2004)
35. Z Li, Y Fu, J Yuan, TS Huang, Y Wu, in *Proceeding of the International Conference on Multimedia and Expo (ICME)*. Query driven localized linear discriminant models for head pose estimation (IEEE, Beijing, China, 2007), pp. 1810–1813
36. L Bretzner, M Krantz, in *Proceeding of the International Conference on Vehicular Electronics and Safety*. Towards low-cost systems for measuring visual cues of driver fatigue and inattention in automotive applications (IEEE, 2005), pp. 161–164
37. MM Trivedi, T Gandhi, J McCall, Looking-in and looking-out of a vehicle: computer-vision-based enhanced vehicle safety. IEEE Trans. Intell. Transpo. Syst. **8**, 108–120 (2007)
38. Smart Eye AntiSleep 4. Date accessed: December 2013 (2013). http://smarteye.se/wpcontent/uploads/2014/12/Product-Sheet-AntiSleep.pdf
39. V Vogelhuber, C Schmid, in *Proceeding of the International Conference on Pattern Recognition (ICPR)*. Face detection based on generic local descriptors and spatial constraints (IEEE, Barcelona, Spain, 2000), pp. 1084–1087
40. L Wiskott, JM Fellous, N Kruger, CVD Malsburg, Face recognition by elastic bunch graph matching. IEEE Trans. Pattern Anal. Mach. Intell. (PAMI). **19**, 775–779 (1997)

41. J Wu, MM Trivedi, A two-stage head pose estimation framework and evaluation. Pattern Recognit. **41**, 1138–1158 (2008)
42. WT Freeman, EH Adelson, The design and use of steerable filters. IEEE Trans. Pattern Anal. Mach. Intell. **13**, 891–906 (1991)
43. C Papageorgiou, T Poggio, A trainable system for object detection. Int. J. Comput. Vis. **38**, 15–33 (2000)
44. H Bay, A Ess, T Tuytelaars, LV Gool, SURF: speeded up robust features. Comput. Vision and Image Underst. **110**, 346–359 (2008)
45. N Dalal, B Triggs, in *Proceeding of the Computer Society Conference on Computer Vision and Pattern Recognition (CVPR), vol. 1*. Histograms of oriented gradients for human detection (IEEE, San Diego, CA, USA, 2005), pp. 886–893
46. C Burges, Tutorial on support vector machines for pattern recognition. Data Min. Knowl. Disc. **2**, 121–167 (1998)
47. SH Park, J Furnkranz, in *Machine Learn. Lecture Notes in Computer Science, vol. 4701*. Efficient pairwise classification (Springer, Warsaw, Poland, 2007), pp. 658–665
48. J Black, M Gargesha, K Kahol, P Kuchi, S Panchanathan, in *Internet Multimedia Systems II (ITCOM)*. A framework for performance evaluation of face recognition algorithms (SPIE, Boston, USA, 2002)
49. ML Cascia, S Sclaroff, V Athitsos, Fast, reliable head tracking under varying illumination: an approach based on registration of textured-mapped 3D models. IEEE Trans. Pattern Anal. Machine Intell. (PAMI). **22**, 322–336 (2000)
50. LP Morency, A Rahimi, T Darrell, in *Proceeding of the Computer Society Conference on Computer Vision and Pattern Recognition (CVPR)*. Adaptive view-based appearance model (IEEE, Madison, USA, 2003), pp. 803–810

6

Robust surface normal estimation via greedy sparse regression

Mingjing Zhang and Mark S. Drew*

Abstract

Photometric stereo (PST) is a widely used technique of estimating surface normals from an image set. However, it often produces inaccurate results for non-Lambertian surface reflectance. In this study, PST is reformulated as a sparse recovery problem where non-Lambertian errors are explicitly identified and corrected. We show that such a problem can be accurately solved via a greedy algorithm called orthogonal matching pursuit (OMP). The performance of OMP is evaluated on synthesized and real-world datasets: we found that the greedy algorithm is overall more robust to non-Lambertian errors than other state-of-the-art sparse approaches with little loss of efficiency. Along with providing an overview of current methods, novel contributions in this paper are as follows: we propose an alternative sparse formulation for PST; in previous PST studies (Wu et al., Robust photometric stereo via low-rank matrix completion and recovery, 2010), (S. Ikehata et al., Robust photometric stereo using sparse regression, 2012), the surface normal vector and the error vector are treated as two entities and are solved independently. In this study, we convert their formulation into a new canonical form of the sparse recovery problem by combining the two vectors into one large vector in a new "stacked" formulation in this domain. This allows for a large repertoire of existing sparse recovery algorithms to be more straightforwardly applied to the PST problem. In our application of the OMP greedy algorithm, we show that greedy solvers can indeed be applied, with this study supplying the first of such attempt at employing greedy approaches to estimate surface normals within the framework of PST. We numerically compare the performance of several normal vector recovery methods. Most notably, this is the first detailed test on complex images of the normal estimation accuracy of our previously proposed method, least median of squares (LMS).

Keywords: Photometric stereo, Robust, Surface normals, OMP

1 Introduction

Shading in 2D images provides a valuable visual cue for understanding the spatial structure of objects. Photometric Stereo (PST) is a powerful technique that exploits shading information to directly estimate the 3D surface orientation, i.e. normal vectors. In the classical PST problem, the input is a set of n images captured from a fixed viewpoint under n different calibrated lighting conditions; hence, there are n observations of luminance at each pixel location. Under the assumption of a Lambertian reflectance model, where the observed luminance is proportional to the cosine of the incident angle and remains constant regardless of the viewing angle, the relationship between n observations $\mathbf{y} \in \mathbb{R}^n$ at each pixel and the collection of n lighting directions $\mathbf{L} \in \mathbb{R}^{n\times 3}$ is formulated as a linear equation group with respect to the normal vector $\mathbf{n} \in \mathbb{R}^3$, i.e.,

$$\mathbf{y} = \mathbf{Ln}. \tag{1}$$

We emphasize that there are indeed such a set of n equation at each pixel. In PST, the linear system Eq. 1 is solved via ordinary least squares (LS). The advantage of PST over 3D laser scanning is that the former provides a very high resolution (depending on the actual resolution of the camera) and therefore can capture the fine details of the surface that may not show up in the scanned model. In addition, PST only requires a simple and inexpensive hardware setup whereas 3D scanning devices are usually costly and less portable. Innovative recent work [3, 4] can reduce PST to a single-shot scenario in a different setup

*Correspondence: mark@cs.sfu.ca
School of Computing Science, Simon Fraser University, Vancouver, BC V5A 1S6, Canada

with spectral multiplexing and more than three colour channels or polarized illumination.

Although the classical PST method almost always guarantees a visually plausible normal map, it in fact suffers from a serious accuracy problem: the simple Lambertian reflectance model adopted in PST does not strictly apply to most real-world textures, which exhibit specular reflection properties to various degrees. Even if the surface is indeed approximately Lambertian, other non-Lambertian errors can be introduced by the interaction of the light and the objects' geometry, resulting in cast shadows, including self-shadowing, as well as interreflections. Attached shadows are also outside the simple shading model. Such non-Lambertian observations, regarded as "outliers" in a Lambertian-based linear model, may severely reduce the accuracy of LS results. Hence, a PST method that is robust to such non-Lambertian effects is needed in order to generate a high-quality normal map.

Many improved PST methods have been proposed since the original PST in an attempt to minimize the effect of non-Lambertian components. These methods either adopt a more sophisticated reflectance model to accommodate non-Lambertian observations as "inliers" (e.g. [5–7]) or rather keep the Lambertian model but use robust statistical methods to rule out or reduce the effect of non-Lambertian outliers (e.g. [8–10]). A typical example of the second category is the least median of squares (LMS) approach used in our previous study [11] (and see [10, 12]), in which the observations outside a certain confidence band are deemed to be outliers. In this study, we again adopt the Lambertian model, but solve for the normal vectors via a sparse representation framework that estimates both the normals and non-Lambertian errors at the same time. This sparse method is more closely related to the statistical-based methods.

1.1 Sparse representation and recovery

It is well understood that ordinary LS fails to unambiguously reconstruct a signal that is passed through an underdetermined linear system, where the number of unknown variables exceeds that of linear equations (Fig. 1b). However, it has been shown that if the signal to be recovered is sparse—having a considerable number of zero or nearly zero entries (Fig. 1c)—then an accurate reconstruction of the signal is still possible via a sparse recovery scheme [13].

The canonical form of a sparse recovery problem can be stated as follows: given an underdetermined linear model $\mathbf{y} = \mathbf{Ax}$, where $\mathbf{A} \in \mathbb{R}^{n \times p}$ is the so-called dictionary matrix ($n < p$), and $\mathbf{y} \in \mathbb{R}^{n \times 1}$ is the vector consisting of n scalar observations, find the unknown sparse signal $\mathbf{x} \in \mathbb{R}^{p \times 1}$ such that

$$\min_{\mathbf{x}} \|\mathbf{x}\|_0 \qquad \text{s.t.} \quad \mathbf{y} = \mathbf{Ax}, \tag{2}$$

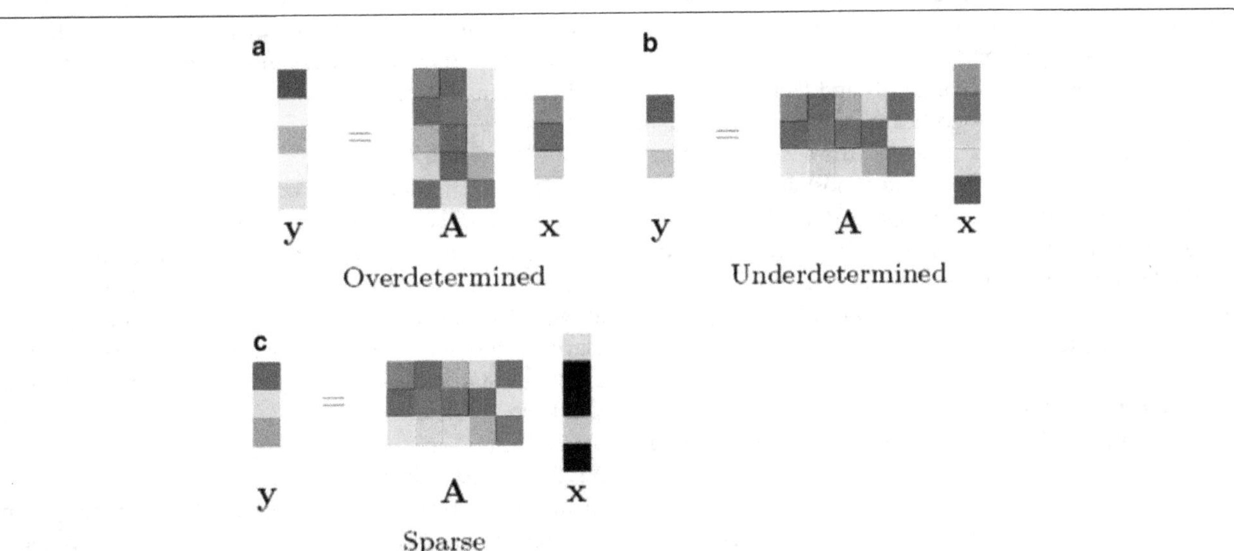

Fig. 1 Stylized visualization of three examples of linear equation systems. **A** and **y** represent the design matrix and observations, respectively; **x** is the unknown signal to be recovered. Positive and negative values are shown as *coloured blocks*, and zero entries are represented by *black blocks*. **a** Overdetermined system, where there are more observations (5) than unknowns (3). **b** Underdetermined system, where there are more unknowns (5) than observations (3). The signal **x** cannot be uniquely determined from such a system. **c** Underdetermined system with sparse signal. It is possible to recover **x** using sparse recovery methods as long as we know that **x** is sparse, even though the system is underdetermined and the exact positions of non-zero entries are not known *a priori*

where $\|\cdot\|_0$ represents ℓ_0 pseudo-norm, the number of non-zero entries.

Equation 2 is generally a non-deterministic polynomial-time (NP)-hard combinatorial problem [14]. In practice, it is more feasible to solve a relaxed form. We will briefly discuss various alternative formulations and corresponding solvers in Section 2.2.

1.2 Photometric stereo and sparse recovery

PST is often formulated as an overdetermined regression problem. The classical PST adopts three lights (hence three observations of luminance at each pixel location) [15] to solve for the 3D normal vectors. Later methods use more lights ranging from four to hundreds [6, 10, 16, 17]. Recently, a few attempts have been made to represent PST as an underdetermined system; firstly, in the case of calibrated lighting directions [1, 2], as addressed here, as well as for the alternative case of unknown lighting conditions [18, 19], not studied in this report. Reconfiguring PST as an underdetermined system means explicitly modelling the non-Lambertian error for each observation as additional unknowns. Suppose there are n lights (hence n equations for each pixel): the per-pixel number of unknowns would be $n + 3$ (three normal vector components and n non-Lambertian errors). As was already pointed out, such a system cannot be unambiguously solved through ordinary LS. Fortunately, if we make an assumption that the majority of luminance observations are approximately Lambertian, then the error vector is essentially a sparse vector with a large number of zero or approximately zero entries. Now that we have a sparse representation of the PST problem, we can solve it using a sparse recovery algorithm.

It has been shown by Wu et al. [1] and Ikehata et al. [2] that sparse PST behaves significantly more robustly than the classical PST method. However, the accuracy is contingent on the solver. At present, the most accurate solver for the sparse formulation is sparse Bayesian learning (SBL) as tested by Ikehata et al. in [2]. In the current study, we employ a modified form of the sparse representation given in [2], but solve it via a different approach—greedy sparse recovery algorithms.

1.3 Novel contributions

The main contributions of the current study are threefold:

1. We propose an alternative sparse formulation for PST (Eqs. 10 and 11). In previous PST studies [1, 2], the surface normal vector and the error vector are treated as two entities and are solved independently. In this study, we convert their formulation into a new canonical form of the sparse recovery problem by combining the two vectors into one large vector. Although such a "stacked" formulation is not novel (e.g. [20]), it is used in the context of surface normal estimation for the first time. The advantage of this formulation is that it allows for a large repertoire of existing sparse recovery algorithms to be more straightforwardly applied to the PST problem.
2. We apply a greedy algorithm called orthogonal matching pursuit (OMP) [21–23], from information theory, to solve the PST problem. It has been previously demonstrated in [1, 2] that PST can be solved by several sparse recovery algorithms that fall into different categories, including augmented Lagrangian rank-minimization [1], ℓ_1 optimization approaches and probability-based methods [2]. However, the possibility of applying greedy solvers, an important category of sparse recovery algorithms, to the PST problem has never been explored. To the best of our knowledge, this study is the first of such attempt at employing greedy approaches to estimate surface normals within the framework of PST.
3. We numerically compare the performance of several normal vector recovery methods. Most notably, it is the first time that the normal estimation accuracy of our previously proposed method—LMS—has been tested and quantitatively demonstrated on complex models.

1.4 Overview

This paper is organized as follows: Section 2 provides a short survey on recent robust PST and sparse recovery methods. In Section 3, we provide a detailed description of our sparse formulation and the OMP algorithm. Experimental results and discussions are presented in Section 4, followed by several possible future research directions discussed in Section 5.

2 Related work

2.1 Robust photometric stereo

This section presents a brief overview of current PST methods. Since the original non-robust Lambertian-based PST [15], many methods have been proposed in an attempt to address non-Lambertian effects such as specularities and shadows. These approaches usually adopt a robust statistical method and/or an improved non-Lambertian reflectance model.

2.1.1 Statistics-based methods

In statistics-based methods, a robust statistical algorithm is employed to detect the non-Lambertian observations as outliers and exclude them from the estimation process in order to minimize their influence on the final result. Early examples include a four-light PST approach in which the values yielding significantly differing albedos are excluded [16, 24, 25]. In a similar five-light PST method [17], the highest and the lowest values, presumably corresponding to highlights and shadows, are simply discarded. Another four-light method [26] explicitly included ambient illumination and surface integrability and adopted an iterative strategy, using current surface estimates to accept or reject each additional light based on a threshold indicating a shadowed value. The problem with these methods is that they rely on throwing away a small number of outlier observation values, whereas our robust sparse methods in the current study reaches the solution based on all observations, by correcting the non-Lambertian error of the outlier observations.

Willems et al. [27] used an iterative method to estimate normals. Initially, the pixel values within a certain range (10–240 out of 255) were used to estimate an initial normal map. In each of the following iterations, error residuals of normals for all lighting directions are computed and the normals are updated based only on those directions with small residuals. Sun et al. [28] showed that at least six light sources are needed to guarantee that every location on the surface is illuminated by at least three lights. They proposed a decision algorithm to discard only doubtful pixels, rather than throwing away all pixel values that lie outside a certain range. However, the validity of their method is based on the assumption that out of the six values for each pixel, there is at most one highlight pixels and two shadowed pixels. Mallick et al. [29] introduced a method based on colour space transformation to separate specular and diffuse components. Holroyd et al. [30] exploited the symmetries in the 2D slices of bidirectional reflectance distribution function (BRDF) obtained at each pixel to recover surface normal and tangent vectors. Both [29] and [30] can be applied to a great variety of surface reflectance, but they do not provide enough focus on the robustness against shadow pixels. Julià et al. [31] utilized a factorization technique to decompose the luminance matrix into surface and light source matrices. They consider the shadow and highlight pixels as missing data, with the objective of reducing the influence of these pixels on the final result.

Some recent studies utilize probability models as a mechanism to incorporate the handling of shadows and highlights into the PST formulation. Tang et al. [32] model normal orientations and discontinuities with two coupled Markov random fields (MRF). They proposed a tensorial belief propagation method to solve the *maximum a posteriori* (MAP) problem in the Markov network. Chandraker et al. [33] formulate PST as a shadow labelling problem where the labels of each pixel's neighbours are taken into consideration, enforcing the smoothness of the shadowed region, and approximate the solution via a fast iterative graph-cut method. Another study [8] employs a maximum likelihood (ML) imaging model for PST. In this method, an inlier map modelled via MRF is included in the ML model. However, the initial values of the inlier map would directly influence the final result, whereas our sparse method does not depend on the choice of any prior.

A few other studies employ random-sampling-based methods. Using three-light datasets, Mukaigawa et al. [34] adopt a random sample consensus (RANSAC)-based approach to iteratively select random groups of pixels from different regions of the image, and the sampled group whose pixels are all taken from diffuse regions are used to calculate the coefficients in the linear equation. RANSAC is also used in a multiview context [9] as a robust fitting approach to select the points on a certain 3D curve. Drew et al. [10, 12] and Zhang and Drew [11] employ a LMS method. Instead of taking samples from different regions on the image, they use a denser image set (50 lights) and sample only from the observations at each pixel location. Non-Lambertian observations are rejected as outliers and excluded from the following LS step. Based on [33], Miyazaki et al. [35] used a median filtering approach similar to LMS but also considering neighbouring pixels. Instead of taking random samples, they simply compare all the three combinations of observations, which is feasible for the small number of lights used in their study. Although guaranteeing a high statistical robustness, these methods are computationally heavy since they usually rely on a large number of samples to take effect.

2.1.2 Non-Lambertian reflectance modelling

Instead of statistically rejecting non-Lambertian effects as outliers, another way to minimize their negative influence on surface normal recovery is to incorporate a more sophisticated reflectance model to directly account for the non-Lambertian components.

Tagare and de Figueiredo [36] constructed an m-lobed reflectance map model to approximate diffuse non-Lambertian surface-light interactions. In [25], a Torrance-Sparrow model is employed to estimate the roughness of the surface that is divided into different areas. Similarly, Nayar et al. [37] adopt a Torrance-Sparrow and Beckmann-Spizzichino hybrid reflectance model. Georghiades [38] applied Torrance-Sparrow model to handle the uncalibrated photometric stereo problem. Other mathematical models to encode surface reflectance include polynomial texture mapping (PTM) [39] and spherical harmonics (SH) [40]. Drew et al. [10] proposed a radial basis function (RBF) interpolation to handle the rendering of specularities and shadows.

Other studies use reference objects to facilitate the estimation of surface properties. In [41], an object with simple, known 3D geometry and approximately Lambertian reflectance (for instance, a white matte sphere) is present in the captured images. A look-up table is established that relates luminance observations at each pixel location and the surface orientation. Then, the surface properties of other objects with similar reflectance as the reference object can simply be inferred from the look-up table. This method, however, only applies to isotropic materials. Hertzmann and Seitz [5] later revisited the idea of including reference material. By adopting an orientation-consistency cue assumption that two points on the surface with the same orientation have the same observed light intensity, they effectively cast PST as a stereoptic correspondence problem. This approach is capable of handling a wider range of anisotropic materials with a small number of reference objects, usually one or two. Similar to [5], an appearance-clustering method proposed by Koppal and Narasimhan [42], also adopting the orientation consistency cue, focuses on finding iso-normals across frames in a captured image sequence, and a classical PST approach may be applied later to obtain the accurate value of the surface normals. Although their method does not rely on the presence of a reference object, it does require the image sequence to be densely captured on a continuous path.

Recent studies attempt to solve a more complicated problem where neither shape nor material information of the object surface is available. Goldman et al. [43] employed an objective function that contains terms for both shape and material and proposed an iterative approach where the reflectance and shape are alternately optimized. The estimation of the material is an inseparable part of the reconstruction process so an explicit reference object is no longer needed. Alldrin et al. [6] also adopt a similar iterative approach that updates shapes and materials alternately. Their formulation is non-parametric and data-driven, and as such is capable of capturing an even wider range of reflectance materials. Ackermann et al. [7] proposed an example-based multi-view PST method which uses the captured object's own geometry as reference.

Yang et al. [44] include a dichromatic reflection model into PST for both estimating surface normals as well as separating the diffuse and specular components, based on a surface chromaticity invariant. Their method is able to reduce the specular effect even when the specular-free observability assumption (that is, each pixel is diffuse in at least one input image) is violated. However, this method does not address shadows and fails on surfaces that mix their own colours into the reflected highlights, such as metallic materials. Moreover, their method also requires knowledge of the lighting chromaticity—they suggest a simple white-patch estimator—whereas in our method, we have no such requirement. Kherada et al. [45] proposed a component-based mapping (CBM) method. They decompose the captured images into direct components (single bounce of light from a surface) and global components (illumination onto a point that is interreflected from all other points in the scene). They then model matte, shadow and specularity separately within each component. This method depends on a training phase, requires accurate disambiguation of direct and global contributions and has a high computational load. Shi et al. [46] introduced a bi-polynomial representation to model the low-frequency component of reflectance and used only the low-frequency information to recover shape and estimate reflectance.

The problem with these methods is that they usually do not work well against non-Lambertian effects that are not accounted for by the surface reflectance alone, such as cast shadows. In our current sparse method, we make no assumption of the surface reflectance property and treat all non-Lambertian effects (specularity and shadow) equally.

2.1.3 Sparse formulation

Recently, a few studies began to adopt sparse representation into PST. Wu et al. [1] model the matrix of all luminance observations as a linear combination of Lambertian and non-Lambertian components and represent the non-Lambertian error as an additive sparse noise vector. Under the assumption that most pixel observations approximately follow the Lambertian reflectance model, they obtain the solution by finding a sparse vector such that the rank of the Lambertian component matrix is minimized. The formulation is known as robust principal component analysis (R-PCA) in the field of sparse recovery. Specifically, they adopted a fast and scalable algorithm suitable for handling a large amount of data points, i.e. the augmented Lagrange multiplier method [47]. However, this method requires a shadow mask to be specified explicitly. Later, Ikehata et al. [2] reconsider PST as a pure sparse regression problem and aim to minimize the number of entries (i.e. the ℓ_0 pseudo norm) in the error matrix. They also add an ℓ_2 relaxation term to account for cases when the sparse assumption is violated. In order to avoid the difficult combinatorial problem involved in the minimization of ℓ_0 norm, they introduced two possible algorithms. One is to relax the ℓ_0 pseudo norm into ℓ_1 norm, as justified in [13, 48], and the solution is obtained via iteratively reweighted L1 minimization (IRL1) [49]. The other method is a hierarchical Bayesian approach called SBL [50]. It has been shown that SBL has an improved accuracy over IRL1 at the expense of lower efficiency, and both IRL1 and SBL perform better than R-PCA [2]. Independently of [1, 2], a similar, but quite complex, schema called alternating direction method of multipliers (ADMM) is developed in a recent paper by Adler et al. [51]. That paper in part applies sparse coding to PST and thus provides a motivating source for the present paper; however, our paper makes use of a radically different formulation as well as ties the PST problem into well-studied sparse solver algorithms. Moreover, that paper is aimed at producing a matte image with specularity and shadows attenuated, whereas the present work is aimed at accurate normal-vector recovery. Whereas our use of OMP uses a greedy algorithm where each component is picked one at a time, in contrast, the ADMM approach adjusts all the components in each iteration. The present study is the first to use greedy approaches to surface normal estimation within a PST framework.

Sparse methods have also found their use in uncalibrated PST, where the lighting directions are not known (but note that in this study we do assume known lighting directions so that these works are somewhat peripheral). Favaro et al. [18] incorporate the rank-minimization algorithm proposed in [1] into the uncalibrated PST problem as a pre-processing step to remove shadow and specularity effects. Argyriou et al. [19] recently also adopt a sparse representation framework to decide the weights for finding the best illuminants to use, again with the lighting directions unknown.

2.2 Sparse recovery methods

As was pointed out in Section 1.1, the canonical form of the sparse recovery problem (Eq. 2) is NP-hard [14] and cannot be solved efficiently as-is. In this section, we summarize alternative formulations to Eq. 2 and several types of solvers.

The first type of approach is convex ℓ_1 relaxation. It has been shown that for a dictionary matrix $\mathbf{A}$ that satisfies a certain restriction, Eq. 2 is likely to be equivalent to an ℓ_1 minimization problem [13, 48]:

$$\min_{\mathbf{x}} \|\mathbf{x}\|_1 \quad \text{s.t.} \quad \mathbf{y} = \mathbf{A}\mathbf{x}, \tag{3}$$

which can be solved via convex optimization techniques such as interior-point (IP) methods [52], gradient projection [53], IRL1 [49] and so forth.

Alternatively, sparse recovery can be achieved via greedy algorithms. The basic idea of such an algorithm is employing an iterative method to find the collection of non-zero entries, or *support*, of the signal $\mathbf{x}$, and then recover $\mathbf{x}$ via LS using only the observations in the support.

One of the most notable greedy algorithms is OMP [21–23], an improvement over the simple matching pursuit (MP) algorithm [54]. In OMP, a column $\mathbf{a}_j$ in $\mathbf{A}$ is iteratively chosen such that $\mathbf{a}_j$ is most greatly correlated with the current residual $\mathbf{r}$. Then, $\mathbf{r}$ is updated by taking into consideration the contribution of $\mathbf{a}_j$. The algorithm is terminated as a fixed number of non-zero entries are recovered or other stopping criteria are met. Then, a simple LS is performed only on a submatrix of $\mathbf{A}$ consisting of the columns chosen by OMP, and the regressed result will be assigned only to the signal entries corresponding to the selected columns. The columns that are not selected by OMP, on the other hand, will not be used in the final LS step, and their corresponding signal entries are simply set to zero.

In fact, OMP approximately solves the following k-sparse recovery problem:

$$\min_{\mathbf{x}} \|\mathbf{y} - \mathbf{A}\mathbf{x}\|_2 \quad \text{s.t.} \quad \|\mathbf{x}\|_0 \leq k. \tag{4}$$

Many state-of-the-art greedy algorithms nowadays are based on OMP. Examples include regularized OMP (ROMP) [55, 56], stagewise OMP (StOMP) [57], compressive sampling matching pursuit (CoSaMP) [58], probability OMP (PrOMP) [59], look ahead OMP [60], OMP with replacement (OMPR) [61], A* OMP [62] etc.

Another type of solvers employ a thresholding step to iteratively refine the recovered support, i.e. the selection/rejection of an entry at each step, is decided by whether the value of a certain function dependent on this entry falls below a given threshold. Algorithms in this category include iterative hard thresholding (IHT) [63], subspace pursuit (SP) [64], approximate message passing (AMP) [65], two-stage thresholding (TST) [66], algebraic pursuit (ALPS) [67] etc.

The fourth category is probability-based algorithms. These methods assume the signal to be recovered follows a specific probability distribution and solve the sparse recovery problem with statistical methods such as ML or MAP estimation. SBL [50] is one of the major algorithms in this category and has already been applied in the context of PST [2].

3 Sparse regression

3.1 Sparse formulation for photometric stereo

In this section, we explore the possibility of formulating and solving PST as a sparse regression problem. Since only the normal recovery is studied in this paper, we omit the albedo α from all equations in this and the following sections for simplicity and always use $\mathbf{n}$ to represent the unnormalised surface normal vector unless otherwise specified.

Here, we assume a Lambertian reflectance model with an additional term $e \in \mathbb{R}$ to account for the non-Lambertian error. Hence, the observed luminance y can be expressed as:

$$y = \boldsymbol{l} \cdot \mathbf{n} + e, \tag{5}$$

where $\boldsymbol{l} \in \mathbb{R}^3$ and $\mathbf{n} \in \mathbb{R}^3$ represent the lighting direction and surface normal, respectively. For each pixel, we have n observations $\mathbf{y} = (y_1, y_2, \ldots y_n)^T \in \mathbb{R}^n$. Now, let us write Eq. 5 in vector form

$$\mathbf{y} = \mathbf{L}\mathbf{n} + \mathbf{e}, \tag{6}$$

where $\mathbf{L} = (\boldsymbol{l}_1, \boldsymbol{l}_2, \ldots \boldsymbol{l}_n)^T \in \mathbb{R}^{n\times 3}$ and $\mathbf{e} = (e_1, e_2, \ldots e_n)^T \in \mathbb{R}^n$.

Equation 6, containing n linear equations but $n + 3$ unknowns (n components in $\mathbf{e}$ and three components in

n), is effectively an underdetermined problem and as such cannot be solved unambiguously. However, if the error **e** is a sparse matrix, i.e. most or at least a great percentage of its elements are zero, then it is still possible to recover **e** exactly or almost exactly by solving the following sparse regression problem:

$$\min_{\mathbf{n},\mathbf{e}} \|\mathbf{e}\|_0 \quad \text{s.t.} \quad \mathbf{y} = \mathbf{L}\mathbf{n} + \mathbf{e}. \tag{7}$$

In Eq. 7, $\|\cdot\|_0$ represents the ℓ_0 pseudo-norm or the number of non-zero elements in **e**. This formulation, however, has two major issues: (1) it is an NP-hard combinatorial problem and (2) real-world scenes may contain a large variety of materials that are only poorly approximated by the Lambertian reflectance model. For those materials, it is very likely that **e** is not strictly sparse. Thus, the equality constraint is very hard to be satisfied. Instead, it is more realistic to use an inequality constraint with a user-defined error tolerance ϵ

$$\min_{\mathbf{n},\mathbf{e}} \|\mathbf{e}\|_0 \quad \text{s.t.} \quad \|\mathbf{y} - \mathbf{L}\mathbf{n} - \mathbf{e}\|_2 \leq \epsilon. \tag{8}$$

Alternatively, if we care more about how much the reconstructed luminance approximates real observation rather than the sparsity of **e**, then it would be more natural to reformulate Eq. 8 as

$$\min_{\mathbf{n},\mathbf{e}} \|\mathbf{y} - \mathbf{L}\mathbf{n} - \mathbf{e}\|_2 \quad \text{s.t.} \quad \|\mathbf{e}\|_0 \leq s, \tag{9}$$

where the scalar s is the sparsity of vector **e**. To further simplify Eq. 9, we propose merging **n** and **e** into one large vector and treating them as one entity, i.e.

$$\begin{aligned} \mathbf{y} &= \mathbf{L}\mathbf{n} + \mathbf{e} \\ &= \mathbf{L}\mathbf{n} + \mathbf{I}\mathbf{e} \\ &= (\mathbf{L}, \mathbf{I}) \begin{pmatrix} \mathbf{n} \\ \mathbf{e} \end{pmatrix} \\ &= \mathbf{A}\mathbf{x}, \end{aligned} \tag{10}$$

where $\mathbf{I} \in \mathbb{R}^{n \times n}$ is an $n \times n$ identity matrix, $\mathbf{A} = (\mathbf{L}, \mathbf{I}) \in \mathbb{R}^{n \times (n+3)}$ is a new merged dictionary matrix and $\mathbf{x} = (\mathbf{n}^{\mathrm{T}}, \mathbf{e}^{\mathrm{T}})^{\mathrm{T}} \in \mathbb{R}^{(n+3) \times 1}$ is the combined vector of all the unknown variables. Hence, Eq. 9 can be rewritten as

$$\min_{\mathbf{n},\mathbf{e}} \|\mathbf{y} - \mathbf{A}\mathbf{x}\|_2 \quad \text{s.t.} \quad \|\mathbf{x}\|_0 \leq s. \tag{11}$$

The stacked formulation was inspired by the work of Wright et al. [20, Eq. 20]. However, in [20], both the signal and the noise are assumed sparse, whereas in our case, the signal (normal vector) has only three components and is not at all sparse.

By formulating our problem in the form of Eq. 11, we can now take advantage of existing algorithms to efficiently achieve an accurate solution. One such solver is a greedy algorithm known as OMP [21–23], which is known for its high accuracy and low time-complexity. We will describe this algorithm in Section 3.2 in detail.

Previously, Ikehata et al. [2] proposed a different formulation to Eq. 11. They expressed the PST problem in a so-called Lagrangian form, i.e.

$$\min_{\mathbf{n},\mathbf{e}} \|\mathbf{y} - \mathbf{L}\mathbf{n} - \mathbf{e}\|_2^2 + \lambda \|\mathbf{e}\|_1 \tag{12}$$

and applied two solving algorithms: IRL1 minimization and SBL. They showed that SBL provides a more accurate estimation but is more computationally expensive. Later, in Section 4, we will show that our OMP solver produces a more accurate result than SBL with comparable efficiency to IRL1.

3.2 Orthogonal matching pursuit

Sparse recovery problems like Eq. 11 can be solved via many different methods (see Section 2.2 for a brief overview). Here, we choose to apply the classical greedy OMP to our surface normal recovery problem. Given the

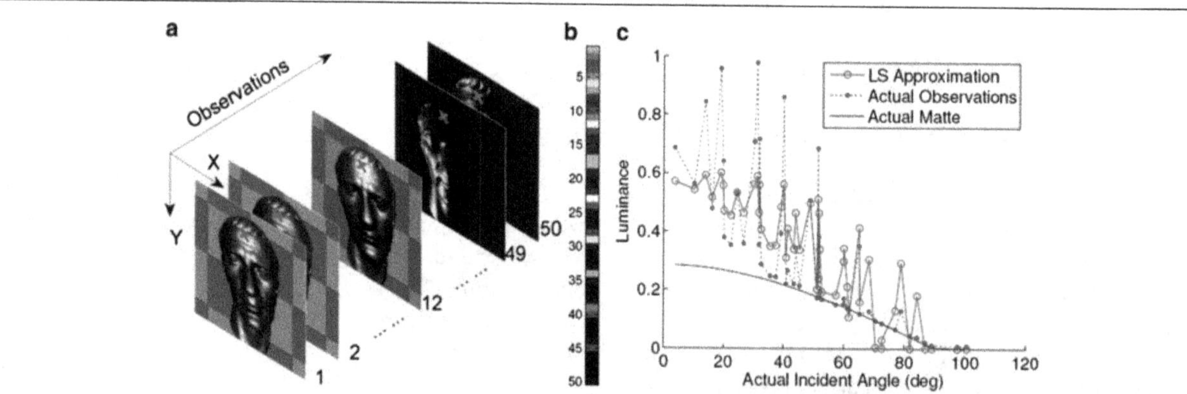

Fig. 2 Visualization of all observations of one pixel from dataset *Caesar*. **a** The pixel studied is marked with *blue crosses* at the same location ($X = 90, Y = 39$) on all images numbered from 1 to 50. **b** Luminance observations arranged by the image index 1 to 50. **c** Luminance observations sorted by incident angle. *Blue dotted line* shows the actual 50 observations; *red circled line* shows the approximated luminance using least squares; *black solid line* represents the ground truth matte (Lambertian) luminance

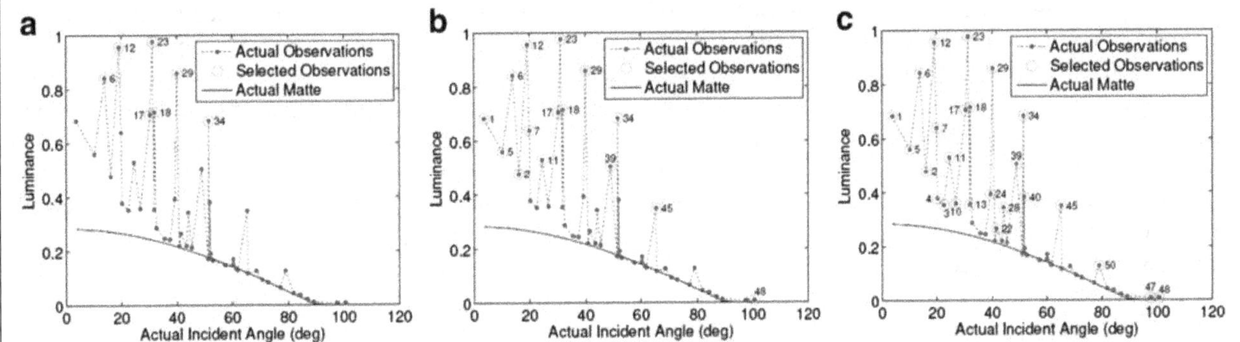

Fig. 3 Outliers identified by orthogonal matching pursuit. **a** Outliers with great non-Lambertian error (*red circles*) detected in iterations 4–10. **b** Outliers with medium to great error (*red circles*) detected in iterations 4–18. **c** All outliers (*red circles*) detected as of iteration 28. *Blue dotted lines* show actual luminance observations in all three plots

linear model in Eq. 10, the basic idea of OMP is to iteratively select columns of the dictionary matrix **A** that are most closely correlated with the current residuals, then project the observation **y** to the linear subspace spanned by the columns selected until the current iteration. We denote each column of **A** as $\mathbf{A}_j$.

Let i be the current number of iterations and $\mathbf{r}_i$ and c_i the residuals and the subset of selected columns in **A** at the ith iteration, respectively. Let $\mathbf{A}(c_i)$ and $\mathbf{x}(c_i)$ represent the columns indexed by c_i in **A** and the entries indexed by c_i in the signal **x** to be recovered, respectively. The OMP algorithm [23], as we here apply to our PST problem, can be summarized as follows:

Algorithm Orthogonal matching pursuit

1. Normalise each column of the dictionary matrix **A** and denote the resulting matrix as $\mathbf{A}'$, i.e. $\|\mathbf{A}'_j\|_2 = 1$ for $j = 1, 2, \ldots, p$. Initialize the iteration counter $i = 1$, residual $\mathbf{r}_0 = \mathbf{y}$ and $c_0 = \varnothing$.
2. Find a column $\mathbf{A}'_t$ ($t \in \{1, 2, \ldots, p\} - c_{i-1}$) that is most closely correlated with the current residual. Equivalently, solve the following maximization problem:

$$t = \underset{j}{\operatorname{argmax}} \|\mathbf{A}_j'^T \mathbf{r}_{i-1}\| \tag{13}$$

3. Add t to the selected set of columns, i.e. update $c_i = c_{i-1} \cup t$, and use $\mathbf{A}'(c_i)$ as the current selected subset of $\mathbf{A}'$.
4. Project the observation **y** onto the linear space spanned by $\mathbf{A}'(c_i)$. The projection matrix is calculated as follows:

$$\mathbf{P} = \mathbf{A}'(c_i)(\mathbf{A}'(c_i)^T\mathbf{A}'(c_i))^{(-1)}\mathbf{A}'(c_i)^T. \tag{14}$$

5. Update the residuals with respect to the new projected observation

$$\mathbf{r}_i = \mathbf{y} - \mathbf{P}\mathbf{y}. \tag{15}$$

6. Increment i by 1. If $i > n/2 + 3$ (= 28 for our typical datasets of 50 images), then proceed to step 7, otherwise go back to step 2.
7. Solve only for the entries indexed by c_i in signal **x** using the original, **unnormalised** design matrix **A**, and simply set the rest of the entries to 0, i.e.

$$\mathbf{x}(c_i) = \mathbf{A}(c_i)^{\dagger}\mathbf{y} \tag{16}$$

and

$$x_j = 0 \qquad \text{for each} \quad j \notin c_i. \tag{17}$$

8. Take the first three entries in **x** as the solutions for the x, y and z component of the normal vector, respectively,

$$\mathbf{n} = (x_1, x_2, x_3). \tag{18}$$

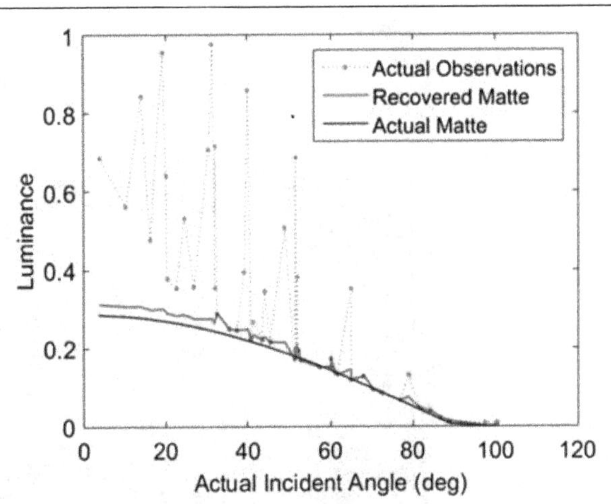

Fig. 4 Matte component recovered by orthogonal matching pursuit. *Blue dotted line* represents actual luminance observations; *black solid line* shows the actual matte model; *red solid line* shows the matte recovered by OMP

In our formulation Eq. 10, we merge the normal and the errors into a large vector, so the components of the two vectors are treated equally by OMP. In each iteration, which column in the dictionary matrix is to be chosen purely depends on its correlation with the current residuals. Thus, there is no strict mathematical guarantee that the normal vector components will be selected in the first s iterations. Indeed, this failure could happen if the non-Lambertian error vector accounts for most of the observations. However, since the observed luminance is usually a function of the surface normal, it is expected that the normal vector components are more closely correlated to the observations than the sporadic non-Lambertian errors. In our experiments, all normal vector components are usually selected within the first few iterations (<10). On the other hand, if one or two components of the surface normal are rather small, then they might not be selected by our algorithm. However, since they are very close to zero anyway, simply treating them as zero would not negatively impact the accuracy of our estimation.

One of the biggest advantages of OMP is its low computational cost and straightforward implementation. We have found that it is significantly faster than LMS as well as other state-of-the-art robust regression methods that

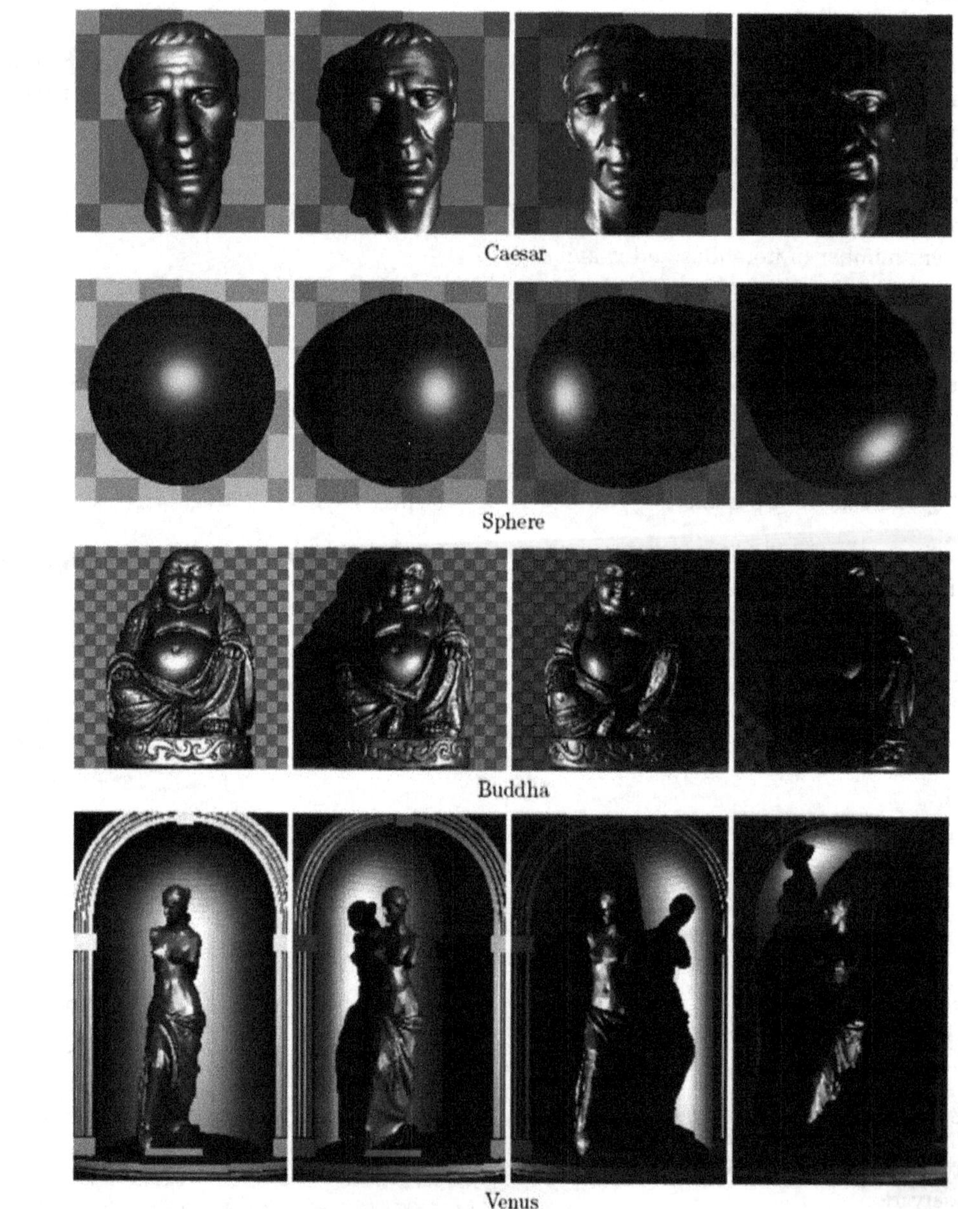

Fig. 5 Sample images from four synthesized datasets rendered with POV-Ray. From top row to bottom row: *Caesar*, *Sphere*, *Buddha* and *Venus*

have been applied in the context of PST (see Section 4.3 and Fig. 21 for more details). Note that for our particular choice of the design matrix **A**, the correlation between any column in the identity matrix and the residual **r** can be simply represented by one element in **r**. Therefore, the inner product in Eq. 13 may be reduced to finding the maximum entry in **r**. This observation allows for an even more efficient implementation. In this work, however, we still implement OMP according to Eq. 13 for generality.

3.2.1 Normalization and orthogonality

As a requirement of the standard OMP algorithm, we used the column normalised version of the design matrix $\mathbf{A}'$ in the column selecting process. After normalization, the first three columns in $\mathbf{A}'$ no longer hold the correct value of lighting vector components. In other words, it appears that the lighting directions are modified by normalization. However, this observation does not negatively affect our results. In step 2 of OMP, the column most correlated with the current residual vector is selected. Normalization only makes sure one column does not have a numerical advantage over another simply because it has a greater L2 norm. Therefore, normalization does not interferes with the selection of the outliers. On the contrary, it enforces the correctness of selection. It is also important to note that after the outliers are selected, we use the original unnormalised dictionary matrix **A**, instead of $\mathbf{A}'$, to make sure that the normal vector are recovered on the actual lighting directions.

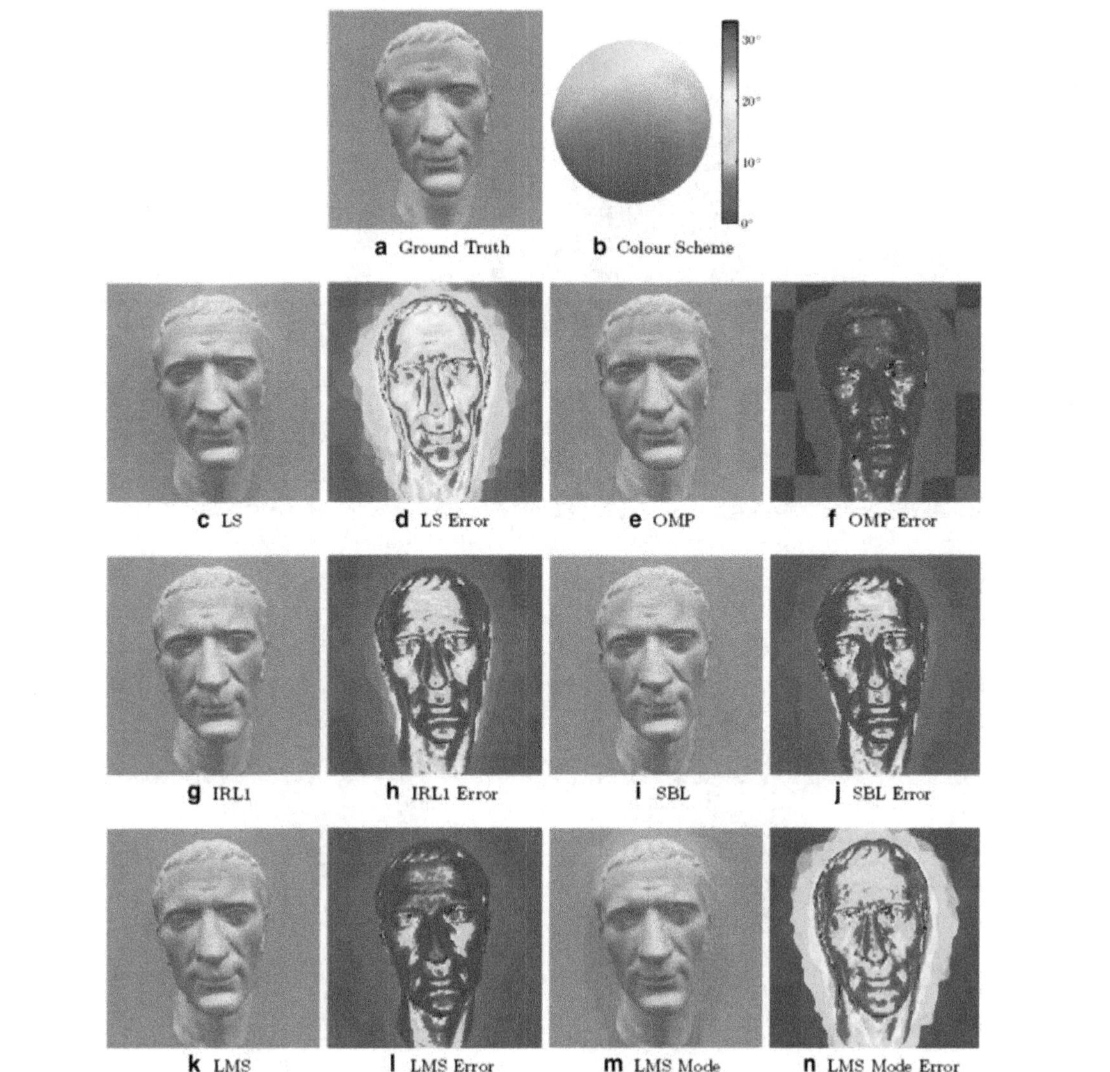

Fig. 6 Normal maps of a head statue of Caesar recovered using various methods. **a** Ground truth normal map; **b** colour wheel and colour bar used for normal and angular error visualization, respectively. Angular error is measured in degrees. **c**, **e**, **g**, **i**, **k**, **m** Normal map recovered using LS, OMP, IRL1, SBL, LMS and LMS mode, respectively. **d**, **f**, **h**, **j**, **l**, **n** Angular error of normal maps recovered using the aforementioned methods

Another issue worth noting is the orthogonality of the dictionary matrix **A**. It has been shown that if **A** satisfies a restricted isometry property (RIP), then the exact recovery of signal **x** may be possible [68, 69]. Essentially, RIP specifies a near-orthonormal condition for **A**. Although our dictionary matrix **A** unfortunately does not obey the RIP property in its general form, we still argue that this matrix is near-orthogonal: with our uniform light distribution, the first three columns are indeed near-orthogonal (the dot products between column 1–2, 1–3 and 2–3 are 8.85×10^{-8}, -1.04×0^{-6} and -5.17×10^{-7}, respectively). The rest of **A** is a large identity matrix **I**, which itself is orthonormal. Also, due to the large number of zeros in **I**, the dot product of any of the first three columns and any column in **I** is a rather small number ($10^{-3} - 10^{-2}$ scale on average). Thus, although it is yet to be strictly proven, we speculate that the dictionary matrix **A** in near-orthogonal enough for our purpose of recovering the three components of normal out of the 53-element signal. As our results show, OMP indeed achieves highly precise recovery of surface normal for most of the pixel locations (see Section 4). We also show that even with a very biased lighting distribution (such that the orthogonality of the first three column are greatly reduced), OMP still provides an accepted recovery with higher accuracy than other sparse methods (see Section 4.2.4).

3.2.2 Stopping criterion

For simplicity, here we set the stopping criterion as a fixed number (s) of iterations. We make a conservative

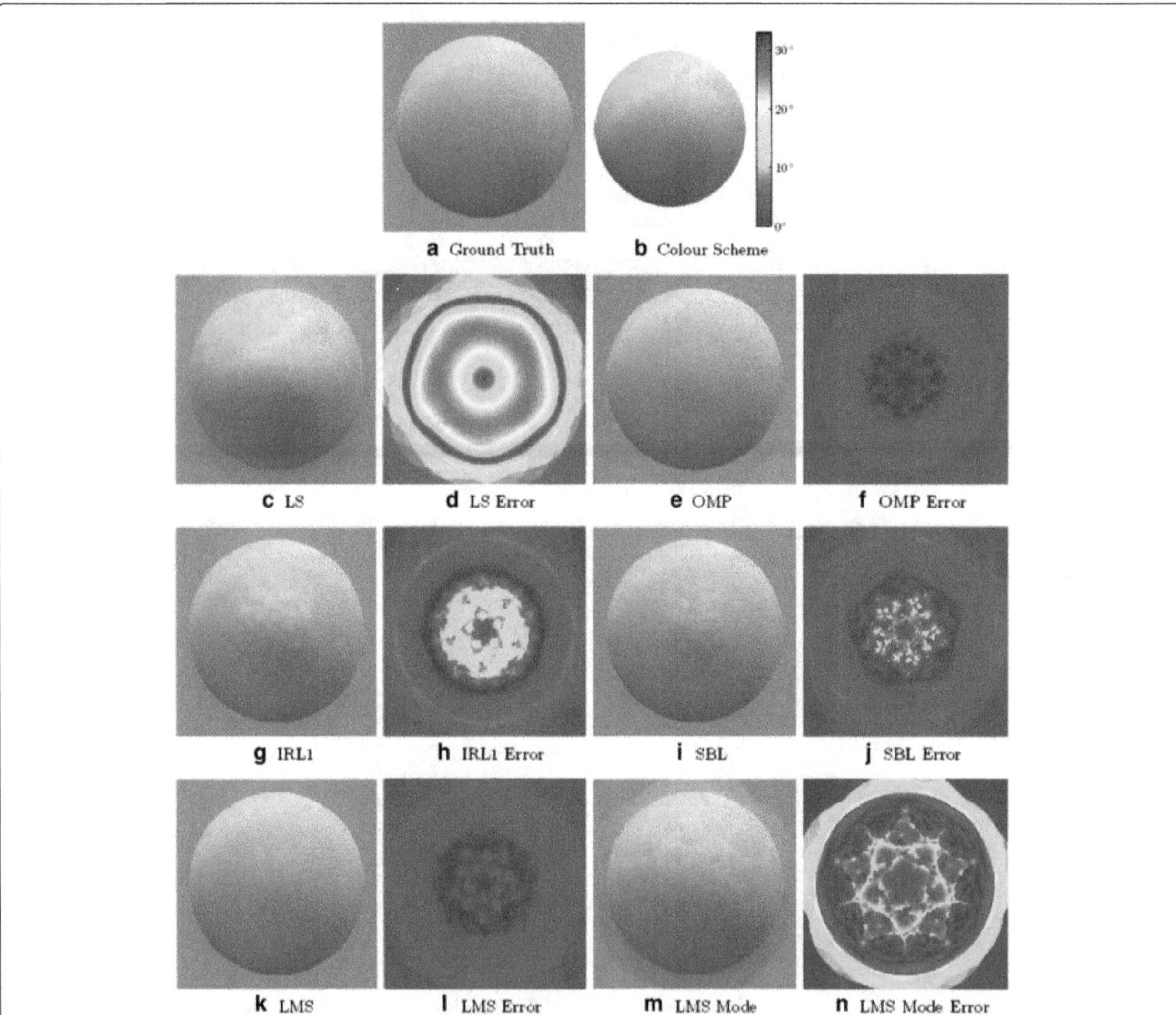

Fig. 7 Normal map of an ideal sphere recovered using various methods. **a** Ground truth normal map; **b** colour wheel and colour bar used for normal and angular error visualization, respectively. Angular error is measured in degrees. **c**,**e**, **g**, **i**, **k**, **m** Normal map recovered using LS, OMP, IRL1, SBL, LMS and LMS mode, respectively. **d**, **f**, **h**, **j**, **l**, **n** Angular error of normal maps recovered using the aforementioned methods

assumption that 50 % of the observed pixels are polluted by non-Lambertian noise. Thus, for our typical datasets of $n = 50$ images and normal vectors with three components, the stopping criterion is $i > s = n/2 + 3 = 28$. This criterion ensures that there is always a moderate number of observations (25) available for regression.

An alternative choice of stopping criterion is based on the residual, i.e. $|\mathbf{r}| <$ threshold. It has been proved that in the matching pursuit and its orthogonal version, OMP, are guaranteed to converge [21, 54]. Thus, such a stopping criterion is theoretically viable. Indeed, we have validated that in our OMP-based method, the residual converges on all pixels used in our datasets. However, we have also noticed in our tests that setting a hard threshold for all pixels results in a slightly decreased accuracy than using the sparsity-based criterion (results not shown). Therefore, in the current study, we will continue using the sparsity-based criterion, i.e. $i >= 28$.

3.3 Visual demonstration

In this section, we demonstrate how OMP enforces robustness onto the normal recovery process by using a simple example. Particularly, we use a synthesized dataset *Caesar* (see Section 4.1.1 for more information) and study all the 50 observations of one pixel (marked by blue crosses in Fig. 2a) at location $(X = 90, Y = 39)$ where the ground truth normal vector is $\mathbf{n}_{gt} = (-0.0780, 0.1828, 0.9801)$. The luminance profile of these observations, sorted by incident light angle, are shown in Fig. 2c (blue dotted line), along with the actual matte

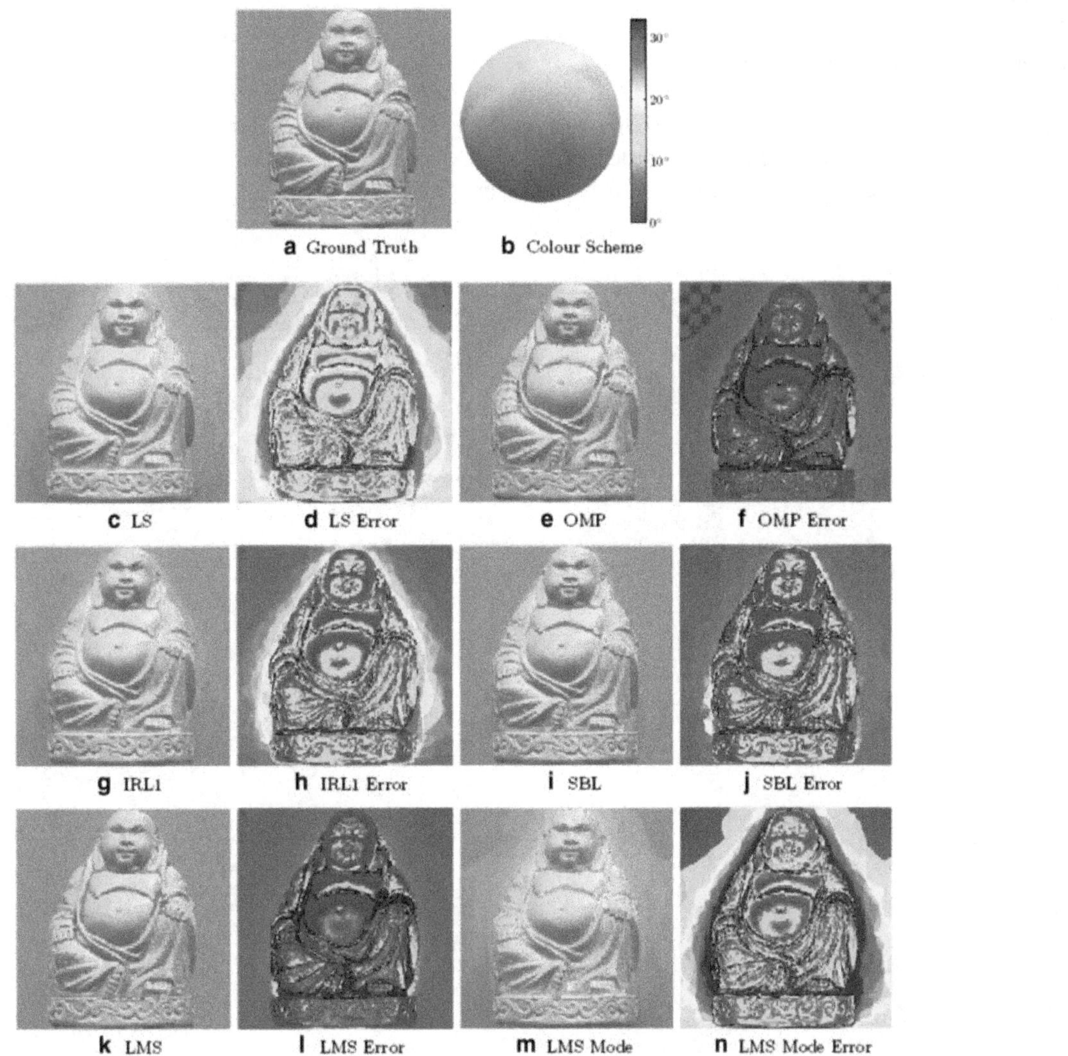

Fig. 8 Normal maps of a head statue of Buddha recovered using various methods. **a** Ground truth normal map; **b** colour wheel and colour bar used for normal and angular error visualization, respectively. Angular error is measured in degrees. **c, e, g, i, k, m** Normal map recovered using LS, OMP, IRL1, SBL, LMS and LMS mode, respectively. **d, f, h, j, l, n** Angular error of normal maps recovered using the aforementioned methods

model curve (black solid line), i.e. the theoretical values of luminance, if the surface is purely Lambertian. It is obvious from Fig. 2c that due to the existence of specular reflection, a good percentage of observations (especially when the incident angle is small) deviate from the values predicted by a matte model.

The naive LS regression, when applied to this pixel, attempts to approximate the values of all observations without taking the actual matte model into consideration (Fig. 2c, red line marked with circles). Naturally, the LS result $\mathbf{n}_{\text{LS}} = (-0.1578, 0.4852, 0.8600)$ deviates greatly from the ground truth $(-0.0780, 0.1828, 0.9801)$.

On the other hand, OMP first attempts to identify s entries, one in each iteration, from the stacked signal $\mathbf{x} = (x_1, x_2, \ldots x_{n+3})^T \in \mathbb{R}^{n+3}$ (see Eq. 10). Usually, these s entries include three components for normal vectors (x_1, x_2, x_3) and $(s-3)$ components from the remaining n entries $(x_4, x_5, \ldots, x_{n+3})$ that correspond to error values. When the OMP algorithm as described in Section 3.2 is applied to this pixel, it behaves as follows:

Iterations 1–3: The entries that correspond to normal vectors, x_3, x_2 and x_1, are selected in the order listed. This is not a coincidence since the first three columns of the dictionary matrix $\mathbf{A}$ are overall more strongly correlated to the observations than any of the rest of the columns that correspond to noise values. Also, we noticed that these entries are in fact selected in order of the absolute value of their corresponding normal components. For instance, the third component of the ground truth normal $(-0.0780, 0.1828, 0.9801)$ is greater than the other two components. Therefore, x_3 gets selected in the first iteration.

Iterations 4–10: Entries x_{26}, x_{15}, x_{32}, x_9, x_{37}, x_{21} and x_{20} are selected sequentially. These entries correspond to non-Lambertian errors at observation #23, #12, ...#17, respectively (marked with red circles in Fig. 3a). Note that the indices of observations mentioned here (23, 12, ...) are equal to the entry indices found (26, 15, ...) minus 3, since the first three elements in $\mathbf{x}$ do not represent errors. We notice that the corresponding observations of these selected entries all have very high error values. As in iterations 1–3, these error entries are also selected in order of their absolute value. For instance, observation #23 (incident angle $\approx 32°$) has the greatest non-Lambertian error; therefore, its corresponding error entry x_{26} is selected in iteration 4, before other entries.

Table 1 Statistics for the angular error between the normal maps recovered for various methods and the ground truth, for three synthesized datasets. All numbers are shown in degrees

	25 % Quantile	Median	Mean	75 % Quantile	STD
Dataset Caesar					
LS	4.669	8.586	9.672	14.32	5.875
LMS mode	1.313	6.769	8.148	13.46	7.620
LMS	1.335	2.692	3.700	4.305	4.242
IRL1	1.854	3.339	5.276	6.440	5.080
SBL	1.576	2.441	4.487	4.488	5.158
OMP	0.7571	2.624	2.424	2.690	3.332
Dataset Sphere					
LS	7.255	12.23	13.55	20.79	7.566
LMS mode	1.658	4.939	6.211	9.907	4.901
LMS	2.588	3.601	3.261	4.030	1.103
IRL1	3.045	4.253	4.960	5.662	3.119
SBL	2.762	3.620	3.671	4.239	1.822
OMP	2.617	3.051	3.179	3.991	0.9324
Dataset Buddha					
LS	9.039	14.03	15.29	20.19	8.25
LMS mode	7.304	14.70	17.42	25.03	13.79
LMS	1.985	4.736	8.454	6.391	15.35
IRL1	4.066	7.354	10.08	13.28	9.178
SBL	2.412	5.295	8.034	9.333	9.479
OMP	1.289	3.049	7.065	5.495	14.17

Iterations 11–18: Another eight entries x_4, x_{10}, x_{42}, x_8, x_{14}, x_{51}, x_{48} and x_5 are selected sequentially. Their corresponding observations have medium error values (Fig. 3b).

Iterations 19–28: Select the rest of the error entries x_{43}, x_{27}, x_{50}, x_{31}, x_{16}, x_7, x_{13}, x_6, x_{53} and x_{25}. The corresponding observations have small error values (Fig. 3c).

Through the 28 iterations above, we have obtained 28 indices; 3 of them correspond to the normal-vector components and the remaining 25 represent the observations that have significant non-Lambertian effect, i.e. non-zero values in signal $\mathbf{x}$ in the sparse regression problem $\mathbf{y} = \mathbf{Ax}$ (Eq. 10). Suppose the indices of 25 selected non-Lambertian outliers are collectively represented as $c_{\text{out}} \subset \{1, 2, \ldots, 50\}$, we can obtain the normal vector $\mathbf{n}$ and an error vector $\mathbf{e}$ by solving the following equation (which is essentially the same as 16):

$$\mathbf{y} = (\mathbf{L}, \mathbf{I}(c_{\text{out}})) \begin{pmatrix} \mathbf{n} \\ \mathbf{e}(c_{\text{out}}) \end{pmatrix}. \tag{19}$$

For our sample pixel, the above equation gives $\mathbf{n}_{\text{OMP}} = (-0.0877, 0.2282, 0.9697)$, which well approximates the ground truth $\mathbf{n}_{\text{gt}} = (-0.078, 0.1828, 0.9801)$ compared to the naive LS result $\mathbf{n}_{\text{LS}} = (-0.1578, 0.4852, 0.8600)$.

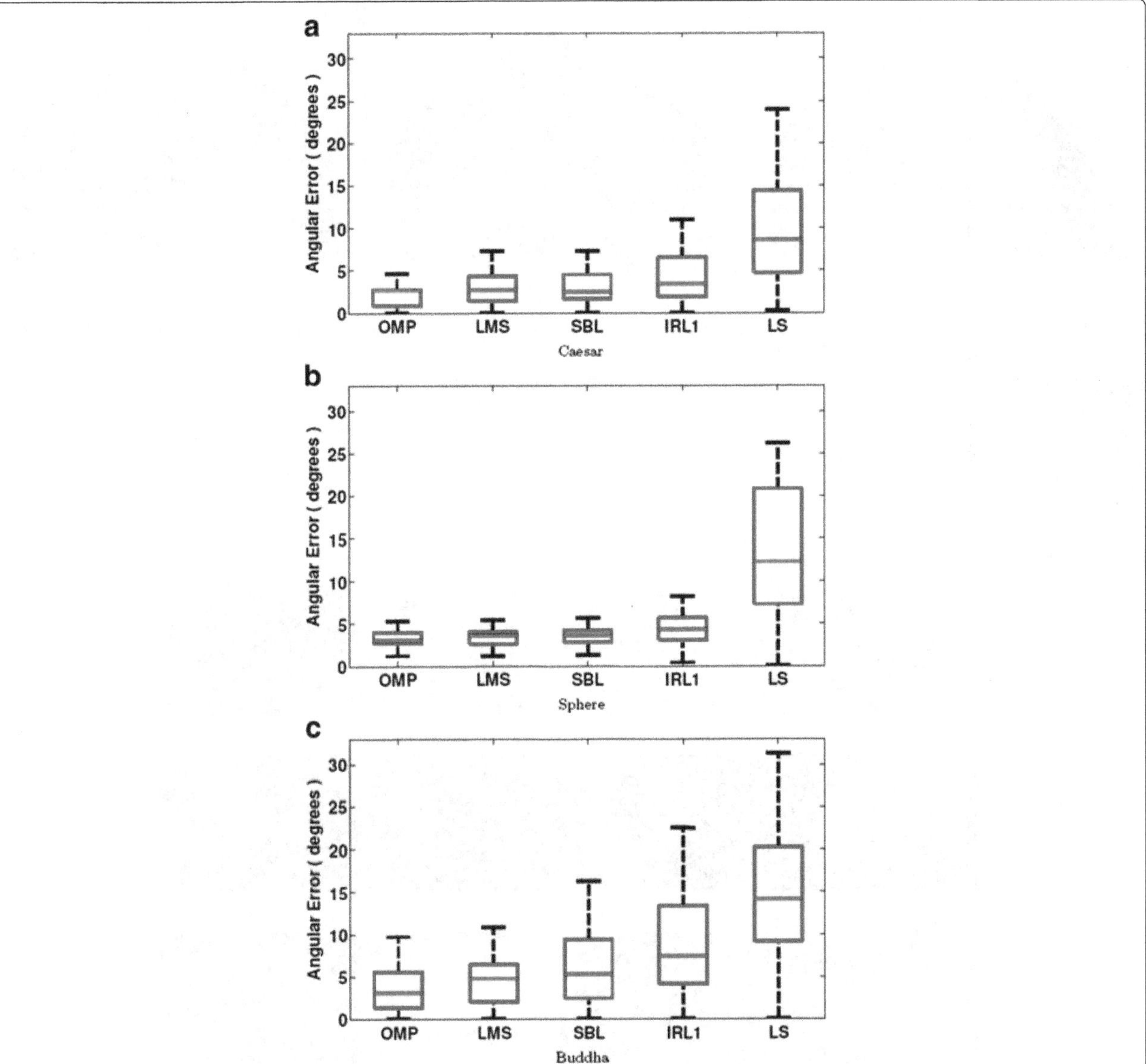

Fig. 9 Angular error of normal maps recovered using various methods (**a–c**). *Red horizontal lines* indicate medians. Upper and lower border of the *blue boxes* represent third (Q3) and first quartile (Q1), respectively. Upper and lower whiskers show 1.0 interquartile range (the difference between the upper and lower quartiles, the IQR) extended from Q3 and Q1, respectively. The whiskers show the quantile Q1 and Q3 extended by 1.0 IQR (as opposed to the typically shown 1.5 IQR)

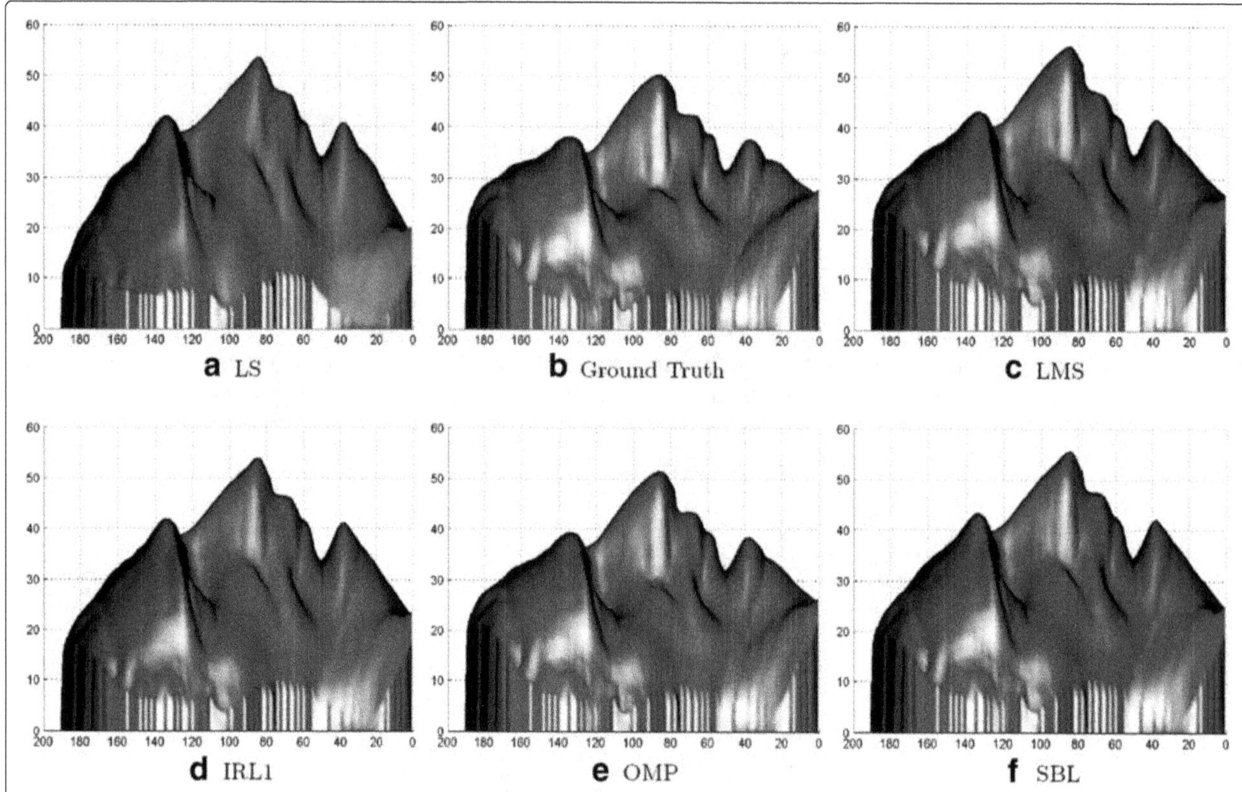

Fig. 10 Three-dimensional surfaces reconstructed from normal maps. **b** Depth map recovered with the ground truth normal map. **a**, **d**, **e**, **c**, **f** Depth maps recovered using LS, IRL1, OMP, LMS and SBL, respectively

Gold

Elba

Frag

Fig. 11 Sample images from three datasets of real-world scenes. From top row to bottom row: *Gold*, *Elba* and *Frag*

Additionally, we can directly recover the matte components by subtracting the error vector **e** from the actual luminance observation **y**. Note that the matte components obtained this way (Fig. 4, red solid line) almost coincide with the ground truth matte model, exhibiting a high degree of robustness.

4 Results and discussion

In this chapter, we present our experimental results and observations on synthesized and real datasets. All experiments were carried out on a Dell Optiplex 755 computer equipped with an Intel Core Duo E6550 CPU and 4 GB RAM, running Windows 7 Enterprise 64 bit. All algorithms were implemented in MATLAB R2014a 64 bit.

4.1 Normal map recovery

We first examine the angular error of normal maps recovered by different methods on both synthesized and real datasets. For synthesized datasets, we quantitatively inspect the difference between the ground truth normal map and the recovered normal maps. For real datasets without a ground truth map, on the other hand,

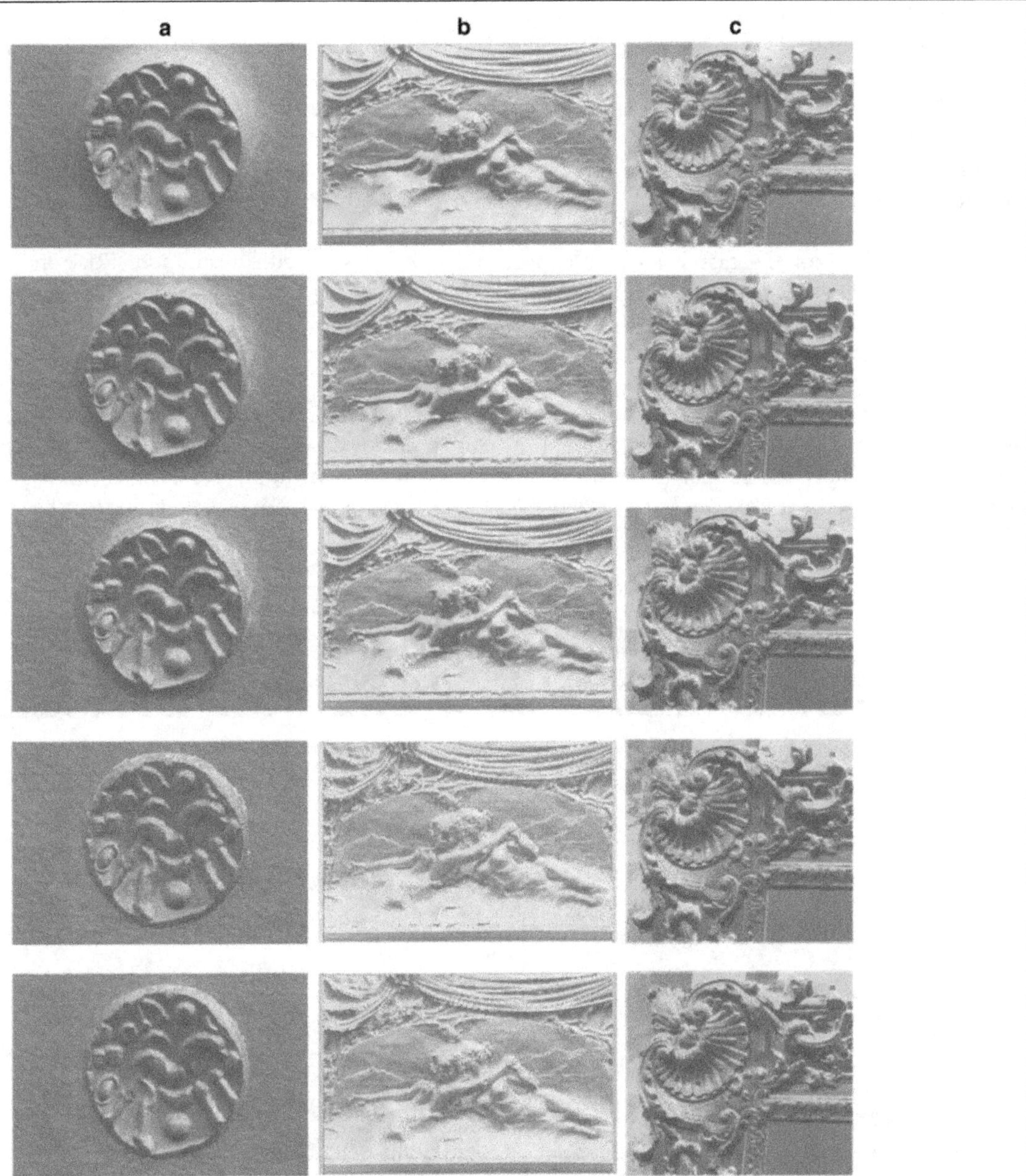

Fig. 12 Normal maps for three real-world dataset recovered using various methods. Columns **a**–**c** represent dataset *Gold*, *Elba* and *Frag*, respectively. Rows *1*–*5* show results using LS, IRL1, SBL, OMP, and LMS, respectively

the recovered normal maps are examined visually and qualitatively.

4.1.1 Synthesized datasets

Four 3D objects are used for our synthesized datasets in this study: *Sphere, Caesar, Buddha* and *Venus*. All 3D models are either created programmatically as geometrical primitives (*Sphere*) or downloaded from the AIM@SHAPE Shape Repository (*Caesar, Buddha*) [70] and the INRIA Gamma research database (*Venus*) [71]. For each object, 50 images are rendered under various lighting directions using raytracing software (POV-Ray 3.6) at a resolution of 200 × 200 (except for *Venus*, whose resolution is 150 × 250). Global illumination is enabled to ensure a highly photorealistic appearance. All scenes feature significant specularity and large areas of cast shadow. *Caesar, Buddha* and *Venus* are rendered with the specular highlight shading model provided by POV-Ray (a modified version of the Phong model) [72], and *Sphere* is rendered with a pure Phong model. A checkered plane is intentionally included in the rendered scene as background to (1) allow for the cast shadow to appear and (2) add further challenges to the algorithms since it introduces local fluctuation in luminance while the surface normals remain constant. Sample images for these datasets are shown in Fig. 5.

For each image set, the normal map is estimated using the OMP method [21–23] as proposed in this study. For comparison, we show the results for two other state-of-the-art sparse recovery methods—IRL1 and SBL [2]. Another two of our previously proposed outlier detection-based methods, LMS [10] and LMS mode finder [11], are also applied and compared. Then, the angular error between the normal map recovered using each method and the ground truth is quantitatively measured. Note that only results for *Caesar, Sphere* and *Buddha* are shown in this section. The fourth dataset *Venus* is reserved for later in Section 4.2.2 as a failure case.

We found these methods exhibit similar relative performance to each other on all the three datasets tested in this section—*Sphere, Caesar* and *Buddha*. In *Caesar*, the normal maps recovered using OMP (Fig. 6e) have a higher quality than those by IRL1 (Fig. 6g) and SBL (Fig. 6i) both qualitatively and quantitatively.

We observe that IRL1 and SBL, although much more robust than LS, still produce a considerable error at highly specular regions, most notably the cheek and the forehead. As a result, the faces on IRL1 and SBL normal maps appear to be more protruding than the ground truth. Also, some fine details on these two normal maps, such as the wrinkles on the forehead, are not well preserved. In addition, IRL1 and SBL fail to handle the regions right beside the neck which are heavily shadowed.

On the other hand, OMP shows a higher degree of robustness than previous sparse methods at specularity-affected regions (cheek, forehead and nose) as well as shadowed regions (areas around the neck on the checkered background), resulting in a normal map closer to the original. For example, the forehead appears flat on

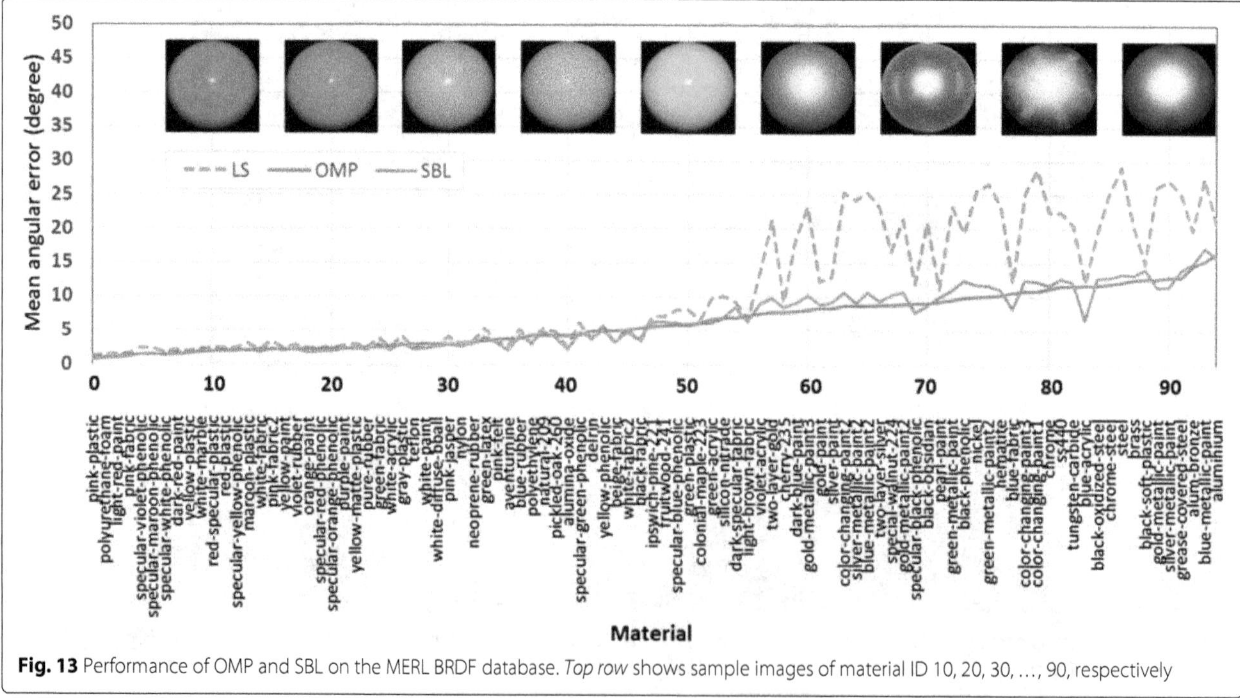

Fig. 13 Performance of OMP and SBL on the MERL BRDF database. *Top row* shows sample images of material ID 10, 20, 30, ..., 90, respectively

OMP normal maps, closely resembling the ground truth. The wrinkles are almost perfectly recovered. However, OMP appears to be confused by the checkered pattern of background, producing a small angular error in these flat regions.

The LMS result (Fig. 6k) is better than IRL1 and SBL but worse than OMP.

The 1D version of LMS—finder—produces a poorer visual result (Fig. 6m) compared to the other robust methods, although it does give a statistically more reliable result than LS. We will exclude this method from future discussion but still show its result for reference.

The effect of specularity on normal map recovery can be further seen from the results for the *Sphere* dataset in Fig. 7. Again, IRL1 and SBL results are noisy in the specularity-affected areas, whereas OMP gives much cleaner results. LMS performs similarly to OMP. Interestingly, a pentagon-shaped pattern is visible on each error map because there are exactly five lights at each elevation angle.

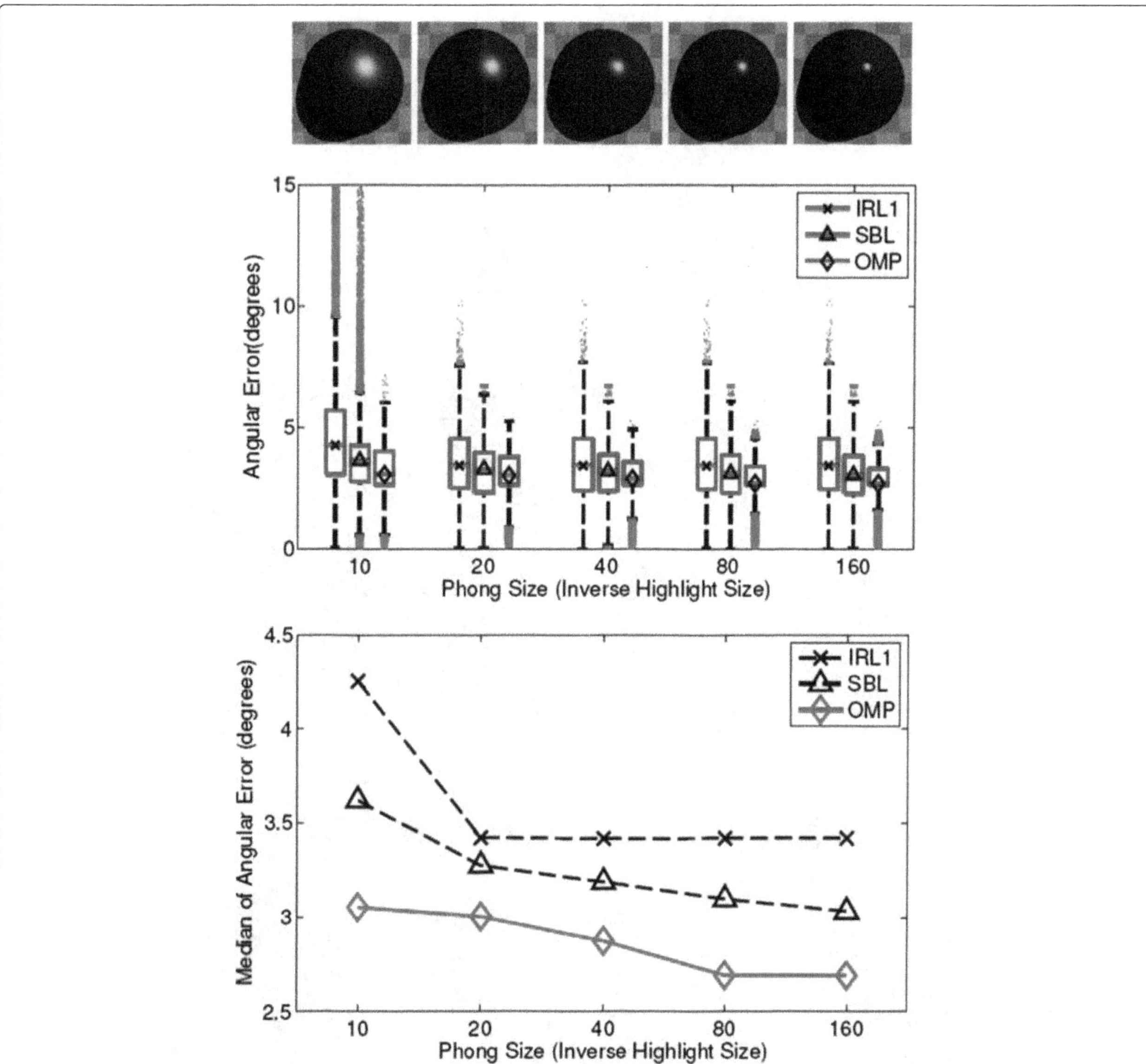

Fig. 14 Statistics for the angular error of normal maps recovered using sparse methods. *Upper row*: sample rendered images with *phong_size* 10–160, respectively, from left to right. *Middle figure*: boxplot of the angular error of normal maps under various degrees of specularity. *Upper and lower border of blue boxes* represent third (Q3) and first quartile (Q1), respectively. *Upper and lower whiskers* show 1.0 interquartile range (IQR) extended from Q3 to Q1, respectively. *Red bars in blue boxes* represent medians. At each specularity level, four sparse recovery methods are compared, symbolized by different markers on the median bar: IRL1 (*cross*), SBL (*triangle*) and OMP (*diamond*). Red dots are outliers distributed out of the error bar range. *Lower figure*: the medians of the angular error

For *Buddha* (Fig. 8), OMP again produces a better overall result than IRL1 and SBL. However, the relatively poorer performance of the greedy methods in shadowed concave regions now becomes a more significant problem due to the prevalence of concave regions such as creases on the clothes. The angular error distribution of the LMS result is similar to that of OMP, though with slightly greater overall error.

From the normal map recovery results obtained on the three datasets, we can see that OMP generally performs better than IRL1 and SBL on convex objects and are more resistant to specularities and cast shadows. The statistical result of the angular error of the normal maps recovered with different methods are listed in Table 1. OMP result has the lowest mean, median, 25 % and 75 % quantiles, as well as standard deviation for all three datasets mentioned above. The LMS result is better than IRL1 and SBL,

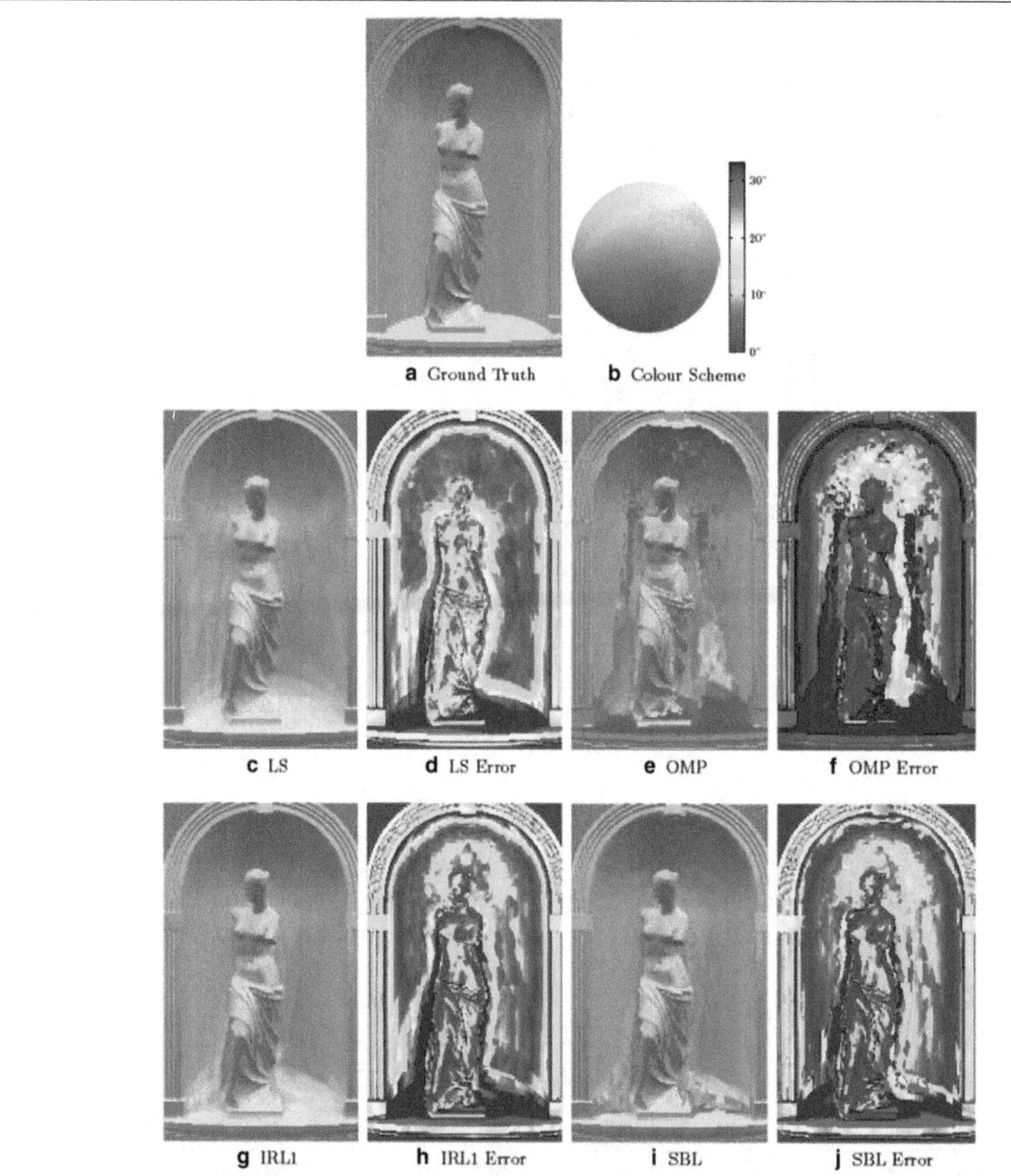

Fig. 15 Normal maps of the *Venus* dataset recovered using sparse methods. **a** Ground truth normal map. **b** Colour wheel and colour bar used for normal and angular error visualization, respectively. Angular error is measured in degrees. **c**, **e**, **g**, **i** Normal maps recovered using LS, OMP, IRL1 and SBL, respectively. **d**, **f**, **h**, **j** Angular error of normal maps recovered using the aforementioned methods

but worse than OMP. These results are also depicted in Fig. 9. Curiously, we notice that the estimation accuracy is generally lower on *Sphere* than *Caesar*, despite the simple geometry of the former. This may be jointly caused by the unique lighting model, surface colour and material that *Sphere* is rendered with. The exact explanation for this observation requires further investigation in the future.

As is witnessed on the *Buddha* dataset, OMP performs less optimally than IRL1 and SBL on small concave regions that are rarely illuminated. This problem also occurs for *Caesar* on the medial side of the eyes and under the eyebrows. It is a lesser concern for objects that are generally convex such as *Caesar* and *Buddha* but may exert a strong negative influence on a scene that contains large concave areas. We will demonstrate the result for such a scene using *Venus* in Section 4.2.2.

4.1.2 Comparison via reconstructed surfaces

Using the normal maps recovered with various methods, we also reconstructed the 3D surface with the Frankot-Chellapa method [73] for direct comparison of the shape. Here, only the reconstruction result for *Caesar* is used

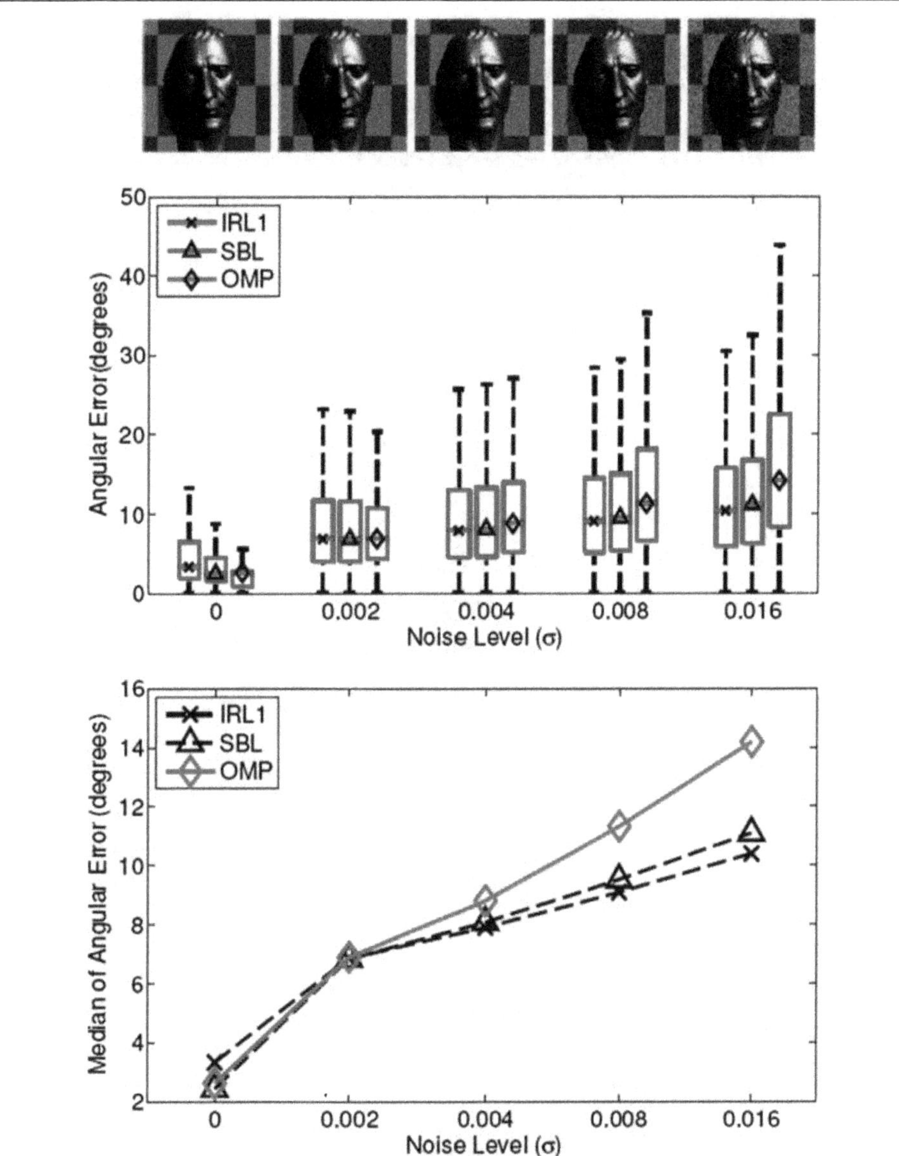

Fig. 16 Robustness of sparse recovery methods against Gaussian noise. *Upper tow*: sample images rendered with Gaussian noise of mean = 0.5 and STD = 0–0.016 , respectively, from left to right. *Middle figure*: boxplot of the angular error of normal maps under various degrees of Gaussian noise. Upper and lower border of *blue boxes* represent the third (Q3) and first (Q1) quartiles, respectively. *Upper and lower whiskers* show 1.0 IQR extended from Q3 and Q1, respectively. *Red bars in blue boxes* represent medians. At each noise level, three sparse recovery methods are compared, symbolized by different markers on the median bar: IRL1 (*cross*), SBL (*triangle*) and OMP (*diamond*). *Lower figure*: the medians of angular error

for demonstration. It is apparent from Fig. 10 that in the LS, IRL1, and SBL results, the overall shape of the face appear to be more protruding then it actually is, especially at the eyebrow ridge and the nose, whereas the OMP manages to preserve the shape accurately. Again, the LMS result appears to be less protruding than IRL1 and SBL results, although still not as accurate as OMP. We speculate that the exaggerated convexity originates from the inaccurately estimated normal vectors at highlight areas, such as the forehead and the nose. Since our greedy algorithm generally provides a better recovery in those regions, they naturally yield a more accurate shape recovery.

4.1.3 Real datasets

Three datasets of real-world scenes are tested in this study: *Gold*, an ancient golden coin, *Elba*, an Italian high-relief sculpture, and *Frag*, a much-decorated golden frame (that surrounds a painting by Fragonard). Sample images of the three datasets are shown in Fig. 11.

The advantages and disadvantages of the methods we found using synthesized datasets are also observed in the

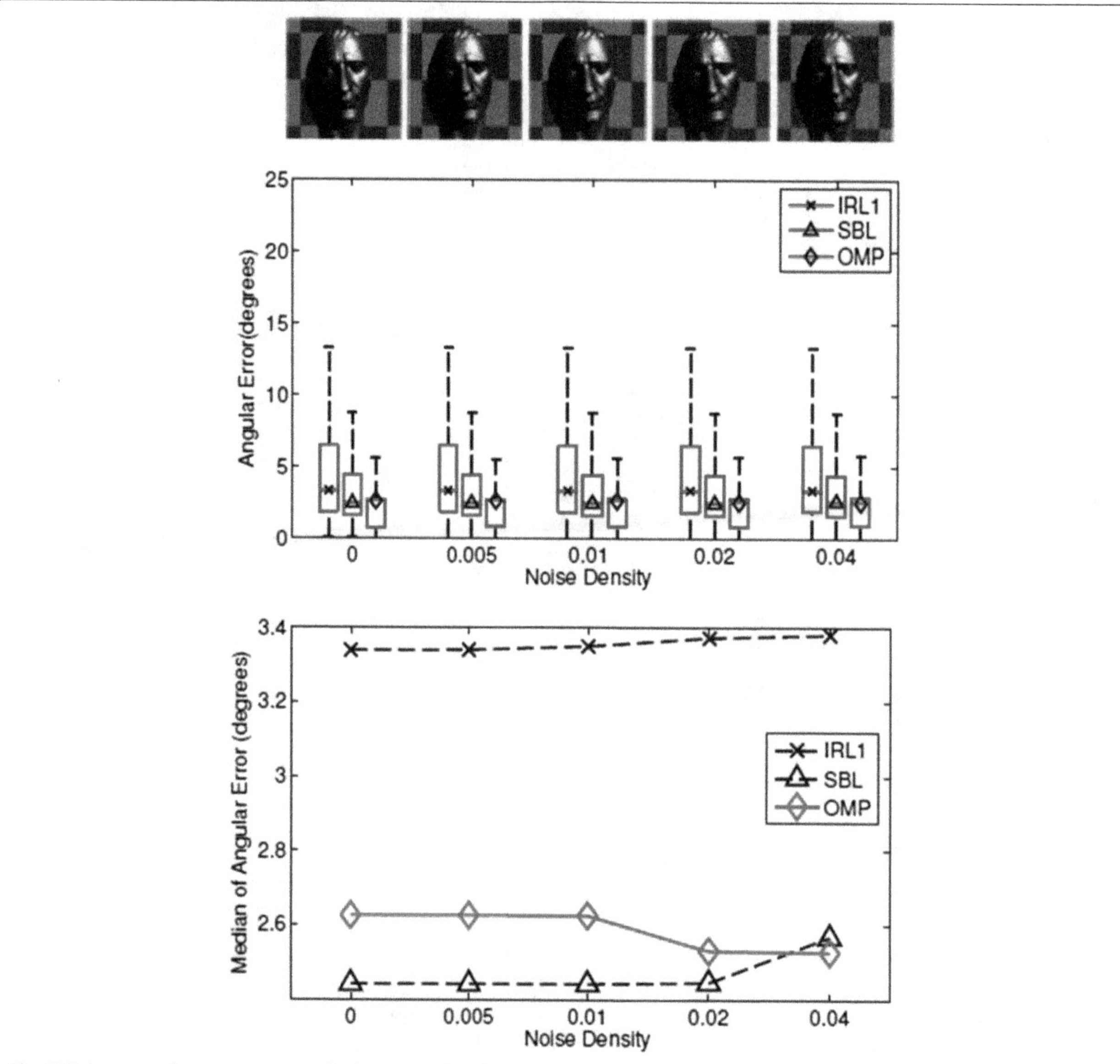

Fig. 17 Robustness of sparse recovery methods against salt and pepper noise. *Upper row*: sample images rendered with salt and pepper noise of density = 0–4 %, respectively, from left to right. *Middle figure*: boxplot of the angular error of normal maps under various degrees of salt and pepper noise. Upper and lower border of *blue boxes* represent third (Q3) and first (Q1) quartiles, respectively. *Upper and lower whiskers* show 1.0 IQR extended from Q3 and Q1, respectively. *Red bars in blue boxes* represent medians. At each noise level, four sparse recovery methods are compared, symbolized by different markers on the median bar: IRL1 (*cross*), SBL (*triangle*) and OMP (*diamond*). *Lower figure*: the medians of angular error

real datasets. Most images in dataset *Gold* have a large area of cast shadow. The influence of shadow can be clearly seen on the normal maps recovered by LS, IRL1 and SBL (Fig. 12a (1–3)) but is completely eliminated by OMP (Fig. 12a (4)). As for *Elba*, the scene contains a great number of small concave regions such as the pleats on the curtain. As expected, the greedy algorithm fail at these regions. Again, we notice that the LS, IRL1 and SBL results are more protruded than greedy results for both *Gold* and *Elba* (Fig. 12a (1–5), b (1–5)). Although there is not a ground truth normal map to support our speculation, it is reasonable to argue that the non-greedy algorithms exaggerate the convexity for *Elba*, as was the case for *Caesar* (Fig. 10). The complex geometry of the object in our third dataset—*Frag*—accounts for the noisy estimates observed in concave regions in the greedy normal maps (Fig. 12C4 and C5). Note that the non-greedy results also show a large degree of inaccuracy in these regions (Fig. 12C1–C3), but in a less noticeable manner since these artefacts are usually smoothly blended into less-affected areas.

4.1.4 MERL database

We also tested the performance of OMP on 95 materials from the MERL BRDF database [74]. Each material is rendered on a sphere at a resolution of 200 × 200 on 50 images of various lighting directions. The performance

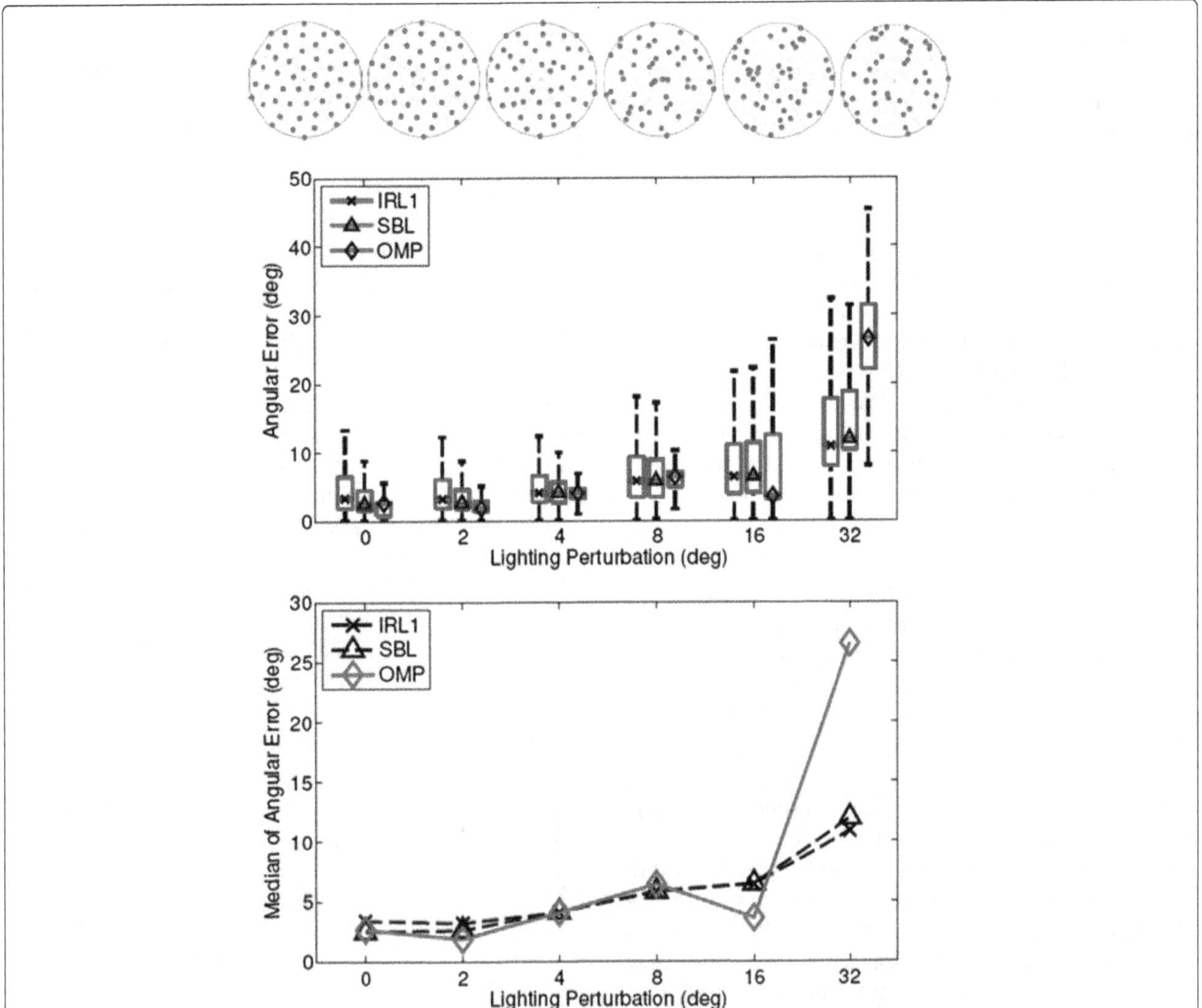

Fig. 18 Robustness of sparse recovery methods against light calibration error. *Top row*: leftmost plot shows actual light positions used for generating the dataset. Remaining plots show miscalibrated light positions with angular perturbations (2°–32°, from left to right) from the actual light positions at random directions. *Middle figure*: boxplot of the angular error of normal maps under various degrees of light calibration error. Upper and lower border of *blue boxes* represent third (Q3) and first (Q1) quartiles, respectively. *Upper and lower whiskers* show 1.0 IQR extended from Q3 and Q1, respectively. *Red bars in blue boxes* represent medians. At each angular perturbation level, four sparse recovery methods are compared, symbolized by different markers on the median bar: IRL1 (*cross*), SBL (*triangle*) and OMP (*diamond*). *Bottom figure*: the medians of angular error

pattern of OMP is very similar to SBL (Fig. 13): both methods are good at handling materials with an insignificant specular component (e.g. #10, *red-specular-plastic*). On the other hand, they both show decreased accuracy on shiny, strongly non-Lambertian metallic materials (e.g. #90, *silver-metallic-paint*) probably due to the violation of the sparsity assumption. Overall, the mean angular error of OMP over all 95 materials is 6.3174°, on par with SBL (6.5370°), and both methods significantly outperform the naive LS (10.8027°).

4.2 Robustness

To further understand of how well these methods behave in the presence of non-Lambertian effects, we tested their performance on *Sphere* with varying degrees of specularity and on *Venus*, where a large portion of the scene is concave, and as such, is heavily polluted by cast shadow. To find out the robustness of these methods against external error introduced by the experimental setup, we also tested the methods with additive image noise and light calibration error.

4.2.1 Specularity

We rendered five datasets of the same object *Sphere* with various sizes of highlight area (Fig. 14, top row) and tested how the size of the specular region affects the performance of our sparse regression methods. The size of the highlight is controlled by the *phong_size* parameter in POV-Ray [72]. We found that although the accuracy of all three methods compared (IRL1, SBL, OMP) decreases as the specular size increases, the greedy algorithm is less affected (Fig. 14, middle and bottom figures).

4.2.2 Shadow and concavity: a failure case

In Section 4.1.1, we have already noticed the possibility that the performance of our greedy algorithm may be negatively affected at shadowed concave regions. Here, we use the *Venus* dataset to further demonstrate this observation. In *Venus* (Fig. 5, bottom row), the convex foreground (the Venus statue) and the concave background (the dome) are well separated, allowing us to clearly inspect the performance of algorithms on different regions.

The result is shown in Fig. 15. As speculated, OMP shows robustness in shiny, convex regions such as the outer rim of the dome, and on the statue itself, but fails on the heavily shadowed background. The other three methods (LS, IRL1, SBL), on the contrary, suffer from noticeable angular error in convex areas. However, they are less severely affected by shadow and concavity on the background than greedy methods. Overall, the normal map recovered with the greedy approach are less smooth for *Venus* due to the inaccurate estimation of normal vectors in the concave regions.

4.2.3 Image noise

We tested three sparse algorithms (IRL1, SBL and OMP) against Gaussian noise as well as salt and pepper noise. For Gaussian noise (Fig. 16), the accuracy of all three methods drastically decreases as the noise level increases, although OMP appears to be slightly more adversely affected. On the other hand, all sparse methods are quite insensitive to salt and pepper noise (Fig. 17).

4.2.4 Lighting

There might be cases when the lighting directions are not properly calibrated. That is, the assumed lighting directions deviate from their actual values. In this test, we introduce for every assumed lighting vector a fixed angular perturbation, ranging from 2° to 32°, at a random direction, while keeping the actual arrangement of lights unchanged.

We tested the performance of the sparse methods under various degrees of light calibration error on the *Caesar* dataset. The actual arrangement of lights is displayed in Fig. 18 (leftmost plot on the top row). As an increasingly greater random perturbation is added to the assumed lighting directions, the angular error gradually increases for all sparse methods. Note that OMP appears to be susceptible to the random calibration error the most, especially when the perturbation reaches 32°.

Also note that in Fig. 18 (bottom), the median of the angular error produced by OMP slightly decreases at 16° compared to previous conditions. We believe that this is a fluctuation caused by the particular arrangement of lights at this condition. Despite this decrease in the median of error, the widths of the error distributions steadily increase at 16° for all three methods, as can be clearly seen from Fig. 18 (middle).

It was reported that the number of lights has a large impact on the accuracy of sparse photometric stereo recovery [2]. We found that our OMP-based method also shows a similar but somewhat greater dependency on the

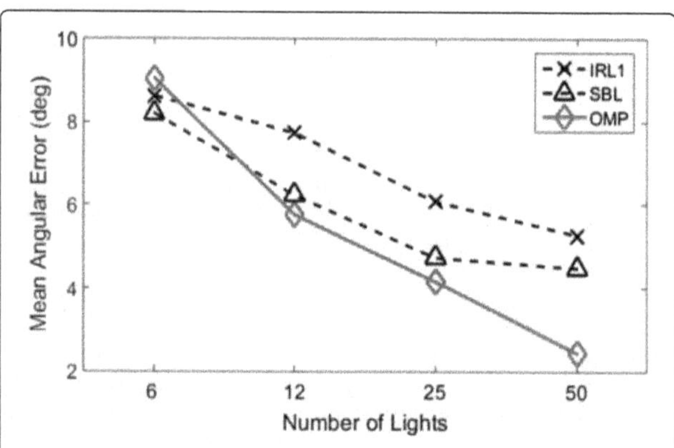

Fig. 19 Mean angular error of recovered normal maps under different numbers of lights

number of lights (Fig. 19). This observation indicates that the OMP works best when a large number of lights are present.

We also investigated the performance of sparse methods under a highly biased lighting distribution. We used 25 lights; 23 of them is located on the left or upper-left hemisphere and the other two on the right (Fig. 20). Under such a biased lighting, OMP still has the best mean angular error (8.2140°) compared with IRL1 (9.1173°), SBL (8.6561°) and LS (12.0726°).

4.3 Efficiency

The actual per-pixel processing time for the MATLAB implementation of the algorithms tested in this study is reported in Fig. 21. The maximum number of iterations for IRL1 and SBL are set to 100 although iteration will be terminated as soon as another stopping criterion is met; OMP always terminates after exactly 28 iterations for our datasets of 50 images; for LMS, the number of iterations is fixed at 1500.

In OMP, the operation with the highest asymptotic complexity is the inversion of a $k \times k$ matrix (where k is the number of selected columns) in Eq. 14. With a naive Gauss-Jordan elimination method, the inversion takes $O(k^3)$, which is asymptotically $O(n^3)$ since $k \leq n/2 + 3$. Since the above operation is repeated $n/2 + 3$ times, the overall time complexity of our OMP algorithm is $O(n^4)$. In our current implementation, the running time of OMP (4.823 ms/pixel) is comparable to IRL1 (3.338 ms/pixel). LMS is the slowest (57.48 ms/pixel), though it can be made faster with fewer iterations at the expense of accuracy.

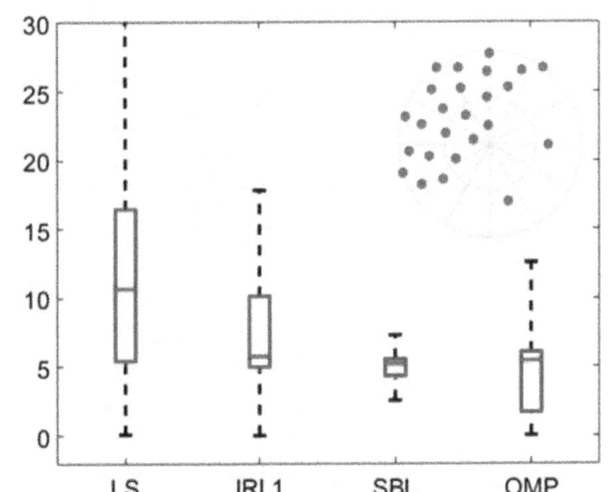

Fig. 20 Recovery accuracy under highly biased lighting distribution. Polar plot shows the arrangement of lights. Upper and lower border of *blue boxes* represent third (Q3) and first (Q1) quartiles, respectively. *Upper and lower whiskers* show 1.0 IQR extended from Q3 and Q1, respectively. *Red bars in blue boxes* represent medians

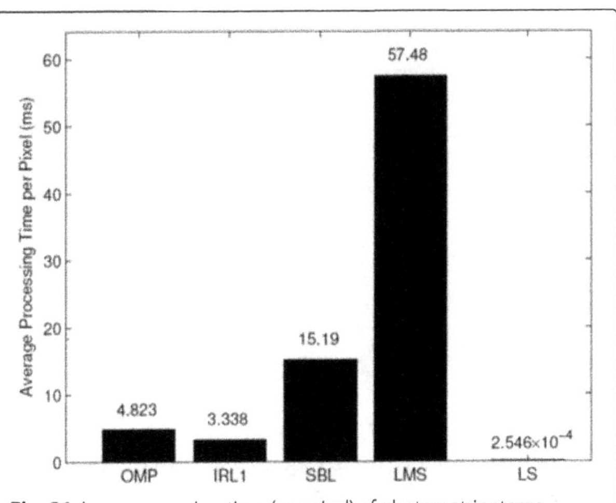

Fig. 21 Average running time (per pixel) of photometric stereo algorithms measured in milliseconds

4.4 Summary

Based on the experimental results above, we have come to the conclusion that our greedy algorithm overall has a higher accuracy than L1 minimization and SBL with a comparable efficiency, though OMP may be less robust in poorly illuminated regions. LMS is close to the greedy sparse algorithm in accuracy, despite its low efficiency. The algorithms tested in this chapter are summarized and compared in Table 2.

5 Conclusions

In this study, the classical PST is reformulated in terms of the canonical form of sparse recovery, and a greedy algorithm—OMP—is applied to solve the problem. Our formulation is different from previous ones [1, 2] in that the former incorporates normal vector components and non-Lambertian errors in one combined vector, allowing for the straightforward application of OMP. In order for OMP to obtain normal estimations, the normal vector components have to be selected before the iteration stops. Although it is not theoretically guaranteed, we observed that the normal components are always selected within the first few iterations in the datasets we tested, unless some components are indeed zero or very close to zero. We also speculate that the dictionary matrix in our formulation is near-orthonormal and satisfies the conditions required by OMP to achieve exact recovery.

We found that, in general, our greedy method OMP outperforms other state-of-the-art sparse solvers such as IRL1 and SBL [2] with comparable efficiency. In particular, OMP provides a more numerically accurate estimation of normal vectors in the presence of common non-Lambertian effects such as highlights and cast shadows, although it may occasionally fail at concave areas that are poorly illuminated. In addition, all sparse methods

Table 2 Qualitative comparison of photometric stereo algorithms

	Robustness				Smoothness	Efficiency
	Overall	Highlight	Shadow	Concavity		
LS	Very low	Very low	Very low	Very low	Very high	Very high
LMS	Very high	Very high	Very high	–	Low	Very low
LMS mode	Low	Low	Low	–	Medium	Very high
IRL1	High	High	High	Medium	High	High
SBL	High	High	High	Medium	High	Low
OMP	Very high	Very high	Very high	Low	Medium	High

The performance is evaluated on a five-level scale: 'Very low', 'low', 'medium', 'high' and 'very high'. Fields that are not available are indicated by a '–' sign

tested are reasonably robust against additive image noise and lighting calibration error.

Another two outlier-removal based methods—LMS and LMS mode finder—are also tested in this study for comparison. LMS results are overall statistically more accurate than IRL1 and SBL but less so than OMP. LMS mode finder, the 1D simplification of LMS, shows some robustness against non-Lambertian errors, especially highlights but performs poorly against cast shadows.

This study opens up many possible directions for future research. First, a great number of sparse recovery algorithms have already been proposed in the past few decades, each designed for a specific formulation. Even within the domain of greedy algorithms, there are many other potential candidates aside from OMP that may be directly applied to the PST problem. It would be interesting to explore this large repertoire of sparse formulations and recovery algorithms to find an optimal method.

It has been shown that sparse methods such as IRL1 and SBL can be used to estimate the lighting directions in the context of uncalibrated PST [19]. It is highly possible that greedy algorithms such as OMP can also be extended to be applied for such a purpose. Future studies may reveal more applications of greedy algorithms in different aspects of the PST framework.

Competing interests
The authors declare that they have no competing interests.

Acknowledgements
The authors would like to thank the anonymous referees for their insightful comments and suggestions, which help improve the quality of the current study substantially.

References
1. L Wu, A Ganesh, B Shi, Y Matsushita, Y Wang, Y Ma, in *Proceedings of Asian Conference of Computer Vision*. Robust photometric stereo via low-rank matrix completion and recovery (Springer, Berlin Heidelberg, 2010), pp. 703–717
2. S Ikehata, D Wipf, Y Matsushita, K Aizawa, in *IEEE Conference on Computer Vision and Pattern Recognition*. Robust photometric stereo using sparse regression (IEEE Computer Society, Washington DC, USA, 2012), pp. 318–325
3. G Fyffe, X Yu, P Debevec, in *Int. Conf. on Computational Photog.* Single-shot photometric stereo by spectral multiplexing (IEEE Computer Society, Washington DC, USA, 2011), pp. 1–6
4. G Fyffe, P Debevec, in *Int. Conf. on Computational Photog.* Single-shot reflectance measurement from polarized color gradient illumination (IEEE Computer Society, Washington DC, USA, 2015)
5. A Hertzmann, SM Seitz, in *IEEE Computer Society Conference on Computer Vision and Pattern Recognition*. Shape and materials by example: a photometric stereo approach, vol. 1 (IEEE Computer Society, Washington DC, USA, 2003), pp. 533–540
6. N Alldrin, T Zickler, D Kriegman, in *IEEE Conference on Computer Vision and Pattern Recognition*. Photometric stereo with non-parametric and spatially-varying reflectance (IEEE, Anchorage AK, 2008), pp. 1–8
7. J Ackermann, M Ritz, A Stork, M Goesele, in *Trends and Topics in Computer Vision*, Lecture Notes in Computer Science, ed. by KN Kutulakos. Removing the example from example-based photometric stereo (Springer, Berlin Heidelberg, 2012), pp. 197–210
8. F Verbiest, L Van Gool, in *2008 IEEE Computer Society Conference on Computer Vision and Pattern Recognition*. Photometric stereo with coherent outlier handling and confidence estimation (IEEE Computer Society, Washington DC, USA, 2008), pp. 1–8
9. C Hernandez, G Vogiatzis, R Cipolla, Multiview photometric stereo. IEEE Trans. Pattern Anal. Mach. Intell. **30**(3), 548–554 (2008)
10. MS Drew, Y Hel-Or, T Malzbender, N Hajari, Robust estimation of surface properties and interpolation of shadow/specularity components. Image Vis. Comput. **30**(4–5), 317–331 (2012)
11. M Zhang, MS Drew, in *CPCV2012: European Conference on Computer Vision Workshop on Color and Photometry in Computer.* Vision Lecture Notes in Computer Science. Robust luminance and chromaticity for matte regression in polynomial texture mapping, vol. 7584/2012 (Springer, Berlin Heidelberg, 2012), pp. 360–369
12. MS Drew, N Hajari, Y Hel-Or, T Malzbender, in *Proceedings of the British Machine Vision Conference*. Specularity and shadow interpolation via robust polynomial texture maps, (2009), pp. 114–111411
13. DL Donoho, Compressed sensing. IEEE Trans. Inf. Theory. **52**(4), 1289–1306 (2006)
14. BK Natarajan, Sparse approximation solutions to linear systems. SIAM J. Comput. **24**(2), 227–234 (1995)
15. RJ Woodham, Photometric method for determining surface orientation from multiple images. Opt. Eng. **19**(1), 139–144 (1980)
16. S Barsky, M Petrou, The 4-source photometric stereo technique for three-dimensional surfaces in the presence of highlights and shadows. IEEE Trans. Pattern Anal. Mach Intell. **25**(10), 1239–1252 (2003)
17. H Rushmeier, G Taubin, A Guéziec, in *Proceedings of Eurographics Workshop on Rendering*. Applying shape from lighting variation to bump map capture, (1997), pp. 35–44
18. P Favaro, T Papadhimitri, A closed-form solution to uncalibrated photometric stereo via diffuse maxima. IEEE Conf. Computer Vision and Pattern Recognit. **157**(10), 821–828 (2012)
19. V Argyriou, S Zafeiriou, B Villarini, M Petrou, A sparse representation method for determining the optimal illumination directions in photometric stereo. Signal Process. **93**(11), 3027–3038 (2013)

20. J Wright, AY Yang, A Ganesh, SS Sastry, Y Ma, Robust face recognition via sparse representation. IEEE Trans. Pattern Anal. Mach. Intell. **31**(2), 210–227 (2009)
21. YC Pati, R Rezaiifar, PS Krishnaprasad, Orthogonal matching pursuit: recursive function approximation with applications to wavelet decomposition. Proceedings of 27th Asilomar Conference on Signals, Systems and Computers. **1**, 40–44 (1993)
22. JA Tropp, AC Gilbert, Signal recovery from random measurements via orthogonal matching pursuit. IEEE Trans. Inf. Theory. **53**(12), 4655–4666 (2007)
23. TT Cai, L Wang, Orthogonal matching pursuit for sparse signal recovery with noise. IEEE Trans. Inf. Theory. **57**(7), 4680–4688 (2011)
24. EN Coleman Jr., R Jain, Obtaining 3-dimensional shape of textured and specular surfaces using four-source photometry. Comput. Graph. Image Process. **18**, 309–328 (1982)
25. F Solomon, K Ikeuchi, Extracting the shape and roughness of specular lobe objects using four light photometric stereo. IEEE Trans. Pattern Anal. Mach. Intell. **18**, 449–454 (1996)
26. A Yuille, D Snow, in *IEEE Computer Society Conference on Computer Vision and Pattern Recognition*. Shape and albedo from multiple images using integrability (IEEE Computer Society, Washington DC,USA, 1997), pp. 158–164
27. G Willems, F Verbiest, W Moreau, H Hameeuw, K Van Lerberghe, L Van Gool, in *Short and Project Papers Proceedings of 6th International Symposium on Virtual Reality, Archaeology and Cultural Heritage*. Easy and cost-effective cuneiform digitizing (Eurographics Association, Geneve, Switzerland, 2005), pp. 73–80
28. J Sun, M Smith, L Smith, S Midha, J Bamber, Object surface recovery using a multi-light photometric stereo technique for non-Lambertian surfaces subject to shadows and specularities. Image Vis. Comput. **25**, 1050–1057 (2007)
29. S Mallick, T Zickler, D Kriegman, P Belhumeur, in *IEEE Comp. Soc. Conf. on Comp. Vis. and Patt. Rec. 2005*. Beyond lambert: reconstructing specular surfaces using color, vol. 2, (2005), pp. 619–6262
30. M Holroyd, J Lawrence, G Humphreys, T Zickler, A photometric approach for estimating normals and tangents. ACM Trans. Graph. (Proceedings of SIGGRAPH Asia 2008). **27**(5), 32–39 (2008)
31. C Julià, F Lumbreras, AD Sappa, A factorization-based approach to photometric stereo. Int. J. Imaging Syst. Technol. **21**(1), 115–119 (2011)
32. K-L Tang, C-K Tang, T-T Wong, in *IEEE Computer Society Conference on Computer Vision and Pattern Recognition*. Dense photometric stereo using tensorial belief propagation, vol. 1, (2005), pp. 132–139
33. M Chandraker, S Agarwal, D Kriegman, in *IEEE Computer Society Conference on Computer Vision and Pattern Recognition*. ShadowCuts: photometric stereo with shadows, (2007), pp. 1–8
34. Y Mukaigawa, Y Ishii, T Shakunaga, Analysis of photometric factors based on photometric linearization. J. Opt. Soc. Am. A, Optics, image science, and vision. **24**(10), 3326–3334 (2007)
35. D Miyazaki, K Hara, K Ikeuchi, Median photometric stereo as applied to the Segonko Tumulus and museum objects. Int. J. Comput. Vision. **86**(2–3), 229–242 (2010)
36. HD Tagare, RJP de Figueiredo, A theory of photometric stereo for a class of diffuse non-Lambertian surfaces. IEEE Trans. Pattern Anal. Mach. Intell. **13**(2), 133–152 (1991)
37. SK Nayar, K Ikeuchi, T Kanade, Determining shape and reflectance of hybrid surfaces by photometric sampling. IEEE Trans. Robot. Autom. **6**(4), 418–431 (1990)
38. AS Georghiades, in *9th IEEE International Conference on Computer Vision*. Incorporating the torrance and sparrow model of reflectance in uncalibrated photometric stereo, (2003), pp. 816–8232. doi:10.1109/ICCV.2003.1238432
39. T Malzbender, D Gelb, H Wolters, in *Proceedings of the 28th Annual Conference on Computer Graphics and Interactive Techniques*. SIGGRAPH '01. Polynomial texture maps (ACM, New York, USA, 2001), pp. 519–528
40. R Basri, D Jacobs, I Kemelmacher, Photometric stereo with general, unknown lighting. Int. J. Comput. Vision. **72**, 239–257 (2007)
41. WM Silver, *Determining shape and reflectance using multiple images*. (Massachusetts Institute of Technology, Cambridge MA, 1980)
42. SJ Koppal, SG Narasimhan, in *IEEE Conference on Computer Vision and Pattern Recognition*. Clustering Appearance for Scene Analysis, vol. 2, (2006), pp. 1323–1330
43. D Goldman, B Curless, A Hertzmann, S Seitz, in *10th IEEE International Conference on Computer Vision*. Shape and spatially-varying BRDFs from photometric stereo, vol. 1, (2005), pp. 341–348
44. Q Yang, N Ahuja, Surface reflectance and normal estimation from photometric stereo. Comput. Vision and Image Underst. **116**(7), 793–802 (2012)
45. S Kherada, P Pandey, A Namboodiri, in *IEEE Workshop on Applications of Computer Vision*. Improving realism of 3D texture using component based modeling (IEEE Computer Society, Washington DC, USA, 2012), pp. 41–47
46. B Shi, T Ping, Y Matsushita, K Ikeuchi, Bi-polynomial modeling of low-frequency reflectances. IEEE Trans. Pattern Anal. Mach. Intell. **36**(6), 1–1 (2013)
47. Z Lin, M Chen, Y Ma, *The augmented Lagrange multiplier method for exact recovery of corrupted low-rank matrices*. (Technical report, UIUC (UILU-ENG-09-2215), 2009)
48. E Candès, J Romberg, l1-magic: recovery of sparse signals via convex programming. URL: http://statweb.stanford.edu/~candes/l1magic/downloads/l1magic.pdf 4(2005)
49. EJ Candès, MB Wakin, SP Boyd, Enhancing sparsity by reweighted L1 minimization. J. Fourier Anal. Appl. **14**(5–6), 877–905 (2008)
50. DP Wipf, BD Rao, Sparse Bayesian learning for basis selection. IEEE Trans. Signal Process. **52**(8), 2153–2164 (2004)
51. A Adler, M Elad, Y Hel-Or, E Rivlin, Sparse coding with anomaly detection. J. of Signal Proc. Systems. **79**, 179–188 (2015)
52. SP Boyd, L Vandenberghe, *Convex Optimization*. (Cambridge University Press, Cambridge, UK, 2004), p. 744
53. MAT Figueiredo, RD Nowak, SJ Wright, Gradient projection for sparse reconstruction: application to compressed sensing and other inverse problems. IEEE J. Selected Topics in Signal Process. **1**(4), 586–597 (2007)
54. SG Mallat, Z Zhang, Matching pursuits with time-frequency dictionaries. IEEE Trans. Signal Process. **41**(12), 3397–3415 (1993)
55. D Needell, R Vershynin, Uniform uncertainty principle and signal recovery via regularized orthogonal matching pursuit. Found. Comput. Math. **9**(3), 317–334 (2009)
56. D Needell, R Vershynin, Signal recovery from incomplete and inaccurate measurements via regularized orthogonal matching pursuit. IEEE J. Selected Topics in Signal Process. **4**(2), 310–316 (2010)
57. DL Donoho, Y Tsaig, I Drori, J-L Starck, Sparse solution of underdetermined systems of linear equations by stagewise orthogonal matching pursuit. IEEE Trans. Inf. Theory. **58**(2), 1094–1121 (2012)
58. D Needell, JA Tropp, CoSaMP: iterative signal recovery from incomplete and inaccurate samples. Commun. ACM. **53**(12), 93–100 (2010)
59. A Divekar, O Ersoy, *Probabilistic matching pursuit for compressive sensing*. (Technical report, Purdue University, West Lafayette, IN, USA, 2010)
60. S Chatterjee, D Sundman, M Skoglund, in *IEEE International Conference on Acoustics, Speech and Signal Processing (ICASSP)*. Look ahead orthogonal matching pursuit, (2011), pp. 4024–4027
61. P Jain, A Tewari, IS Dhillon, in *Proceedings of Neural Information Processing Systems*. Orthogonal matching pursuit with replacement, (2011), pp. 1215–1223
62. NB Karahanoglu, H Erdogan, A* orthogonal matching pursuit: best-first search for compressed sensing signal recovery. Digital Signal Process. **22**(4), 555–568 (2012)
63. T Blumensath, ME Davies, Iterative hard thresholding for compressed sensing. Appl. Comput. Harmonic Anal. **27**(3), 265–274 (2009)
64. W Dai, O Milenkovic, Subspace pursuit for compressive sensing signal reconstruction. IEEE Trans. Inf. Theory. **55**(5), 2230–2249 (2009)
65. DL Donoho, A Maleki, A Montanari, Message-passing algorithms for compressed sensing. Proc. Natl. Acad. Sci. **106**(45), 18914–18919 (2009)
66. A Maleki, DL Donoho, Optimally tuned iterative reconstruction algorithms for compressed sensing. IEEE J. Sel. Topics in Signal Process. **4**(2), 330–341 (2010)
67. V Cevher, in *IEEE International Conference on Acoustics, Speech and Signal Processing (ICASSP)*. An ALPS view of sparse recovery, (2011), pp. 5808–5811
68. EJ Candès, T Tao, Decoding by linear programming. IEEE Trans. Inf. Theory. **51**(12), 4203–4215 (2005)
69. EJ Candès, JK Romberg, T Tao, Stable signal recovery from incomplete and inaccurate measurements. Commun.Pure and Appl. Math. **59**(8), 1207–1223 (2006)

70. AIM@SHAPE shape repository. http://visionair.ge.imati.cnr.it/ontologies/shapes/
71. 3D meshes research databse by INRIA Gamma Group. http://www-roc.inria.fr/gamma/gamma/download/download.php
72. POV-Ray 3.6.0 documentation online. http://www.povray.org/documentation/view/3.6.0/347/
73. RT Frankot, R Chellapa, A method for enforcing integrability in shape from shading algorithms. IEEE Trans. Pattern Anal. Mach. Intell. **10**(4), 439–451 (1988)
74. W Matusik, H Pfister, M Brand, L McMillan, A data-driven reflectance model. ACM Trans. Graph. **22**(3), 759–769 (2003)

7

A novel approach for handedness detection from off-line handwriting using fuzzy conceptual reduction

Somaya Al-Maadeed*, Fethi Ferjani, Samir Elloumi and Ali Jaoua

Abstract

A challenging area of pattern recognition is the recognition of handwritten texts in different languages and the reduction of a volume of data to the greatest extent while preserving associations (or dependencies) between objects of the original data. Until now, only a few studies have been carried out in the area of dimensionality reduction for handedness detection from off-line handwriting textual data. Nevertheless, further investigating new techniques to reduce the large amount of processed data in this field is worthwhile. In this paper, we demonstrate that it is important to select only the most characterizing features from handwritings and reject all those that do not contribute effectively to the process of handwriting recognition. To achieve this goal, the proposed approach is based mainly on fuzzy conceptual reduction by applying the Lukasiewicz implication. Handwritten texts in both Arabic and English languages are considered in this study. To evaluate the effectiveness of our proposal approach, classification is carried out using a K-Nearest-Neighbors (K-NN) classifier using a database of 121 writers. We consider left/right handedness as parameters for the evaluation where we determine the recall/precision and F-measure of each writer. Then, we apply dimensionality reduction based on fuzzy conceptual reduction by using the Lukasiewicz implication. Our novel feature reduction method achieves a maximum reduction rate of 83.43 %, thus making the testing phase much faster. The proposed fuzzy conceptual reduction algorithm is able to reduce the feature vector dimension by 31.3 % compared to the original ⊠BEST OF ALL COMBINED FEATURES⊠algorithm.

Keywords: Index terms-handwriting, Fuzzy binary relation, Left/right identification, Feature, Lukasiewicz implication, Galois connections, Closure of Fuzzy Galois connections

1 Introduction

Handwriting recognition is the ability of a computer to receive and interpret intelligible handwritten input from sources such as paper documents, photographs, touch-screens, and other devices. The image of the written text may be sensed "off-line" from a piece of paper by optical scanning (optical character recognition) or intelligent word recognition. Alternatively, the movements of the pen tip may be sensed "on-line", for example by a pen-based computer screen surface. Handwriting recognition principally entails optical character recognition. However, a complete handwriting recognition system also handles formatting, performs correct segmentation into characters and determines the most plausible words.

Handwriting recognition has been one of the fascinating and challenging research areas in the field of image processing and pattern recognition [1, 2]. It contributes immensely to the advancement of an automation process and can improve the interface between human beings and machines in numerous applications. Several research works have been focusing on new techniques and methods that would reduce the processing time while providing higher recognition accuracy [3].

In general, handwriting recognition is classified into two types of methods: off-line and on-line handwriting recognition methods. In off-line recognition, the writing is usually captured optically by a scanner, and the complete text is available as an image. In the on-line system, the two-dimensional coordinates of successive points are

*Correspondence: s_alali@qu.edu.qa
Department of Computer Science and Engineering, Qatar University, Doha, Qatar

represented as a function of time, and the order of strokes made by the writer is also available. The on-line methods have been shown to be superior to their off-line counterparts in recognizing handwritten characters due to the temporal information available with the former [4, 5]. Several applications, including mail sorting, bank processing, document reading, and postal address recognition, require off-line handwriting recognition systems. As a result, off-line handwriting recognition continues to be an active area of research toward exploring newer techniques that would improve recognition accuracy [6].

In our current study, we focus on off-line handwriting and use a K-Nearest Neighbors (K-NN) classifier to make classifications based on many parameters, such as gender, age, handedness, and nationality, to measure the performance of our proposed algorithm. This type of classification has several applications. For example, in the forensic domain, handwriting classification can help investigators to focus on a certain category of suspects. Additionally, processing each category separately leads to improved results in writer identification and verification applications.

There are few studies in the literature that investigate the automatic detection of gender, age, and handedness from handwritings. Bandi et al. [7] proposed a system that classifies handwritings into demographic categories using the "macro-features" introduced in [8]. These features focus on measurements such as pen pressure, writing movement, stroke formation, and word proportion. The authors reported classification accuracies of 77.5, 86.6, and 74.4 % for gender, age and handedness classification, respectively. However, in this study, all the writers produced the same letter. Unfortunately, this is not always the case in real forensic caseworks. Moreover, the dataset used in this study is not publicly available.

Liwicki et al. [9] also addressed the classification of gender and handedness in the on-line mode (which means that the temporal information about the handwriting is available). The authors used a set of 29 features extracted from both on-line information and its off-line representation and applied support vector machines and Gaussian mixture models to perform the classification. The authors reported a performance of 67.06 % for gender classification and 84.66 % for handedness classification. In [10], the authors separately reported the performance of the off-line mode, the on-line mode, and their combination. The performance reported for the off-line mode was 55.39 %, which is slightly better than chance.

In this paper, we propose a novel approach to the detection of the handedness of the writer of a handwritten document based on the Lukasiewicz implication where a set of features was proposed and evaluated to predict the handedness of the writer. These features are combined using a K-Nearest Neighbors (K-NN) classifier under the *Rapidminer* platform [11]. This method is evaluated using the QUWI database, which is the only publicly available dataset containing annotations regarding gender, age range, and nationality.

The rest of the paper is organized as follows. In section 2, we review some basic definitions from relational algebra, the mathematical background related to fuzzy set theories and useful for this research paper. In section 3, the state-of-the-art in writer identification for the English and Arabic languages is presented in detail. The evaluation was made using larger amounts of text and may not produce acceptable results when limited amounts of text are available. Writer recognition from short handwritten texts is therefore an interesting area of study. Section 4 gives a description of the system overview and the database used for carrying out the experimental evaluations. Next, we describe the proposed features and the method in which they are extracted. Then, we present the utilized (K-NN) classifier followed by the detailed results and an analysis of the experimental evaluations. Finally, we conclude the paper with some discussion on future research directions on the subject.

2 Key settings and new definitions

The domains of computer science, relational algebra, formal concept analysis, and lattice theory have seen important advances in research [12]. This research has contributed enormously to the search for original solutions for complex problems in the domains of knowledge engineering, data mining, and information retrieval. Relational algebra and formal concept analysis may be considered as useful mathematical foundations that unify data and knowledge in information retrieval systems.

2.1 Binary relations

In the following, we review some basic definitions from relational algebra. Let us consider two sets $\mathcal{X}$ and $\mathcal{Y}$ and two elements e and e', where $e \in \mathcal{X}$ and $e' \in \mathcal{Y}$.

- A relation $\mathcal{R}$ is a subset of the Cartesian product of two sets $\mathcal{X}$ and $\mathcal{Y}$.
- An element $(e, e') \in \mathcal{R}$, where e' denotes the image of e by $\mathcal{R}$.
- A binary relation identity $\mathcal{I}(\mathcal{A}) = \{(e, e) | e \in \mathcal{A}\}$.
- The relative product or composition of two binary relations $\mathcal{R}$ and $\mathcal{R}'$ is $\mathcal{R} \circ \mathcal{R}' = \{(e, e') | \exists t \in \mathcal{Y} : ((e, t) \in \mathcal{R}) \& ((t, e') \in \mathcal{R}')\}$.
- The inverse of the relation $\mathcal{R}$ is $\mathcal{R}^{-1} = \{(e, e') | (e', e) \in \mathcal{R}\}$.
- The set of images of e is defined by $e.\mathcal{R} = \{e' | (e, e') \in \mathcal{R}\}$.
- The set of antecedents of e' is defined by $\mathcal{R}.e' = \{e | (e, e') \in \mathcal{R}\}$.

- The cardinality of $\mathcal{R}$ is defined by $Card(\mathcal{R})$ = the numbers of pairs $(e, e') \in \mathcal{R}$.
- The complement of the relation $\mathcal{R}$ is $\overline{\mathcal{R}} = \{(e, e') | (e, e') \notin \mathcal{R}\}$.
- The domain of $\mathcal{R}$ is defined by $Dom(\mathcal{R}) = \{e | \exists e' : (e, e') \in \mathcal{R}\}$.
- The range or codomain of $\mathcal{R}$ is defined by $Cod(\mathcal{R}) = \{e' | \exists e : (e, e') \in \mathcal{R}\}$.

2.2 Formal concept analysis

Formal Concept Analysis (FCA) is the mathematical theory of data analysis using formal contexts and concept lattices [12–14]. It was introduced by Rudolf Wille in 1984 and builds on applied lattice and order theory, which were developed by Birkhoff et al. [15]

Definition 1. A formal context
A formal context (or an extraction context) is a triplet $\mathcal{K} = (\mathcal{X}, \mathcal{Y}, \mathcal{R})$, where $\mathcal{X}$ represents a finite set of objects, $\mathcal{Y}$ is a finite set of attributes (or properties), and $\mathcal{R}$ is a binary (incidence) relation, (i.e., $\mathcal{R} \subseteq \mathcal{X} \times \mathcal{Y}$). Each couple $(x, y) \in \mathcal{R}$ expresses that the object $x \in \mathcal{X}$ verifies property y belonging to $\mathcal{Y}$.

Definition 2. Formal concept in fuzzy binary relation
Let $\mathcal{X}$ be a set called the universe of discourse. Elements of $\mathcal{X}$ are denoted by lowercase letters. A fuzzy set $E = \{x_1/v_1, x_2/v_2, \cdots, x_n/v_n\}$ is defined as a collection of elements $x_i \in \mathcal{X}, i = 1 : n$, which includes a degree of membership v_i for each element x_i [16, 17].

A fuzzy binary context (or fuzzy binary relation) is a fuzzy set defined on the product of two sets $\mathcal{O}$ (set of objects) and $\mathcal{P}$ (set of properties). Hence, $\mathcal{X} = \mathcal{O} \times \mathcal{P}$.

Definition 3. Galois connection
Let $(\mathcal{X}, \mathcal{Y}, \mathcal{R})$ be the formal context, and let $A \subseteq \mathcal{X}$ and $B \subseteq \mathcal{Y}$ be two finite sets. We define two operators f(A) and g(B) on A and B as follows:

1. $f(A) = \{e' | \forall e \in A, (e, e') \in \mathcal{R}\}$,
2. $g(B) = \{e | \forall e' \in B, (e, e') \in \mathcal{R}\}$.

Operator f defines the properties shared by all elements of A, and operator g defines objects sharing the same properties included in set B. The operators f and g define a Galois connection between the sets $\mathcal{X}$ and $\mathcal{Y}$ with respect to the binary context $(\mathcal{X}, \mathcal{Y}, \mathcal{R})$ [12, 18].

Definition 4. Fuzzy set
In classical set theory, elements fully belong to a set or are fully excluded. However, a fuzzy set $\mathcal{A}$ in universe $\mathcal{X}$ is the set whose elements also partially belong to $\mathcal{X}$. The grade of belonging of each element is determined by a membership function μ_A given by [16]

- $\mu_A : \mathcal{X} \to [0, 1]$,
- *A finite fuzzy set can be denoted as* $A = \{\mu_A(x_1)/x_1, \mu_A(x_2)/x_2, \ldots, \mu_A(x_n)/x_n\}$, *for any* $x_i \in \mathcal{X}$.

Example 1. *Let us consider the fuzzy relation $\mathcal{F}_{BR}$ depicted in Table 1. $\mathcal{F}_{BR}$ contains five objects O_1, O_2, O_3, O_4, and O_5 and six properties {**a**, **b**, **c**, **d**, **e**, **f**}, where the values have been set randomly.*

Definition 5. *Fuzzy Galois connection*
Let $\mathcal{F}_{BR}$ be a fuzzy binary relation defined on $\mathcal{X}$. For two sets A and B such that $A \subseteq \mathcal{O}$, B is a fuzzy set defined on $\mathcal{P}$, and $\delta \in [0, 1]$ [17, 19–21]. *We define the operators $\mathcal{F}$ and $\mathcal{H}_\delta$ as follows:*

- $\mathcal{F}(A) = \{d/\alpha | \alpha = min\{\mu_{\mathcal{F}_{BR}}(g, d) | g \in A, d \in \mathcal{P}\}\}$
- $\mathcal{H}_\delta(B) = \{g | d \in \mathcal{P} \Rightarrow (\mu_B(d) \to_L \mu_{\mathcal{F}_{BR}}(g, d) \geqslant \delta\}$,

where $\to_L$ denotes the Lukasiewicz implication. For example, for $x_i, x_j \in [0, 1]$,

$$x_i \to_L x_j = min(1, 1 - x_i + x_j). \tag{1}$$

Note that $\mu_{\mathcal{F}_{BR}}(g, d)$ denotes the weight of the pair (g,d) in the fuzzy relation $\mathcal{F}_{BR}$.

Definition 6. A fuzzy closure operator
For two sets A and B such that $A \subseteq \mathcal{O}$, B is a fuzzy set defined on $\mathcal{P}$, and $\delta \in [0, 1]$. We define $Closure(A) = \mathcal{H}_\delta(\mathcal{F}(A)) = A'$ and $Closure(B) = \mathcal{F}(\mathcal{H}_\delta(B)) = B'$.

The composition $f \circ g$ defines the closure of the Galois connection. Let A_i , A_j be subsets of objects $\mathcal{O}$, and B_i , B_j fuzzy subsets defined on $\mathcal{P}$. The operators f and g have the following properties [12]:

1. $A_i \subseteq A_j \Rightarrow f(A_i) \supseteq f(A_j)$;
2. $B_i \subseteq B_j \Rightarrow g(B_i) \supseteq g(B_j)$;
3. $A_i \subseteq g \circ f(A_i)$ and $B_i \subseteq f \circ g(B_i)$;
4. $A \subseteq g(B) \Leftrightarrow B = f(A)$; and
5. $f = f \circ g \circ f$ and $g = g \circ f \circ g$.

Fuzzy data reduction To manage the large amount of features, it is important to select the most pertinent ones.

Table 1 Fuzzy binary relation

	a	b	c	d	e	f
O_1	0.5	0.2	0.6	0.4	0.7	0.5
O_2	0.7	0.3	0.2	0.3	0.2	1
O_3	1	0.4	0.6	1	0.7	0.5
O_4	0.5	0.2	0.6	0.4	0.8	0.6
O_5	0.7	0.3	0.7	0.4	0.6	0.7

In this paper, we use fuzzy conceptual reduction applied to the original data. Fuzzy conceptual data reduction methods have the main objective of minimizing the size of data while preserving the content of the original document. Unfortunately, most of the methods presented in the literature are based on heuristics and are not accurate. Moreover, reducing fuzzy data becomes a difficult problem because the handling of imprecision and uncertainty may cause information loss and/or deformation. In this work, we develop a fuzzy conceptual approach based on Lukasiewicz fuzzy Galois. This method is based on fuzzy formal concept analysis, which has been recently developed by several researches and applied for learning, knowledge acquisition, information retrieval, etc. The Lukasiewicz implication based on the fuzzy Galois connection is mainly used in this paper. It allows one to consider different precision levels according to the value of δ in the definition of fuzzy formal concepts.

The advantage of reduced data is that it can be used directly as a prototype for making decisions, for supervised learning, or for reasoning. For that purpose, we first prove that some rows can be removed from the initial fuzzy binary context at a given precision level (value given to δ by application of the Lukasiewicz implication). It is primordial to assess that there is an equivalence between an object and a set of objects. Second, we define a solution for data reduction in the case of fuzzy binary relations.

Equivalence between an object and a subset of other objects An object x is equivalent (for a given value of δ for δ varying from 0 to 1) to a set of objects S_x, relative to a fuzzy binary context $\mathcal{F}_{BR}$, if and only if $\{x\} \cup S_x$ is a domain of a concept of $\mathcal{F}_{BR}$, and the closure $(S_x) = \{x\} \cup S_x$, where $x \notin S_x$. As intuitive justification, x is equivalent to S_x means that $S_x \rightarrow x$ within some precision δ.

3 A review of related works

Handwriting refers to the style of writing textual documents with a writing instrument such as a pen or pencil by a person. Characteristics of handwriting include the following: (1) specific shapes of letters, e.g., their roundness or sharpness; (2) regular or irregular spacing between letters; (3) the slope of the letters; (4) the rhythmic repetition of the elements or arrhythmia; (5) the pressure to the paper; and (6) the average size of letters. Because each person's handwriting is unique, it can be used to verify a document's writer. Therefore, writer identification has been recently studied in a wide variety of applications, such as security, financial activity, and forensics and has been used for access control. Writer identification is the task of determining the writer of a document among different writers. Writer identification methods can be categorized into two types: text-dependent methods and text-independent methods. In text-dependent methods, a writer has to write the same fixed text to perform identification, but in text-independent methods, any text may be used to establish the identity of the writer. These methods can be performed on-line, where dynamic information about the writing is available, or off-line, where only a scanned image of the writing is available. Recently, different approaches for writer identification have been proposed. A scientific validation of the individuality of handwriting was performed in [22]. In that study, handwriting samples from 1500 individuals, representative of the US population with respect to gender, age, ethnic groups, etc., were obtained. The writer can be identified based on macro-features and micro-features that were extracted from handwritten documents. The authors in [23] proposed a global approach based on multi-channel Gabor filtering, where each writer's handwriting is regarded as a different texture. Bensefia et al. [24] used local features based on graphemes extracted from segmentations of cursive handwriting. In addition, writer identification has been performed by a textual-based information retrieval model. The work in [25] presented a new approach using a connected-component contour codebook and its probability density function. In addition, combining connected-component contours with an independent edge-based orientation and curvature PDF yields very high correct identification rates. Schlapbach and Bunke [26] proposed an HMM-based approach for writer identification and verification. In [27], the authors used a combination of directional, grapheme, and run-length features to improve writer identification and verification performance. Other studies used chain code and global features for writer identification [28]. Both proposed methods are applicable to cursive handwriting and have practical feasibility for writer identification. Other applications including such as off-line handwriting recognition systems for different languages reached up to 99 % for handwritten characters [29]. For handedness detection our earlier study proved that it can work [30]. Authors in [25] evaluated the performance of edge-based directional probability distributions as features in comparison to a number of non-angular features. [31] extracted a set of features from handwritten lines of text. The features extracted correspond to visible characteristics of the extracted feature score writing, such as the width, slant, and height of the three main writing zones. In [32], a new feature vector was employed by means of morphologically processing the horizontal profiles of the words. Because of the lack of a standard database for writer identification, a comparison of the previous studies is not possible. Because our purpose is to introduce an automatic method and because no limitation on handwriting is considered, methods that need no segmentation or connected-component analysis are regarded. Most previous studies are based on English documents with the assumption that the written text is fixed (text-dependent methods), and no research has been reported on English and Arabic texts or Arabic documents. In this paper, we propose a new method that is text-independent for off-line writer identification based on English/Arabic handwriting. Based on the idea that was presented in [23], we assume handwriting as a texture image, and a set of new features are extracted from preprocessed images of documents.

4 System overview of proposed generic approach for combined features

In this section, we present a system overview of our proposed approach. We then describe the dataset that was utilized to obtain the results. In the following, we give a description of the feature extraction and subsequently detail our proposed algorithm. The main experiment is discussed in this paper.

4.1 System overview

Figure 1 shows the handedness recognition system where the features are extracted from training and testing documents. The (K-NN) classifier [11] is then used to predict the handedness of the writer.

4.2 Description of the dataset

The dataset contains samples from 121 writers, which half of them are left-handed and the other half are right-handed. The dataset is a subset of the QUWI dataset, which was described in [33] in which 475 writers produced four handwritten documents: the first page contains an Arabic handwritten text that varies from one writer to another, the second page contains an Arabic handwritten text that is the same for all the writers, the third page contains an English handwritten text that varies from one writer to another, and the fourth page contains an English handwritten text that is the same for all the writers. left-handed writers were less represented in QUWI dataset. Only 121 writers were selected from this dataset to have evenly sampled data over the right- and left-handed writers. Figures 2 and 3 illustrate an example of the two pages. The images have been acquired using an EPSON GT-S80 scanner with a 600-DPI resolution. The images were provided in a JPG uncompressed format. The whole dataset of images and the corresponding features can be downloaded from the web site[1].

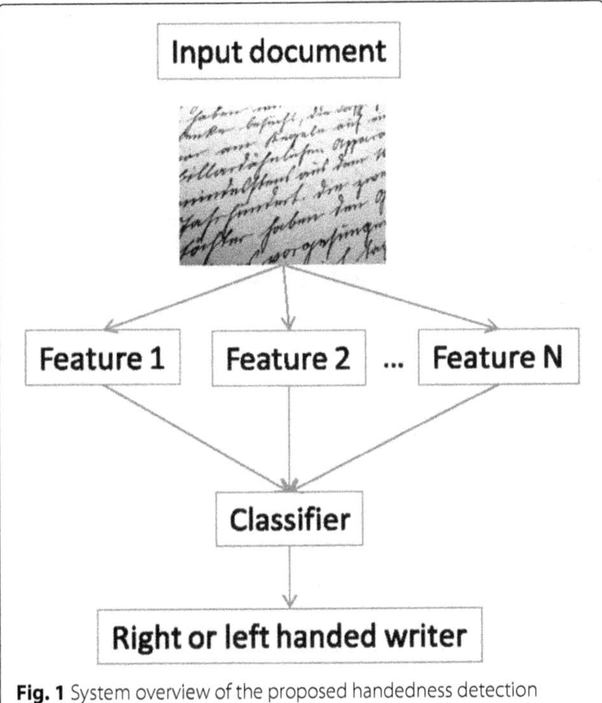

Fig. 1 System overview of the proposed handedness detection approach

4.3 Feature extraction

In this step, the characterizing features are extracted from the handwriting data. As a preprocessing step for feature f1, we calculate the Zhang skeleton of the binarized image. This algorithm is popular for not creating parasitic branches compared to most skeletonization algorithms. For some other features, the contours were calculated using Freeman chain codes. We then continue the extraction of features, which measure the direction of writing (f1), curvature (f2), tortuosity (f3), chain code feature (f4, 6, and 16), and edge-based directional features (f16), in order to compare the results. In addition, edge-based directional features using the whole window computed for size 7 (f18, whose PDF size is 112) and size 10 (f26, whose PDF size is 220) are also extracted. These features enable us to discriminate between left- and right-handed writers as will be explained in the following sections. Figures 2 and 3 show handwriting examples of left- and right-handed writers. Figure 4 shows some examples of feature extraction from preprocessed handwritten character.

To make the system independent of the pen, images are first binarized using the Otsu thresholding algorithm [34]. The following subsections describe the features considered in this study. It is to be noted that these features do not correspond to a single value but are defined by a probability distribution function (PDF) extracted from the handwriting images to characterize the writer's individuality [35, 36]. The PDF describes the relative likelihood for a certain feature to take on a given value.

4.3.1 *Directions (f1)*

We move along the pixels of the obtained segments of the skeleton using a predefined order favoring the four connectivity neighbors. For each pixel p, we consider the

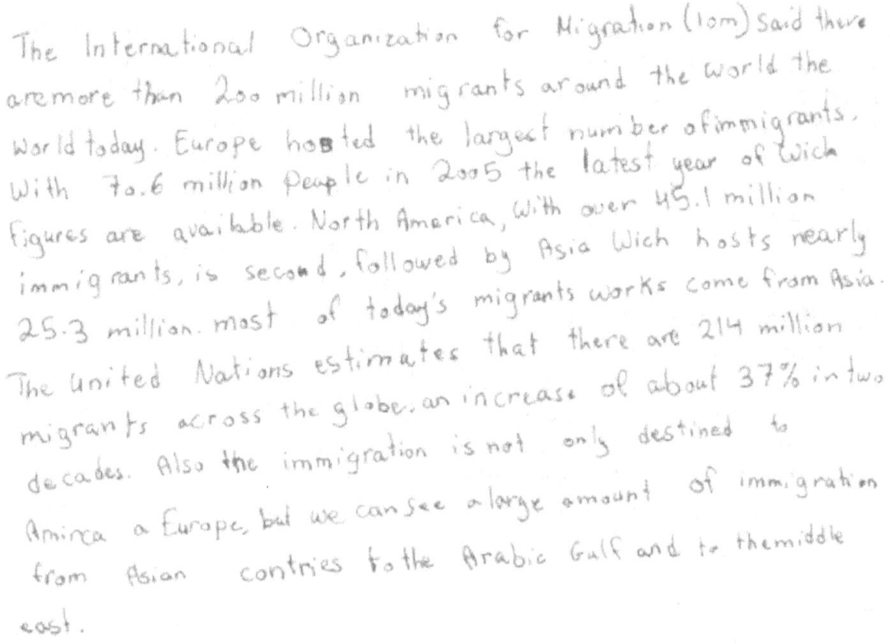

The International Organization for Migration (Iom) Said there
aremore than 200 million migrants around the world the
World today. Europe hosted the largest number ofimmigrants.
With 70.6 million People in 2005 the latest year of Wich
figures are available. North America, With over 45.1 million
immigrants, is second, followed by Asia Wich hosts nearly
25.3 million. most of today's migrants works come from Asia.
The United Nations estimates that there are 214 million
migrants across the globe. an increase of about 37% in two
decades. Also the immigration is not only destined to
Amirca a Europe, but we can see a large amount of immigration
from Asian contries to the Arabic Gulf and to themiddle
east.

Fig. 2 Typical example of right-handed writer

$2 \times N + 1$ neighboring pixels centered at p. The linear regression of these pixels is calculated to give the tangent at the pixel p [36].

The PDF of the resulting directions is computed as a probability vector for which the size has been empirically set to 10.

4.3.2 Curvatures (f2)

For each pixel p belonging to the contour, we consider a neighboring window, which has a size t. We compute the number of pixels n_1 inside this neighboring window belonging to the background and the number of pixels n_2 representing the foreground. Obviously, the difference $n_1 - n_2$ increases with the local curvature of the contour. We then estimate the curvature as being $C = \frac{n_1 - n_2}{n_1 + n_2}$. The PDF of the curvatures is computed as a vector whose size has been empirically set to 100 (s pixels in each side) [36].

4.3.3 Tortuosity (f3)

This feature makes it possible to distinguish between fast writers who produce smooth handwriting and slow writers who produce "ortuous"/twisted handwriting [36] by finding the longest line in the middle of the character shape. This feature has a PDF vector of 10. The PDF of the angles of the longest traversing segments are produced in a vector whose size has been set to 10.

4.3.4 Chain code features (f4-f7)

Chain codes are generated by browsing the contour of the text and assigning a number to each pixel according to its location with respect to the previous pixel. A chain code might be applied in different orders:

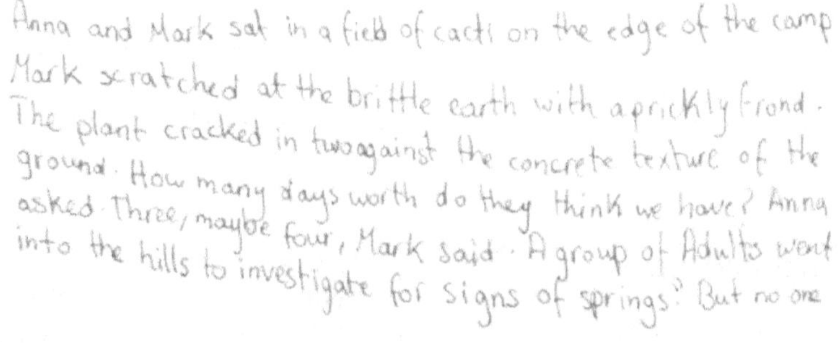

Anna and Mark sat in a field of cacti on the edge of the camp
Mark scratched at the brittle earth with a prickly frond.
The plant cracked in twoagainst the concrete texture of the
ground. How many days worth do they think we have? Anna
asked. Three, maybe four, Mark said. A group of Adults went
into the hills to investigate for signs of springs? But no one

Fig. 3 Typical example of left-handed writer

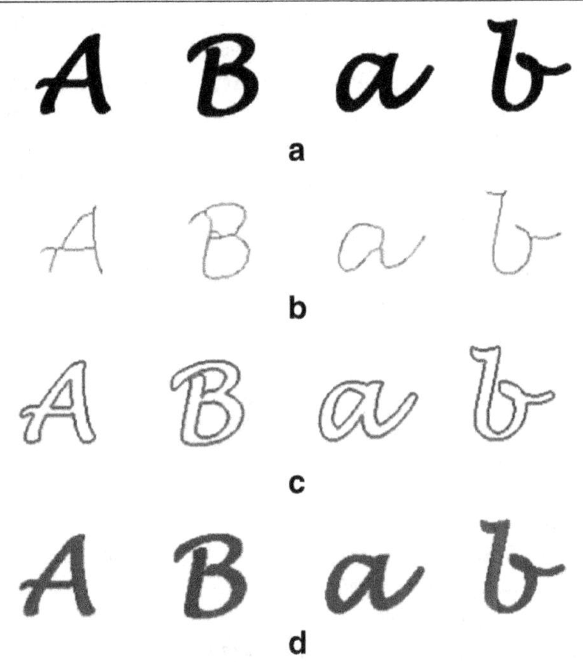

Fig. 4 Examples of some of feature extractions. **a** Original binary handwritten characters. **b** Direction feature f1 where the *red* color corresponds to a $\pi/2$ tangent, and the *blue* color corresponds to a zero tangent. **c** Curvature feature f2 highlighted in the binary image; *red* corresponds to the maximum curvature, and *blue* corresponds to the minimum curvature. **d** Tortuosity feature f3 where *red* corresponds to the maximum line segment that traverses a point and *blue* to the minimum length

f4: PDF of i patterns in the chain code list such that $i \in 0, 1, \ldots, 7$. This PDF has a size of 8.

f5: PDF of (i, j) patterns in the chain code list such that $i, j \in 0, 1, \ldots, 7$. This PDF has a size of 64. Similarly, **f6** and **f7** correspond to a PDF of (i, j, k) and (i, j, k, l) in the chain code list, where $i, j, k, l \in 0, 1, \ldots, 7$. Their respective sizes are 512 and 4096.

4.3.5 *Edge-based directional features (f8-f26)*

In this paper, this feature has been computed from size 1 (**f8**, whose PDF size is 4) to size 10 (**f17**, whose PDF size is 40). We have also extended these features to include the whole window. This feature has been computed from size 2 (**f18**, whose PDF size is 12) to size 10 (**f26**, whose PDF size is 220) (ref. Table 2).

Each contour $Contour_i$, being a sequence of consecutive boundary points, is computed as follows:

$Contour_i = \{p_j | j \leqslant M_i, p_1 = p_{Mi}\}$, where M_i is the length of $contour_i$.

In the following section, we will present the classifier used to predict the class of the set of features. Then, we define the different steps of the proposed algorithm.

Table 2 Overview of the implemented features

Feature	Description	Dimension
f1	Run-length distribution of white pixels in four directions	10
f2	Run-length distribution of black pixels in four directions	100
f3	Run-length distribution of white and black pixels in four directions	10
f4	Edge-direction distribution using 16 angles	8
f6	Polygon-based features	512
f16	Chain-code-based local features	36
f23	Codebook-based features	112
f26	AR-coefficient-based features	220

4.4 Proposed classifier

4.4.1 *(K-NN) Classifier*

The K-Nearest-Neighbor (K-NN) classifier is one of the most basic classifiers for pattern recognition and data classification. The principle of this method is based on the intuitive concept that data instances of the same class should be closer in the feature space. As a result, for a given data point x of an unknown class, we can simply compute the distance between x and all the data points in the training data and assign the class determined by the K nearest points of x. Due to its simplicity, K-NN is often used as a baseline method in comparison with other sophisticated approaches utilized in pattern recognition. The K-Nearest-Neighbor classification divides data into a test set and a training set. In our case, we choose $K = 5$ to be used in a sample of 128 writers for both English and Arabic texts.

The main task of classification is to use the feature vectors provided by the feature extraction algorithm to assign the object to a category [37]. In our work, we use the K-Nearest Neighbors (K-NN) for the classification of the

Table 3 Detailed accuracy for left and right handwriting: *fuzzy 2-combinations*

		Left and right handedness			
# of Att	δ	% Reduction	Precision	Recall	F-measure
f1-f2	0.995	17.7 %	65.69 %	66.86 %	66.27 %
f1-f3	0.85	*31.3 %*	69.47 %	69.40 %	*69.43 %*
f1-f4	0.92	20.0 %	62.81 %	62.80 %	62.80 %
f1-f23	0.995	34.3 %	64.55 %	64.43 %	64.48 %
f1-f6	0.995	5.9 %	63.55 %	61.93 %	62.73 %
f1-f26	0.998	36.9 %	67.23 %	66.89 %	67.05 %
f1-f16	0.993	23.8 %	63.85 %	63.58 %	63.71 %

extracted features. K-NN running on the Rapidminer platform [11] classifier classifies an unknown sample based on the known classification of its neighbors [38, 39]. Given unknown data, the K-Nearest-Neighbor classifier searches the pattern space for the K training data that are closest to the unknown data. These K training tuples are the K "Nearest Neighbors" of the unknown data. Typically, we normalize the values of each attribute. This helps to prevent attributes with initially large ranges from outweighting attributes with initially smaller ranges (such as binary attributes).

In this step, the features previously presented are used to predict the handedness of the writer of each document. When performing the classification, each element of the feature vector will be used as a separate input for the classifier (for example, *f1* will be an input vector of 10 elements for the classifier, as shown in Table 2). We have combined these features using a K-Nearest-Neighbor classifier [40]. A description of the combination of features using the (K-NN) classifier is given below.

4.4.2 *Proposed algorithm*

In this section, we give a description of the proposed algorithm based on a new heuristic approach for obtaining the best feature combination by applying the Lukasiewicz implication with variations of different values of δ. Only the best value of δ that provides the highest F-measure score was retained. The proposed approach is split into three sub-modules: (1) the main algorithm, which takes into consideration the input fuzzy binary relation for the features (f1, f2, f3, f4, f6, f16, f23 and f26) denoted $\mathcal{F}_{BR}$ (these features are presented in excel files in the form of fuzzy binary relations, where the rows represent the different writers based on their left/right handedness and the columns represent the values of features); (2) the remaining feature module processing that identifies the features to be rejected and the features that will be maintained according to the best value of δ that provides a high score of the F-measure and a considerable improvement in the data reduction percentage (which is accomplished by computing the closure of the Galois connection); and (3) the third sub-module, which determines the computed closure of the remaining attributes. In this case, we consider the fuzzy binary matrix relevant to a given feature F_i, where $\mathcal{O}_i$ represents an object and $A_0, A_1, \ldots, A_n$ represent the corresponding attributes. The rows represent the different writers (left and right handedness). We use 121 writers. The columns represent the measured values of the features. For instance, feature f26 describes the measured values of the "Polygon-based features" using 512 measures (i.e., 0.019815 represents a measure).

Step 1: Performing a matrix transposition; transpose the rows to the attributes and columns to the objects. Each row represents a different feature F_i, and the objects corresponding to 121 different writers with both left and right handedness are represented as columns.

Step 2: Choosing different arbitrary values of δ, we compute the closure of the attributes by using the following formula: $h_\delta \circ f(A) = \{A_0, A_1, \ldots, A_n\}$. The discovered redundant attributes are removed. Intuitively, an attribute is redundant if we can regenerate it by association from other attributes. Finally, we keep the last subset that contains the reduced subset of *Objects* $\times$ *Attributes* with the highest values of δ in terms of precision, recall and F-measure.

Algorithm 1: MAIN ALGORITHM

Input: FUZZY WORKING RELATION $\mathcal{F}_{BR}$
Output: REMAINEDFEATURES
begin
 Initialize the input fuzzy binary relation to the fuzzy working relation relation
 $\mathcal{F}_{BR} \leftarrow$ INITIAL CONTEXT
 We denote that R_i is denoted by x
 for *each feature x in the domain* $\mathcal{F}_{BR}$ **do**
  LISTCLOSE $\leftarrow$ COMPUTECLOSURE(x)
  //Compute the closure of x
  $S_x \leftarrow$ LISTCLOSE.CLOSURE ({X}) $- \{x\}$
  //Compute the closure of Sx $- \{x\}$
  $CL_{Sx} \leftarrow$
  LISTCLOSE.CLOSURE ($S_x - \{x\}) - \{x\}$
  if ($CLS_x \equiv S_x$) **then**
   //Remove feature x

Algorithm 2: COMPUTING CLOSURE

Input: VECTOR LISTSX
Output: LISTVECTOR
begin
 Compute the Galois connection of ListSx According to different values of δ
 VECTORMIN $\leftarrow$ GALOISF(*ListSx*)
 compute the closure of Galois connection of ListSx LISTVEC $\leftarrow$ GALOISH(*VectorMin*, δ)

In general, the sub-modules are composed of the following steps:

1. In the main algorithm, we determine for each row
   - The closure list "ListClose", which is denoted S_x and computed using the following formula: $\mathcal{H} \circ \mathcal{F}(x)$.

- The next step consists of removing from Listclose the redundant feature: The values of $\mathcal{F}_{\mathcal{BR}}$ (*i.e.,* $CL_{Sx} \leftarrow$ LISTCLOSE.CLOSURE $- (x)$.
- If CL_{Sx} is equivalent to S_x, then feature x is removed.

2. The second sub-module (Algorithm 2) consists of computing the Galois connection of $ListS_x$ according to the specified value of δ (i.e., CL_{Sx})
3. Another module may be added in order to update the context if CLS_x is equal to S_x.

To summarize, the algorithm falls into the following detailed steps:

- Compute the F-measure for each feature.
  1. Vary different values of δ (for example, 0.95), and generate the features that satisfy the Lukasiewicz implication according to the fixed value of δ;
  2. Choose the best results of the features and determine the percentage of reduction;
  3. Combine the feature with the highest score (e.g., the F-measure) with all the other features;
  4. Compute the F-measure for the combined two selected features; and
  5. Retain only the combined feature with the highest score.
- Repeat the steps above in a similar way for combinations of the next levels (3, 4 and so on) until no improvement is obtained.
- Select the combination of features with the highest F-measure score.

4.5 Results and their analysis

To evaluate our approach, we use the QUWI dataset presented earlier in this paper. We use standard evaluation metrics, including the precision, recall and F-measure (which is derived from the precision and recall [41]). In our experiments, we conduct the evaluation of the features after applying the feature reduction process using the Lukasiewicz implication for different values of δ (varying from 0..1). Recall that the precision of a class i is defined as

$$Precision = \frac{\#documents\ Correctly\ Classified\ into\ Class\ i}{\#of\ documents\ classified\ into\ Class\ i}$$

and the recall of class i is defined as:

$$Recall = \frac{\#documents\ Correctly\ Classified\ into\ Class\ i}{\#of\ documents\ that\ are\ truly\ in\ class\ i}$$

and then the F-measure, which reflects the relative importance of the recall versus precision, is defined as

$$\text{F-measure} = \frac{2 \times Precision \times Recall}{Precision + Recall}$$

$Accuracy = \frac{\alpha}{\beta}$ such that $\alpha = \alpha_1 + \alpha_2$ and $\beta = \alpha_1 + \alpha_2 + \beta_1 + \beta_2$ where

- $\alpha_1 =$ # of true documents correctly classified into $Class_i$
- $\alpha_2 =$ # of true documents incorrectly classified into $Class_i$
- $\beta_1 =$ # of false documents correctly classified into $Class_i$
- $\beta_2 =$ # of false documents incorrectly classified into $Class_i$

The precision, recall and F-measure metrics are used to evaluate our approach using English and Arabic texts. In this work, we have chosen the K-Nearest-Neighbors (K-NN) algorithm, which is widely used for classification, machine learning, and pattern recognition by data miners [42].

In the (K-NN) classifier, we have used a cross validation which is defined as follows:

- Divide training examples into two sets, a training set (95 %) and a validation set (5 %);
- Predict the class labels for the validation set by using the examples in the training set; and
- Choose the number of neighbors $K = 5$ that maximizes the classification accuracy.

Table 4 Detailed accuracy for left and right handwriting: *fuzzy 3-combinations*

		Left and right handedness			
# of Att	δ	% Reduction	Precision	Recall	F-measure
f1-f3-f2	0.98	32.3 %	67.81 %	67.75 %	67.77 %
f1-f3-f4	0.94	16.7 %	70.27 %	70.26 %	70.26 %
f1-f3-f23	0.995	*33.3 %*	71.99 %	71.87 %	*71.93 %*
f1-f3-f6	0.997	9.9 %	68.91 %	68.58 %	68.59 %
f1-f3-f26	0.997	19.6 %	71.99 %	71.87 %	71.93 %
f1-f3-f16	0.994	24.4 %	71.99 %	71.87 %	71.93 %

Table 5 Detailed accuracy for left and right handwriting: *fuzzy 5-combinations*

		Left and right handedness			
# of Att	δ	% Reduction	Precision	Recall	F-measure
f1-f3-f23-f16-f2	0.995	8.9 %	66.94 %	66.94 %	66.94 %
f1-f3-f23-f16-f4	0.992	32.4 %	66.12 %	66.12 %	66.12 %
f1-f3-f23-f16-f6	0.990	11.7 %	70.27 %	70.23 %	70.25 %
f1-f3-f23-f16-f26	0.995	*31.3 %*	71.99 %	71.87 %	*71.93 %*

Table 6 Detailed accuracy for left and right handwriting: *fuzzy 6-combinations*

Left and right handedness					
# of Att	δ	% Reduction	Precision	Recall	F-measure
f1-f3-f23-f16-f26-f2	0.990	29.7 %	66.15 %	66.09 %	66.12 %
f1-f3-f23-f16-f26-f4	0.997	18.8 %	66.97 %	66.95 %	66.96 %
f1-f3-f23-f16-f26-f6	0.992	*20.6 %*	69.47 %	69.40 %	*69.43 %*

4.5.1 Experiments with no feature reduction

The classification is carried out separately for the Arabic and English languages in a first step and jointly in a second step. The results are reported for the case of similar texts written by all the writers and different texts for each writer. In the following, we present the results of the classification at the end of each iteration:

First iteration: we compute the F-measure of the features separately. We present (1) the reduction percentage and (2) the improvement of the F-measure through application of the Lukasiewicz implication.

Second iteration: it is clear that feature f1 has the highest F-measure (i.e., 70.73 %), with a 20 % data reduction percentage. Therefore, we have combined f1 with each feature, a combination of two features (i.e., f6, f4, f3 and so on) and the results obtained are shown in Table 3. The combination strategy proves that combining a feature with a low recognition rate can provide a good result while reducing the amount of data. When we combine f1 with f3, the recognition rate drops by one point, but a data reduction of 31.3 % is obtained. This is almost 13 % improvement than the first result with no combination.

The highest F-measure score was 69.43 % for the two combined features f1 and f3 with a data reduction of 31.3 %, Table 3. Therefore, we have combined the latter with the other remaining features (e.g., f2 and f4). This prpocess is continued until the end where we only select the combination having the highest F-measure with the respective high-reduction percentage. The highest selected F-measure is marked in italics, as shown in the next Tables 4, 5, 6, 7, and 8.

The remaining classification rates obtained in each evaluation are given in Tables 4 through 8 for three to eight combinations, respectively. With the three combinations

Table 7 Detailed accuracy for left and right handwriting: *fuzzy 7-combinations*

Left and right handedness					
# of Att	δ	% Reduction	Precision	Recall	F-measure
f1-f3-f23-f16-f26-f4-f2	0.995	*4.6 %*	67.01 %	66.91 %	*66.96 %*
f1-f3-f23-f16-f26-f4-f6	0.990	30.7 %			66.94 %

Table 8 Detailed accuracy for left and right handwriting: *fuzzy 8-combinations*

Left and right handedness					
# of Att	δ	% Reduction	Precision	Recall	F-measure
f1-f3-f23-f16-f26-f4-f6-f2	0.995	*5.2 %*	68.77 %	68.55 %	*68.66 %*

as depicted in Table 9, it can be seen that further improvements are achieved in the F-measure rates, with the best performance obtained when we combine f1, f3, and f23. Furthermore, the reduction rate improved by 3 % compared to the best result using two combinations. In the case of four combinations, (e.g., Table 4), the result of the F-measure remains almost the same as when using three combinations but with further reductions using f1-f3-f23-f4 and f1-f3-f23-f16.

The combined features f1-f3-f23-f16-f26 yield the highest score (F-measure is approximately equivalent to 71.93 %). Therefore, this feature will be combined with the other features (i.e., f26-f4, f26-f23, f26-f16, f26-f3, f26-f6, and f26-1). We continue the combination process until reaching the point where no possible combination of features that obtain a higher F-measure score is possible. In the previous tables, this is shown by the recognition rates and the reduction rates which start to decrease. Finally, the F-measure slightly improves with the 8-combination with only a 5 % reduction rate.

4.5.2 Summary of experiments using the Lukasiewicz implication

The scores, as shown in Table 5, are computed. The top scores for the combined features approach were Precision = 71.99 %, Recall = 71.87 %, F-measure = 71.93 %, and Accuracy = 71.86 +/- 7.74 % (71.90 %). The data reduction percentage was 31.3 %. This represents a considerable data reduction ability relative to the whole volume of data with an interesting improvement in the precision and recall metrics. Interestingly, computing confidence intervals for these results yields a confidence interval of (+/-5 %). Thus, we conclude that our approach

Table 9 Detailed accuracy for left and right handwriting: *fuzzy 4-combinations*

Left and right handedness					
# of Att	δ	% Reduction	Precision	Recall	F-measure
f1-f3-f23-f2	0.995	12.1 %	66.96 %	66.93 %	66.94 %
f1-f3-f23-f4	0.992	46.8 %	67.02 %	66.97 %	66.99 %
f1-f3-f23-f6	0.995	33.3 %	71.99 %	71.87 %	71.93 %
f1-f3-f23-f26	0.997	16.2 %	71.99 %	71.87 %	71.93 %
f1-f3-f23-f16	0.992	*35.0 %*	71.99 %	71.87 %	*71.93 %*

Table 10 Different combinations

Left and right handedness	
Combination	Features selected
1-combination	f6
2-combination	f1-f3
3-combination	f1-f3-f23
4-combination	f1-f3-f23-f16
5-combination	f1-f3-f23-f16-f26
6-combination	f1-f3-f23-f16-f26-f6
7-combination	f1-f3-f23-f16-f26-f4-f2
8-combination	f1-f3-f23-f16-f26-f4-f6-f2

improves the quality of the obtained features and contributes enormously to the reduction of the number of features. Figure 5 results of different combined features.

4.5.3 *Key findings*

In the following, we provide a summary of the obtained results. We show the best results for each combination. We then graph these results in an appropriate figure. Finally, we comment on the results.

Discussions: Based on the conducted experiments, we provide the following remarks. The most striking feature is that according to the obtained scores, the highest number of labels is explored using the combined features f1-f3-f23-f16-f26 (the F-measure is approximately equal to 71.93 %).

- The accuracy is reasonable (69.78 % on average) in all approaches except for the combination f1-f3 (69.47 %). This is due to the quality of the features (the accuracy of f1 was 70.58 %, and the accuracy of f3 was 62.85 %). It reaches its maximum for the following combination (approximately 71.99 %): f1-f3-f23-f16-f26 (Table 10).
- The recall attains its highest values (71.87 %) for the combination f1-f3-f23-f16 and the lowest value (69.60 %) for the combination f1-f3-f23-f16-f26-f4-f2. It is clear that the features f26-f4-f2 did not improve the score. On average, the score was 69.60 %.
- The F-measure reaches its highest values (71.93 %) for the combination f1-f3-f23-f16 and the lowest value (69.69 %) for the combination f1-f3-f23-f16-f26-f4-f2. It is clear that the same features f26-f4-f2 did not improve the score. On average, the score was 69.69 %.
- The reduction percentage reaches its maximum (83.43 %) for the feature F6 alone and its minimum of 4.57 % for the feature f1-f3-f23-f16-f26-f4-f2, while on average, the reduction percentage was 30.36 %); and
- Finally, if one takes into consideration the highest F-measure with an improvement in the reduction percentage, it is clear that using f1-f3-f23-f16 with the F-measure (71.93 %) results in a percent reduction of 31.3 %, which emphasized a considerable improvement in dimensionality reduction of the features. Tables 11, 12 and 13.

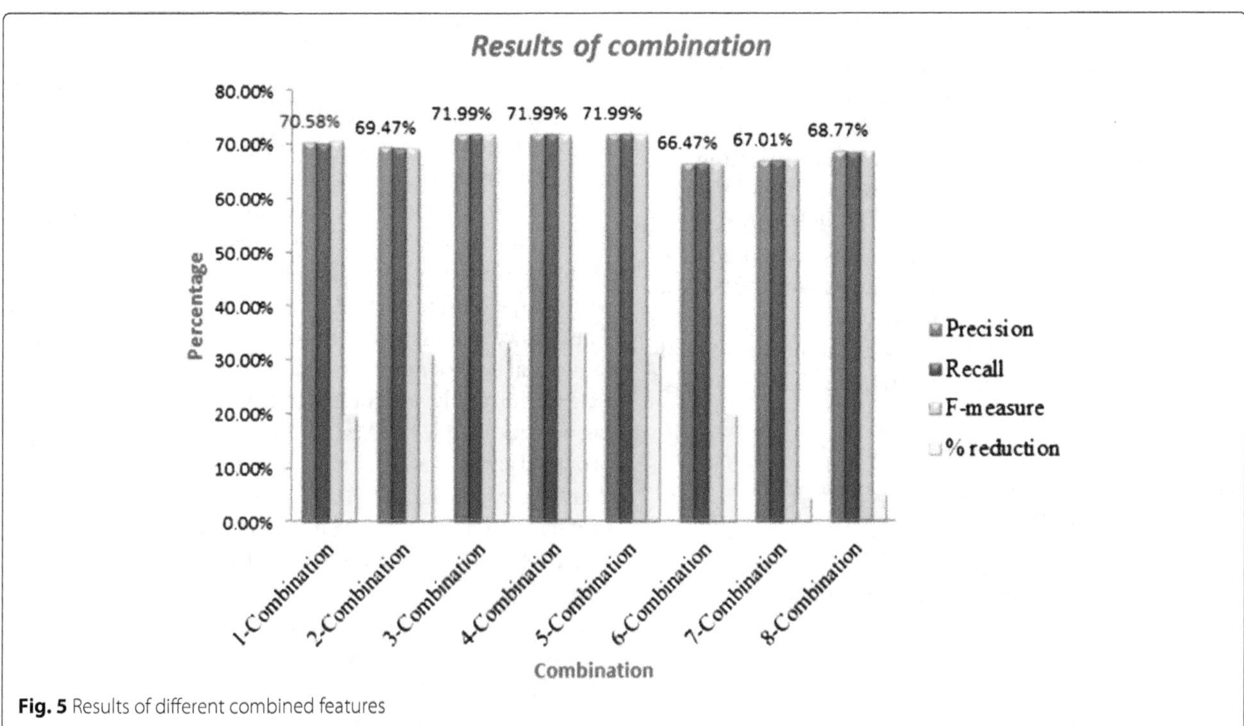

Fig. 5 Results of different combined features

Table 11 Detailed accuracy for left- and right-handwriting: *fuzzy 1-combination*

Left and right handedness					
# of Att	δ	% Reduction	Precision	Recall	F-measure
f26	0.997	74.6 %	66.12 %	66.12 %	66.12 %
f6	0.992	83.43 %	70.07 %	69.35 %	69.71 %
f4	0.92	33.3 %	68.67 %	68.57 %	68.61 %
f3	0.851	33.3 %	62.85 %	62.83 %	62.84 %
f16	0.998	5.6 %	59.55 %	59.47 %	59.50 %
f1	0.885	*20.0 %*	70.58 %	70.19 %	*70.73 %*
f23	0.999	13.4 %	68.00 %	67.72 %	67.86 %
f2	0.995	32.6 %	45.40 %	45.43 %	45.41 %

Table 13 Summary of the highest obtained results using different combinations

Left and right handedness					
Combination	δ	% Reduction	Precision	Recall	F-measure
1-combination	0.885 %	20 %	70.58 %	70.19 %	70.73 %
2-combination	0.85 %	31.3 %	69.47 %	69.40 %	69.43 %
3-combination	0.995 %	33.3 %	71.99 %	71.87 %	71.93 %
4-combination	0.992 %	35 %	71.99 %	71.87 %	71.93 %
5-combination	0.995 %	31.3 %	71.99 %	71.87 %	71.93 %
6-combination	0.992 %	20 %	66.47 %	66.40 %	69.43 %
7-combination	0.995 %	4.6 %	67.01 %	66.91 %	66.96 %
8-combination	0.995 %	5.2 %	68.77 %	68.55 %	68.66 %

4.6 Computational complexity reduction

We consider the features f1, f2, f3, f4, f6, f16, f23, and f26 for the computation of the complexity analysis of our approach. Thus, the computational complexity of the determination of the best combined features, for the previous n features, is determined as follows, as the objects correspond to features and attributes to writers. So, the time complexity is definitively $\mathcal{O}(n \star m^2)$ where n is the number of writers and m the number of features. As a matter of fact, we are calculating the closure of each one of the m features, where the closure requires m x n comparisons, where n is the number of writers.

5 Conclusion and future work

We have proposed a new generic approach for combined feature extraction based on a successive combination of the best feature with the highest score with every other feature. We plan to apply this approach to many applications including gender, age, and nationality prediction. The goal of this research consists of investigating the text-independent identification of a script writer. We have employed a set of features (e.g., f1, f2, f3, f4, f6, f16, f23, and f26), which have shown promising results on a database of handwritten documents in two different languages: Arabic and English. The evaluations were carried out on the only existing database of its type, containing short writing samples from 121 different writers. The results of the determination of the best combined feature identification are very encouraging (the F-measure is approximately equal to 71.08 %). They reflect the effectiveness of the run-length features in a text-independent script environment and validate the hypothesis put forward in this research, i.e., that the writing style remains approximately the same across different scripts. It is also worth mentioning that, unlike most of the studies that use complete pages of text, our results are based on a limited amount of handwritten text, which is more realistic. Another interesting aspect of this study was the evaluation and comparison of a number of state-of-the-art methods on this dataset. The features used in these methods naturally show a decrease in performance when exposed to different script scenarios. In all cases, the run-length features outperform these features. Finally, for the comparison of the proposed method with other methods, the average correct handedness detection results are over 83.43 %, which exceeds the results reported in [30] for off-line gender identification (70 %) on the same dataset. The results also compare well with the 73 %; 55.39 % reported for gender classification in [36, 43] on different datasets. It would be interesting to evaluate these features on a larger dataset with a large number of writers and many scripts per writer. This, however, involves the challenging task of finding individuals who are familiar with multiple scripts. To extend this study, we intend to utilize a database including writing samples in Arabic, French, and other languages provided by the same writer. In addition, classifiers other than those discussed in this paper can be evaluated to find out how they perform in a many-script environment. The proposed approach can also be extended to include a rejection threshold to reject writers that are not a part of the database. Finally, it would be interesting to apply a feature selection strategy to reduce the dimension of the proposed feature set and to study which subset of features is the most discriminative in characterizing the writers.

Table 12 Detailed accuracy for left and right handwriting using the Lukasiewicz implication

Left and right handedness					
# of Att	δ	% Reduction	Precision	Recall	F-measure
f1-f3 f23-f16 f26	0.995	*31.3 %*	71.99 %	71.87 %	*71.93 %*

The application of Latent Semantic Analysis (LSA) techniques seems promising regarding the reduction of the volume of data. These aspects will constitute the focus of our future research on writer recognition. With regard to data reduction, in our future work, we would like to investigate reducing the high-dimensional data of features gathered from user-cognitive loads, which results from the density of data to be visualized and mined, and reducing the dimensionality of the dataset while associations (or dependencies) between objects as applied to writer identification. This dimensionality reduction will be based on fuzzy conceptual reduction through the application of the Lukasiewicz implication.

Endnote

[1]https://www.kaggle.com/c/icdar2013-gender-prediction-from-handwriting.

Competing interests

The authors declare that they have no competing interests.

Acknowledgments

This publication was made possible by a grant from the Qatar National Research Fund NPRP 09-864-1-128. Its contents are solely the responsibility of the authors and do not necessarily represent the official views of the QNRF.

References

1. A Hassaine, S Al Maadeed, J Aljaam, A Jaoua. ICDAR 2013 Competition on Gender Prediction from Handwriting, Twelfth International Conference on Document Analysis and Recognition ICDAR2013 (IEEE, Washington, DC, 2013). https://www.computer.org/csdl/-abs.html
2. S Impedovo, L Ottaviano, S Occhinegro, Optical character recognition. Int.J. Pattern Recognit. Artif. Intell. **5**(1-2), 1–24 (1991)
3. VK Govindan, AP Shivaprasad, Character recognition a review. Pattern Recognit. **23**(7), 671–683 (1990)
4. R Plamondon, SN Srihari, On-line and off-line handwritten character recognition: a comprehensive survey. IEEE Trans. Pattern Anal. Mach. Intell. **22**(1), 63–84 (2000)
5. N Arica, F Yarman-Vural, An overview of character recognition focused on off-line handwriting. IEEE Trans. Syst. Man Cybernet. Part C: Appl. Rev. **31**(2), 216–233 (2001)
6. U Bhattacharya, BB Chaudhuri, Handwritten numeral databases of Indian scripts and multistage recognition of mixed numerals. IEEE Trans. Pattern Anal. Mach. Intell. **31**(3), 444–457 (2009)
7. K Bandi, SN Srihari, in *Proceedings of the International Graphonomics Society Conference (IGS)*. Writer demographic identification using bagging and boosting (Publisher International Graphonomics Society (IGS), 2005), pp. 133–137. http://www.graphonomics.org/publications.php
8. S Srihari, SH Cha, H Arora, S Lee, in *Proceedings of the Sixth International Conference on Document Analysis and Recognition*. Individuality of handwriting: a validation study (IEEE, 2001), pp. 106–109
9. M Liwicki, A Schlapbach, P Loretan, H Bunke, in *Proceedings of the 13th Conference of the International Graphonomics Society*. Automatic detection of gender and handedness from on-line handwriting (Publisher International Graphonomics Society (IGS), 2007), pp. 179–183. http://www.graphonomics.org/publications.php
10. M Liwicki, A Schlapbach, H Bunke, Automatic gender detection using on-line and off-line information. Pattern. Anal. Appl. **14**, 87–92 (2011)
11. KDnuggets, Data integration, analytical ETL, data analysis, and reporting, rapid miner journal (2012). Software available at, http://sourceforge.net/projects/rapidminer/
12. B Ganter, R Wille, *Formal Concept Analysis*. (Springer-Verlag, Berlin Heidelberg, 1999), p. 283
13. L Wang, in *Proceedings Part I of the Second International Conference, (FSKD 2005), Changsha, China*. Fuzzy Systems and knowledge discovery (Springler-Verlag Berlin, Heidelberg, 2005), pp. 515–519. ISBN 10 3-540-28312-9
14. B Ganter. *Two basic algorithms in concept analysis*. Preprint 831, Technische, (Hochschule Darmstadt, Germany, 1984)
15. G Birkhoff. Lattice Theory First edition, Providence: American. Mathemathics Society (Springler-Verlag Berlin, Heidelberg, 2005), pp. 515–519. ISBN 10 3-540-28312-9
16. LA Zadeh, Fuzzy sets, information and control. **8**, 338–353 (1965)
17. S Elloumi, J Jaam, A Hasnah, A Jaoua, I Nafkha, A multi-level conceptual data reduction approach based on the Lukasiewicz implication. Inf. Sci. **163**, 253–262 (2004)
18. J Riguet, Lattice Theory First edition, Relations binaires, fermetures et correspondances de Galois. Bull.Soc. Math. France. **78**, 114–155 (1948)
19. R Belohlavek, Fuzzy Galois connections. Math. Logic Quart. **45**, 497–504 (1999)
20. R Belohlavek, Lattices of fixed points of Galois connections. Math.Logic Quart. **47**, 111–116 (2001)
21. A Frascella, Lattice Theory First edition, Fuzzy Galois connections under weak conditions, fuzzy sets and systems. **172**, 33–50 (2011)
22. SN Srihari, H Arora, SH Cha, S Lee, Individuality of handwriting. J.Forensic Sci. **47**(40), 1–17 (2002)
23. HE Said, TN Tan, KD Baker, Personal identification based on handwriting. Pattern Recognit. **33**(1), 149–160 (2000)
24. A Bensefia, T Paquet, L Heutte, A writer identification and verification system. Pattern Recognit. Lett. **26**(13), 2080–2092 (2005)
25. M Bulacu, L Schomaker, L Vuurpijl, in *Seventh International Conference on Document Analysis and Recognition*. Writer identification using edge-based directional features (IEEE, 2003), pp. 937–941
26. A Schlapbach, H Bunke, in *9th Int.Workshop on Frontiers in Handwriting Recognition*. Using HMM based recognizers for writer identification and verification (IEEE, 2004), pp. 167–172
27. M Bulacu, L Schomaker, Text-independent writer identification and verification using textural and allographic features. IEEE Trans. Pattern Anal. Mach. Intell. **29**(4), 701–717 (2007)
28. I Siddiqi, N Vincent, Text independent writer recognition using redundant writing patterns with contour-based orientation and curvature features. Pattern Recognit. **43**(11), 3853–3865 (2010)
29. U Pal, T Wakabayashi, F Kimura, in *Ninth International conference on Document Analysis and Recognition ICDAR 07*. Handwritten numeral recognition of six popular scripts, vol. 2 (IEEE Computer Society, Washington, DC, USA, 2007), pp. 749–753. ISBN:0-7695-2822-8
30. S Al Maadeed, F Ferjani, S Elloumi, Hassaine Ai, Jaoua A, in *2013 IEEE GCC Conference and exhibition, November 1720, Doha, Qatar*. Automatic handedness detection from off-line handwriting (IEEE, 2013), pp. 119–124. ISBN: 978-1-4799-0722-9
31. UV Marti, R Messerli, H Bunke, in *proceedings of Sixth International Conference on Document Analysis and Recognition*. Writer identification using text line based features (IEEE Computer Society, Washington, DC, USA, 2001), pp. 101–105
32. EN Zois, V Anastassopoulos, Morphological waveform coding for writer identification. Pattern Recognit. **33**, 385–398 (2000)
33. S Al-Madeed, W Ayouby, A Hassaine, J Aljaam. QUWI: An Arabic and English handwriting dataset for off-line writer identification, *International Conference on Frontiers in Handwriting Recognition* (IEEE, 2012), pp. 746–751. ISBN: 978-1-4673-2262-1
34. N Otsu, A threshold selection method from gray-level histograms. Automatica. **11**, 285–296 (1975)
35. A Hassaine, S Al-Maadeed, J AlJaam, A Jaoua, A Bouridane. The ICDAR2011 Arabic writer identification Contest, *Proc. Eleventh International Conference on Document Analysis and Recognition,Beijing, China* (IEEE, 2011)
36. S Al Maadeed, A Hassaine, Automatic prediction of age, gender, and nationality in offline handwriting. EURASIP J Image Video Process. **2014**, 10 (2014)
37. RO Duda, PE Hart, DG Stork, *Pattern Classification*, second edition. (John Wiley & Sons Inc., New York, 2000)

38. BV Dasarathy, *Nearest Neighbor: Pattern Classification Techniques*. (IEEE Computer Society Press, New York, 1990)
39. Y Yang, in *Proc.17th Annual Intl. ACM SIGIR Conf. Research and Development in Information Retrieval, Dublin (Ireland)*. Expert network: effective and efficient learning from human decisions in text categorization and retrieval (ACM, 1994), pp. 13–22
40. L Breiman, Random forests. Mach. Learn. **45**(11), 5–32 (2001)
41. G Salton, J Michael, *An Introduction to Modern Information Retrieval*. (McGraw-Hill, New York, 1983)
42. KDnuggets, Data integration, analytical ETL, data analysis, and reporting, rapid miner (2012). Software available at http://sourceforge.net/projects/rapidminer/
43. M Liwicki, A Schlapbach, H Bunke, Automatic gender detection using on-line and off-line information. Anal. Appl. **14**, 87–92 (2011)

8

Adaptive dualISO HDR reconstruction

Saghi Hajisharif*, Joel Kronander and Jonas Unger

Abstract

With the development of modern image sensors enabling flexible image acquisition, single shot high dynamic range (HDR) imaging is becoming increasingly popular. In this work, we capture single shot HDR images using an imaging sensor with spatially varying gain/ISO. This allows all incoming photons to be used in the imaging. Previous methods on single shot HDR capture use spatially varying neutral density (ND) filters which lead to wasting incoming light. The main technical contribution in this work is an extension of previous HDR reconstruction approaches for single shot HDR imaging based on local polynomial approximations (Kronander et al., Unified HDR reconstruction from raw CFA data, 2013; Hajisharif et al., HDR reconstruction for alternating gain (ISO) sensor readout, 2014). Using a sensor noise model, these works deploy a statistically informed filtering operation to reconstruct HDR pixel values. However, instead of using a fixed filter size, we introduce two novel algorithms for adaptive filter kernel selection. Unlike a previous work, using adaptive filter kernels (Signal Process Image Commun 29(2):203–215, 2014), our algorithms are based on analyzing the model fit and the expected statistical deviation of the estimate based on the sensor noise model. Using an iterative procedure, we can then adapt the filter kernel according to the image structure and the statistical image noise. Experimental results show that the proposed filter de-noises the noisy image carefully while well preserving the important image features such as edges and corners, outperforming previous methods. To demonstrate the robustness of our approach, we have exploited input images from raw sensor data using a commercial off-the-shelf camera. To further analyze our algorithm, we have also implemented a camera simulator to evaluate different gain patterns and noise properties of the sensor.

Keywords: HDR reconstruction, Single shot HDR imaging, DualISO, Statistical image filtering

1 Introduction

The range of radiance intensities found in most real-world scenes, spanning from the sun or direct light sources to areas in shadow, typically exceeds, by orders of magnitude. It is very difficult to accurately capture this wide range using a digital sensor in a single image or video frame. This limitation has spurred the development of techniques for capture of *high dynamic range* (HDR) images and video; for an overview, see [26].

We present two algorithms for HDR image reconstruction based on a single input image where the pixel gain is varied over the sensor [4, 10]. Similar to [34, 35], we use the per-pixel gain of the analog signal, pixel measurements, to increase the dynamic range in the captured image. The analog pixel gain is proportional to the *ISO* setting found on most cameras. The input to our algorithm is a RAW sensor image consisting of pixels with either a high or a low gain setting, for example, varying the gain by every other two rows. The low gain setting enables the capturing of high-intensity region without saturation, while the high-gain setting enables us to capture image with a high signal-to-noise ratio in darker areas of the scene. Without loss of generality, we assume that color is captured using a *color filter array* (CFA), e.g., a Bayer pattern overlaid on the image sensor. Figure 1 illustrates two different distributions of per-pixel gain settings overlaid onto a raw CFA image. This approach to HDR capture is very robust and can be applied to off-the-shelf consumer cameras [4]. It does not suffer from, e.g., the various motion blurs or ghosting artifacts found in the commonly used exposure bracketing methods [7, 12]. Compared to multi-sensor cameras, e.g., [16, 31], it

* Correspondence: saghi.hajisharif@liu.se
Linköping University, Norrköping, Sweden

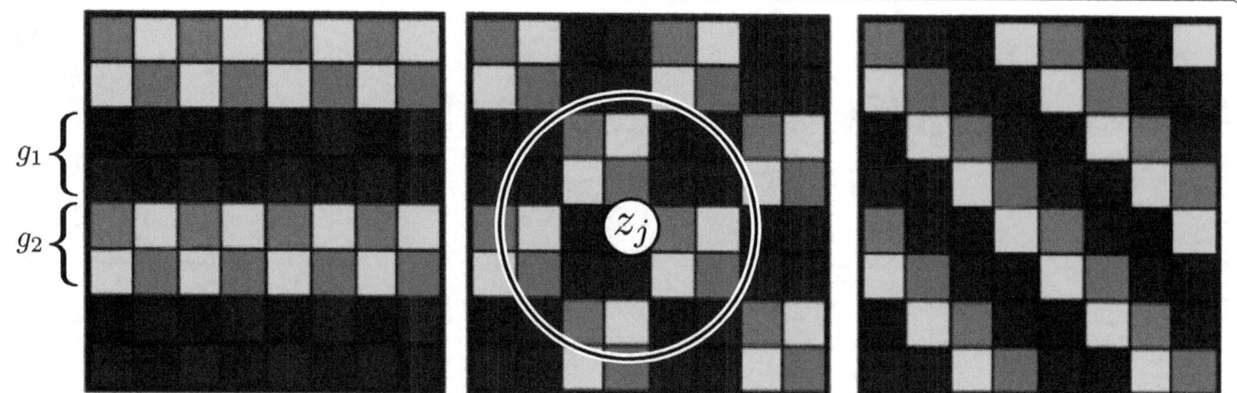

Fig. 1 Illustrates three different gain patterns, with two different gain settings (ISO), for a sensor with a Bayer pattern CFA, and (*middle*) how the multiple gain pixels are filtered to reconstruct the HDR output value z_j at pixel location X_j. The different gains, g_1 and g_2, corresponding to, e.g., 1× and 16×, amplification of the analog readout enables the capture of a wider range of intensities and extends the dynamic range in the final image

does not require costly specialized hardware and removes the requirement of careful geometric sensor calibration and the risk of misalignment between the exposures.

The main contribution in this paper is an extension of the previous statistical reconstruction method for dualISO data developed in [10, 15], using two novel algorithms for adapting the scale of the filtering window. In contrast to previous works [10, 15], the window support is adapted both to the statistical properties of the image noise as well as the underlying signal structure contained in the image. We show that the novel scale selection results in increased image quality in several examples.

2 Background

Since the seminal work by Devebec and Malik [7], a large body of work has developed more robust and higher quality HDR capture and reconstruction methods; for a complete overview, see, e.g., [23, 26]. In this section, we give an overview of the previous work most closely related to the methods proposed in this paper.

2.1 HDR capture

High-quality HDR capture using off-the-shelf image sensors can currently be performed with three distinct approaches.

The traditional approach captures a sequence of images with varying exposure times and then merges these into an HDR image [7, 12]. For dynamic scenes, non-rigid registration of the individual exposures is necessary; and for moving objects, general de-ghosting algorithms are necessary to apply for high-quality results. While there has been a large body of work improving these approaches, see, e.g., the survey [33], they still cannot robustly handle moving cameras and objects in general scenes.

The second approach to HDR capture is based on using beam splitters to project incident light onto multiple sensors with different exposures. The different exposures can be achieved by using varying neutral density (ND) filters in front of the sensors [1, 8, 16, 19] or by clever setups of semi-transparent beam splitter arrangements [31]. These systems offer a major advantage over exposure time fusion methods in that they robustly handle motion of the camera and objects in the scene by using the same exposure time for each sensor.

The third approach, which is most closely related to this work, is spatial multiplexing of the image to achieve HDR capture. Here, a single sensor image is used where the response to incident light varies over the sensor. Most previous works achieved this by placing a spatially varying array of ND filters in front of the sensor [2, 24, 25, 27]. Its most familiar application is color imaging via a color filter array (e.g., the Bayer pattern [6]). By avoiding the need for more than one sensor, this design provides a cost-effective solution to achieve robust HDR capture. However, most existing methods still suffer from noise as large portions of the incident light are wasted in the ND filters. By instead focusing on spatially multiplexing the response to incident light using the gain/ISO setting, we can use the entire incident light for high-quality HDR reconstruction.

2.2 HDR reconstruction

To reconstruct HDR images from a set of images with different exposures, the traditional method is to compute a per-pixel weighted average of the low dynamic range (LDR) measurements. The weights, often based on heuristics, are chosen to suppress image noise and remove saturated values from processing [3, 7, 20]. Mann and Picard [20] assigned weights according to the derivative of the inverse camera response, and later Debevec and Malik [7] used a simple double ramp function that excludes values close to the saturation point or the black level. Later work derived weight functions based on more sophisticated camera noise models. Mitsunaga and Nayar [22] derived a weight function that maximizes SNR assuming signal-independent additive noise, and Kirk and Andersen [13] derived a weight function inversely proportional to the temporal variance of the digital LDR values. Granados et al. [9] extended this approach to include both spatial and temporal camera noises. While most previous methods consider only a single pixel at a time from each LDR exposure, Tocci et al. [31] presented an algorithm that incorporates a neighborhood of LDR samples in the reconstruction.

The vast majority of previous HDR reconstruction algorithms treat the complete imaging pipeline from raw pixel measurements to a full HDR image in a series of steps [7, 9, 31], either performing demosaicing after or before HDR fusion and denoising. In this work, we instead treat all of these operations in a single joint filtering operation. This enables us to take sensor noise into account in a systematic fashion while also improving the reconstruction speed. Recently, Heide et al. [11] proposed a framework for joint demosaicing, denoising, and HDR assembly by solving an inverse problem with different global image priors and regularizers using convex optimization methods. While providing impressive results, their method does not incorporate a well-founded model of the heterogeneous sensor noise, and despite GPU implementations, their implementation is still computationally expensive which requires solving a global optimization problem. Instead, we take a local approach, enabling rapid parallel processing, while also incorporating a well-founded statistical noise model.

Our statistically motivated locally adaptive filtering framework is inspired by recent methods in image processing. The last two decades have seen an increased popularity of image processing operations using locally adaptive filter weights, for applications in, e.g., interpolation, denoising, and upsampling. Examples include normalized convolution [14], the bilateral filter [32], and moving least squares [17]. Recently, deep connections have been shown [21, 29] between these methods and traditional non-parametric statistics [18]. In this paper, we extend the earlier framework for HDR reconstruction developed in [10, 15, 16] based on fitting local polynomial approximations (LPA) [5] to irregularly distributed samples around output pixels using a localized maximum likelihood estimation [30] to incorporate the heterogeneous noise of the samples. In contrast to the previous works [10, 15, 16], we propose a novel adaptation of the filter kernel size that allows the filter extent to adapt not only to local image structure but also the sensor noise in the region.

3 DualISO capture and reconstruction—overview

The goal of the algorithm presented in this paper is to generate an HDR image based on input data in which the per-pixel gain (ISO) is varying over the sensor. This means that the analog readouts are amplified differently between segments of pixels on the sensor. Figure 1 illustrates three different gain patterns with two different gain values, g_1 and g_2, using a sensor with a Bayer pattern *color filter array* (CFA). The unity gain, g_1, pixel segments capture the high-intensity regions in the scene while the amplified segments, g_2, capture low-intensity regions. g_2 pixels may lie well below the acceptable noise floor for g_1 pixels.

The key benefit of using a varying per-pixel gain, g_i, is that the dynamic range in the final output will be extended using a single image as an input [10, 11]. However, accurate reconstruction of the output HDR image is a challenging filtering problem. The different gain settings lead to a loss of data in the spatial domain due to the fact that the amplified pixels, using gain g_2, saturate faster. For high-quality reconstruction, it is also necessary to take into account the heterogeneous image noise, which for a specific camera and exposure setting, varies with both intensity and the choice of gain settings.

The method presented here extends the statistical HDR reconstruction developed by [15, 16] to include reconstruction kernels which adapts to both the image content and the heterogeneous measurement noise. We assume that the input data is a raw CFA sensor image with per-pixel gain settings varying between pixel segments as described in Fig. 1 (middle). Each pixel value, z_j, at a pixel coordinate, X_j, in the output HDR image is, for each color channel, reconstructed by filtering the input pixels within a neighborhood around X_j. Our statistical approach first estimates the variance, or measurement noise, for each input sample in the raw image using a noise model. The input samples are then weighted using the estimated variances and an adaptive Gaussian kernel in the spatial domain. The weights, computed from the variances, ensure that low noise samples are weighted higher than noisy samples, and the Gaussian filter gives lower weights to samples further away from the reconstruction point, X_j. The HDR pixel value z_j at location X_j is then reconstructed iteratively by adjusting the shape of the Gaussian kernel to the weighted input samples. In the first iteration, the Gaussian kernel is very small. The spatial support of the kernel is gradually increased until a statistically informed threshold based on the variances and image content is reached. The final HDR pixel value, z_j, is then estimated by fitting a polynomial to the weighted input samples. Our method performs noise reduction, color interpolation, and HDR fusion in a single operation.

The detailed presentation of the algorithm is laid out as described below. Section 4 first describes the camera noise model used to estimate sample variances, and Section 5 describes how each HDR pixel value is reconstructed using our statistical HDR reconstruction framework. The novel methods for filter scale selection for HDR reconstruction are presented in Section 6. Finally, in Section 7, we describe how the parameters for the noise model are calibrated and in Section 8, we show example results and evaluation of our reconstruction method.

4 Sensor noise model

The camera sensor electronics convert the incident radiant power f, which for convenience we express as the number of photo-induced electrons collected per unit time, to a measured digital value y_i at a pixel i. The samples, y_i, contain measurement noise that is dependent on sensor characteristics such as readout noise, gain/ISO setting, and the inherent Poisson shot noise in the incident illumination.

To model the dependence of the measured digital pixel value on the incident radiant power and the camera parameters, we use a well-established radiometric model derived from previous works [9, 15]. Using this model, the non-saturated pixel values are modeled as random variables following a normal distribution:

$$y_i \sim N\left(g_i a_i t f_i + \mu_R, g^2{}_i a_i t f_i + \sigma_R^2(g_i)\right), \tag{1}$$

where t is the exposure time, g_i is the pixel gain/ISO, a_i is a pixel non-uniformity, μ_R is the mean of the read out noise, and σ_R^2 is the variance of the read out noise. An example showing the standard deviation of the read out noise, σ_R, for varying gain/ISO using a Canon Mark III sensor (saturation around 1600) is shown in Fig. 2.

In order to compute an estimate of the incident radiant power, $\hat{f}_i$, from the noisy digital input sample values y_i, we use the following estimator:

$$\hat{f}_i = \frac{y_i - b_i}{g_i t a_i}, \tag{2}$$

where b_i is obtained from a bias frame captured with no light reaching the sensor.

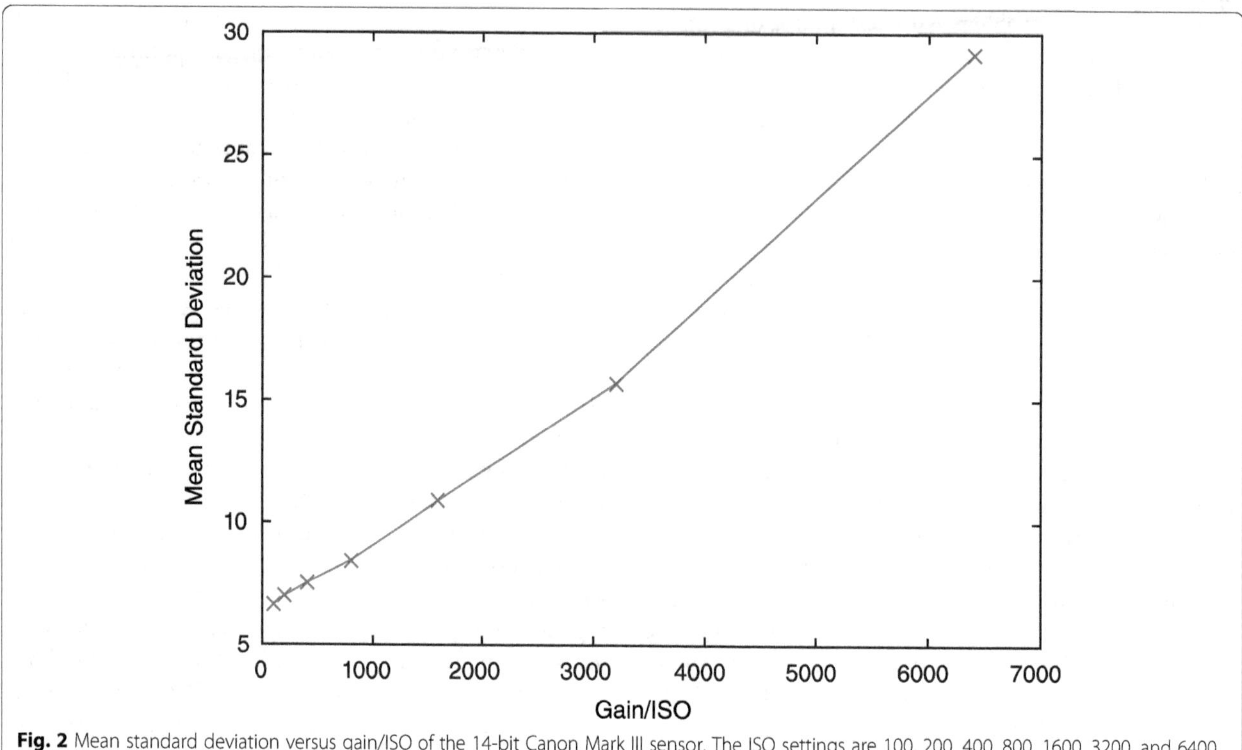

Fig. 2 Mean standard deviation versus gain/ISO of the 14-bit Canon Mark III sensor. The ISO settings are 100, 200, 400, 800, 1600, 3200, and 6400

Similarly, we approximate the variance of this estimator by

$$\hat{\sigma}^2_{\hat{f}_i} = \frac{g_i^2 t a_i \hat{f}_i + \hat{\sigma}^2_R(g_i)}{g_i^2 t^2 a_i^2}, \quad (3)$$

where g_i, a_i, and $\hat{\sigma}^2_R(g_i)$ are found through calibration; see Section 7. We do not include the effect of pixel cross-talk, and the variances, $\hat{\sigma}^2_{\hat{f}_i}$, are assumed to be independent of each other.

5 Adaptive HDR reconstruction

To estimate an HDR pixel value z_j at a location X_j on the sensor, we use a LPA [5] to fit the observation samples of the incident radiant power in the local neighborhood. The same framework is also known as kernel regression [29].

5.1 Local polynomial approximation

To estimate the radiant power, $f(x)$, at an output pixel, we use a generic local polynomial expansion of the radiant power around the output pixel location $X_j = [x_1, x_2]^T$. Assuming that the radiant power $f(x)$ is a smooth function in a local neighborhood around the output location X_j, an Mth order Taylor series expansion is used to predict the radiant power at a point X_i close to X_j as follows:

$$\tilde{f}(X_i) = C_0 + C_1(X_i - X_j) + C_2 tril\left\{(X_i - X_j)(X_i - X_j)^T\right\} + \ldots, \quad (4)$$

where *tril* lexicographically vectorizes the lower triangular part of a symmetric matrix and

$$C_0 = f(X_j) \quad (5)$$

$$C_1 = \nabla f(X_j) = \left[\frac{\partial f(X_j)}{\partial x_1}, \frac{\partial f(X_j)}{\partial x_2}\right] \quad (6)$$

$$C_2 = \frac{1}{2}\left[\frac{\partial^2 f(X_j)}{\partial x_1^2}, 2\frac{\partial^2 f(X_j)}{\partial x_1 \partial x_2}, \frac{\partial^2 f(X_j)}{\partial x_2^2}\right]. \quad (7)$$

Given the fitted polynomial coefficients, $C_{1:M}$, we can thus estimate the radiant power and the HDR pixel value, z_j, at the output location X_j by $z_j = C_0 = f(X_j)$.

5.2 Maximum localized likelihood fitting

To estimate the coefficients, we maximize a localized likelihood function [30] defined using a Gaussian smoothing window centered around X_j

$$\mathcal{W}_h(X_j) = \frac{1}{2\pi h^2} \exp\left\{\frac{-(X_k - X_j)^T (X_k - X_j)}{h}\right\}, \quad (8)$$

where h is a local scale parameter (see Section 6) which determines the shape of the filtering kernel. In Section 6, we discuss how the size of the window function can be selected adaptively depending on the features at each location in the image.

We denote the observed pixel samples (radiant power estimates, $\hat{f}_i(X_j)$ at position X_j) in the support of the local neighborhood window by f_k with a linear index $k = 1 \ldots K$. Note that these are obtained from the digital pixel values using Eq. 2 derived from the sensor noise model.

Using the assumption of normally distributed radiant power estimates, f_k, the polynomial coefficients, $\tilde{C}$, maximizing the localized likelihood function is found by the weighted least squares estimate

$$\tilde{C} = \left(\Phi^T W \Phi\right)^{-1} \Phi^T W \bar{f}, \quad (9)$$

where

$$\bar{f} = [f_1, f_2, \ldots f_K]^T$$

$$W = \mathrm{diag}\left[\frac{\mathcal{W}_h(X_1)}{\hat{\sigma}^2_{f_1}}, \frac{\mathcal{W}_h(X_2)}{\hat{\sigma}^2_{f_2}}, \ldots, \frac{\mathcal{W}_h(X_K)}{\hat{\sigma}^2_{f_k}}\right]$$

$$\Phi = \begin{bmatrix} 1 & (X_1 - X_j) & tril^T\left\{(X_1 - X_j)(X_1 - X_j)^T\right\} & \ldots \\ 1 & (X_2 - X_j) & tril^T\left\{(X_2 - X_j)(X_2 - X_j)^T\right\} & \ldots \\ \vdots & \vdots & \vdots & \vdots \\ 1 & (X_K - X_j) & tril^T\left\{(X_K - X_j)(X_K - X_j)^T\right\} & \ldots \end{bmatrix}. \quad (10)$$

The operator *tril* lexicographically vectorizes the lower triangular part of a symmetric matrix.

Using this maximum likelihood approach, we can efficiently solve for the polynomial coefficients $C_{1:M}$ and estimate the final HDR pixel value z_j at a pixel location X_j for a given smoothing parameter h. However, in order to enable a good trade-off between bias and variance, i.e., between image sharpness and noise reduction, it is necessary to locally adapt the smoothing parameter h to image features and image noise. If h is globally fixed over the image, reconstruction may lead to a noisy final image for small h and blurry result for a high h value. The best trade-off between image sharpness and denoising is achieved by adapting the smoothing parameter h to local image features.

In the next section, we describe the iterative reconstruction method and two algorithms for selecting the locally best smoothing parameter, h, for each HDR pixel estimate, z_j, individually.

Algorithm 1 Adaptive HDR reconstruction

procedure HDR RECONSTRUCTION
  for each color channel in R, G, B **do**
    for each HDR pixel estimate z_j **do**
      $h_l = h_{min}$
      z_j = estimate $\hat{z}_{j,h_l}$ using LPA with degree M
      for each $h_l < h_{max}$ **do**
        $h_l = h_l + h_{inc}$
        estimate $\hat{z}_{j,h_l}$ using LPA with degree M
        estimate $\hat{z}_{j,h_l}$ variance $\tilde{\sigma}_{\hat{z}_{j,h_i}}$ and the reconstruction error ϵ (for EVS)
        apply update rule (ICI or EVS) based on $\hat{z}_{j,h_l}$ and
        if variation in $\hat{z}_{j,h_l}$ can be explained by $\tilde{\sigma}_{\hat{z}_{j,h_l}}$ **then**
          $z_j = \hat{z}_{j,h_l}$
        else
          break
        end if
      end for
    end for
  end for
end procedure

6 Adaptive scale selection

The size of the window function introduces a trade-off between bias and variance. A large window will reduce the variance but can lead to overly smoothed images (bias). Ideally, it is desirable to have large window supports in regions where the smooth polynomial model, used for the reconstruction, is a good fit to the underlying signal, while keeping the window size small close to the edges or important image features. The size of the smoothing window is determined by the smoothing parameter h. Figure 3 illustrates how a signal value, the black point, is being estimated using a kernel with a gradually increasing smoothing parameter, h. When the smoothing parameter h is increased from h_0, the h_1, i.e., a higher degree of smoothing, the variance in the estimated value can be explained by the signal variance. When the smoothing parameter is increased from h_1 to h_2, the kernel reaches the step in the signal and the estimation at the black point can no longer be explained by the signal variance. Smoothing parameter h_1 thus produces a better estimate.

The adaptation of the smoothing parameter, h, scale selection is carried out iteratively. The goal of the adaptation is to gradually increase h, and find an optimal

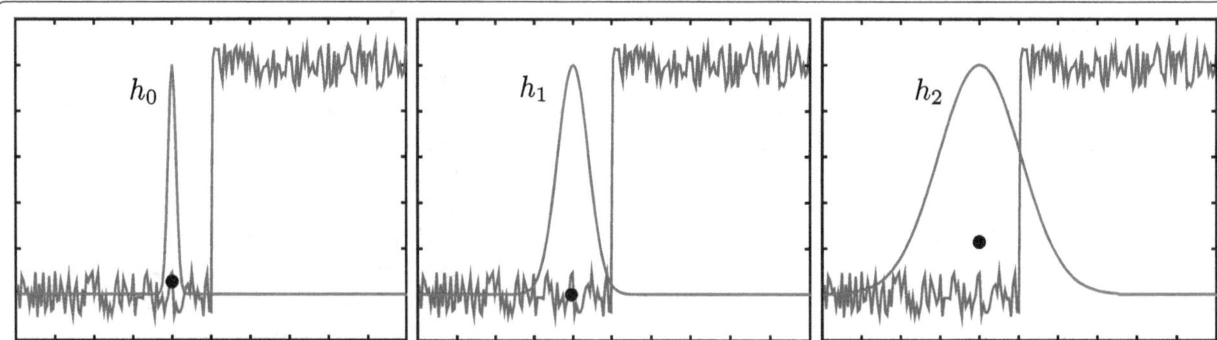

Fig. 3 Illustrating how a signal value, the *black point*, is estimated using a kernel with an iteratively increasing smoothing parameter, h. Increasing from h_0 to h_1, i.e., a higher degree of smoothing, the variance in the estimated value can be explained by the variance in the original signal. However, when the smoothing parameter is increased from h_1 to h_2, the kernel reaches the step in the signal and the estimate at the black point can no longer be explained by the signal variance

Fig. 4 Reconstructed from dualISO data with ISO100–1600 captured with Canon 5D Mark III. Reconstructed with ICI $M = 2$, $h \in [0.6, 5.0]$, and $\Gamma = 1.0$

h such that the variation in the estimated value between iterations can be explained by the signal variance and the smoothing applied. Denoting each iteration by l and the corresponding smoothing parameter by h_1, Algorithm 1 describes the outline of the HDR pixels z_j reconstructed by adapting the smoothing parameter h_1. In each iteration, we estimate the signal value and its variance. We then apply an update rule which determines whether the h value used is valid or not. This is repeated until the update rule does not hold or the maximum h value, $h_{\max}$, is reached. In Sections 6.1 and 6.2, we describe how the variance of the pixel is estimated in detail with the two different update rules.

6.1 Update rule 1: error of estimation versus standard deviation (EVS)

The first update rule is built on the intuition that if the weighted mean reconstruction error is larger than the weighted mean standard deviation, i.e., the difference between the data and the fit cannot be explained by the expected signal variation due to noise, the polynomial model does not provide a good fit to the underlying image data. As described in Algorithm 1, the smoothing parameter, h_1, is iteratively increased with an increment h_{inc}. In each iteration, l, the EVS update rule computes the weighted reconstruction error e_1 as

$$e_l = \sqrt{\sum_k W^2(k,k)(\tilde{f}(X_k) - \hat{f}_k)^2}, \tag{11}$$

where k indexes the pixels in the neighborhood and W is the weights including both the variance of the original pixels and the spatial Gaussian kernel as described in Eq. 10. The weighted standard deviation, $\tilde{\sigma}_{\hat{z}_{j,h_l}}$, of this estimate can be obtained from the covariance matrix M_C for the fitted polynomial coefficients, $\tilde{C}$, which is given by

$$M_C = \left(\Phi^T W \Phi\right)^{-1} \Phi^T W \Sigma W^T \Phi \left(\Phi^T W^T \Phi\right)^{-1}, \tag{12}$$

where $\Sigma = diag[\sigma^2_{f_1}, \sigma^2_{f_2}, \ldots, \sigma^2_{f_k}]$ is the variance of the observation. The variance of estimated radiant power z_j, $\tilde{\sigma}_{\hat{f}_j}$, at the output location X_j, is thus given by the element $\tilde{\sigma}_{\hat{z}_{j,h_l}} = M_C(0,0)$ in M_C. During the iterations, the smoothing parameter, h_l, is updated to $h_{l+1} = h_l + h_{\text{inc}}$ as long as the weighted reconstruction error, $_l$, is smaller than the standard deviation $\epsilon_l < \Gamma \tilde{\sigma}_{\hat{z}_{j,h_l}}$, where Γ is a user-

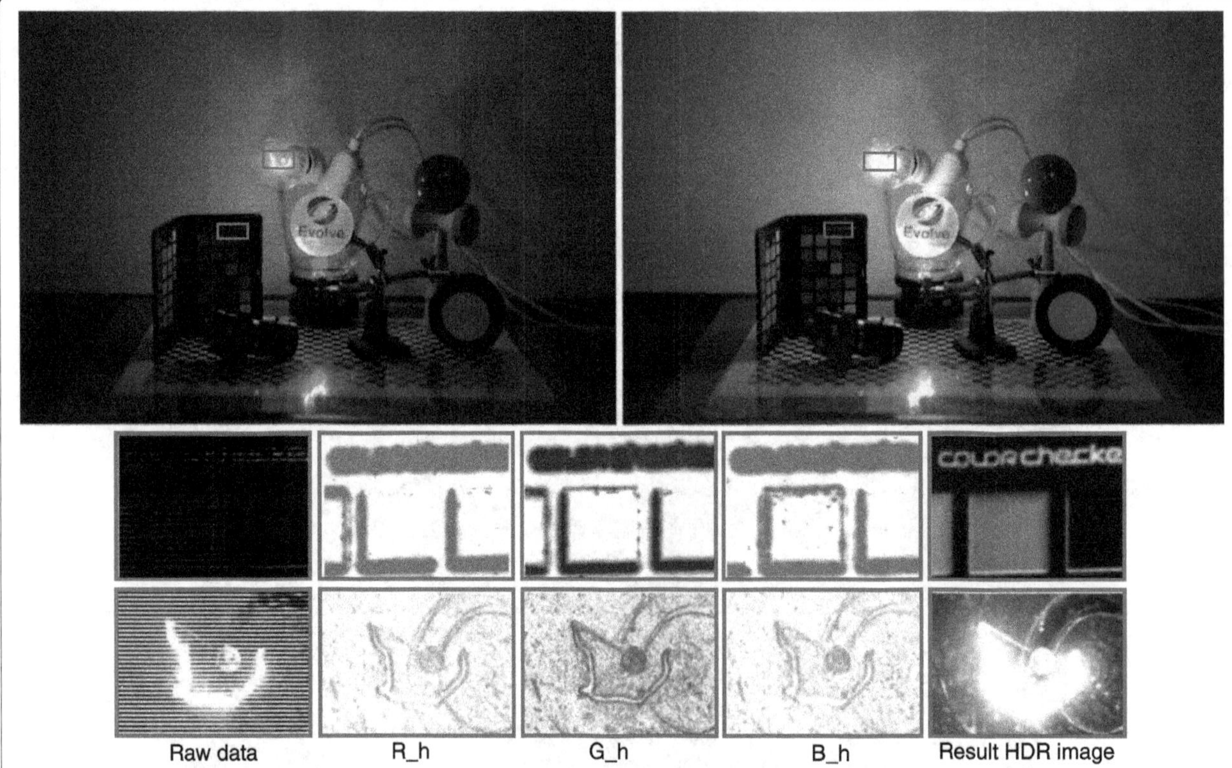

Fig. 5 Reconstruction process of one sample raw image. *Top left* shows the raw input image with CFA Bayer pattern and dualISO row pattern. *Top right* indicates the resulted tone mapped HDR reconstructed image with EVS rule. *Bottom rows* extracted images from *left* to *right*: cutout of the raw image, scaling parameter image for R, G, B color channels with $\Gamma = 1.0$, and the cutout of the reconstructed HDR image

Fig. 6 Lamp scene with different methods for comparison: **a** LPA $M = 2$ from *left* to *right*: $h = 0.6, 1.4,$ and5.0; **b** SKR $M = 2$ from *left* to *right* $h = 0.6, 1.4,$ and5.0; **c** our method with ICI $M = 2$ from *left* to *right* $\Gamma = 0.6, 1.0,$ and1.4; **d** our method with EVS $M = 2$ from *left* to *right* $\Gamma = 0.6, 1.0,$ and1.4

specified parameter controlling the trade-off between levels of denoising applied by the kernel.

6.2 Update rule 2: intersection of confidence intervals (ICI)

The second update rule is based on the ICI algorithm [5]. The main purpose of this algorithm is to obtain the largest scaling parameter in the local neighborhood of the estimation point under the constraint that the polynomial model remains a likely fit to the underlying data. As described in Algorithm 1, the smoothing parameter, $h_{\min} \leq h_l \leq h_{\max}$, is iteratively increased. For each iteration, l, the ICI rule determines a confidence interval, $D_l = [L_l, U_l]$:

$$L_l = \hat{z}_{j,h_l}(x) - \Gamma \tilde{\sigma}_{\hat{z}_{j,h_l}}, \tag{13}$$

$$U_l = \hat{z}_{j,h_l}(x) + \Gamma \tilde{\sigma}_{\hat{z}_{j,h_l}}, \tag{14}$$

where $\hat{z}_{j,h_l}(x)$ is the estimated radiant power given the scaling parameter h_l and $\tilde{\sigma}_{\hat{z}_{j,h_l}}$ is the weighted standard deviation of this estimate computed using Eq. 12. Γ is a scaling parameter controlling how wide the intersection interval is. During adaptation, h_l is increased as long as there is an overlap between the confidence intervals, i.e., h_l is updated to $h_{l+1} = h_l + h_{\text{inc}}$ if there is an overlap between D_l and D_{l+1}. In practice, we utilize Γ as a user parameter, enabling an intuitive trade-off between image sharpness and denoising. A detailed overview of the ICI rule and its robustness can be found in [28].

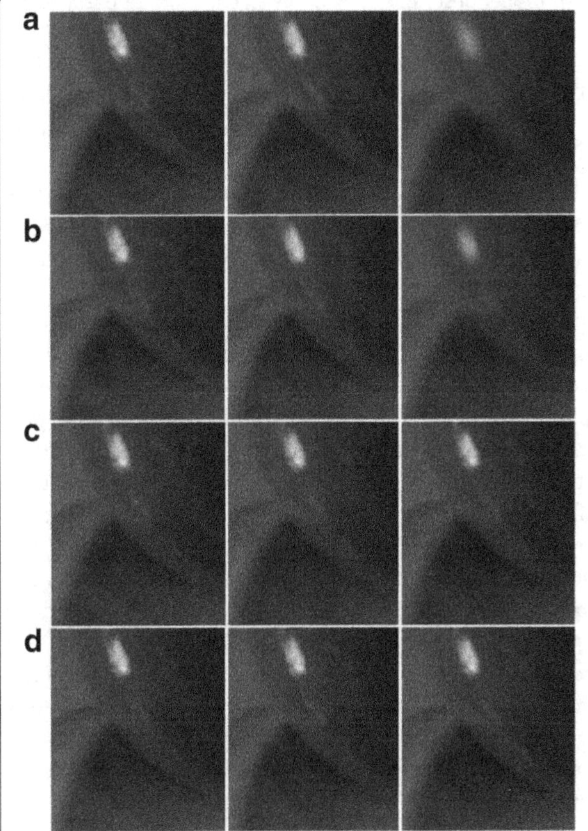

Fig. 7 Another comparison of a cutout of the simulated lamp scene for different methods: **a** LPA $M = 2$ from *left* to *right*: $h = 0.6$, 1.4, and 5.0; **b** SKR $M = 2$ from *left* to *right* $h = 0.6$, 1.4, and 5.0; **c** our method with ICI $M = 2$ from *left* to *right* $\Gamma = 0.6$, 1.0, and 1.4; **d** our method with EVS $M = 2$ from *left* to *right* $\Gamma = 0.6$, 1.0, and 1.4

7 Camera parameter calibration

The variance of the readout noise, the sensor gain, bias, and the sensor saturation point are calibrated once for each sensor. The bias frame, b, and readout noise variance, $\text{Var}[r_i(g_i, t)]$, are calibrated as the per-pixel mean and the variance, respectively. This calibration is done over a set of black images captured with the lens covered, so that no photons reach the sensor. The sensor gain, g_i, can be calibrated using the relation,

$$\frac{\text{Var}[y_i] - \text{Var}[b_i]}{E[y_i] - E[b_i]} = \frac{g_i^2 \text{Var}[e_i]}{g_i E[e_i]} = g_i, \tag{15}$$

where the second equality follows from e_i being Poisson distributed shot noise with $E[e_i] = \text{Var}[e_i]$. In addition, $E[y_i]$ and $E[b_i]$ can be estimated by averaging flat fields and the bias frame, respectively, and $\text{Var}[b_i]$ as described above. The per-pixel non-uniformity, a_i, can be estimated using a flat field image computed as the average over a large sequence of non-saturated images.

8 Results and evaluation

The proposed algorithm has been evaluated on two different sets of images. One synthetic image data set with known ground truth computed using a camera simulator and one set of images captured using a Canon 5D Mark III running the Magic Lantern firmware with the dualISO module installed. The synthetic data is generated using a camera simulation framework which takes a noise-free HDR image as input and applies noise based on the camera

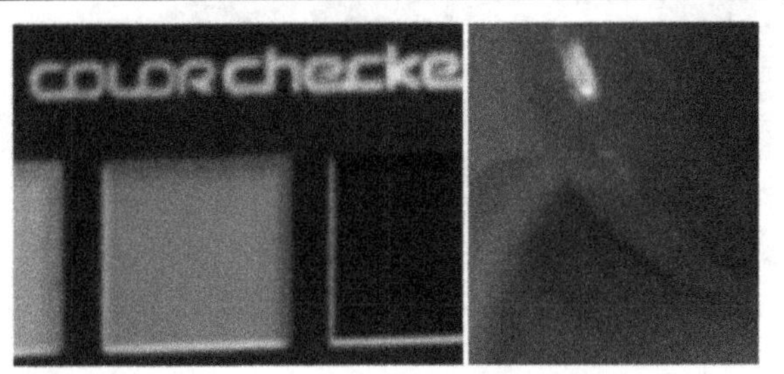

Fig. 8 The ground truth reference images for the cutouts compared in Figs. 6 and 7

noise model described in Section 4. The camera parameters estimated from real cameras as described in Section 7 are used for simulating dualISO sample data. The noise-free HDR images (ground truth) were captured as a set of carefully calibrated exposure brackets one *f*-stop apart covering the dynamic range of the scene. Each of the different exposures in the bracketing sequence was captured as the average of 100 calibrated RAW images with the same exposure settings. The test images used exhibit a very large dynamic range, were selected to be representative for challenging scenes, and include features such as dark and bright image regions, high- and low-frequency regions, image noise, and strong local contrasts. In our evaluation, we compare three different gain patterns as shown in Fig. 1. We have tested our algorithm for a polynomial degree of $M = 0, 1, 2$ and a range of different parameter settings for Γ. In all tests (except for the non-adaptive fixed h comparisons), the smoothing parameter, h, is allowed to vary between $h = 0.6$ and $h = 5.0$. Figure 4 shows an image captured with a Canon 5D Mark III running the Magic Lantern dualISO module and reconstructed by the proposed method. The image shows that our algorithm performs well in the reconstruction by keeping image sharpness while allowing high-quality noise reduction.

Figure 5 shows a high contrast scene simulating a Canon 5D camera with dualISO settings of ISO100 and ISO1600 alternating in pairs of rows on the sensor as shown in Fig. 1 (left). Figure 5 shows the input raw CFA Bayer image, and three images in the bottom row show the locally adapted h values for the red, green, and blue color channels, respectively. The EVS update rule adapts the smoothing parameter h to both the image features and the image noise. The parameter h becomes smaller as we get closer to edges and textured regions and larger in homogeneous areas.

In Figs. 6 and 7, we focus on the trade-off between image sharpness and denoising. We compare our algorithm using both the ICI and EVS update rules to LPA using non-adaptive filtering kernels, [10], with $h = 0.6$, 1.4, and 5.0, and the widely used steering kernel regression (SKR) method [29]. The images compare two cut-out regions of the lamp scene from Fig. 5. The two regions have been chosen to display the performance of our algorithm in a dark region, Fig. 6, and a highlight region, Fig. 7. The ground truth reference images of the cutouts are displayed in Fig. 8. In both images, the comparisons are ordered as follows: (a) non-adaptive LPA $M = 2$ from left to right with $h = 0.6, 1.4,$ and5.0, (b) SKR [29] $M = 2$ from left to right with $h = 0.6, 1.4,$ and5.0, (c)

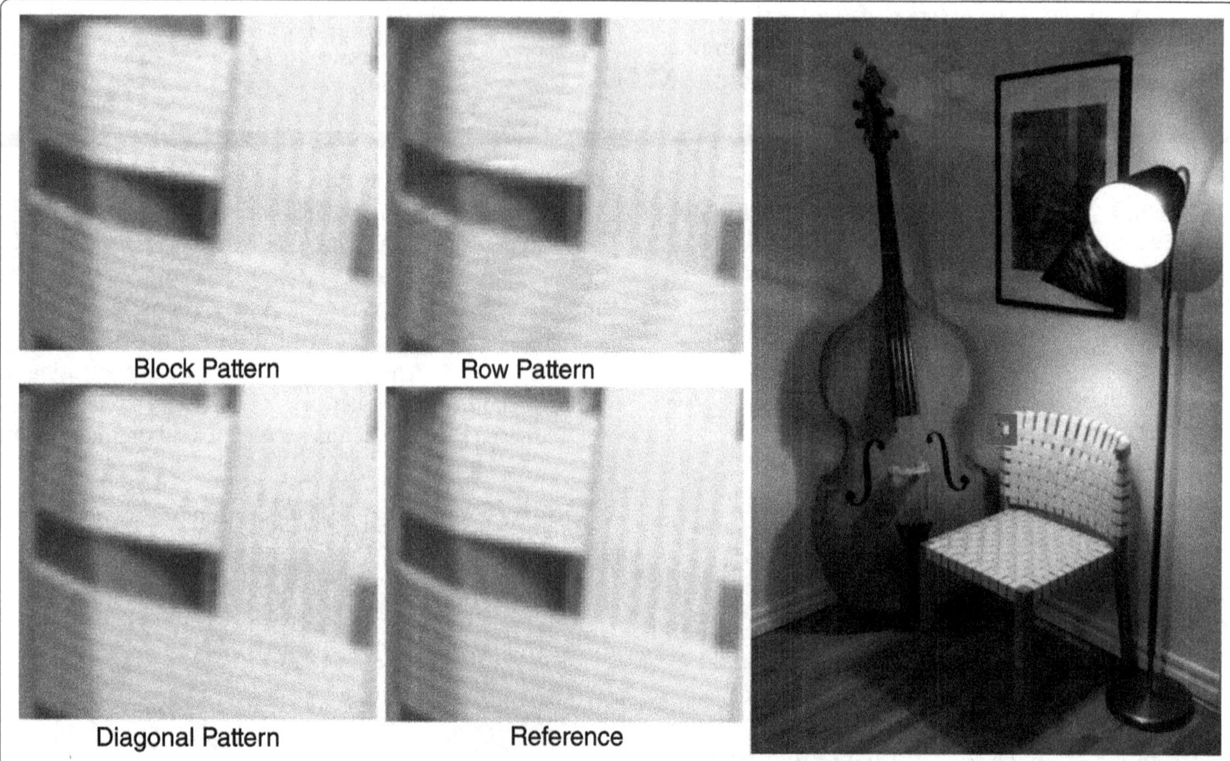

Fig. 9 Living room scene. Comparison of our method with EVS rule $M = 2$; $\Gamma = 1.0$ for different gain patterns: block pattern, row pattern, and diagonal pattern

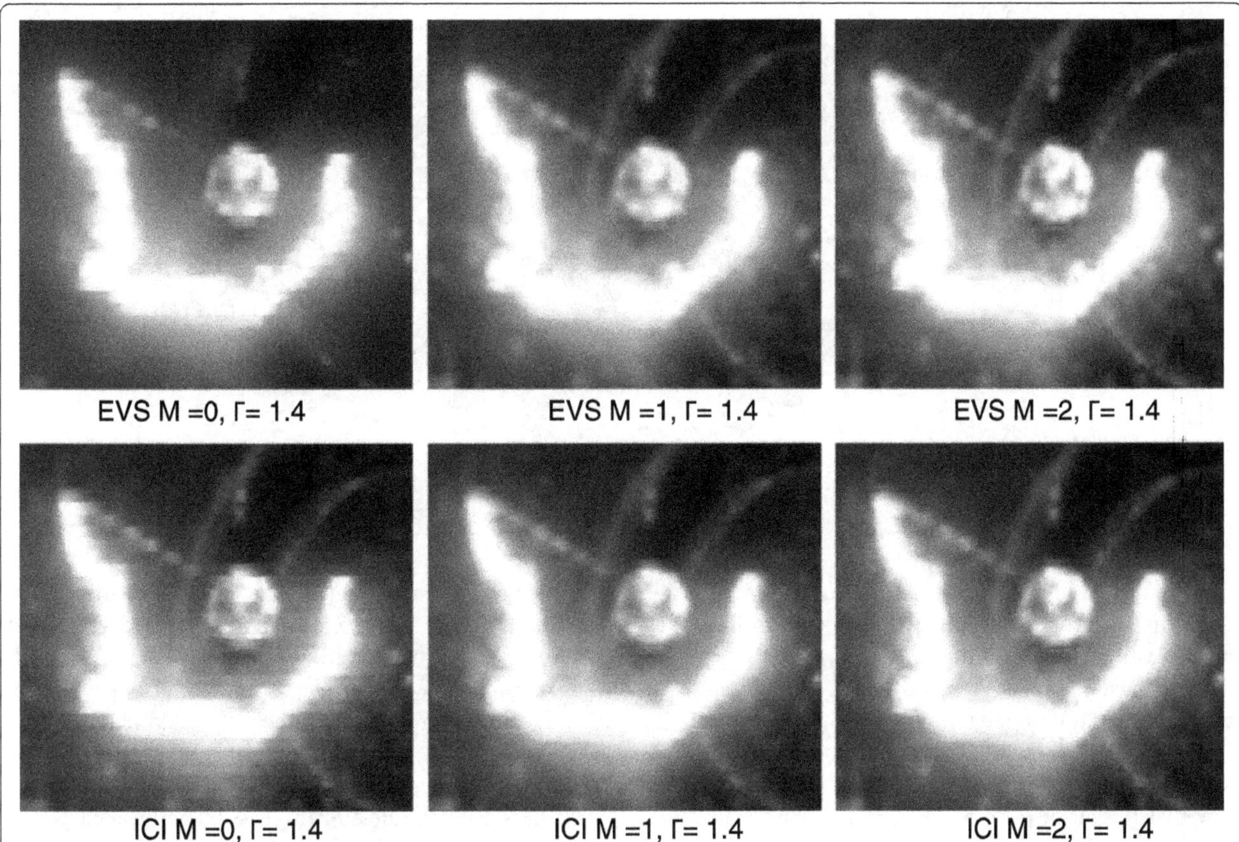

Fig. 10 Lamp scene, evaluation of EVS, and ICI method for different degrees of polynomial for dualISO 100–1600 with row pattern from *left* to *right* with $M = 0, 1,$ and 2

our method with ICI rule for local adaptation of scale parameter, $M = 2$, from left to right: $\Gamma = 0.6, 1.0, \text{and} 1.4$, and (d) our method with EVS rule, $M = 2$, from left to right: $\Gamma = 0.6, 1.0, \text{and} 1.4$. From Fig. 6, it is evident that the non-adaptive method in (a) [10] does not perform well. SKR produces good results for $h = 1.4$ but cannot fully adapt the smoothing parameter as artifacts from the noise filtering are visible (zoom in). Both ICI- and EVS-based algorithms keep sharpness while reducing the image noise more than the other methods. In Fig. 7, SKR with $h = 1.4$ produces a sharp image without color artifacts; however, it also smooths the reflection on the red toy. ICI and EVS produce a similar result, but EVS leads to less smoothing around the highlight areas of the scene compared to ICI. The images show that our algorithms using ICI and EVS update rules produce high-quality images. In general, the EVS update rule allows for a higher degree of smoothing and denoising while keeping higher contrast edges intact. However, in dark regions, the EVS update leads to a loss of detail compared to ICI rule. Another important difference is that although the EVS update rule may produce better results in some cases, it is built on the heuristic argument that the reconstruction error should be smaller than the

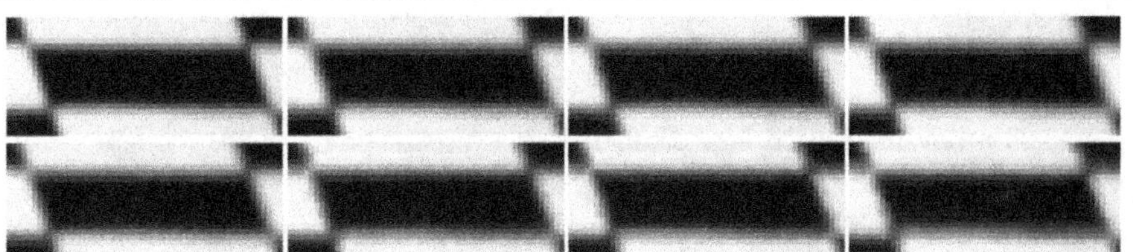

Fig. 11 Cutouts of the checkerboard in the lamp scene, evaluation of ISO settings for EVS $M = 2$, $\Gamma = 1.0$. (*Top row*) From *left* to *right*: reference, dualISO 100–400, dualISO 100–800, and dualISO 100–1600. (*Bottom row*) From *left* to *right*: dualISO 100–3200, dualISO 100–6400, dualISO 100–12800, and dualISO 100–25600

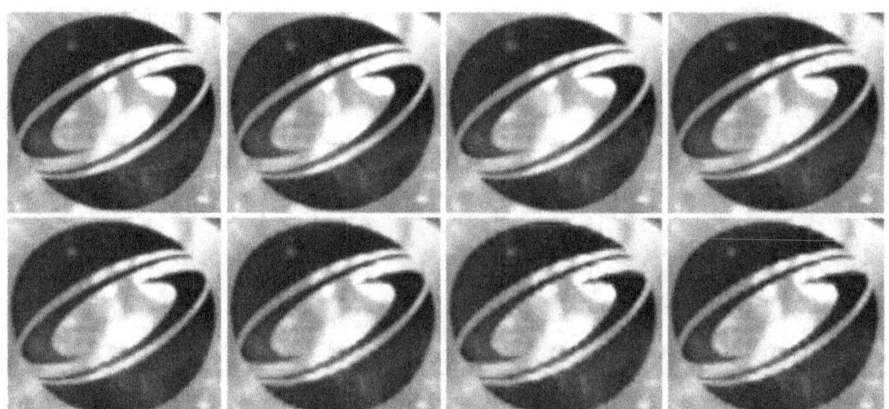

Fig. 12 Cutouts of the glass in the lamp scene, evaluation of ISO settings for EVS $M = 2$, $\Gamma = 1.0$. (*Top row*) From *left* to *right*: reference, dualISO 100–400, dualISO 100–800, and dualISO 100–1600. (*Bottom row*) From *left* to *right*: dualISO 100–3200, dualISO 100–6400, dualISO 100–12800, and dualISO 100–25600

standard deviation in the filtered region. While ICI rule is statistically motivated and designed to minimize the estimate variance.

In Fig. 9, we demonstrate how our algorithms perform using the three different gain patterns illustrated in Fig. 1. This particular image region is selected as it contains slanted edges in different directions. The comparisons show that the block pattern and diagonal pattern in some cases produce better results. However, the reconstruction quality depends on how the image features are oriented and the statistically optimal configuration of the gain pattern is out of scope of this paper. Figure 10 shows a cutout of the lamp scene simulated with row pattern and reconstructed using a varying polynomial degree of $M = 0$, 1, and 2. As expected, $M = 0$ produces a blocky result, and $M = 1$ and $M = 2$ produce increasingly more accurate reconstructions.

In Figs. 11 and 12, we show the effect of increasing the ISO separation in the dualISO image using a simulated 14-bit Canon 5D sensor. The dualISO settings are varied from ISO100–ISO200 to ISO100–ISO25600. As the separation between the ISO settings increase, the number of overlapping bits in the two exposures decrease. The image shows that our algorithm works well up to ISO100–ISO6400, i.e., a separation of six *f*-stops and an overlap of 8 bits. By increasing the separation further, artifacts start to appear along the edges.

9 Conclusions

In this paper, we presented a novel approach for adaptive unified HDR image reconstruction that includes the sensor noise model and error of the estimation for a more robust and accurate reconstruction of single shot spatial multiplexing raw data. The method handles severe noise, especially in the darker regions while it keeps the error of the estimation low to prevent over-smoothing of the image. To the best of our knowledge, none of the previous methods have considered sensor noise model and estimated error and variance in order to adapt the reconstructed kernel for each local region of the image. The robustness of our approach for noise reduction and HDR reconstruction has been experimentally verified on both real data and simulated camera images. While being a simple method to implement, our results demonstrate a relatively good performance.

Competing interests

The authors declare that they have no competing interests.

Acknowledgements

This project was funded by the Swedish Foundation for Strategic Research (SSF) through grant IIS11-0081, Linköping University Center for Industrial Information Technology (CENIIT), and the Swedish Research Council through the Linnaeus Environment CADICS.

References

1. M Aggarwal, N Ahuja, Split aperture imaging for high dynamic range. Int. J. Comput. Vis. **58**(1), 7–17 (2004)
2. C Aguerrebere, A Almansa, Y Gousseau, J Delon, P Musé, Single shot high dynamic range imaging using piecewise linear estimators, in *ICCP*, 2014
3. AO Akyüz, E Reinhard, Noise reduction in high dynamic range imaging. J. Vis. Commun. Image Represent. **18**(5), 366–367 (2007)
4. Alex. Dynamic range improvement for Canon DSLR with 8-channel sensor read-out by alternating iso during sensor readout. Technical documentation, url: http://acoutts.com/a1ex/dual_iso.pdf, July 2013.
5. J Astola, V Katkovnik, K Egiazarian. *Local Approximation Techniques in Signal and Image Processing*. SPIE- International Society for Optical Engineering, 2006
6. B Bayer. Color imaging array. US Patent 3 971 065, 1976.
7. P Debevec, J Malik, Recovering high dynamic range radiance maps from photographs, in *SIGGRAPH*, 1997, pp. 369–378
8. J Froehlich, S Grandinetti, B Eberhardt, S Walter, A Schilling, H Brendel. Creating cinematic wide gamut HDR-video for evaluation of tone mapping operators and HDR-displays. In *SPIE Electronic Imaging*, pages 90230X-90230X. International Society for Optics and Photonics, SPIE digital library, 2014.
9. M Granados, B Ajdin, M Wand, C Theobalt, H Seidel, H Lensch, Optimal hdr reconstruction with linear digital cameras, in *CVPR*, 2010

10. S Hajisharif, J Kronander, J Unger, HDR reconstruction for alternating gain (iso) sensor readout, in *Eurographics 2014 Short Papers*, ed. by MW Eric Galin, 2014
11. F Heide, M Steinberger, YT Tsai, M Rouf, D Pajak, D Reddy, O Gallo, J Liu, W Heidrich, K Egiazarian, J Kautz, L Pulli. FlexISP: a flexible camera image processing framework. *ACM Transactions on Graphics (Proceedings SIGGRAPH Asia 2014)*, 33(6), December 2014
12. S Kang, M Uyttendaele, S Winder, R Szeliski, High dynamic range video. ACM Transactions on Graphics (Proceedings of SIGGRAPH 2003) **22**(3), 319–325 (2003)
13. K Kirk, H Andersen, Noise characterization of weighting schemes for combination of multiple exposures, in *Proc. British Machine Vision Conference (BMVC)*, 2006, pp. 1129–1138
14. H Knutsson, CF Westin, Normalized and differential convolution, in *CVPR*, 1993
15. J Kronander, S Gustavson, G Bonnet, A Ynnerman, J Unger, A unified framework for multi-sensor HDR video reconstruction. Signal Processing: Image Communications **29**(2), 203–215 (2014)
16. J Kronander, S Gustavson, G Bonnet, J Unger, Unified HDR reconstruction from raw CFA data, in *IEEE International Conference on Computational Photography (ICCP)*, 2013
17. P Lancaster, K Salkauskasr, Surfaces generated by moving least squares methods. Math. Comput. **87**, 141–158 (1981)
18. C Loader, *Local regression and likelihood* (Springer, New York, 1999)
19. A Manakov, JF Restrepo, O Klehm, R Hegedüs, E Eisemann, HP Seidel, I Ihrke, A reconfigurable camera add-on for high dynamic range, multi-spectral, polarization, and light-field imaging. ACM Transaction (Proc. SIGGRAPH 2013) **32**(4), 1–47 (2013)
20. S Mann, RW Picard, On being 'undigital' with digital cameras: extending dynamic range by combining differently exposed pictures, in *IS&T*, 1995
21. P Milanfar, A tour of modern image filtering: new insights and methods, both practical and theoretical. IEEE Signal Process. Mag. **30**(1), 106–128 (2013)
22. T Mitsunaga, SK Nayar, Radiometric self calibration, in *CVPR*, 1999, pp. 374–380
23. K Myszkowski, R Mantiuk, G Krawczyk. *High Dynamic Range Video*. Synthesis lectures on computer graphics and animation, a publication in Morgan and Claypool, 2008
24. SG Narasimhan, SK Nayar, Enhancing resolution along multiple imaging dimensions using assorted pixels. IEEE Transaction on Pattern Analysis and Machine Intelligence **27**(4), 518–530 (2005)
25. S Nayar, T Mitsunaga, High dynamic range imaging: spatially varying pixel exposures, in *CVPR*, 2000
26. E Reinhard, W Heidrich, S Pattanaik, P Debevec, G Ward, K Myszkowski. High dynamic range imaging: acquisition, display and image-based lighting (Morgan Kaufmann Publishers Inc., San Francisco, CA, USA, 2005)
27. M Schoberl, A Belz, J Seiler, S Foessel, A Kaup, High dynamic range video by spatially non-regular optical filtering, in *Image Processing (ICIP), 2012 19th IEEE International Conference*, 2012, pp. 2757–2760
28. L Stankovic, Performance analysis of the adaptive algorithm for bias-to-variance tradeoff. IEEE Transaction on Signal Processing **52**(5), 1228–1234 (2004)
29. H Takeda, S Farsiu, P Milanfar, Kernel regression for image processing and reconstruction. IEEE Trans. Image Process. **16**(2), 349–366 (2007)
30. R Tibshirani, T Hastie, Local likelihood estimation. J. Am. Stat. Assoc. **82**(398), 559–567 (1987)
31. MD Tocci, C Kiser, N Tocci, P Sen, A versatile HDR video production system. ACM Transactions on Graphics(Proceedings of SIGGRAPH 2011) **30**(4), 1–41 (2011)
32. C Tomasi, R Manduchi, Bilateral filtering for gray and color images, in *ICCV*, 1998, pp. 839–846
33. OT Tursun, AO Akyüz, A Erdem, E Erdem. The state-of-the-art in HDR deghosting: a survey and evaluation. *Computer Graphics Forum (Proceedings of Eurogprahics STARs)*, 2015
34. J Unger, S Gustavson, High-dynamic-range video for photometric measurement of illumination, in *SPIE Electronic Imaging*, 2007
35. J Unger, S Gustavson, M Ollila, M Johannesson, A real time light probe, in *Proceedings of the 25th Eurographics Annual Conference, volume Short Papers and Interactive Demos*, 2004, pp. 17–21

9

Stopping criterion for linear anisotropic image diffusion: a fingerprint image enhancement case

Tariq M. Khan[1*†], Mohammad A. U. Khan[2†], Yinan Kong[1] and Omar Kittaneh[2]

Abstract

Images can be broadly classified into two types: isotropic and anisotropic. Isotropic images contain largely rounded objects while anisotropics are made of flow-like structures. Regardless of the types, the acquisition process introduces noise. A standard approach is to use diffusion for image smoothing. Based on the category, either isotropic or anisotropic diffusion can be used. Fundamentally, diffusion process is an iterated one, starting with a poor quality image, and converging to a completely blurred mean-value image, with no significant structure left. Though the process starts by doing a desirable job of cleaning noise and filling gaps, called under-smoothing, it quickly passes into an over-smoothing phase where it starts destroying the important structure. One relevant concern is to find the boundary between the under-smoothing and over-smoothing regions. The spatial entropy change is found to be one such measure that may be helpful in providing important clues to describe that boundary, and thus provides a reasonable stopping rule for isotropic as well as anisotropic diffusion. Numerical experiments with real fingerprint data confirm the role of entropy-change in identification of a reasonable stopping point where most of the noise is diminished and blurring is just started. The proposed criterion is directly related to the blurring phenomena that is an increasing function of diffusion process. The proposed scheme is evaluated with the help of synthetic as well as the real images and compared with other state-of-the-art schemes using a qualitative measure. Diffusions of some challenging low-quality images from FVC2004 are also analyzed to provide a reasonable stopping rule using the proposed stopping rule.

Keywords: Isotropic diffusion, Fingerprint enhancement, Entropy, Squared-difference, Stopping criterion

1 Introduction

In image processing problems, many times one comes across the task to enhance flow-like structures, for instance, the automatic assessment of wood surfaces or fabrics, fingerprint image analysis, scientific image processing in oceanography [1], seismic image analysis [2], or sonogram image interpolated for Fourier analysis [3]. All images as mentioned above have one thing common; they contain elongated structures [4–6]. Such images can be referred to as *anisotropic*. The isotropic, by contrast, is an image category having largely round objects. The isotropic as well as anisotropic images, once acquired from their respective sources are mostly noisy. The noise treatment is different based on the category they belong. The case of noise smoothing for anisotropic images is more interesting and is the focus of research presented here.

Classifying images into their category will help to devise a proper noise removal strategy for them. The authors in [7] suggested to use local anisotropy strength as a measure for an image to classify as anisotropic or isotropic. They later extended their anisotropy strength definition to construct a complete flow-coordinate system for anisotropic images. Their proposed anisotropy strength measure computation can be summarized as follows. First, the image $L(x,y)$ is smoothed with a Gaussian of small standard deviation. The result $C(x,y)$ is then differentiated in x- and y- direction to form $C_x(x,y)$ and

*Correspondence: tariq045@gmail.com
†Equal contributors
1 Department of Engineering, Macquarie University, Balaclava Rd, 2109 Sydney, Australia
Full list of author information is available at the end of the article

$C_y(x,y)$, respectively. Next the covariance matrix components $J_1(x,y) = 2C_x(x,y)$ and $J_2(x,y) = C_x^2(x,y) - C_y^2(x,y)$, and $J_3(x,y) = \sqrt{C_x^2(x,y) + C_y^2(x,y)}$ are computed. The components are smoothed again with a larger Gaussian. The local orientations and their anisotropy strength measure are computed as

$$\theta(x,y) = \frac{\arctan(\frac{J_1(x,y)}{J_2(x,y)})}{2}, \tag{1}$$

and

$$\chi(x,y) = \frac{\sqrt{J_1^2(x,y) + J_2^2(x,y)}}{J_3(x,y)}. \tag{2}$$

Applying this definition to our test images, that is, Blackball and Curves image, will result in a graphical display as shown in Fig. 1. The local flow directions are depicted by the orientations of the small needles superimposed on the image. The length of each needle is drawn proportional to the amount of local anisotropy at that pixel point. It is noted that the Blackball image is largely isotropic with no preferred local directions, whereas the Curves image showed a profound anisotropic character, largely in the vicinity of the elongated structures. This justifies labeling Blackball image as isotropic and the Curves image as anisotropic.

The rest of this paper is organized as follows. In Section 3, a discrete image as a spatial distribution is discussed. The spatial entropy of linear isotropic diffusion process is described in Section 4. Section 5 talks about spatial entropy of a linear anisotropic diffusion process followed by results and discussion in Section 6. Finally, the paper is concluded in Section 7.

2 Related work

The research concerned here is to smooth noise present in fingerprint images (a representative of anisotropic class) without affecting their ridge/valley pattern. This aim can be conveniently served in a *scale-space* construction. A scale-space framework describes a noisy image as a stack of progressively evolving many smooth images, each one with their corresponding scale [8]. The stack is ordered in increasing smoothness scale, where the scale varies in fine-to-coarse. The fine-to-coarse transformation is implemented, in general, by a linear isotropic diffusion process, governed by a partial differential equation (PDE) as follows.

Let $L(x,y)$ denote a noisy grayscale input image and $L(x,y;t)$ be an evolving image at scale t, initialized with $L(x,y;0) = L(x,y)$. Then, the linear isotropic diffusion process can be defined by the equation

$$\frac{\partial L}{\partial t} = \nabla \cdot (c\nabla L) = c\nabla^2 L. \tag{3}$$

This equation appears in many physical processes [9, 10]. In the context of heat transfer, it is referred to as the famous *heat equation*. For image processing, the amount of heat is replaced with the intensity value at a certain location. The diffusivity parameter c is constant across the image, making it a *linear* isotropic equation. The linear isotropic equation has an elegant solution $L(x,y;t) = G_{\sqrt{2ct}}(x,y) * L(x,y)$, where $G_\sigma = \frac{1}{2\pi\sigma^2}\exp\left(-\frac{x^2+y^2}{2\sigma^2}\right)$. This solution provides the required interpretation in the form of low-pass filtering. Due to low-pass nature of this diffusion, as it progresses from fine scale images to coarser images, the blurring intensifies and may result in removing significant image structure, typically edges, lines, or other details, well before it had taken care of the noise. To protect the structure in a diffusion process, the diffusivity

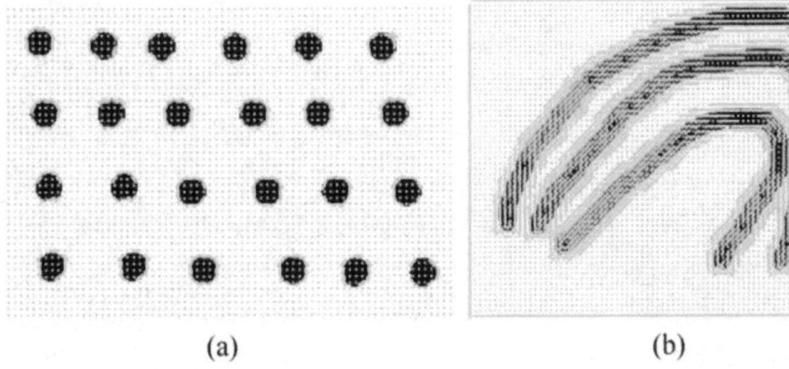

(a) (b)

Fig. 1 Anisotropy strength measure. The figure reveals anisotropy strength measure in the form of an array of needles on top of the image. The length of needles is representative of anisotropy strength and the needle direction is an estimation of local flow. The blackball image is largely isotropic with little amount of anisotropy at almost all the points in the image. However, the curve image on the right is largely anisotropic, with a strong strength measure appearing around the elongated feature of interest. **a** represents the black ball image. **b** is the curve image

parameter should be made dependent on some characterization of image structure. This results in the famous nonlinear isotropic diffusion process, proposed by [11]. The diffusivity now becomes a function of gradients, so at the edge point the diffusion is completely inhibited and in smooth regions diffusion is allowed. However, computing gradients for a noisy image is an ill-posed problem. A remedy was pointed out by [12], that suggests the use of Gaussian smoothing before computing gradients. This modification lays the foundation for a well-behaved *non-linear* isotropic diffusion process. Later on, instead of inhibiting diffusion at edge points, it was thought of to steer the diffusion in the direction parallel to the edge [13–16] rather than across it. This paved the way for the use of the diffusion matrix. This evolved the current form of *non-linear* anisotropic diffusion. The diffusion matrix-based equation is defined as

$$\partial_t L = \nabla (D \nabla L), \tag{4}$$

where D is the 2×2 diffusion matrix. The eigenvectors of the diffusion matrix provide the required steering while the eigenvalues as a function of gradients, add the non-linearity character. In our wish to keep connected with the Gaussian convolution interpretation that provides a mathematical tractability to the whole process, the research reported here is restricted to the linear anisotropic diffusion case. For that, the eigenvalues of the diffusion matrix are kept fixed. It is found that the Gaussian convolution connection is also useful for linking anisotropic diffusion with its earlier counterpart isotropic diffusion in a more natural way. The support for this modification came from the argument made in [14], that a non-uniform Gaussian can act as a solution of the Anisotropic Gaussian scale-space as long as the diffusion matrix is *spatially constant*, i.e., it does not depend on (x, y) spatial location. Keeping in line with this argument, only spatially-invariant diffusion matrix is used; however, the steering was allowed. This leaves us with the so-called *linear anisotropic diffusion* process. The constant eigenvalues are responsible for the linear part of the name, while the steering of the eigenvectors is what provided the word anisotropic in the nomenclature. The linear anisotropic diffusion equation has a convolution solution with a non-uniform Gaussian of the form:

$$G_{\lambda_u,\lambda_v}(u,v) = \frac{1}{\sqrt{2\pi}\lambda_u} \exp\left(-\frac{u^2}{2\lambda_u^2}\right) \frac{1}{\sqrt{2\pi}\lambda_v} \exp\left(-\frac{v^2}{2\lambda_v^2}\right), \tag{5}$$

where (u, v) are the rotated coordinates obtained using eigenvectors of the diffusion matrix. The eigenvalues λ_u, λ_v represent the standard deviations of the Gaussian in u and v direction, respectively. Normally, for noisy images, one of the eigenvalues is set to be much smaller than the other one, resulting in a non-uniform Gaussian function with more generalized elliptical support.

Searching for a suitable linear anisotropic diffusion strategy for noisy images in literature, we stumble upon considerable activity regarding the impact of a non-linear anisotropic diffusion equation on noisy images. The non-linear anisotropic literature is used as a stepping stone to reach a linear anisotropic diffusion strategy. The idea of non-linear anisotropic diffusion was pioneered by Nitzberg et al. [17] and Cottet et al. [12]. Later on, Weickert [3] put forward a formal method for enhancing the elongated structure, referred to as coherence-enhanced diffusion (CED). The CED works by steering the diffusion process in a particular direction with the help of a spatially varying diffusion matrix. The design was further generalized by adopting a diffusion matrix to learn the local structure iteratively [18]. Since smoothing elongated structure is desired, the CED procedure comes in handy. The CED is adopted as it is, but with one major modification. That is, the eigenvalues are forced to be independent of spatial position without disturbing the eigenvectors. Thus, our proposed linear anisotropic diffusion process will steer the non-uniform Gaussian to lay along the structure, but its size will remain constant regardless of the position. Towards the end, we will desribe another variant of CED, where even the steering part of the diffusion matrix will also be precomputed and kept constant throughout the evolution process. This is referred to as the linear-oriented diffusion process.

The suggested linear anisotropic process for anisotropic images are confronted with one basic problem: when to stop the diffusion. For the case of a noisy image, the diffusion process initializes with an under-smooth situation that ultimately turns into an over-smooth one (the mean-value image at the end with no structure). Over-estimating stopping time will result in an over-smoothed blurry image while under-estimating may leave significant noise in the image. Therefore, it is crucial that an appropriate time is selected in an automatic way. The literature activity in this respect can be divided into two broad categories. One that deals with stopping criterion selection in additive noise model setting. These methods adopt the stopping time by treating the noisy image as the result of a noise addition, where the correlation between the diffused image and the initial noisy image minimized [3]. The authors in [19] introduced a multigrid algorithm using a normalized cumulative periodogram. A frequency approach to the problem was presented in [20]. Whereas, [21] uses the extent of noise smoothing in every iteration as a stopping parameter for diffusion. Later on, a spatially-varying stopping method was introduced that increased the computational cost significantly [22]. By identifying it as a Lyapunov functional of a large class of scalar-valued nonlinear diffusion filters, Weickert [23] introduced decreasing the variance of an evolving image as a stopping tool.

Since additive noise model may break down for some real-world images, where noise manifests itself in the form of gaps in regular ridge structures. Therefore, a second category of stopping rule was evolved. The category deals with examining entropy profile of the diffused image and proposed stopping criterion for the evolving image entropy distance from that of the entropy of the original noisy image [3]. The idea of local image entropy was introduced in [24], where the measure of local entropy defines the segmentation boundaries in multiple-object images. Local image entropy definition can be extended to define a global characteristic of the scale-space image, that is spatial entropy [25].

The research work reported here takes an investigative look at the stopping rule concerning the change in spatial entropy of an image as it goes through diffusion process. The connection, between last peak in spatial entropy curve and the size of the image structure, is found to be related to the start of significant information loss. This observation paves the way to the hypothesis that peak entropy change will happen at the time instant on diffusion time axis when dominant image structures just start blending with the background right at their boundaries. This finding, substantiated by extensive empirical evidence provided here, motivated us to put forward the idea that a maximum entropy change may well be posed as a good stopping time for the diffusion process.

3 A discrete image as spatial distribution

Consider a discrete fingerprint image $L(x, y)$, where x is the row index and y is the column index. This discrete image can be realized as spatially distribution light intensity [26]. Each spatial location that is (x, y) in the image registers the number of light quantum-hit. In this way, we may define

$$p(x, y) = \frac{L(x, y)}{\sum_x \sum_y L(x, y)}. \tag{6}$$

This spatial probability perspective was found to correspond very nicely with the theory of scale-space [27]. As we move higher in scale-space for an image, and the spatial smoothing is high, or equivalently, the spatial uncertainty increases. In the limit, the spatial distribution, becomes close to uniform distribution. The spatial entropy of an image is given as

$$H_t(L) = -\sum_x \sum_y \frac{L(x, y; t)}{\sum_x \sum_y L(x, y; t)} \log \left(\frac{L(x, y; t)}{\sum_x \sum_y L(x, y; t)} \right). \tag{7}$$

As stated in [26], the spatial entropy of the image increases monotonically towards an equilibrium state $\log N$, where N is dimension $N = rows \times columns$.

4 Spatial entropy of linear isotropic diffusion process

The linear diffusion process implemented by so-called heat equation is the oldest and well-investigated noise-smoothing process in the image processing domain. The linear diffusion process can be visualized as an evolution process with an artificial variable t denoting the *diffusion time*, where the noisy input image is repeatedly smoothed at a constant rate in all directions. No preference to any direction is what justifies the name *isotropic*. This evolution results in *scale space* representation of the noisy image. As we move up to coarser scales, the evolving images become more and more simplified since the diffusion process removes the image structures present at finer scales. In the process, noise also gets smoothed as it is considered a smaller size object while diffusion just reaches the point of touching the boundaries of the large dominating structure.

During the process of diffusion from fine-scale image to the higher coarser scale images, the mean of the resulting image remains constant with a monotonic decrease in variance (a second-order statistic [13]). Later on, it was found that spatial entropy associated with linear isotropic diffusion process also rises smoothly in a monotonic fashion [25]. Motivated by the smoothness of the spatial entropy graph for the diffusion process, the first derivative of the entropy function on natural scale parameter $\tau = \log(t)$ was investigated. It was shown that entropy change graph do show important peaks related to dominating structures present in the original fine scale image. However, their experiments did not involve smoothing noisy images, and the authors fell short of suggesting to use these peaks as stopping criterion. The empirical evidence is provided here to show that once a linear isotropic diffusion process is involved in smoothing noisy images, these peaks will come at a much later stage in diffusion time. Therefore, most of the noise being low size structure already wiped by the process, and thus the peaks could be regarded as a suitable stopping time. This proposition is tested by tracking experimental data.

To provide a quantitative measure for checking our test results, two binary statistical measures are used: sensitivity and specificity. This is due to the use of a binary image as input test, and the final diffused image is thresholded to come up with the final binary output image. Since we are dealing with binary images, the two measures suit us. The measures deal with comparing the output binary image A with a standard ground truth image B. Let us first define four related quantities: true positive (TP) (the black pixels in image A are also black in image B), false positive (FP) (the black pixels in image A are white pixels in image B), false negative (FN) (the black pixels of image B are identified as white in image A, that is we missed the true black pixels), and true negative (TN) (the white pixels in image A are same as white pixels in image B). Sensitivity is given by

$$Sensitivity = \frac{number\ of\ TP}{number\ of\ TP + number\ of\ FN}. \qquad (8)$$

Specificity is more concerned with

$$Specificity = \frac{number\ of\ TN}{number\ of\ TN + number\ of\ FP} \qquad (9)$$

First, a linear isotropic diffusion process is conducted for the image without noise. Figure 2b shows the entropy curve with natural scale parameter. The monotonic behavior of entropy curve is noted. The curve starts increasing from a low value and moves onwards to an almost stable asymptotic value on a much larger scale. The regularity of the entropy curve motivates us to compute its derivative on the natural scale parameter. The entropy change curve for this image diffusion process is depicted in Fig. 2c. One clear peak in the graph is observed, corresponding well with the radius of the black balls. If the linear diffusion process is stopped at a scale where the peak in entropy-change happens, then output resulting diffused image is displayed in Fig. 2d. It is observed that diffused image is still intact with all the black balls showing their characteristic black colors, with diffusion just started at the boundaries of these balls. Hoping that this peak in entropy change will remain fixed at this scale with the noise added to the image, the best possible stopping time will be the scale of the peak. The sensitivity and specificity numbers for the comparison of the output diffused binary image with the original are 88 and 96 %.

To investigate the shape and location of the peaks in entropy change with noise-added images, we start with lower SNR images. The black balls image is considered with Gaussian noise added, such that its SNR reduces to 2. The black ball image with SNR = 2 dB is depicted in Fig. 3a. The linear diffusion process was conducted for this noisy image to mitigate the effect of Gaussian noise. The resulting entropy change graph is displayed in Fig. 3b. We see two peaks in the graph. The first peak is largely the contribution of the noise added to the image. The second peak is due to the presence of black balls, at the same location where we saw it before in the clean image entropy-change graph. This validated the claim made in [25], that peaks in entropy change graphs are representative of the corresponding sizes of the structures present in the

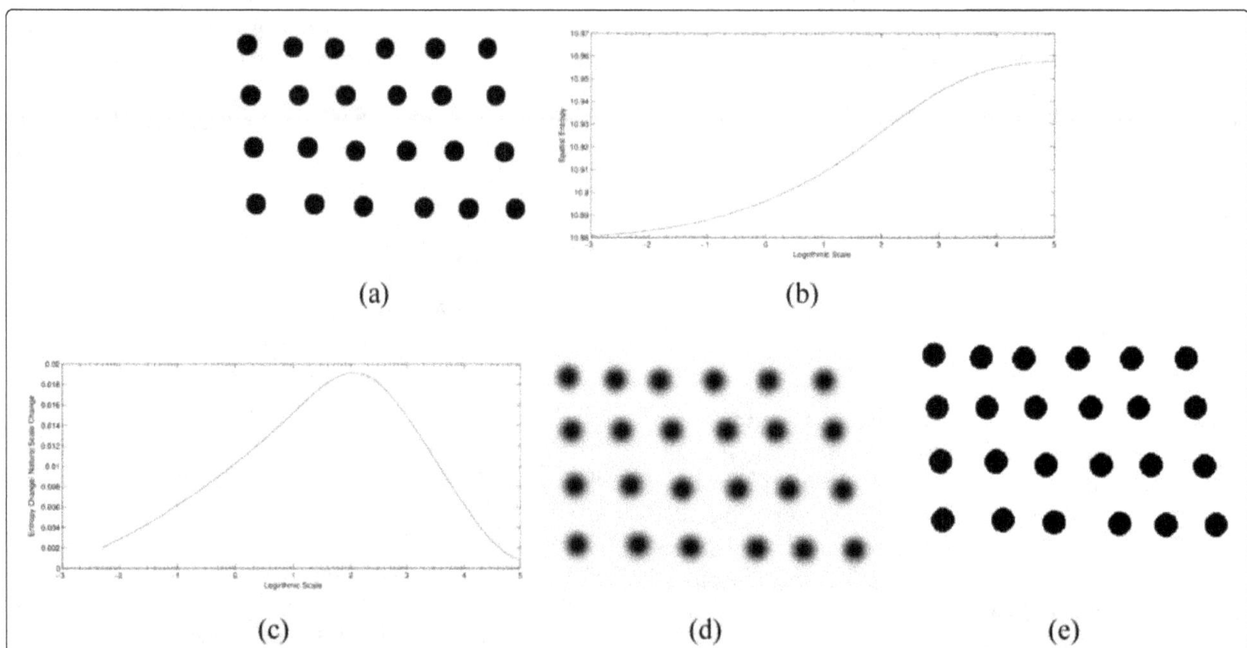

Fig. 2 Linear isotropic diffusion process. **a** shows a black ball test image with white background. The features present in the image are isotropic in shape with a constant radius of two pixels. **b** shows the smooth spatial entropy graph resulting from diffusion process on natural scale parameter. The entropy change with natural scale change is displayed in (**c**), where the peak corresponds to the size of the black balls. The diffused image resulting from stopping the diffusion process at the location of the peak in entropy change is shown in (**d**). The diffused image is converted to binary image using Otsu optimal threshold of 0.63. The final binary image is displayed as (**e**)

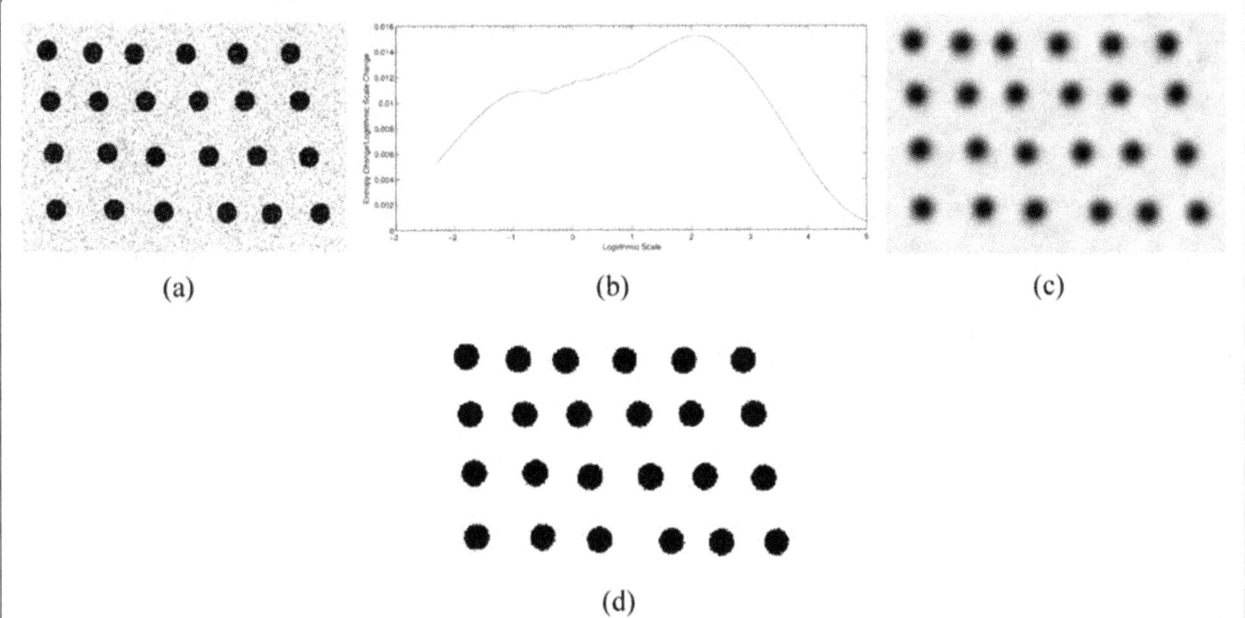

Fig. 3 Linear isotropic diffusion process for noisy image. **a** shows a noisy black ball test image with white background. The zero mean Gaussian Noise added such that SNR reduces to 2 dB. **b** shows the spatial entropy-change graph resulting from diffusion process on the natural scale parameter for a noisy image. Two peaks can be observed, where the first peak is the result of adding Gaussian noise, and the second peak is representing the characteristic size of the black balls. The diffused image resulting from stopping the diffusion process at the location of the second peak in entropy change is shown in **c**. Binarized image as a result of the threshold, set to the mean value of the diffused image results in (**d**)

images. The linear diffusion process can be stopped at the location of the second peak, the resulting output diffused image is shown in Fig. 3c. The image clearly shows a diffused image where largely the noise is smoothed with the black balls still intact. The diffused image can be binarized by using its histogram, clearly showing a valley between black and white bars. Doing so, the image of Fig. 3d is reached, with sensitivity and specificity numbers being 85 and 91 %.

To further investigate the entropy change graph of a noisy image, the black ball imaged are severely degraded with a large amount of Gaussian noise till its SNR drops to −3 dB. The noisy black ball image is depicted in Fig. 4a. The linear diffusion process is applied to this noisy image, with the resulting entropy change graph displayed in Fig. 4b. The presence of two peaks is observed, as previously did in a less noisy image. However, this time, the peak associated with noise is much large in amplitude to the peak of the black balls. This clearly is the outcome of a large amount of noise added to the image pixels. The second peak, though small in amplitude, is still present at the same location as that of clean image entropy-change graph. By stopping the linear diffusion process at the second peak location, we get the diffused image is shown in Fig. 4c. By converting this diffused image by selecting a threshold from its histogram, we reach the binary result as displayed in Fig. 4d, having sensitivity and specificity numbers as 78 and 88 %.

5 Spatial entropy of a linear anisotropic diffusion process

In this section, spatial entropy analysis is carried out for the anisotropic diffusion process. What we are looking for is the finding whether we will get a smooth spatial entropy increasing function, and then will we get a distinct peak in the entropy change curve for the anisotropic diffusion process.

The anisotropic scale-space for the image $L(x, y)$ can be constructed by the diffusion equation

$$\frac{\partial L}{\partial t} = \nabla (D \nabla L), \tag{10}$$

where D is the 2×2 diffusion matrix, adapted to the local image structure, via a structural descriptor, called the second-moment matrix μ, defined as

$$S = \begin{pmatrix} s_{11} & s_{12} \\ s_{12} & s_{22} \end{pmatrix} = \begin{pmatrix} L^2_{x,\sigma} & L_{x,\sigma} L_{y,\sigma} \\ L_{x,\sigma} L_{y,\sigma} & L^2_{y,\sigma} \end{pmatrix}, \tag{11}$$

where L^2_x, $L_x L_y$, and L^2_y represent the second order Gaussian-derivative filters, in the x and y directions. This symmetric 2×2 matrix has two eigenvalues λ_1 and λ_2, given by:

$$\begin{aligned} \mu_1 &= 1/2\,(s11 + s12 + \alpha) \\ \mu_2 &= 1/2\,(s11 + s12 - \alpha), \end{aligned} \tag{12}$$

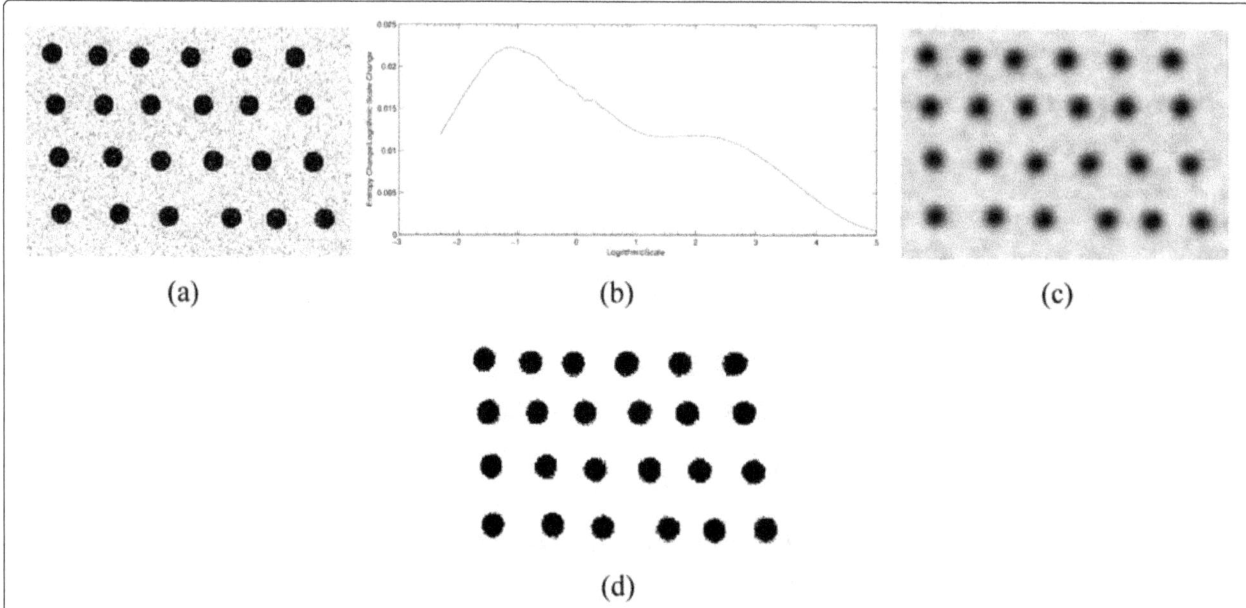

Fig. 4 Linear isotropic diffusion process for noisy image. **a** shows a noisy black ball test image with white background. The zero mean Gaussian Noise added such that SNR reduces to -3dB. **b** shows the spatial entropy-change graph resulting from diffusion process with respect to natural scale parameter for noisy image. Two peaks can be observed, where the first peak is much larger in amplitude than the second peak. The diffused image resulting from stopping the diffusion process at the location of the second peak in entropy change, is shown in **c**. Binarized image as a result of threshold, set to the mean value of the diffused image is resulted in (**d**)

where

$$\alpha = \sqrt{(s11 - s22)^2 + 4s12^2} \tag{13}$$

The second-moment matrix comes with two eigenvectors. The first normalized eigenvector can be written as $(\cos\theta,\ \sin\theta)^T$, and the second orthogonal eigenvector comes out to be as $(-\sin\theta,\ \cos\theta)^T$. One of these eigenvectors is parallel, and the other is perpendicular to the structure. The parameter θ represents the local orientations of the given image. What is observed here is that eigenvalues are dependent on the local structure. In order to transform CED process into a linear anisotropic process, fixed values are assigned to the eigenvalues. Specifically, the eigenvalue associated with eigenvector that goes parallel to the structure has given a larger value than that of the eigenvalue of an eigenvector that is perpendicular to the structure boundary. Our specific choice of λ_1 and λ_2 for this experiment are

$$\begin{aligned} \lambda_1 &= 0.1 \\ \lambda_2 &= 1 - 0.1, \end{aligned} \tag{14}$$

with a step size of 0.01 to provide a stable diffusion process.

The diffusion matrix D can now be reconstructed with help of its structure-invariant eigenvalues and structure-dependent eigenvectors as

$$\begin{aligned} d11 &= \lambda_1 \cos^2\theta + \lambda_2 \sin^2\theta \\ d12 &= (\lambda_1 - \lambda_2) \sin\theta \cos\theta \\ d22 &= \lambda_1 \sin^2\theta + \lambda_2 \cos^2\theta \end{aligned} \tag{15}$$

Once the diffusion matrix is constructed, the evaluation process is set to start. The diffusion process proceeds in four steps.

1. Calculate the second-moment matrix for each pixel.
2. Construct the diffusion matrix for each pixel.
3. Calculate the change in intensity for each pixel as $\nabla\,(D\nabla L)$.
4. Update the image using the diffusion equation as

$$L^{t+\Delta t} = L^t + \Delta t \times \nabla\,(D\nabla L)\,. \tag{16}$$

This monotonic decreasing behavior of the image variance is also evident in the graph depicted in Fig. 5 when we are diffusing our fingerprint image shown in Fig. 5. What can be seen from the graph is that it is fast decreasing in the beginning, but towards the end, it becomes saturated, providing convergence. Thus, by bounding the relative change in the variance, one can define the diffusion stopping rule. However, this rule does not guarantee an optimal time to stop the process. It is based on the user-defined ratio of diffused image variance to that of initial image variance. This ratio might be useful if we want to compare various diffusion schemes. Its utility to provide a well-diffused image with all the important structure cleaned but intact may be limited.

Under the CED process, the fingerprint image becomes strongly coherent as the number of iterations increased. In other words, as the scale increases, the image becomes

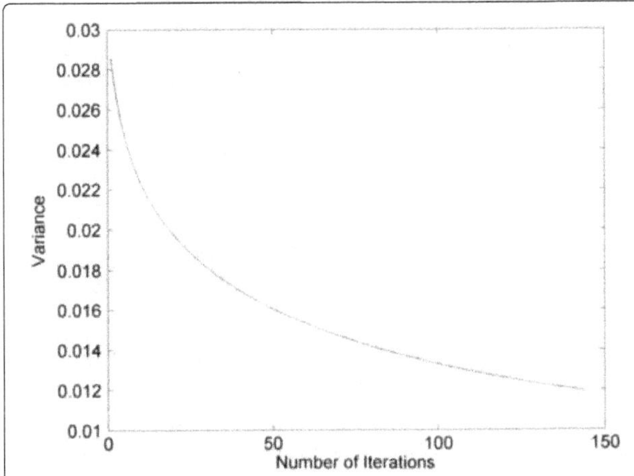

Fig. 5 This graph shows the monotonic decreasing behavior of the variance of the image for coherence enhanced diffusion (CED)

diffused with a corresponding change in its spatial distribution. Taking $p_t(x,y) = \frac{L(x,y;t)}{\sum_x \sum_y L(x,y;t)}$ and $C = \sum_x \sum_y L(x,y;t)$, we get

$$H_t(L) = -\sum_x \sum_y p_t(x,y) \log p_t(x,y). \quad (17)$$

Now, we track the change in entropy with respect to natural scale parameter $\tau = \log t$. The natural scale parameter is defined in [28]. The entropy change is thus,

$$\frac{dH_t(L)}{d\tau} = -\sum_x \sum_y \frac{d}{d\tau}\left(p_t(x,y) \log p_t(x,y)\right). \quad (18)$$

After some mathematical manipulations, reach to the equation

$$\frac{dH_t(L)}{d\tau} = -\sum_x \sum_y \left[1 + \log p_t(x,y)\right] \frac{d}{d\tau} p_t.(x,y). \quad (19)$$

Using chain rule $\tau = \log t$ and $d\tau = \frac{1}{t}dt$

$$\frac{dH_t(L)}{d\tau} = -\sum_x \sum_y \left[1 + \log p_t(x,y)\right] \left(\frac{d}{dt} p_t(x,y)\right) t. \quad (20)$$

Now, as $p_t(x,y) = \frac{L(x,y;t)}{\sum_x \sum_y L(x,y;t)} = \frac{L_t(x,y)}{C}$

$$\frac{dp_t(x,y)}{dt} = \frac{1}{C} \frac{dL_t(x,y)}{dt} \quad (21)$$

$$\frac{dp_t(x,y)}{dt} = \frac{1}{C} \nabla\left(D \nabla L_t(x,y)\right) \quad (22)$$

$$\frac{dp_t}{dt} = \frac{1}{C} \nabla D \nabla L_t. \quad (23)$$

The Eq. (20) lends itself now as

$$\frac{dH_t}{d\tau} = -t \sum_x \sum_y \left(1 + \log \frac{L_t}{C}\right) . \frac{1}{C} \nabla D \nabla L_t \quad (24)$$

$$\frac{dH_t}{d\tau} = -t \sum_x \sum_y (1 - \log C + \log L_t) . \frac{1}{C} \nabla D \nabla L_t \quad (25)$$

$$\frac{dH_t}{d\tau} = -\frac{t}{C} \sum_x \sum_y (k + \log L_t) . \nabla D \nabla L_t. \quad (26)$$

The rate of change in the entropy for the linear isotropic diffusion case is the special case of 26, and this happens when the diffusion matrix D is replaced by a scalar diffusivity, say c. Spatial entropy change for linear isotropic diffusion process is given by

$$\frac{dH_t}{d\tau} = -\frac{ct}{C} \sum_x \sum_y (k + \log L_t) . \nabla^2 L_t. \quad (27)$$

For both, anisotropic as well as isotropic cases, the spatial entropy change equation contains the same constant $k = 1 - \log C$.

The same tests, as were performed earlier for linear isotropic diffusion process, are conducted for *linear anisotropic diffusion* process. The test anisotropic image for this purpose consists of three curves, as shown in Fig. 6. At the heart of the anisotropic process is the construction of diffusion matrix D. The diffusion matrix handles steering the elliptical Gaussian to go around the structure. The geometric visualization in the form of ellipses corresponding to point-wise diffusion matrix is displayed in Fig. 6, where it can be seen that they align well with the local flow of the curve. The diffusion parallel to the edges is enabled due to the large eigenvalue while

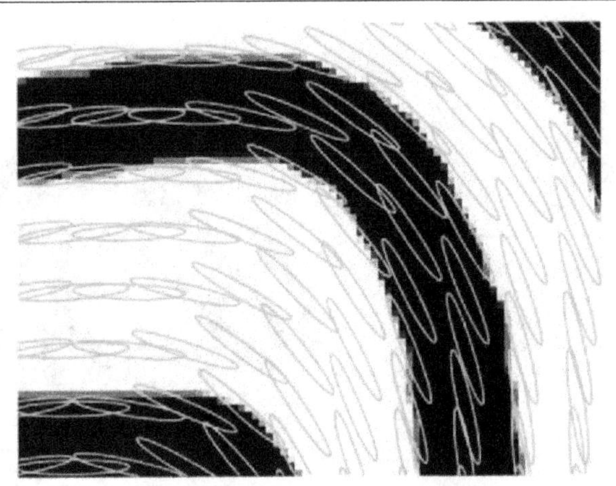

Fig. 6 Geometric interpretation of diffusion matrix. The figure shows part of the anisotropic curve image. The diffusion matrix associated with each pixel is depicted as ellipses on top of the image. It is observed that ellipses are steered to follow with the curve flow direction

avoiding the cross-over edge problems due to small eigenvalues. The linear anisotropic diffusion character is made evident by having constant eccentricity for all the ellipses across the image. The term anisotropic used here is related to changing direction of the ellipse at each pixel due to the diffusion matrix eigenvector adaptability with the given local structure. Therefore, with each iteration, the ellipse does grow without changing the eccentricity ratio and for a given diffusion time, the size of the ellipse remains constant throughout the image. Since the major axis of the ellipse is parallel to the edge of the curve, so no harm in increasing it. The minor axis of the ellipse is aligned with the width of the curve. So increasing the ellipse minor axis will eventually make the ellipse protrude outside the boundary of the curve, and the disturbed structure is obtained, and that is precisely where the diffusion should stop eventually.

First, linear anisotropic diffusion process was applied to a clean curve image. The entropy and entropy change graphs as depicted in Fig. 7b,c. Both graphs are smooth and well-behaved, validating the notion that the linear anisotropic diffusion process is a lot like their isotropic counterparts. A prominent peak is located at $\tau = 4$ in the entropy change graph, representing the characteristic width of the curves present in the image. By stopping the diffusion process by that peak location, the diffused image is shown in Fig. 7d. The image is largely undisturbed with small diffusion effects at the boundaries and ends of the curves. The quantitative measures, of sensitivity and specificity, for the output image, are computed as 82 and 89 %. The peak in entropy change graph, thus, presents itself as a suitable stopping time for the linear anisotropic diffusion process.

The experiment for linear anisotropic diffusion process was also conducted for an extremely noise situation. A Gaussian noise is added to the original curve image such that the resulting SNR is lowered to become -10 dB. The noisy curve image is displayed in Fig. 8a. After the completion of the linear anisotropic diffusion process, the entropy change graph is obtained as depicted in Fig. 8b,c, respectively. It is clearly observed that the curve for entropy change is steeply coming down in the beginning and then hits a bottom. After the minimum is reached, it rises again to display a peak at the characteristic width of the curves in the noisy image. The noise can be largely curtailed by stopping the diffusion process

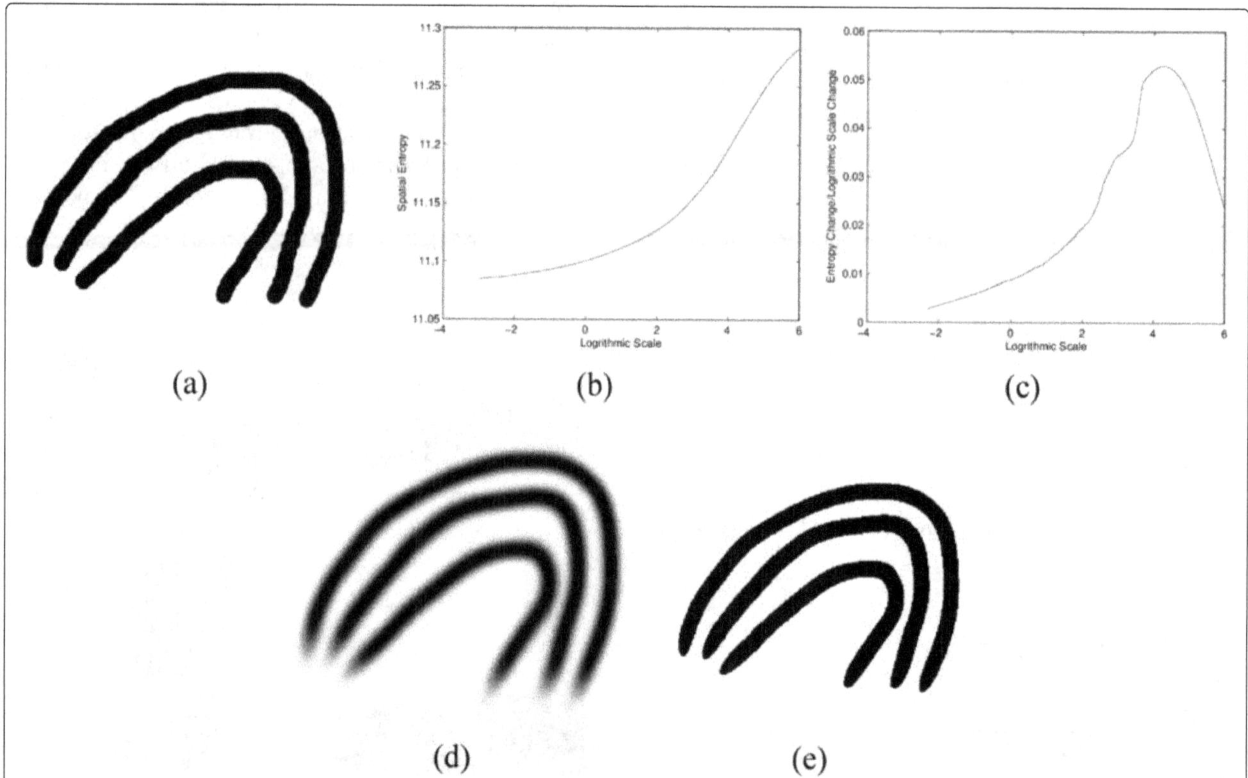

Fig. 7 Linear anisotropic diffusion process. **a** shows a flow-like test image having three black curves with a white background. The features present in the image are elongated in shape with a constant width of two pixels. **b** shows the smooth spatial entropy graph resulting from diffusion process with respect to the natural scale parameter. The entropy change with natural scale change is displayed in (**c**), where the peak corresponds to the width of the curves. The diffused image resulting from stopping the diffusion process at the location of the peak in entropy change is shown in (**d**). The diffused image is converted to the binary image using Otsu optimal threshold of 0.63, as shown in (**e**)

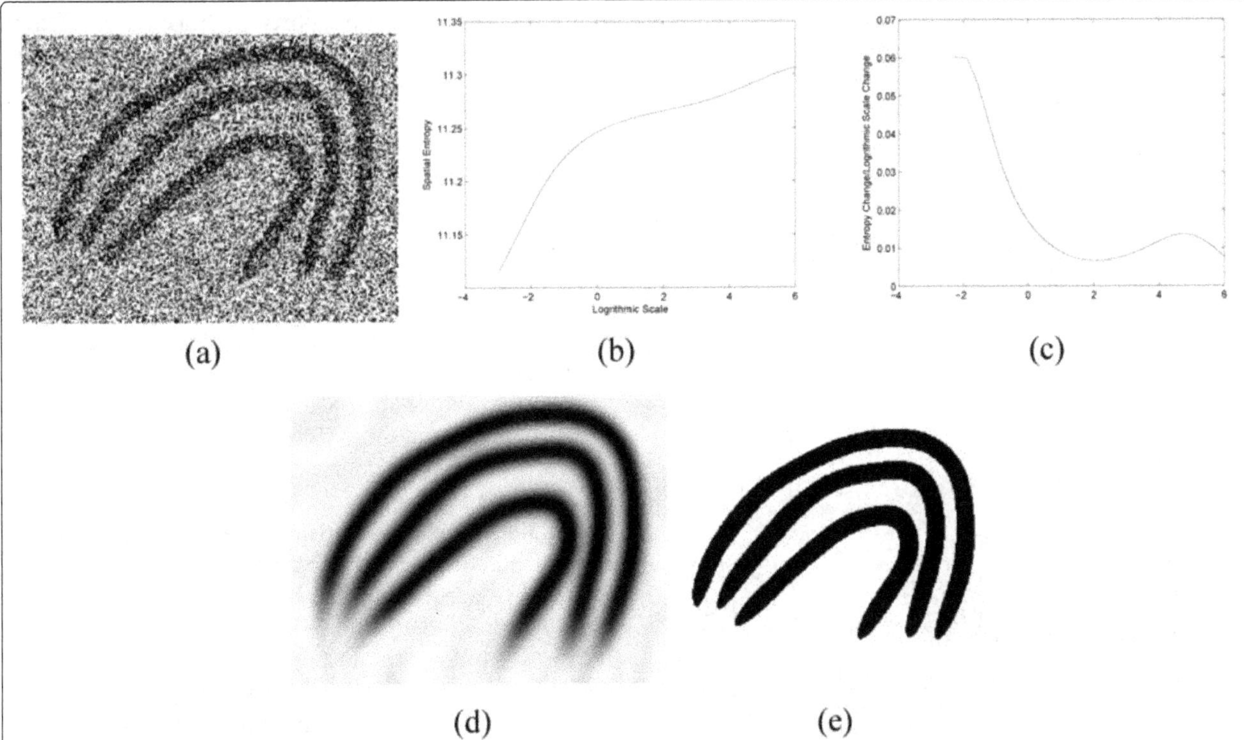

Fig. 8 Linear anisotropic diffusion process for noisy image. **a** shows a flow-like noisy test image having three black curves with a white background. The Gaussian noise is added to bring down the SNR of the resulting image to be −10 dB. **b** shows the smooth spatial entropy graph resulting from diffusion process on the natural scale parameter. The entropy change with natural scale change is displayed in (**c**), where a distinct peak is still observable. The diffused image resulting from stopping the diffusion process at the location of the peak in entropy change is shown in (**d**). The diffused image is converted to binary image using Otsu optimal threshold of 0.55, as shown in (**e**)

at the peak. The diffused image stopped by the peak is shown in Fig. 8d. The image does show a large smoothing of the noise with minimum disturbance to the structure of interest. Thresholding the image by Otsu method, a final binarized image is obtained, as shown in Fig. 8e. The quantitative measures of sensitivity and specificity for the binarized output image are recorded as 75 and 84 %.

6 Results and discussion for real fingerprint images

This section deals with real fingerprint images. We look into their acquisition process then process them for uniform background and later investigate their spatial entropy characteristic as the image evolves under linear anisotropic process. The first test that we performed is to check the anisotropic strength measure for the acquired fingerprint. Figure 9 shows the results of the test. It is observed that the regular ridge/valley pattern found in the fingerprint image is largely anisotropic in nature. This justifies the employment of linear anisotropic diffusion process for smoothing these images.

The acquired fingerprint images often show important illumination variations, poor contrast in some areas, and gaps in ridge/valley regions. To reduce the illumination imperfections and generate images more suitable for enhancement and minutia extraction, a preprocessing comprising the non-uniform illumination correction is applied. It occurs due to the very process of scanning a finger. The middle finger surface is thicker as compared to the surrounding region. This results in blocking the light in the middle while the outer surface is fairly highly illuminated. The fingerprint scanner registers this uneven illumination. Consequently, background variation will add bias for different regions of the same image to disturb the ridge/valley contrast. Since the ridge/valley pattern is identified and classified by its gray-level profile, this effect may worsen the performance of diffusion and disturb our spatial entropy analysis. With the purpose of removing this disturbing factor from our experimental analysis, a homomorphic filtering approach is adopted. The process is described below.

In basic terms, homomorphic filtering assumes that an image can be represented in terms of product of illumination and reflectance. That is

$$L(x,y) = i(x,y) \times r(x,y), \tag{28}$$

where $L(x,y)$ is the fingerprint image, $i(x,y)$ is the background illumination image, and $r(x,y)$ is the reflectance

Fig. 9 Anisotropic strength measurement for real fingerprint images. This figure shows an acquired digital fingerprint with local anisotropy strength displayed as length of the needles on top of the image. We observe a large presence of significant anisotropy in the image

image [29]. Reflectance r arises due to the object itself, but the illumination image i is independent of the object, is a pure representation of lighting conditions at the time of the image capture. To compensate for the non-uniform illumination, the illumination image part has to be made constant. Illumination is assumed to be slowly varying lending itself in the low-frequency region as compared to the reflectance image that contains abrupt changes, showing a considerable high-frequency attitude.

For implementing homomorphic filtering, we first transform the multiplicative model of image formation to additive model by moving to the log domain.

$$ln(L(x, y)) = ln(i(x, y)) + ln(r(x, y)). \tag{29}$$

Then, a low-frequency filter is used with an appropriate cutoff to get a background illumination image $i(x, y)$ estimate. The difference $d(x, y)$ between original image $L(x, y)$ and background illumination $i(x, y)$ is calculated for every pixel,

$$d(x, y) = L(x, y) - i(x, y). \tag{30}$$

To this respect, literature reports illumination-correction methods based on the subtraction of the background illumination image from the original image [30–32]. The background image is shown in Fig. 10. After subtraction, a grayish look image is obtained, as depicted in Fig. 10b. Finally, an illuminated-corrected image is obtained by transforming linearly new image pixels into the whole range of possible gray levels [0–1] using the linear stretch. Figure 10c shows the new image corresponding to stretched and uniformly illuminated image. The proposed illumination correction algorithm is observed to reduce background intensity variations and enhance contrast in the middle region than the original fingerprint image. The method was validated for all the images that were processed in the database.

To validate the effect of the homomorphic filtering, the histogram analysis is investigated before and after homomorphic filtering stage. Histogram of an image represents the relative occurrences of the gray-level present in an image. According to [29, 33], the non-uniform illumination will modify the histogram of an image in a way that it can not be binarized by a single global threshold. For this purpose, the Otsu method [34] is used, which chooses the threshold to minimize the intraclass variance of the background and foreground, to compute the binary threshold for the original fingerprint and that of the uniformly illuminated image. The results are displayed in Fig. 11.

The uniformly illuminated fingerprint image is now fed to the linear anisotropic diffusion process. The image went through diffusion evolution process from a small scale $\tau = \log(t) = -3$ till $tau = \log(t) = 5$. The normal width of the ridges was found to be 9, with half the width equal to 4.5. The spatial entropy graph is depicted in Fig. 12. We see a smooth curve with ever increasing entropy values. The entropy change graph in Fig. 12b displays a clear peak at $\tau = \log(t) = 1$, that results in $t = 2.13$. The scale value t in fingerprint images is linked to the width of the ridges as proposed in [14]. By stopping the process at $\tau = 1$, a diffused image is obtained as shown in Fig. 12c. If we let the diffusion process continue for long time ($\tau = 5$), we get a mean image as shown in Fig. 12d.

What remains to be tested is the comparison of entropy-change based stopping criterion with that of correlation-based method, presented in [35]. If the unknown additive noise n is uncorrelated with the unknown signal $u(t)$, it could be reasonable to minimize the covariance of the noise $u(0) - u(t)$ with the signal $u(t)$. The covariance is represented by the correlation coefficient and is given by,

$$corr\,(u\,(0) - u\,(t)\,,\,u\,(t)) = \frac{corr(u(0)-u(t), u(t))}{\sqrt{\text{var}(u(0)-u(t))\cdot\text{var}(u(t))}} \tag{31}$$

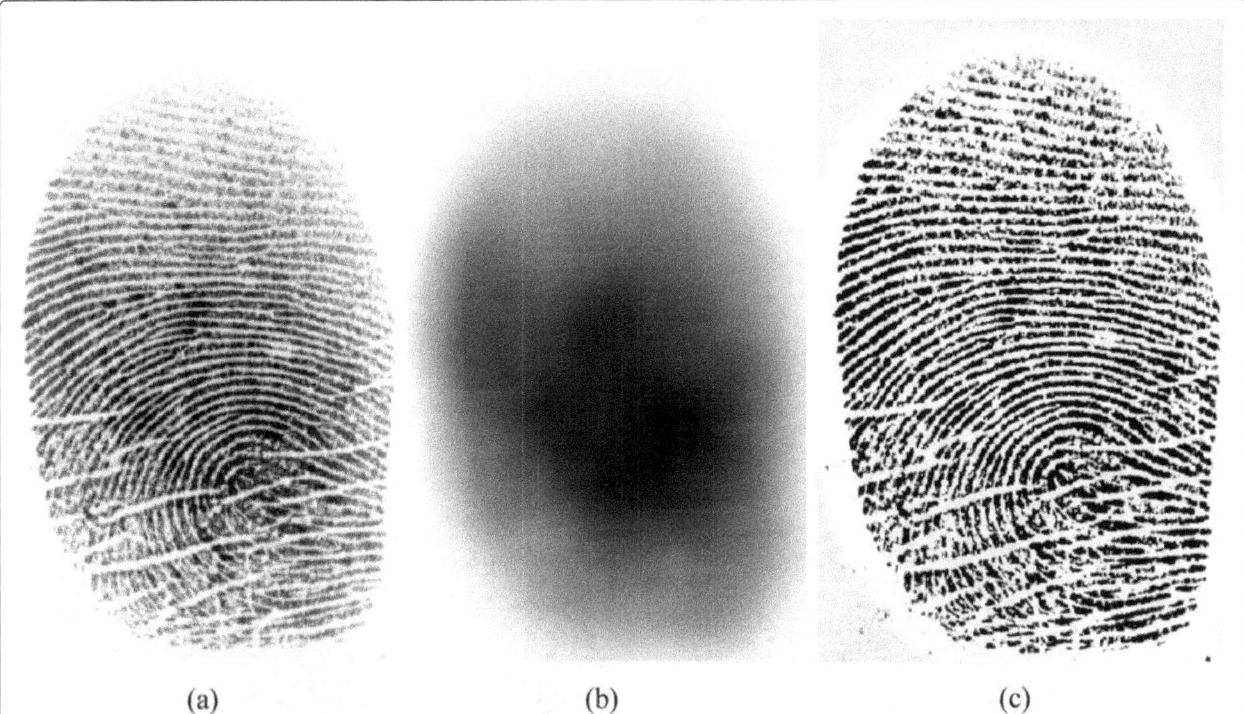

Fig. 10 Non-uniform illumination correction. **a** shows an acquired digital fingerprint. **b** depicts the estimated illumination surface, clearly showing non-uniform background lighting conditions. **c** is an output result after passing the image through homomorphic filtering operation and then linearly stretched. We observe that illumination has been corrected with clear ridge/valley structure

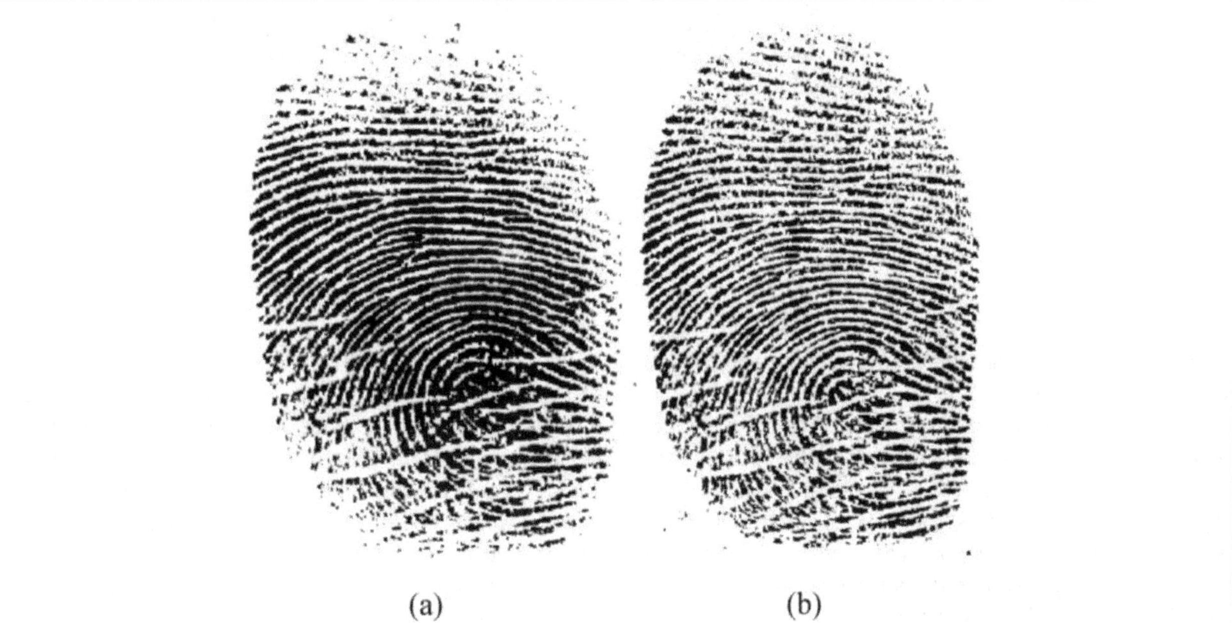

Fig. 11 Validation test for homomorphic filtering output. **a** shows the binarization of an acquired digital fingerprint using the optimal Otsu method. **b** depicts the binarization of the uniformly illuminated fingerprint with homomorphic filtering, also using the optimal Otsu method. We observe that binarization results for filtering output shows all the regions with ridge/valley structure intact

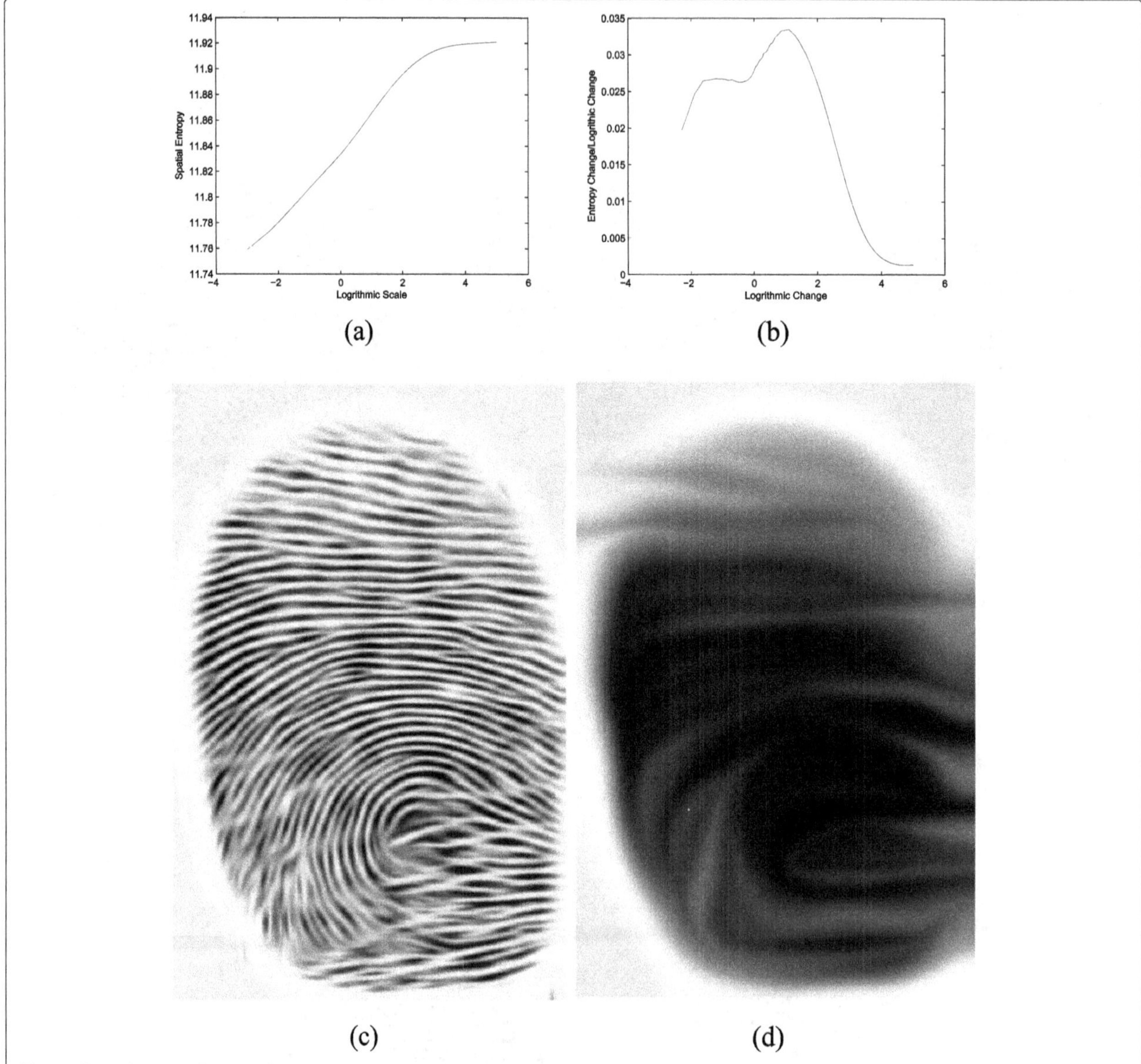

Fig. 12 Spatial Entropy for a real fingerprint image under linear anisotropic diffusion process. **a** displays spatial entropy graph of an acquired digital fingerprint. **b** depicts the entropy-change graph with one clear peak. the peak goes well with half-width of the average ridge present in fingerprint image. The diffused image obtained by peak of entropy-change is depicted in (**c**). While the image shown in (**d**) is the image we will ultimately get if we let the diffusion go on for a long enough diffusion time

and choose the stopping time T so that the expression 31 is as small as possible.

Later on, the authors in [36] proposed to use the quality of the edges in the process of finding the optimal time to stop the diffusion process. To assess the quality of our fingerprint edge structures, the edge contrast measure is used which is defined in [37]. The edge quality index is referred to as the edge based contrast measure (EBCM). The EBCM is based on the observation that human perception mechanisms are very sensitive to contours (or edges). The larger the width of the edge pixels, the larger will be this quality index. In our diffusion process, the edges are larger in width due to the poor image quality, so this EBCM is larger at the beginning of the diffusion process. After certain iterations, the smoothness of the noise happens, and the edges improve with less width and a lower value for the EBCM. After reaching a certain minimum, the edges again starts to widen due to over-smoothing, and the corresponding EBCM values increase. The best stopping time could be the minimum of the EPCM values, as shown in Fig. 13.

Image enhancement for fingerprint images is essentially to raise the contrast of ridge/valley structure, such that enhanced version is more suitable for binarization that

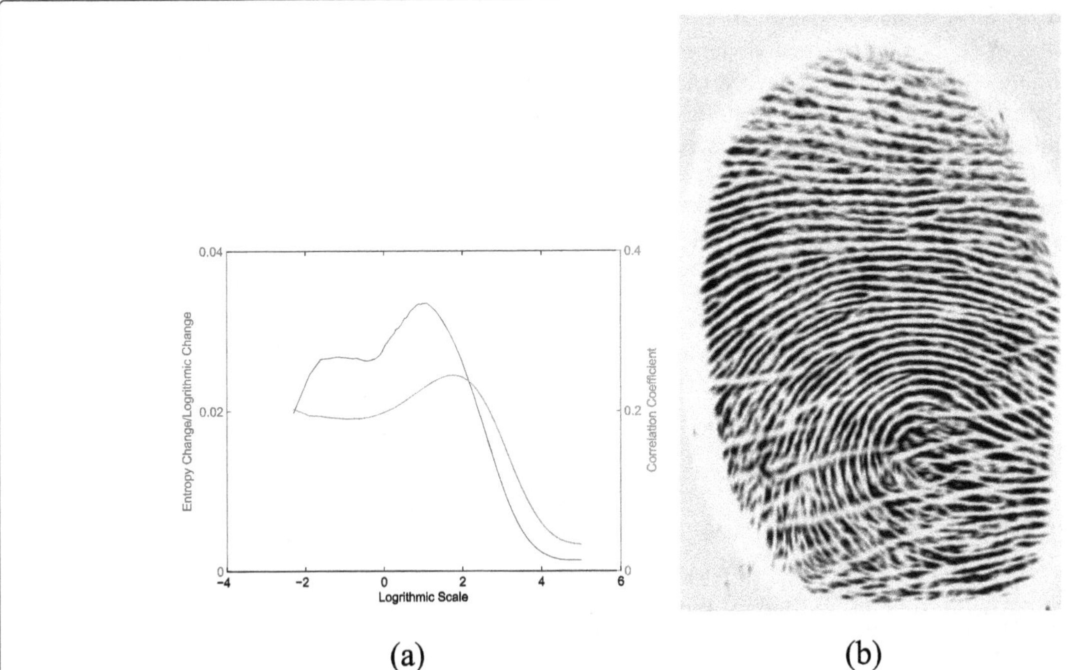

Fig. 13 Comparison between spatial entropy-based and correlation-based stopping rule. **a** displays spatial entropy change graph of an acquired digital fingerprint as black curve and correlation coefficient between (input noisy image - diffused image) and diffused image. **b** shows stopping the diffusion process at the minimum of the correlation coefficient curve. The diffused image still shows signs of interrupted ridges

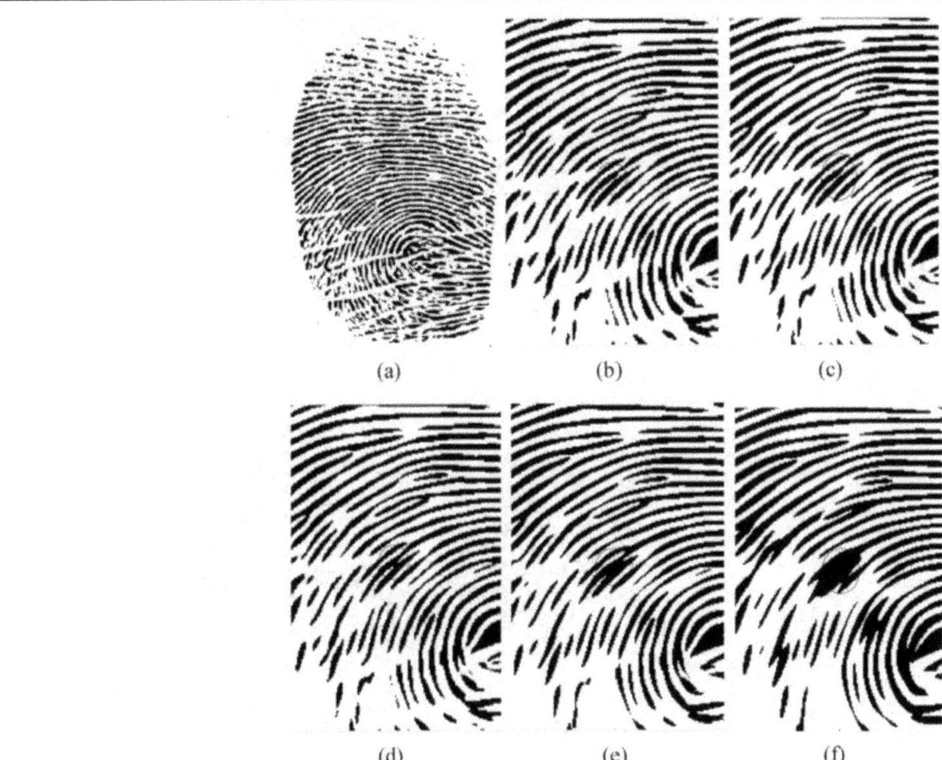

Fig. 14 Progression of diffusion for a fingerprint image. Image binarized using global threshold using the Otsu method with respect to various locations of the entropy-change graph. **a** displays image at $\tau = -1.2$, the location specified by the correlation method. Similarly, (**b**) at $\tau = 0.9$, (**c**) at $\tau = 1$, (**d**) at $\tau = 1.1$, (**e**) at $\tau = 1.4$, and (**f**) at *tau* = 2. We observe that as the diffusion increases, the the gaps within ridges started to fill. However, after a certain limit as $\tau = 1$, the closer ridges started to get merged into one. The ellipse is drawn of the portion of the fingerprint to facilitate observation

will eventually be used for automated identification system. To perform the evaluation of the real fingerprint image after diffusion, the third party minutia extractor as provided in [38] is used. The noisy acquired images were stopped at three different time instants due to correlation method, EPCM, and the proposed entropy-change based, and the resulting three output diffused images were then compared quantitatively. Analysis of the diffused image yields a list of candidate minutiae. However, due to the use of non-optimal stopping time, there are usually a large proportion of false minutiae, i.e. points that have been incorrectly identified as minutiae. This diffusion process directly affects the binarization which creates wrong minutiae, as shown in Fig. 14. Therefore, the total number of candidate minutiae detected in three types of diffused images indicate the relative degree of noisiness

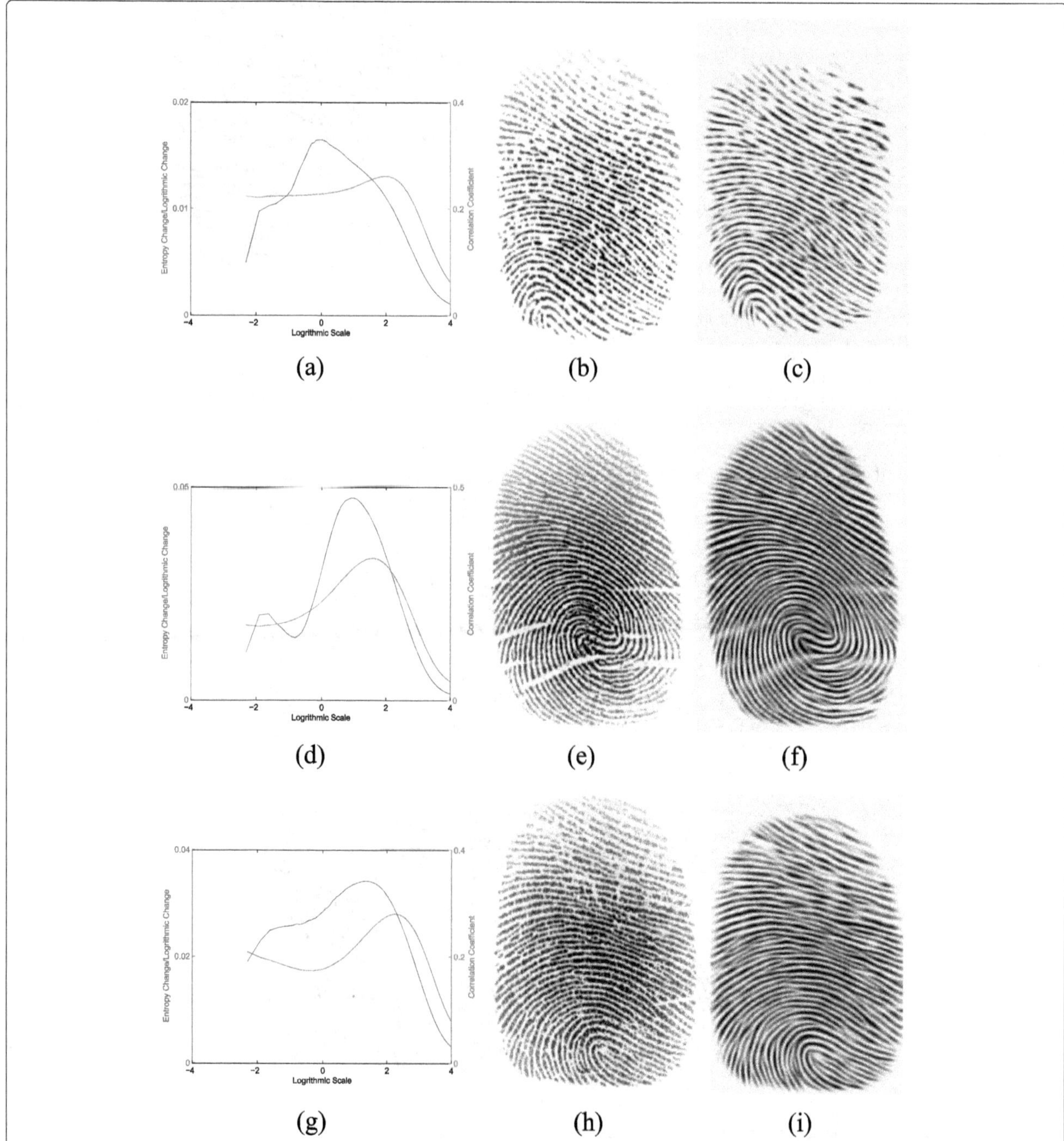

Fig. 15 Comparison of fake minutia point of the proposed method with correlation-based and EBSM methods. **a**, (**d**), and (**g**) are the graph shows the comparison of stopping time for correlation-based in green and proposed method in blue. **b**, (**e**) , (**h**) are the fingerprint images diffused and stopped by the correlation-based optimal stopping method. **c**, (**f**), and (**i**) are the final optimal stopped images for the proposed method

still present in them, and will cause false minutiae. Figure 15 depicts a comparison of fake minutiae of the proposed method with correlation-based and EBSM method. A Table 1 has been generated for the six test images from the university campus students, that indicate that correlation based stopping method and EPSM-based stopping criterion had detected considerable more minutiae, indicating the immature diffusion of the noisy input image. The correlation-based stopping generated on the average 350 minutiae per image (4 times the ground truth image) while EPSM provided 210 minutiae per image (2.4 times the ground truth). The proposed entropy-change generated 145 minutiae per image (1.65 time ground truth).

Another set of experiments was conducted to assess the suitability of proposed stopping criterion for some extremely low-quality fingerprint images present in the FVC2004 database to assess the ultimate strength of the proposed stopping rule. One such challenging image is displayed in Fig. 17c. The fingerprint shows broken ridges, salt and pepper noise, non-uniform illumination, and on top of it a dark square patch right at the center. The image was preprocessed first with small median filter of size 3×3 to tackle salt and pepper noise, and was then made to go through homomorphic filtering to eliminate to a larger extent the non-uniform background variations.

After initial treatments, the image was passed on to a linear diffusion process to join broken ridges while avoiding the mixing of ridge/valley pattern. A modified coherence enhancing diffusion (CED) as suggested earlier in linear anisotropic section proves to be of little success for diffusing low-quality fingerprints. This is due to the finding that our earlier attempts at introducing constant eigenvalues with CED process (to transform CED into a linear anisotropic process) seems to inadequate for low-quality fingerprint image diffusions. The spatial entropy curve was found to be increasing in the beginning but show a dip in spatial entropy values towards the end (large logarithmic scales). A search was conducted to look into some recent robust variant of CEDs while dealing with low-quality fingerprints. The search culminated into a new class of diffusion process that was developed specifically for low-quality challenging fingerprints. The new process deploys a precomputed orientation field to transform the Coherence-enhancing diffusion process into that of linear oriented diffusion process [39], much more robust to the extremely noisy situations. The new process was studied with special care for its spatial entropy behavior while smoothing low-quality fingerprints. The spatial entropy was found to be monotonically growing quantity as a function of increasing logarithmic scale. This desirable behavior was found to be consistent across many database images that were tested here. The large part of the stable behavior for entropy graph can be attributed to the injection of precomputed orientation filed that was extremely helpful to steer the diffusion matrix in right direction in sensitive later stages of diffusion process, where large scales were involved. Specifically, the linear oriented diffusion process was adopted for experimentation here with two fixed eigenvalues as $\lambda_1 = 0.01$ and $\lambda_2 = 1 - 0.01$. The diffusion matrix was constructed as before:

$$d11 = \lambda_1 \cos^2(\theta) + \lambda_2 \sin^2(\theta), \quad (32)$$

$$d12 = (\lambda_1 - \lambda_2) \sin(\theta) \cos(theta), \quad (33)$$

$$d11 = \lambda_1 \sin^2(\theta) + \lambda_2 \cos^2(\theta), \quad (34)$$

but with one major change that is θ is now precomputed orientation field from the use of directional filter bank framework for the image [40]. The orientation field θ was kept constant in the whole evolution process. The diffusion process was evolved starting from scale $\tau_i = \log(t = \exp(-3))$ and reaching final scale$\tau_f = \log(t = \exp(5.5))$ (providing mean value image) with a step size of $t = \exp(-3)$. The spatial entropy was computed along the way and reported to be growing entity with steady value at the end, as depicted in Fig. 16d. The entropy graph contains a multitude of discontinuities corresponding to a small left-over noise particles in the fingerprint after preprocessing. The curve can be smoothed by fitting a piecewise spline while caring for some real big discontinuities. To do so, a smoothing spline function was fitted to the noisy entropy curve with a coarser soothing parameter of value 0.95 on a scale of $[0, 1]$. The entropy change curve is constructed from fitted spline curve and is depicted in Fig. 16e. It shows a number of peaks representing different structures dominating at different scales. There may well be some small broken parts of otherwise long ridges. The last peak at the farthest end represents the largest dominating structure that may be linked tom average ridge width of the fingerprint. By stopping the linear diffusion process at that peak $\tau = 3.2$, the diffused image is displayed in Fig. 16f. The uneven image contrast can be straightforwardly improved using well-known block-based contrast enhancement scheme such as contrast limited adaptive

Table 1 A comparison. Total minutiae found by the detection algorithm enhanced by edge-width-based, correlation-based, and entropy-change-based. The sample images are used from FVC2004 DB2_B 101_1 to 101_6

	Edge width-based	Correlation-based	Entropy-change-based
Image1	220	367	155
Image2	200	333	135
Image3	222	370	150
Image4	208	330	140
Image5	224	380	160
Image6	206	320	130

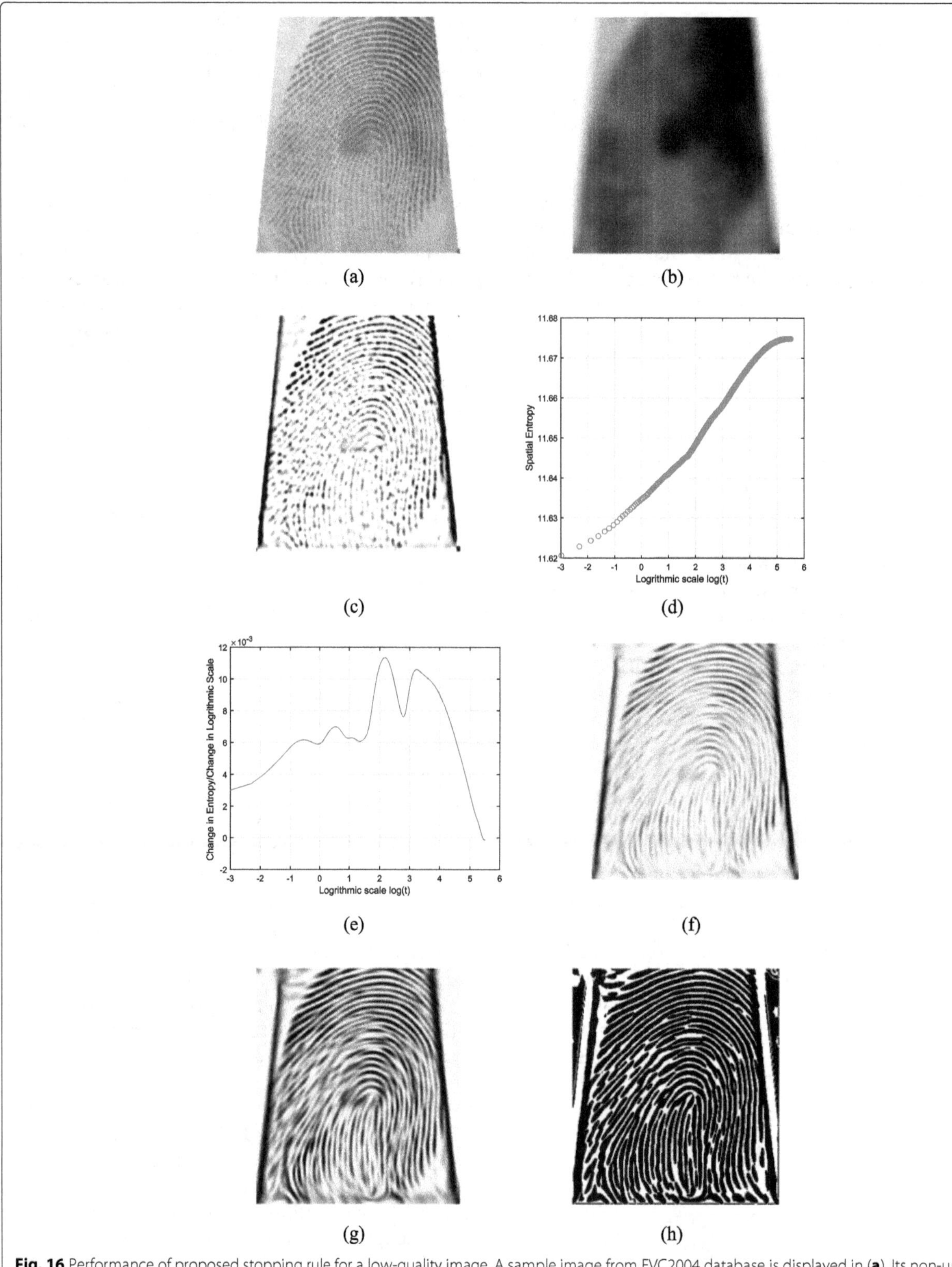

Fig. 16 Performance of proposed stopping rule for a low-quality image. A sample image from FVC2004 database is displayed in (**a**). Its non-uniform illumination image is extracted as shown in (**b**). The (**c**) depicted the uniform image. Spatial entropy points for the uniform image are plotted in (**d**). A piecewise smooth spline was fitted due to noisy nature of the entropy points, and subsequently, its derivative is computed as shown in (**e**), proving a smoothed entropy-change curve with increasing logarithmic scale. The optimally diffused image stopped at the farthest peak in entropy-change curve is displayed in (**f**). The contrast-adjusted image through linear stretch is shown in (**g**). Finally, a 9 × 9 block-based binarization was used to come up with a clean binary image as depicted in (**h**)

histogram equalization (CLAHE) [41], to provide evenly-contrasted image, as in Fig. 16g. The contrast-adjusted image was then binarized with a block-by-block process to result in Fig. 16h. The binarized result shows a clear fingerprint with ridge/valley structure largely intact (minimum mixing of nearby ridges) with greatly diminishing the intensity of noise. Most of the genuine minutia points (ridge ending and bifurcation points) are still valid and can be easily detected by the subsequent extraction process.

To quantitatively assess the performance of proposed stopping rule for image diffusion, a measure goodness index (GI), was adopted from an earlier fingerprint image enhancement [42]. This goodness index (GI) is defined as follows:

$$GI = \frac{\sum_{i=1}^{r} q_i \left[p_i - a_i - b_i \right]}{\sum_{r=1}^{r} q_i t_i}, \qquad (35)$$

where, p represents the paired minutiae (between the manually extracted and machine extracted), a represents the missing minutiae, b represents the spurious minutiae and t represents the true minutiae. The measure is suppose to give a number between 0 and 1. This goodness index is applied on Fig. 17c. The GI without enhancement is found to be 0.34, with enhancing using CED [18] is 0.45 and after applying the proposed method is 0.52. A larger test is performed on the 40 images of FVC2000 DB4_B (101 to 105). The averaged GI without enhancement comes out to be 0.26, with enhancing using CED [18] is 0.37 and after applying the proposed method is 0.43.

The proposed stopping rule being an iterated process can be analyzed with its computation complexity profile. The stopping rule involves three nested loops. First one is the do-while loop that let the process runs till it reaches the farthest peak in the entropy change graph, and the remaining two are FOR loops that span the dimensionality of the fingerprint. Therefore, an estimate of the computational complexity associated with the proposed stopping rule can be described as a product $N \times M \times ITERATIONS$, where N and M represents the rows and columns of the fingerprint and ITERATIONS are the count of repetitions to reach the required peak. Since the peaks represent the dominating structure, which is this case is the width of the ridges, an experiment was conducted to see that linkage more explicitly. A sequence of same dimension fingerprint images was created by increasing zoom values and center cropping the resultant image. For each of these

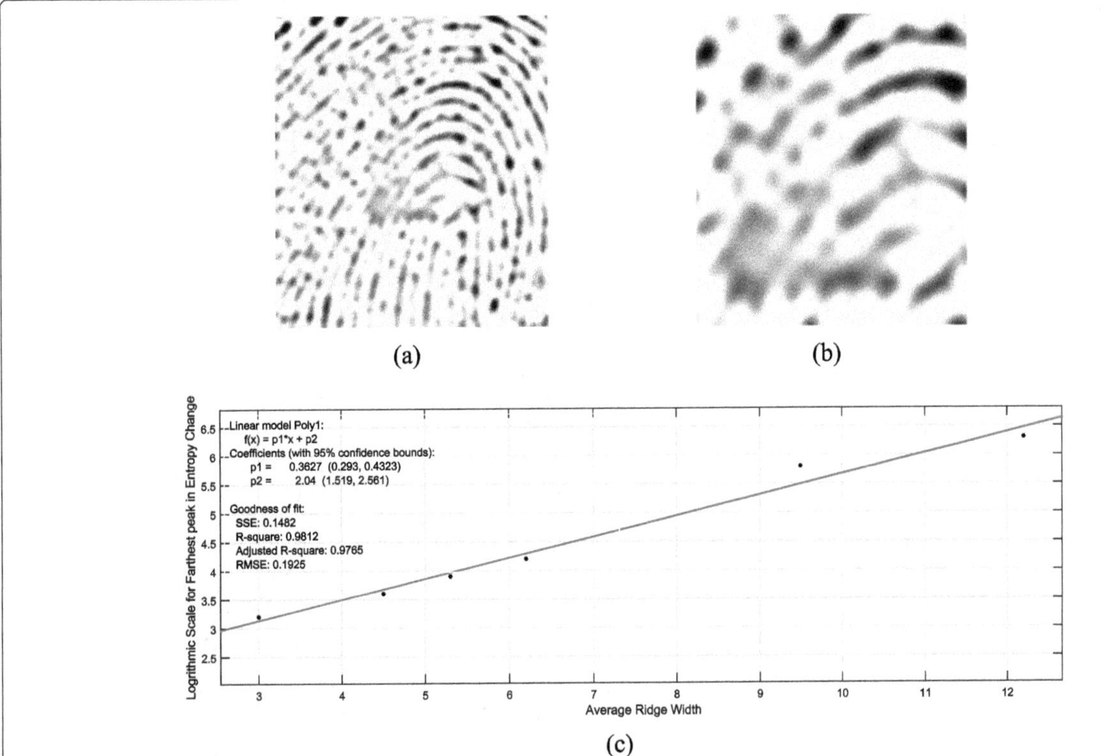

Fig. 17 Linear relationship between stopping point and average ridge width. The figure shows the fitting of a linear curve through some discrete points for the stopping point of the entropy-change versus logarithmic scale curve, corresponding to the furthest peak. For creating increasing large ridge widths, the center-cropped zoomed images of same dimension are being employed. The figure shows the first and the last such zoomed images. Figure (**a**) Shows the first zoomed image and (**b**) displays the last image in zoomed series. **c** gives a comparison of average ridge width with Logarithmic scale for farthest peak in Entropy change

images, an identical linear diffusion scheme with pre-computed orientation filed was run to locate the desired peak in their respective entropy-change graphs. A plot in Fig. 17 is shown connecting logarithmic scale at which the process stopped and the average width of the ridges in the respective zoomed images. The graph in fig shows the dots, obtained from this experiment, and were fitted with a linear curve having 95 % confidence interval. The logarithmic scale, at which the diffusion process stopped, in turn, can provide the number of iterations knowing the step size involved in the diffusion process. Thus, given dimension of the input fingerprint and an estimate of the average ridge width, a reasonable guess at the computation complexity of the proposed stopping rule can be reached.

7 Conclusions

In this paper, the entropy-change for an anisotropic diffusion of a fingerprint image is investigated. a unique peak is found, associated with blurring of the dominant structure. This provides a reasonable stopping rule for the anisotropic diffusion process, whose goal is to smooth the image without disturbing the structural information. The numerical results validated the existence of the boundary between under-smooth and over-smooth regions of anisotropic diffusion.

Competing interests

All the authors of this paper declare that they have no significant competing financial, professional, or personal interests that might have influenced the performance or presentation of the work described in this manuscript.

Acknowledgements

The authors would like to acknowledge the support of the Department of Engineering, Macquarie University, Sydney, Australia for the work presented in this paper.

Author details

[1]Department of Engineering, Macquarie University, Balaclava Rd, 2109 Sydney, Australia. [2]Department of Electrical and Computer Engineering, Biometric and Sensor Lab, Effat University, Jeddah, Saudi Arabia.

References

1. J Bernd, *Spatio-Temporal Image Processing*, 1st, vol. 751. (Springer-Verlag, Berlin Heidelberg, 1993)
2. AZ Averbuch, FG Meyer, JO Stromberg, RR Coifman, A Vassiliou, Low bit-rate efficient compression for seismic data. IEEE Trans. Image Process. **10**(12), 1801–1814 (2001)
3. J Weickert, Coherence enhancing diffusion of colour images. Image Vis. Comput. **17**, 201–212 (1999)
4. TM Khan, MA Khan, Y Kong, Boosting CED using robust orientation estimation. Int. J. Multimedia Appl. **6**(2) (2014)
5. MAU Khan, TM Khan, Fingerprint image enhancement using data driven Directional Filter. Bank. Optik-In. J. Light Electron Optics. **124**(23), 6063–6068 (2013)
6. MAU Khan, A Khan, TM Khan, M Abbas, N Mohammad, in *International Conference on Information and Emerging Technologies (ICIET)*. Fingerprint image enhancement using principal component analysis (PCA) filters (IEEE, Karachi, 2010), pp. 1–6
7. M Kaas, A Witkin, Analyzing oriented patterns. Comput. Vis. Graphics Image Process. **37**, 362–385 (1987)
8. MAU Khan, TM Khan, O Kittaneh, Y Kong, Stopping criterion for anisotropic image diffusion. Optik-Int. J. Light Electron Optics. **127**(1), 156–160 (2016)
9. G Hellwig, *Partial differential equations*. (Teubner, Stuttgart, 1977)
10. IG Petrowsk, *Vorlesungen uber partielle Differentialgleichungen*. (Teubner, Leipzig, 1955)
11. P Perona, J Malik, Scale-space and edge detection using anisotropic diffusion. IEEE Transa. Pattern Anal. Mach. Intell. **12**(7), 629–639 (1990)
12. GH Cottet, L Germain, Image processing through reaction combined with nonlinear diffusion. Math. Comput. **61**, 659–673 (1993)
13. J Weickert, *Anisotropic Diffusion in Image Processing*. (ECMI Series, Teubner-Verlag, Stuttgart, Germany, 1998)
14. A Almansa, LT Indeberg, Fingerprint enhancement by shape adaptation of scale-space operators with automatic scale selection. IEEE Trans. Image Process. **9**, 2027–2041 (2000)
15. TM Khan, MA Khan, Y Kong, Fingerprint image enhancement using multi-scale DDFB based diffusion filters and modified Hong filters. Optik-Int. J. Light Electron Optics. **125**(16), 4206–4214 (2014)
16. MAK Khan, TM Khan, SA Khan, in *7th International Conference on Emerging Technologies*. Coherence enhancement diffusion using Multi-Scale DFB (IEEE, Islamabad, 2011), pp. 1–6
17. M Nitzberg, T Shiota, Nonlinear image filtering with edge and corner enhancement. IEEE Trans. Pattern Anal. Mach. Intell. **14**, 826–833 (1992)
18. J Weickert, Coherence-enhancing diffusion filtering. Int. J. Comput. Vis. **31**, 111–127 (1999)
19. D Chen, MacS Lachlan, M Kilmer, Iterative parameter-choice and multigrid methods for anisotropic diffusion denoising. SIAM J. Sci. Comput. **33**, 2972–2994 (2011)
20. A Ilyevsky, E Turkel, Stopping criteria for anisotropic PDEs in image processing. J. Sci. Comput. **45**, 333–347 (2010)
21. G Gilboa, N Sochen, YY Zeevi, Estimation of optimal PDE-based denoising in the SNR sense. IEEE Trans. Image Process. **15**(8), 2269–2280 (2006)
22. G Gilboa, Nonlinear scale space with spatially varying stopping time. IEEE Trans. Pattern Anal. Mach. Intell. **30**, 2175–2187 (2008)
23. J Weickert, *Scale-space properties of nonlinear diffusion filtering with a diffusion tensor Report No. 110*. (Laboratory of Technomathematics, University of Kaiserslautern, P.O. Box 3049, 67653 Kaiserslautern, Germany, 1994)
24. M Ferraro, G Boccignone, T Caelli, On the representation of image structures via scale space entropy conditions. IEEE Trans. Pattern Anal. Mach Intell. **21**(11), 1199-1203 (1999)
25. J Sporring, in *Proceeding of ICPR'96*. The entropy of scale-space (Springer, Washington DC, 1996), pp. 900–9004
26. J Sporring, *The entropy of scale-space. Department of Computer Science / University of Copenhagen Universitetsparken 1 / DK-2100 Copenhagen East Denmark*, (1996)
27. AP Witkin, in *Proc. of International Joint Conference on Arti?cial Intelligence (IJCAI)*. Scale space filtering, (1983)
28. JJ Koenderink, The structure of images. Biol. Cybernet. **50**(5), 363–370 (1984)
29. RC Gonzalez, RE Woods, *Digital Image Processing*, 3rd ed. (Prentice-Hall, Inc., Upper Saddle River, NJ, USA, 2006)
30. AJ Frame, PE Undrill, MJ Cree, JA Olson, KC McHardy, PF Sharp, JV Forrester, A comparison of computer based classification methods applied to the detection of microaneurysms inophthalmic fluorescein angiograms. Comput. Biol. Med. **28**(3), 225–238 (1998)
31. M Niemeijer, B van Ginneken, JJ Staal, MSASSMD Abramoff, Automatic detection of red lesions in digitalcolor fundus photographs. IEEE Trans. Med. Imaging. **24**(5), 584–592 (2005)
32. B Zhang, X Wu You, Q Li, F Karray, Detection of microaneurysms using multi-scale correlation coefficients. Pattern Recognit. **43**, 2237–2248 (2010)
33. M Sharma, A Kumar, Non uniform background illumination removal (NUBIR) from microscopic images. Int. J. Adv. Res. Technol. Eng. Sci. **1**(2), 23–29 (2014)
34. N Otsu, A threshold selection method from gray-level histograms. IEEE Trans. Syst. Man Cybernet. **9**(1), 62–66 (1979)
35. P Mrazek, Selection of optimal stopping time for nonlinear diffusion filtering. Int. J. Comput. Vis. **52**(2), 189–203 (2003)

36. C Tsiotsios, M Petrou, On the choice of the parameters for anisotropic diffusion in image processing. Pattern Recognit. **46**(5), 1369–1381 (2013). Article in Press
37. A Beghdadi, AL Negrate, Contrast enhancement technique based on local detection of edges. Conpu. Vis. Graphics Image Process. **46**, 162–174 (1989)
38. MATLAB, MATLAB Central; 20015. MATLAB Central - MathWorks
39. C Gottschlich, CBS Nlieb, Oriented diffusion filtering for enhancing low-quality fingerprint images. IET Biometrics. **1**, 105–113 (2012)
40. MAU Khan, K Ullah, A Khan, IU Islam, Robust multi-scale orientation estimation: Directional filter bank based approach. Elsevier J. Appl. Math. Comput. **242**, 814–824 (2014)
41. K Zuiderveld, *Contrast limited adaptive histogram equalization*. (Academic Press Professional, Inc., San Diego, CA, USA, 1994)
42. L Hong, Y Wan, A Jain, Fingerprint image enhancement: algorithm and performance evaluation. IEEE Trans. Pattern Anal. Mach. Intell. **20**, 777–789 (1998)

10

Rotation update on manifold in probabilistic NRSFM for robust 3D face modeling

Chengchao Qu[1,2*], Hua Gao[3] and Hazım Kemal Ekenel[4]

Abstract

This paper focuses on recovering the 3D structure and motion of human faces from a sequence of 2D images. Based on a probabilistic model, we extensively studied the rotation constraints of the problem. Instead of imposing numerical optimizations, the inherent geometric properties of the rotation matrices are taken into account. The conventional Newton's method for optimization problems was generalized on the rotation manifold, which ultimately resolves the constraints into unconstrained optimization on the manifold. Furthermore, we also extended the algorithm to model within-individual and between-individual shape variances separately. Evaluation results give evidence to the improvement over the state-of-the-art algorithms on the Mocap-Face dataset with additive noise, as well as on the Binghamton University A 3D Facial Expression (BU-3DFE) dataset. Robustness in handling noisy data and modeling multiple subjects shows the capability of our system to deal with real-world image tracks.

Keywords: Non-rigid structure from motion, Manifold optimization, Newton's method, PLDA, Face model

1 Introduction

Recovering scene geometry and camera motion from sequences of 2D monocular images has seen significant success for the 3D geometry of static objects. The widely used rigid factorization method was first introduced by Tomasi and Kanade [1]. Orthonormality constraints are adopted on the rotation matrices in order to recover structure and motion in a single step. Unfortunately, most biological objects and natural scenes are deformable. 3D rigid motions, i.e., camera rotation and translation, along with non-rigid deformations, e.g., stretching and bending, are mixed altogether in their image measurements. Hence, extending the existing rigid algorithms to the non-rigid scenario turns out to be a far more challenging task than it appears to be.

It is known that the problem of non-rigid structure from motion (NRSFM) is generally underconstrained and thus intractable, if each point of the object moves arbitrarily. In practice, however, many objects, e.g., faces, deform under certain rules. A possible approach is to learn an application-specific 3D model of non-rigid structure from the training data to constrain deformation [2]. Another possibility is to hard-code and learn a model incrementally [3]. Some approaches [4–7] were proposed from another perspective to remove the need of such a prior model, which is not available in most real-world situations. The shape model, i.e., shape bases, is treated as unknowns to be solved, with only the orthonormality constraints on camera rotations being utilized. Xiao et al. [8] proved that only enforcing the orthonormality constraints is not enough for the factorization-based method; therefore, they introduced the basis constraints to reduce ambiguity.

In this work, two major contributions have been made. We first investigated the geometric properties of the orthonormality constraints and generalized the Newton's optimization method to the underlying manifold of the camera rotation matrices. That means, non-linear optimization can be carried out on the manifold without any imprecise approximations. We used a probabilistic principal component analysis (PPCA)-based framework [9] to model NRSFM as it is more robust to noise than the closed-form factorization techniques. Our second contribution is about dealing with multiple subjects. The current NRSFM algorithms mostly focus on the reconstruction of a single subject. While dealing with

*Correspondence: qu@kit.edu
[1]Vision and Fusion Laboratory (IES), Karlsruhe Institute of Technology (KIT), Adenauerring 4, 76131 Karlsruhe, Germany
[2]Fraunhofer Institute of Optronics, System Technologies and Image Exploitation (Fraunhofer IOSB), Fraunhoferstr. 1, 76131 Karlsruhe, Germany
Full list of author information is available at the end of the article

data containing multiple subjects, no difference is taken into account, when modeling between-individual variation (e.g., face model of different identities) and within-individual variation (e.g., facial expression of the same identity). For that reason, we extended the PPCA-based framework to the probabilistic linear discriminant analysis (PLDA) [10] model to improve reconstruction performance on data with multiple subjects.

The remainder of this paper is organized as follows. Previous research on NRSFM is reviewed in Section 2. Section 3 presents the probabilistic NRSFM model [9] and our novel manifold optimization technique on the orthonormality constraints. Section 4 discusses the experimental results of our algorithm. Finally, we conclude our work in Section 5.

2 Related work

Modern structure from motion (SFM) algorithms employ the factorization method for orthographic camera projection proposed by Tomasi and Kanade [1]. The rank theorem ensures that the input matrix can be factorized into two matrices, one corresponds to the camera motion, and the other represents the shape. Although the resulting matrices from singular value decomposition (SVD) are not unique, they only differ by a linear transformation. By imposing metric constraints, a decent solution of the SFM problem for rigid objects can be achieved.

In the seminal work of Bregler et al. [11] and Torresani et al. [6] for solving NRSFM, they assumed that the 3D shape of an object can be explained as a linear combination of deformation shapes applied to a dominant rigid component. In this way, the non-rigid motion recovery is formulated as a factorization problem and the low rank of the image measurements is analyzed. In general, this model assumes that the number of basis shapes should be known, an inaccurate choice that can lead to performance drop. Theoretically, if the number is underestimated, it is not sufficient to represent all variations of the object; otherwise, the extra degree of freedom is unconstrained and is unlikely to generalize well, which starts fitting to noise.

Using the linear representation, Xiao et al. [8] proposed a closed-form scheme for solving the NRSFM problem. They proved in the previous work that by imposing orthonormality constraints alone on camera rotations, the increased degree of freedom will cause ambiguity. The additional basis constraints will determine the shape bases uniquely. In [12], Xiao and Kanade pointed out that even enforcing both sets of linear metric constraints above could still lead to ambiguity, if there exist degenerate bases, which are not of full rank three. However, by exploiting the rank three constraints inherently, Akhter et al. [13] analytically proved that orthonormality constraints alone are sufficient to recover the exact structure. Ambiguity solely lies in the transformation of linear basis vectors, which does not affect the 3D structure reconstruction. Dai et al. [14] proved this claim by solving the NRSFM problem without any prior using matrix trace norm minimization.

Torresani et al. [9] proposed a probabilistic deformation model based on PPCA and suggested that it reveals better reconstruction result than the conventional linear model. In their work, 3D shapes are drawn from non-uniform probability distribution functions (PDFs) with a Gaussian prior on each shape in the subspace instead of the common linear subspace model, which is a specific usage of PPCA. The parameters of the PDF are unknown in advance, which will be optimized using the expectation-maximization (EM) algorithm together with the 3D shapes and rigid motions. An advantage of PPCA over the simple deterministic subspace model is that degeneracy of closed-form solutions does not occur so that the ambiguity problem figured out by Xiao et al. in [12] does not happen here. However, the rotation matrices are approximated by using a single Gauss–Newton step with a fixed updating step length, which can lead to a considerable performance drop in the rotation reconstruction if no proper metric on the manifold is defined.

Over the last years, more research on NRSFM has also been done using various forms of non-linear optimization techniques to minimize the 3D reprojection error. In order to overcome the degeneracy problem, some additional heuristic constraints were introduced. Shaji and Chandran [15] introduced a canonical Riemannian metric on the product span subspace of the rotation matrices and articulated shape weights. The Newton's algorithm is generalized to the product manifold to recover those parameters, while the Wiberg algorithm is employed to solve the shape update. It differs from our approach in that our framework uses a probabilistic model with a posterior objective function over the latent variables, which is more robust to noise introduced by tracking error or manual labeling. Section 4 shows the robustness of our model with extreme conditions of noise.

Other than recovering the whole 3D shapes and motion parameters like in almost all the existing applications, Rabaud and Belongie [16] presented a manifold learning approach that only focuses on an embedding of frames within the input image sequence. The intuition is as follows: given enough image frames, a non-rigid deformed 3D shape can be observed several times in different view angles. If some of the frames share a low 3D reconstruction error, they are highly likely to represent a similar 3D shape, otherwise it means a poorly matched set of frames. Following this principle, triplets of frames are compared to exploit all repetitions in possible shape deformations. Then the generalized non-metric multi-dimensional scaling framework is used to estimate the weight of each deformation shape. Bundle adjustment is employed as a further optimization step, which minimizes the reprojection error. This closed-form approach can reconstruct accurate 3D shape on a clean synthetic dataset; however, with the amount of noise added, their performance drops very fast and approaches that of PPCA. Tao and Matuszewski [17] also employed manifold learning-based diffusion maps to handle highly deformable objects.

By exploiting the temporal smoothness of the shape trajectories across the images, Akhter et al. [18] addressed the NRSFM problem in trajectory space, which is the dual problem to the conventional spatial shape bases. By describing the 3D point trajectory linearly using object independent discrete cosine transform (DCT) vectors, unknowns in estimation are reduced, and stable reconstruction is achieved as a result. Gotardo and Martinez extended the temporal dependence to iteratively obtain higher-frequency DCT in [19] and explicitly modeled the complementary spaces of rank three in [20]. Valmadre and Lucey [21] formulated the regularization of the trajectory basis with a temporal filter. Recently, Park et al. [22] simplified the global motion estimation of the trajectory basis with the aid of a few stationary points in the scene. Despite the robust performance on various motion capture datasets, limitation of its application is also obvious, i.e., the object deformation should be temporally continuous and smooth. Otherwise, higher-frequency DCT vectors are needed, which significantly increases the rank of the trajectory matrix factorization and will eventually lead to degeneration and unstable performance. In contrast, the primary problem in shape space does not suffer from this.

In this paper, we demonstrate a probabilistic, iterative alternating approach to solve the NRSFM problem. In contrast to Torresani et al. [9], the conventional Newton's method is generalized on the rotation manifold to solve the optimal rotation matrix for each optimization iteration. The orthonormality constraints are naturally guaranteed by the metric update step without the need of being projected back after constrained optimizations on the Euclidean space. Additionally, a generic PLDA model that takes into account the commonness across all subjects, as well as the specific characteristics between the subjects, can be learned. On datasets with more than one subject, better individual reconstruction is achieved even if insufficient number of frames are available for each subject.

3 NRSFM model

Most of the state-of-the-art NRSFM algorithms make use of a linear subspace model to represent the shape model. A linear combination of deformation shapes is thereby applied to a dominant rigid component. Let the $3P \times 1$ matrix $\bar{\mathbf{s}}$ be the mean shape and the $3P \times K$ matrix $\mathbf{V}$ and the K-dimensional vector $\mathbf{z}_t$ be the remaining basis shapes and their weights, respectively, where P is the number of landmarks in each image frame and K the number of articulation shapes apart from the mean shape. The 3D shape of the tth frame is represented as

$$\mathbf{s}_t = \bar{\mathbf{s}} + \mathbf{V}\mathbf{z}_t. \tag{1}$$

Note that shapes are stacked in matrix $\mathbf{V}$ so that each column represents a basis shape. Camera rotation in frame t is denoted by the 2×3 matrix $\mathbf{R}_t$. Due to the inevitable presence of internal and external noise in image tracks or labeling, a zero-mean Gaussian noise $\mathbf{n}_t$ with variance σ^2 is also added. If we align the images to the center and drop the translations, the 2D observation matrix under the orthographic camera model can be factorized into

$$\mathbf{p}_t = \mathbf{R}_t(\bar{\mathbf{s}} + \mathbf{V}\mathbf{z}_t) + \mathbf{n}_t. \tag{2}$$

This probabilistic formulation of the conventional principal component analysis (PCA) was addressed by Tipping and Bishop in [23]. It has a simple linear probabilistic assumption that all marginal and conditional distributions are Gaussian. PPCA is closely related to factor analysis [24], in which a statistical model is used to describe the relation between the observed vector $\mathbf{p}_t$ and the corresponding latent variables $\mathbf{z}_t$.

In Eq. (2), the weight coefficients $\mathbf{z}_t$ are formulated as an independent and identically distributed (i.i.d.) Gaussian prior

$$\mathbf{z}_t \sim \mathcal{N}(0; \mathbf{I}). \tag{3}$$

These unobserved or latent variables are marginalized out instead of being explicitly calculated. Since there only exists linear transformations in Eq. (2), the measurement matrix $\mathbf{p}_i$ is also Gaussian distributed [9] with the form

$$\mathbf{p}_t \sim \mathcal{N}(\mathbf{R}_t\bar{\mathbf{s}}; \mathbf{R}_t\mathbf{V}\mathbf{V}^\top\mathbf{R}_t^\top + \sigma^2\mathbf{I}). \tag{4}$$

3.1 Shape update

The PPCA model can be estimated iteratively by the EM algorithm [9]. In the expectation step (E-step), the posterior distribution over $\mathbf{z}_t$ is defined as

$$\begin{aligned} q(\mathbf{z}_t) &= p(\mathbf{z}_t|\mathbf{p}_t, \mathbf{R}_t, \bar{\mathbf{s}}, \mathbf{V}, \sigma^2) \\ &= \mathcal{N}(\mathbf{z}_t|\boldsymbol{\mu}_t; \boldsymbol{\Sigma}_t). \end{aligned} \tag{5}$$

Over this distribution, the first two moments of $\mathbf{z}_t$ are given

$$\boldsymbol{\mu}_t = \mathbb{E}[\,\mathbf{z}_t], \tag{6}$$

$$\boldsymbol{\phi}_t = \mathbb{E}[\,\mathbf{z}_t\mathbf{z}_t^\top] = \boldsymbol{\Sigma}_t + \boldsymbol{\mu}_t\boldsymbol{\mu}_t^\top. \tag{7}$$

In the following maximization step (M-step), the expected negative log-likelihood function

$$\begin{aligned}\mathcal{L} &= -\mathbb{E}\left[\sum_t \log p(\mathbf{p}_t|\mathbf{R}_t, \bar{\mathbf{s}}, \mathbf{V}, \sigma^2)\right] \\ &= \frac{1}{2\sigma^2}\sum_t \mathbb{E}\left[||\mathbf{p}_t - \mathbf{R}_t(\bar{\mathbf{s}} + \mathbf{V}\mathbf{z}_t)||^2\right] \\ &\quad + JT\log(2\pi\sigma^2)\end{aligned} \tag{8}$$

is minimized. The shape bases $\{\bar{\mathbf{s}}, \mathbf{V}\}$ and the noise parameter σ^2 can be updated individually in closed form by setting their partial derivative to zero [9] with the help of the expectations in Eqs. (6) and (7).

However, the camera rotation parameter $\mathbf{R}_t$ is subject to orthonormality constraints; hence, closed-form update like the other parameters is not possible. Torresani et al. [9] approximated the solution with a single Gauss–Newton step on the Euclidean space, which is inaccurate and has a theoretically low convergence rate. In the upcoming section, we propose our optimization technique on the manifold.

3.2 Motion update on manifold

In [9], a twist vector $\boldsymbol{\xi}$ is employed to hold the result of the single Gauss–Newton step. The exponential map of the skew-symmetric matrix $\hat{\boldsymbol{\xi}}$ is then set as the updating vector $\boldsymbol{\Delta}$. Note that without defining an appropriate metric on the manifold, a manually selected and fixed updating step length is implemented, which declines the performance obviously, when faced complex setups.

3.2.1 Newton's method on SO(3)

As we consider the orthographic camera model, the camera motion matrix $\mathbf{R}_t$ in Eq. (2) is obtained by projecting a 3D rotation matrix to 2D with an orthographic projection matrix

$$\boldsymbol{\Pi} = \begin{bmatrix} 1 & 0 & 0 \\ 0 & 1 & 0 \end{bmatrix}, \tag{9}$$

so that the mapping

$$\mathbf{R} = \boldsymbol{\Pi}\mathbf{Q} \tag{10}$$

from 3D to 2D is satisfied. The rotation matrix $\mathbf{Q}$ is an orthogonal matrix with a determinant one, which lies exactly on the manifold of the special orthogonal group

$$SO(3) = \left\{\mathbf{Q} \in \mathbb{R}^{3\times 3} : \mathbf{Q}^\top\mathbf{Q} = \mathbf{I}, \det(\mathbf{Q}) = 1\right\}. \tag{11}$$

Hence, instead of putting an approximate algebraic or numeric constraint on the Euclidean space $\mathbb{R}^N$ and projecting them back onto the $SO(3)$ manifold, an unconstrained optimization on the manifold is a natural generalization and is expected to perform better.

To start with, we consider to be on the normal Euclidean space. The Newton's method iteratively finds the stationary points of differentiable functions. Provided that $f(x)$ is a twice-differentiable function, the update sequence x_n can be approximated by the Taylor series expansion up to the second order and rewritten as

$$x_{k+1} = x_k - \left[f''(x_k)\right]^{-1} f'(x_k). \tag{12}$$

Given a quadratic function $f(x)$, the optimal point can be found even in a single step. So on the Euclidean space, the first and second order derivatives of the objective function are needed. Edelman et al. [25] proved that for Stiefel manifolds (set of all orthonormal k-frames in $\mathbb{R}^n$, $V_k(\mathbb{R}^n) = \{\mathbf{A} \in \mathbb{R}^{n\times k} : \mathbf{A}^\top\mathbf{A} = \mathbf{I}\}$), e.g., $SO(3)$, their canonical Riemannian structure makes possible to generalize a Riemannian Newton's method on them. Besides the gradient and Hessian, the definition of the update along the geodesic of the manifold must be known to ensure that the update is valid, because unlike on the Euclidean space, the update path is no longer a straight line but rather a geodesic curve, which stays on the surface of the manifold all the time and defines the shortest path between two points on the surface. The update step is illustrated in Fig. 1.

We define the objective function F with respect to the rotation matrix $\mathbf{Q}$ for the manifold optimization as follows

$$F(\mathbf{Q}) = \mathbb{E}\left[\|\mathbf{p} - \boldsymbol{\Pi}\mathbf{Q}(\bar{\mathbf{s}} + \mathbf{V}\mathbf{z})\|_F^2\right]. \tag{13}$$

Since $\mathbf{Q} \in SO(3)$, following [26], its tangent vector $\boldsymbol{\Delta} \in T(SO(3))$ is given by

$$\boldsymbol{\Delta} = \mathbf{Q}\hat{\mathbf{u}}, \tag{14}$$

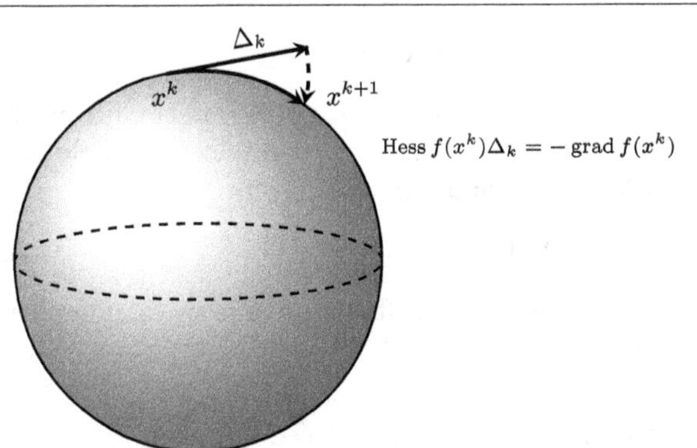

Fig. 1 Generalization of the Newton's method on manifold. Current approximation x^k is updated in the direction of the optimal update vector $\boldsymbol{\Delta}_k$ by a unit distance. Applying the update on the geodesic reveals the new point x^{k+1}

where $\hat{\mathbf{u}}$ is the skew-symmetric matrix of vector $\mathbf{u}$ in the form of

$$\hat{\mathbf{u}} = \begin{bmatrix} 0 & -u_3 & u_2 \\ u_3 & 0 & -u_1 \\ -u_2 & u_1 & 0 \end{bmatrix}. \tag{15}$$

For the Riemannian manifold, the metric can simply be induced from the Euclidean metric as

$$g(\boldsymbol{\Delta}_1, \boldsymbol{\Delta}_2) = \frac{1}{2}\mathrm{tr}(\boldsymbol{\Delta}_1^\top \boldsymbol{\Delta}_2). \tag{16}$$

The explicit formula for geodesics on $SO(3)$ at $\mathbf{Q}$ in direction $\boldsymbol{\Delta}$ is then

$$\begin{aligned} \mathbf{Q}(t) &= \exp(\mathbf{Q}, \boldsymbol{\Delta}t) = \mathbf{Q}\exp(\hat{\boldsymbol{\omega}}t) \\ &= \mathbf{Q}\left(\mathbf{I} + \hat{\boldsymbol{\omega}}\sin(t) + \hat{\boldsymbol{\omega}}^2(1-\cos(t))\right), \end{aligned} \tag{17}$$

where $t \in \mathbb{R}$, $\omega = \mathbf{Q}^\top\boldsymbol{\Delta} \in \mathfrak{so}(3)$ ($\mathfrak{so}(3)$ is the Lie algebra of $SO(3)$). The last equation is called the Rodrigues' rotation formula [27].

3.2.2 Gradient and Hessian

To obtain the gradient and Hessian, we first derive the first and second order derivative for the geodesic $\mathbf{Q}(t)$ with respect to t:

$$\begin{aligned} \left.\frac{\mathrm{d}\mathbf{Q}(t)}{\mathrm{d}t}\right|_{t=0} &= \left.\mathbf{Q}\hat{\boldsymbol{\omega}}\cos(t) + \mathbf{Q}\hat{\boldsymbol{\omega}}^2\sin(t)\right|_{t=0} \\ &= \mathbf{Q}\hat{\boldsymbol{\omega}} \\ &= \mathbf{Q}(\mathbf{Q}^\top\boldsymbol{\Delta}) \\ &= \boldsymbol{\Delta} \end{aligned} \tag{18}$$

$$\begin{aligned} \left.\frac{\mathrm{d}^2\mathbf{Q}(t)}{\mathrm{d}t^2}\right|_{t=0} &= \left.-\mathbf{Q}\hat{\boldsymbol{\omega}}\sin(t) + \mathbf{Q}\hat{\boldsymbol{\omega}}^2\cos(t)\right|_{t=0} \\ &= \mathbf{Q}\hat{\boldsymbol{\omega}}^2 \\ &= \mathbf{Q}(\mathbf{Q}^\top\boldsymbol{\Delta})(\mathbf{Q}^\top\boldsymbol{\Delta}) \\ &= \boldsymbol{\Delta}(\mathbf{Q}^\top\boldsymbol{\Delta}) \\ &= -\boldsymbol{\Delta}\boldsymbol{\Delta}^\top\mathbf{Q} \end{aligned} \tag{19}$$

Note that the last step in Eq. (19) is derived from the property of tangent space on the Stiefel manifold that $\mathbf{Q}^\top\boldsymbol{\Delta}$ is a skew-symmetric matrix with

$$\mathbf{Q}^\top\boldsymbol{\Delta} + \boldsymbol{\Delta}^\top\mathbf{Q} = \mathbf{0}. \tag{20}$$

Given the geodesic definition, we derive the gradient and Hessian in direction $\boldsymbol{\Delta} \in T(SO(3))$:

$$\begin{aligned} \mathrm{d}F(\boldsymbol{\Delta}) &= \left.\frac{\mathrm{d}F(\mathbf{Q}(t))}{\mathrm{d}t}\right|_{t=0} \\ &= \boldsymbol{\Pi}\mathbf{Q}\mathbf{V}\boldsymbol{\phi}\mathbf{V}^\top\dot{\mathbf{Q}}^\top\boldsymbol{\Pi}^\top - \mathbf{p}\boldsymbol{\mu}^\top\mathbf{V}^\top\dot{\mathbf{Q}}^\top\boldsymbol{\Pi}^\top|_{t=0} \\ &= \boldsymbol{\Pi}\mathbf{Q}\mathbf{V}\boldsymbol{\phi}\mathbf{V}^\top\boldsymbol{\Delta}^\top\boldsymbol{\Pi}^\top - \mathbf{p}\boldsymbol{\mu}^\top\mathbf{V}^\top\boldsymbol{\Delta}^\top\boldsymbol{\Pi}^\top \end{aligned} \tag{21}$$

$$\begin{aligned} \mathrm{Hess}\,F(\boldsymbol{\Delta}, \boldsymbol{\Delta}) &= \left.\frac{\mathrm{d}^2F(\mathbf{Q}(t))}{\mathrm{d}t^2}\right|_{t=0} \\ &= \boldsymbol{\Pi}\dot{\mathbf{Q}}\mathbf{V}\boldsymbol{\phi}\mathbf{V}^\top\dot{\mathbf{Q}}^\top\boldsymbol{\Pi}^\top \\ &\quad + \boldsymbol{\Pi}\mathbf{Q}\mathbf{V}\boldsymbol{\phi}\mathbf{V}^\top\ddot{\mathbf{Q}}^\top\boldsymbol{\Pi}^\top \\ &\quad - \mathbf{p}\boldsymbol{\mu}^\top\mathbf{V}^\top\ddot{\mathbf{Q}}^\top\boldsymbol{\Pi}^\top|_{t=0} \\ &= \boldsymbol{\Pi}\boldsymbol{\Delta}\mathbf{V}\boldsymbol{\phi}\mathbf{V}^\top\boldsymbol{\Delta}^\top\boldsymbol{\Pi}^\top \\ &\quad - \boldsymbol{\Pi}\mathbf{Q}\mathbf{V}\boldsymbol{\phi}\mathbf{V}^\top\mathbf{Q}^\top\boldsymbol{\Delta}\boldsymbol{\Delta}^\top\boldsymbol{\Pi}^\top \\ &\quad + \mathbf{p}\boldsymbol{\mu}^\top\mathbf{V}^\top\mathbf{Q}^\top\boldsymbol{\Delta}\boldsymbol{\Delta}^\top\boldsymbol{\Pi}^\top \end{aligned} \tag{22}$$

For any arbitrary pair of vectors $\mathbf{X}, \mathbf{Y} \in T(SO(3))$, polarization [26] helps compute $\mathrm{Hess}\,F(\mathbf{X}, \mathbf{Y})$ with

$$\begin{aligned} \mathrm{Hess}\,F(\mathbf{X}, \mathbf{Y}) = \frac{1}{4}\,&(\mathrm{Hess}\,F(\mathbf{X}+\mathbf{Y}, \mathbf{X}+\mathbf{Y}) \\ &-\mathrm{Hess}\,F(\mathbf{X}-\mathbf{Y}, \mathbf{X}-\mathbf{Y}))\,. \end{aligned} \tag{23}$$

3.2.3 Algorithm summary

With the requirements for generalizing Newton's method being ready, the optimal updating vector on the manifold can be found by modifying the original Newton Eq. (12) to

$$\boldsymbol{\Delta} = -\mathrm{Hess}^{-1}\,\mathbf{G}, \tag{24}$$

assuming that the Hessian is non-degenerate. It is the same as finding a vector $\boldsymbol{\Delta}$ that satisfies for all vector fields $\mathbf{Y}$

$$\mathrm{Hess}\,F(\mathbf{Y}, \boldsymbol{\Delta}) = g(-\mathbf{G}, \mathbf{Y}) = -\mathrm{d}\,F(\mathbf{Y}), \tag{25}$$

where $\mathbf{G} = \nabla F$ stands for the gradient. The Hessian can be uniquely determined by using an orthonormal basis $\{\mathbf{E}^k\}, k = 1, 2, 3$ into Eq. (25) as

$$\mathrm{Hess}\,F(\mathbf{E}^k, \boldsymbol{\Delta}) = -\mathrm{d}\,F(\mathbf{E}^k). \tag{26}$$

For simplicity, the standard basis $\mathbf{e}_k$ for $\mathbb{R}^3$ is chosen so that $\mathbf{E}^k = \mathbf{Q}\hat{\mathbf{e}}_k \in T(SO(3))$. Thus, the 3×3 Hessian matrix $\mathbf{H}$ and the three-dimensional gradient vector $\mathbf{g}$ can be obtained:

$$\mathbf{H}_{kl} = \mathrm{Hess}\,F(\mathbf{E}^k, \mathbf{E}^l), \tag{27}$$

$$\mathbf{g}_k = \mathrm{d}\,F(\mathbf{E}^k),\ k, l = 1, 2, 3 \tag{28}$$

Then,we solve for the vector $\mathbf{u} = [u_1, u_2, u_3]^\top \in \mathbb{R}^3$ using

$$\mathbf{u} = -\mathbf{H}^{-1}\mathbf{g}. \tag{29}$$

Finally, the desired updating vector $\boldsymbol{\Delta} = \mathbf{Q}\hat{\mathbf{u}}$ is obtained. The last step is to update the current rotation along the geodesic in the direction of this vector. The algorithm is summarized in Algorithm 1.

Algorithm 1 Minimize $F(\mathbf{Q}) = \mathbb{E}\left[\|\mathbf{p} - \mathbf{Q}(\bar{\mathbf{s}} + \mathbf{V}\mathbf{z})\|_F^2\right]$

1. At the point $\mathbf{Q}$, compute the optimal update vector $\boldsymbol{\Delta} = -\text{Hess}^{-1}\,\mathbf{G}$
   1: Choose basis tangent vectors $\mathbf{E}^k = \mathbf{Q}\hat{\mathbf{e}}_k \in T(SO(3))$ with $\mathbf{e}_k$ for $1 \le k \le 3$ being the standard basis for $\mathbb{R}^3$
   2: Compute $\mathbf{H}_{kl} = \text{Hess}\,F(\mathbf{E}^k, \mathbf{E}^l)$, $1 \le k, l \le 3$
   3: Compute $\mathbf{g}_k = \mathrm{d}\,F(\mathbf{E}^k)$, $1 \le k \le 3$
   4: Compute $\mathbf{u} = (u_1, u_2, u_3)^\top$ such that $\mathbf{u} = -\mathbf{H}^{-1}\mathbf{g}$
   5: The optimal updating vector $\boldsymbol{\Delta} = -\text{Hess}^{-1}\,\mathbf{G} = \mathbf{Q}\hat{\mathbf{u}}$
2. Update the rotation $\mathbf{Q}$
   1: Move $\mathbf{Q}$ in the direction $\boldsymbol{\Delta}$ along the geodesic to

$$\exp(\mathbf{Q}, \boldsymbol{\Delta}t) = \mathbf{Q}\left(\mathbf{I} + \hat{\omega}\sin(t) + \hat{\omega}^2(1 - \cos(t))\right),$$

where $t = \sqrt{\frac{1}{2}\mathrm{tr}(\boldsymbol{\Delta}^\top\boldsymbol{\Delta})}$ and $\omega = \mathbf{Q}^\top\boldsymbol{\Delta}/t$

3.3 NRSFM with PLDA

PLDA was presented by Prince and Elder in [10] as a probabilistic estimation for deterministic linear discriminant analysis (LDA) [28]. This model separately models the between-individual and within-individual variations among different subjects. Unlike PCA, which only takes into account the whole data distribution, LDA seeks the maximum separability of classes along the direction that has the highest ratio of the variance between the classes to the variance within the classes [29]. Thus, for our PLDA model, the single shape subspace $\mathbf{V}$ in Eq. (1) is replaced by the between-individual subspace $\mathbf{F}$ and the within-individual subspace $\mathbf{K}$ as follows

$$\mathbf{s}_{ij} = \bar{\mathbf{s}} + \mathbf{F}\mathbf{h}_i + \mathbf{K}\mathbf{w}_{ij}, \tag{30}$$

where i denotes the ith individual of the total I subjects and j denotes the jth image of J images belonging to this person. Compared to the original definition in Eq. (1), the latent variables $\mathbf{z}_t$ now consist of two parts. The first part, $\mathbf{h}_i$, indicates the parameter for the between-individual subspace $\mathbf{F}$, which remains constant for all J images of individual i, while the second part $\mathbf{w}_{ij}$ describes how each image varies in the within-individual subspace $\mathbf{K}$. Given this advanced shape model, the latent identity variables $\mathbf{h}_i$ guarantee that a great part of the commonness in the same subject is preserved and taken into account at runtime.

In order to estimate the PLDA parameters, an EM algorithm that is similar to PPCA is presented by Prince and Elder [10] with modifications in the E-step. The main point is to ensure that all J images share the same latent identity variable $\mathbf{h}_i$ despite the image-specific latent variables $\mathbf{w}_{ij}$. Therefore, the calculation of these J images is done in the single step and the corresponding equations in Eq. (1) are stacked up into a composite matrix system

$$\begin{bmatrix} \mathbf{s}_{i1} \\ \mathbf{s}_{i2} \\ \vdots \\ \mathbf{s}_{iJ} \end{bmatrix} = \begin{bmatrix} \bar{\mathbf{s}} \\ \bar{\mathbf{s}} \\ \vdots \\ \bar{\mathbf{s}} \end{bmatrix} + \begin{bmatrix} \mathbf{F} & \mathbf{K} & \mathbf{0} & \dots & \mathbf{0} \\ \mathbf{F} & \mathbf{0} & \mathbf{K} & \dots & \mathbf{0} \\ \vdots & \vdots & \vdots & \ddots & \vdots \\ \mathbf{F} & \mathbf{0} & \mathbf{0} & \dots & \mathbf{K} \end{bmatrix} \begin{bmatrix} \mathbf{h}_i \\ \mathbf{w}_{i1} \\ \mathbf{w}_{i2} \\ \vdots \\ \mathbf{w}_{iJ} \end{bmatrix}, \tag{31}$$

or equivalently

$$\mathbf{s}_i = \bar{\mathbf{s}}' + \mathbf{A}\mathbf{y}_i. \tag{32}$$

Accordingly, the expectations of the new latent variables $\mathbf{y}_i$ for the E-step changes from Eqs. (6) and (7) to

$$\boldsymbol{\mu}_i' = \mathbb{E}[\mathbf{y}_i], \tag{33}$$

$$\boldsymbol{\phi}_i' = \mathbb{E}\left[\mathbf{y}_i\mathbf{y}_i^\top\right] = \boldsymbol{\Sigma}_i + \boldsymbol{\mu}_i'\boldsymbol{\mu}_i'^\top. \tag{34}$$

As for the M-step, most of the existing PPCA updates remain unchanged, if we replace the original shape matrix $\mathbf{V}$ with $[\mathbf{F} \quad \mathbf{K}]$ and each of the latent variables $\mathbf{z}_{ij}$ with $\begin{bmatrix} \mathbf{h}_i \\ \mathbf{w}_{ij} \end{bmatrix}$. Accordingly, the objective log-likelihood function to be minimized in Eq. (8) can be modified to

$$\begin{aligned} \mathcal{L} &= -\mathbb{E}\left[\sum_t \log p(\mathbf{p}_t|\mathbf{R}_t, \bar{\mathbf{s}}, [\mathbf{F} \quad \mathbf{K}], \sigma^2)\right] \\ &= \frac{1}{2\sigma^2}\sum_t \mathbb{E}\left[\|\mathbf{p}_t - \mathbf{R}_t(\bar{\mathbf{s}} + [\mathbf{F} \quad \mathbf{K}]\begin{bmatrix} \mathbf{h}_i \\ \mathbf{w}_{ij} \end{bmatrix})\|^2\right] \\ &\quad + JT\log(2\pi\sigma^2). \end{aligned} \tag{35}$$

We apply the Gauss-Newton step [9] as well as our manifold extension of Newton's method to optimize the objective function.

4 Experiments

In this section, extensive experiments are conducted to validate the proposed approaches. Rotation recovery using the Newton's method on the manifold is first assessed on different datasets. Subsequently, performance of PLDA on generated data with multiple subjects is presented.

4.1 Setup

For our experiments, the evaluation criteria is the same as in [9], i.e., the sum of squared differences between estimated 3D shapes to ground truth depth: $\|\hat{\mathbf{s}}_{1:T} - \mathbf{s}_{1:T}\|_F^2$, with the camera rotation $\mathbf{R}$ also being applied to the 3D shape. As the ground truth for camera rotation is not given, we are not able to measure the absolute performance gain from our algorithm explicitly. However, the decreased reconstruction error implicitly assesses the effectiveness of rotation estimation in our algorithm.

Moreover, additive zero-mean Gaussian noise is imposed to analyze the robustness of reconstruction. The noise level is plotted as the ratio of the noise variance to the norm of the 2D measurements: $JT\sigma^2/\|\mathbf{p}_{1:T}\|_F$. The noise levels range from 0 to 30 % with 2 % step, and the trials for each noise level are averaged over 10 runs. Our test is carried out on two face datasets, i.e., the Vicon motion capture data Mocap-Face [9] and the Binghamton University 3D Facial Expression (BU-3DFE) dataset [30].

The Mocap-Face dataset [9] contains a single video, which captures a single male subject with 40 markers attached to his face. The video contains 316 frames in total. Sample frames from this dataset can be seen in Fig. 2. Throughout the video sequence, the subject made limited changes of facial expression and head pose. Note that the tracking is very accurate using the markers.

The BU-3DFE dataset [30] is originally created for 3D facial expression analysis. The complete dataset consists of 100 subjects, covering different ethnic groups. Seven facial expressions are performed at four intensity levels by each subject. We randomly select 300 frames from 100 subjects for our test, in contrast to the Mocap-Face dataset [9], where there exists only one subject. This is a more practical setup, since in many computer vision datasets, e.g., for face alignment [31, 32], only static images of multiple subjects are provided, where only a few samples of the same subject is available. Separate application of NRSFM for each single subject is then impossible. Random poses are generated by projecting the 3D landmarks to 2D. Note that temporal smoothness of the shapes is not valid in this dataset. The dataset provides manual annotation of 83 marker points as shown in Fig. 3. Due to labeling noise and inconsistency, this dataset contains noise in the original measurements.

4.2 Evaluation of manifold PPCA

We first give quantitative results for the recovery of the 3D camera rotation between our algorithm (manifold PPCA) and the baseline PPCA [9], as well as the other state-of-the-art approaches, point trajectory approach (PTA) [18] and column space fitting (CSF2) [20].

4.2.1 Reconstruction results on Mocap-Face

In the first experiment without noise on Mocap-Face [9], our approach achieves slightly better performance than PPCA, while both having an effective reconstruction result under 3 % error, as is plotted in Fig. 4a. Qualitative results are also shown in Fig. 5a, which yields similar outcome. In comparison, performance of methods in trajectory space is more sensitive to the number of DCT bases. Starting from $K = 8$, both PTA and CSF2 degrade abruptly.

In real life, there are no markers and the automatic point detectors are usually not stable. To assess the performance of the system in a real-world case, it is necessary to test it on the noisy data. We evaluate our system with additive Gaussian noise at different noise levels. As can be observed from Fig. 4c, at the beginning, all algorithms have almost the same error rate up to 6 % noise. With more noise added, PPCA starts to undergo a significantly steeper curve than our approach. Starting from 20 % noise level, our result gets 50 % lower error rate than PPCA. That is most likely because with more noise, it is more difficult for the rotation approximation in PPCA to find the right updating direction. Despite achieving the lowest error in the above noise-free experiment, the state-of-the-art CSF2 surprisingly fails to hold up well against noise, which approaches PPCA as the second worst. The same trajectory-based PTA is more stable, thanks to smoother DCT bases. Shaji and Chandran [15] also evaluated on the Mocap-Face dataset [9] with additive noise. From their plot, the performance degrades very quickly with noise level over 20 %. However, our probabilistic approach does not suffer from this problem. Additionally, the variance of the results of each noise level is also shown in the figure, in which we observe that the manifold PPCA also reduces error variances. That means our approach performs much more stably under noisy circumstances.

4.2.2 Reconstruction results on BU-3DFE

Since the BU–3DFE dataset [30] is a more difficult setup, the performance is lower compared to the test on the Mocap-Face dataset [9]. The purpose of this test is to see how well a generic face model can be generated using different NRSFM approaches. As can be seen in Fig. 4b, the recovered models cannot fit all instances as well as on the Mocap-Face dataset [9]. But again, the error level of our attempt is in overall ca. 8 % lower than that of PPCA regardless of the choice of K, which demonstrates a relative performance gain of 30 to 40 %. As can be observed

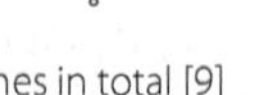

Fig. 2 The Mocap-Face dataset, captured by 40 markers attached to the face of the subject in the left image, 316 frames in total [9]

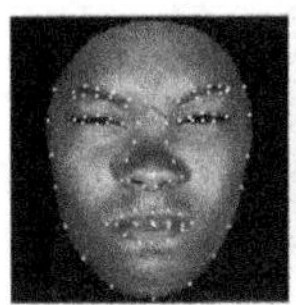 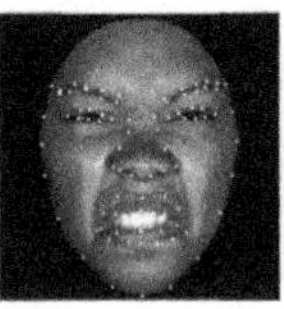 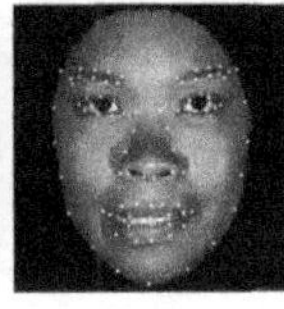 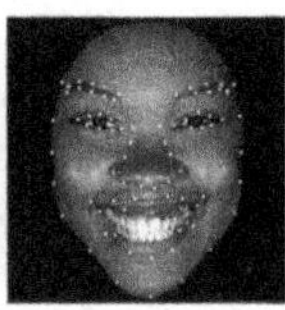 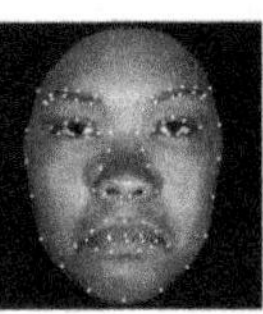 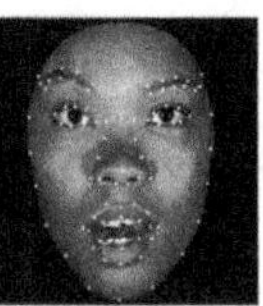

Fig. 3 The BU-3DFE dataset, with 83 points and 300 frames of multiple subjects, plus zero-mean Gaussian random pose changes [30]

qualitatively in Fig. 5b, PPCA's rotation approximation limits its result to getting better rotation estimate in frame 50. It also has difficulties to recover the contour of the faces correctly in frame 250, whereas our system clearly does better. It is also interesting to test the approaches in trajectory bases, where the smooth shape deformation assumption is no longer valid on the BU-3DFE dataset. As expected, CSF2 performs worse than the manifold PPCA, which does not take into account the temporal prior. Selection of K also has no influence to the error rate, unlike in Fig. 4a. If we add Gaussian noise to the data (see Fig. 4d), manifold PPCA and PTA degrades slightly slower than PPCA and CSF2, similar to that on the Mocap-Face dataset [9]. These results reveal that, when modeling more complicated shapes, an optimal rotation estimation using manifold optimization techniques is superior. Another advantage of our manifold-based approach is that it is more robust in noisy environments in general.

4.3 Evaluation of PLDA

In the second part of our experiments, we generate datasets with multiple subjects from the original BU-3DFE dataset [30] to evaluate the PLDA variant of the NRSFM algorithm in comparison with PPCA. We consider two setups with different number of subjects involved. For the first setup, we select six subjects and each subject has 50 images. For the second setup, there are 12 subjects with 25 images, respectively. Thus, the total number of frames is still 300. Similar to the PPCA experiments with additive noise, we randomly generate five input datasets for each setup to obtain statistically significant results.

For all tests in this section, we directly compare the results of PPCA and PLDA as well as the influence of imposing Newton's method on the manifold for recovery of the rotation matrix to both probabilistic frameworks. Since in Section 4.2.2, trajectory-based methods

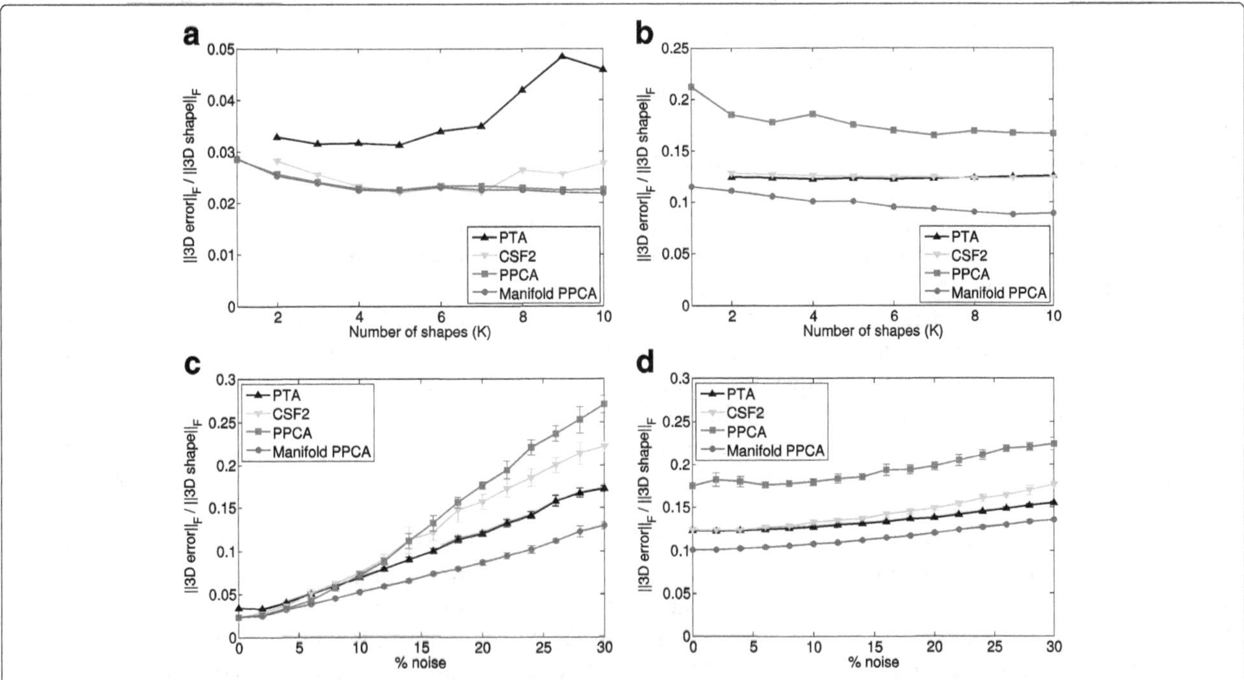

Fig. 4 The first row shows the reconstruction error as a function of bases (K) without adding noise manually on **a** Mocap-Face and **b** BU-3DFE. The second row shows the reconstruction error with additive Gaussian noise up to 30 % on **c** Mocap-Face and **d** BU-3DFE

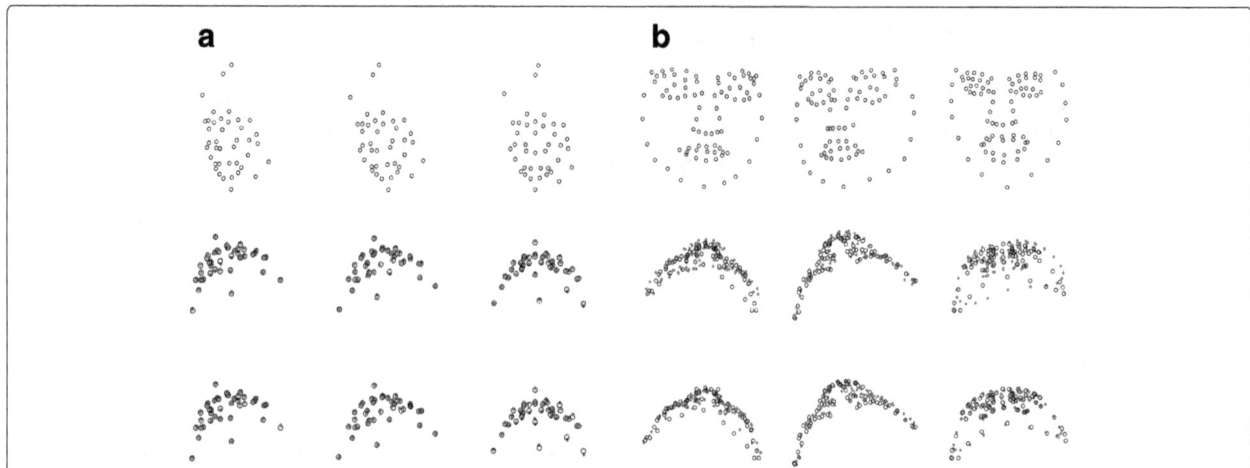

Fig. 5 a Mocap-Face and **b** BU-3DFE 2D tracks in the upper row. Reconstruction results of PPCA and manifold PPCA are *red* and *blue dots* in the *second* and *third rows*, respectively. *Circles* denote the ground truth. Images are captured at frames 50, 150, and 250

are proven not to generalize well when temporal smoothness does not hold, the results of PTA and CSF2 are not included. The curves of both PPCA approaches are plotted in dashed lines while the PLDA results are plotted in solid lines.

In the test case without additive noise, we fix the number of between-subject shape bases $F = 3$ and vary the within-subject bases G in PLDA from 1 to 7, compared to the only shape bases K in PPCA that equals the sum of F and G. We first notice that with the help of PLDA, the performance for both rotation recovery techniques has got further improvement, independently from the input data with 6 subjects (Fig. 6a) or 12 subjects (Fig. 6b). Especially for the baseline PPCA algorithm with Gauss-Newton rotation approximation, there is nearly 10 % less error in the reconstruction in both setups. Moreover, the error variance also drops hugely to an acceptable level. For our Newton's rotation recovery method on the manifold, only little improvement in reconstruction error is observed; however, the performance without PLDA already delivers a satisfactory result. As a result, all of our proposed methods, i.e., manifold PPCA, PLDA and manifold PLDA, manage to make significant performance enhancement. The similar outcome of the manifold PPCA and PLDA, which employ different extensions for the optimization problem, indicates that the achieved result may have approximated to the performance limit of the dataset using the probabilistic framework.

When zero-mean Gaussian noise is added (Fig. 6c, d), although the gaps between the manifold PLDA and PLDA remain close, the lower average error rate and the stability with notably less error deviation at some noise levels again demonstrate the effectiveness and necessity of our better rotation recovery. We also observe that both PLDA-based methods degrade faster than those PPCA-based methods with the amount of imposed Gaussian noise starting from ca. 15 to 20 %. We conclude that its reason is probably because PLDA-based approaches need to estimate more parameters in the E-step than PPCA in each iteration, which makes the additional noise and uncertainty a decisive deficit factor for the approach. But overall, introducing PLDA does help to further decrease the error reconstruction rate with or without additive noise.

Qualitative experiments are also conducted in order to review the effect of applying PLDA on datasets with more than one subject, as can be seen in Fig. 7. We know that from Eq. (30), the between-individual linear shape model of PLDA consists of the global mean shape $\bar{\mathbf{s}}$ plus the subject-specific shape term $\mathbf{F}\mathbf{h}_i$. The frame-specific shape term $\mathbf{K}\mathbf{w}_{ij}$ serves solely as within-individual variance and is therefore omitted in this experiment. Thus, in Fig. 7, the reconstruction of every single subject for each dataset given by the first two terms in Eq. (30) is shown. The 3D ground truth is obtained by averaging all 50 shape vectors for the corresponding subject. As expected, the reconstruction result is fairly satisfactory, thanks to the characteristics of PLDA. Even for some unique faces as in Fig. 7e, g, their contours and facial features are still well modeled, which again proves the capability and tolerance of our approach.

In Fig. 8, we illustrate the reconstructed bases **F** and **K**. The effect of varying the between-individual (**F**) and within-individual (**K**) shape bases learned by PLDA between ± 3 standard deviations from the mean value is analyzed. With the first three bases of **F**, different eye, face contour, and mouth types are modeled, respectively, in Fig. 8b, c, and d. It is interesting to see that the variations in the figures are more related to identification of the subjects than to the expressions. With the bases of **K** instead, different facial expressions present in BU-3DFE

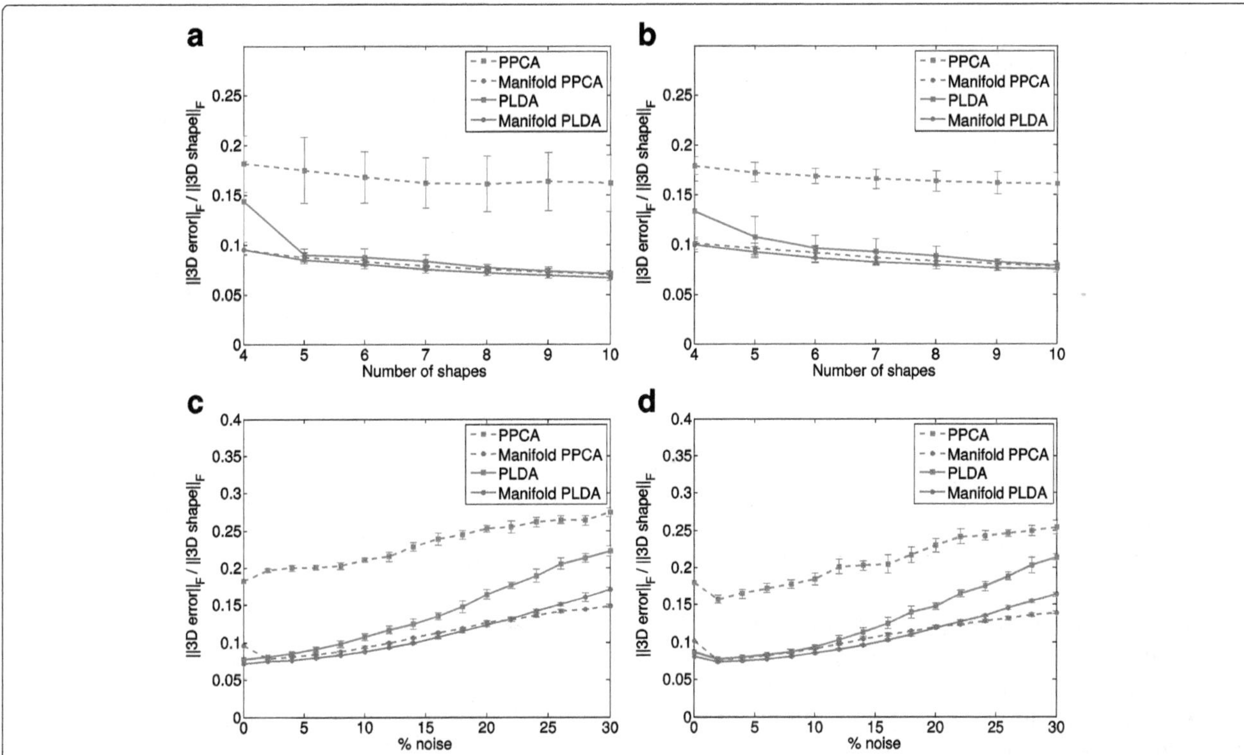

Fig. 6 The *first row* shows the reconstruction error without adding noise manually using PPCA and PLDA. The *second row* shows the reconstruction error with additive Gaussian noise up to 30 %. **a** and **c** in the *left column* are evaluated on datasets with 6 subjects, and **b** and **d** in the *right column* are evaluated on datasets with 12 subjects

[30] are well recognizable. For example, Fig. 8e shows opening and closing mouths. Evolution from angry to fear is illustrated in Fig. 8f and from surprise to happiness in Fig. 8g, respectively. Those results fully meet our expectations and conform to the characteristics of PLDA, which provide optimally and more meaningfully reconstructed shape bases than those given by PPCA.

5 Conclusions

In this work, we have presented a novel solution to unleash the orthonormality constraints of the camera rotation matrix in the NRSFM problem. Without requiring conducting complex approximations, performing rotation update on the $SO(3)$ manifold implicitly ensures the validity of the constraints. In the experiments

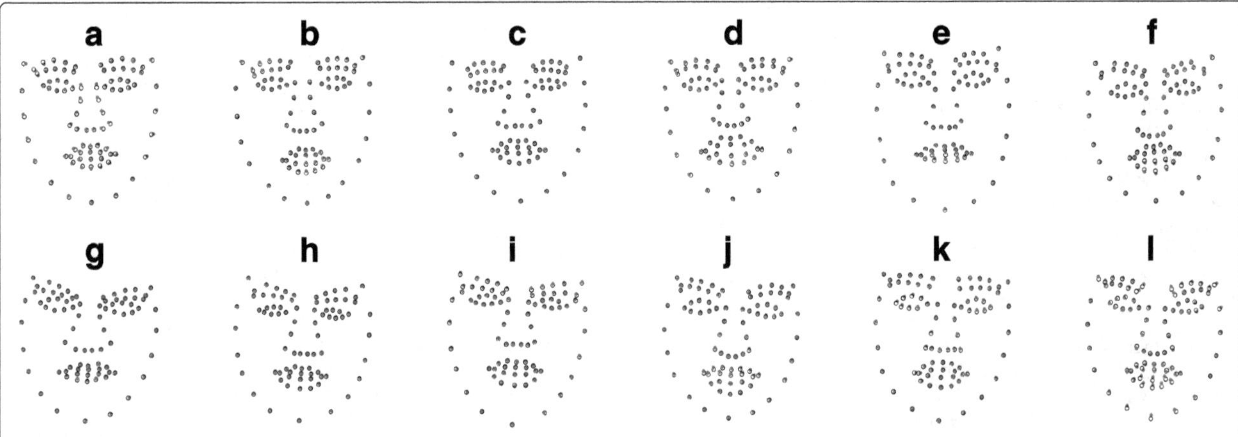

Fig. 7 a–l Qualitative reconstruction results of different subjects using PLDA. Each *row* corresponds to a generated dataset consisting of six subjects based on BU-3DFE

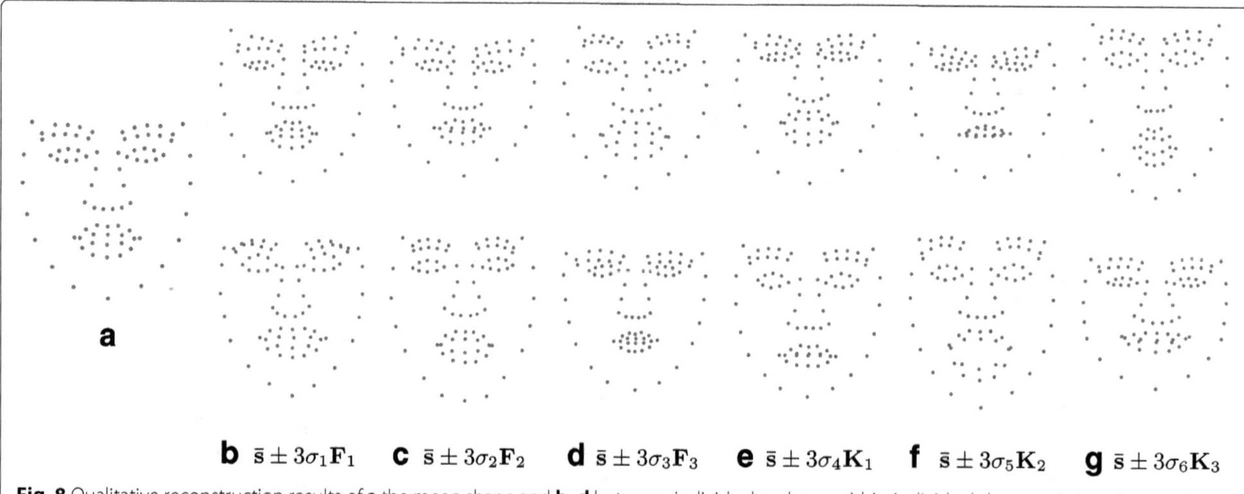

Fig. 8 Qualitative reconstruction results of **a** the mean shape and **b–d** between-individual and **e–g** within-individual shape variations imposed on the mean shape

on the Mocap-Face dataset [9] with additional noise, which contains only one subject, our approach performs significantly better by reducing up to 50 % reconstruction error. Furthermore, the proposed PLDA approach successfully extends the existing probabilistic framework to separately model between-subject and within-subject shape variations during the alternating optimization for datasets with multiple identities. On the BU-3DFE dataset [30] with multiple subjects and manually annotated landmarks, we clearly outperform the baseline approaches in all tests. To conclude, we have shown that the proposed approaches are robust against noise, which indicates that they are more capable of dealing with real-world data. In addition to its robustness, our approaches generalize better on datasets with multiple subjects.

Competing interests
The authors declare that they have no competing interests.

Authors' contributions
In this work, two major contributions have been made. We first investigated the geometric properties of the orthonormality constraints and generalized the Newton's optimization method to the underlying manifold of the camera rotation matrices. That means, non-linear optimization can be carried out on the manifold without any imprecise approximations. We used a PPCA-based framework [9] to model NRSFM as it is more robust to noise than the closed-form factorization techniques. Our second contribution deals with multiple subjects. Current NRSFM algorithms mostly focus on the reconstruction of a single subject. While dealing with data containing multiple subjects, no difference is taken into account, when modeling between-individual variation (e.g., face model of different identities) and within-individual variation (e.g., facial expression of the same identity). For that reason, we extend the PPCA-based framework to the PLDA [10] model to improve reconstruction performance on data with multiple subjects.

Acknowledgements
This work was done when H. Gao was at Computer Vision for Human-Computer Interaction Lab (CV:HCI), Karlsruhe Institute of Technology (KIT). H. K. Ekenel was partially supported by TUBITAK, project no. 113E121 and a Marie Curie FP7 Integration Grant within the 7th EU Framework Programme. We acknowledge support by Deutsche Forschungsgemeinschaft (DFG) and Open Access Publishing Fund of KIT.

Author details
[1]Vision and Fusion Laboratory (IES), Karlsruhe Institute of Technology (KIT), Adenauerring 4, 76131 Karlsruhe, Germany. [2]Fraunhofer Institute of Optronics, System Technologies and Image Exploitation (Fraunhofer IOSB), Fraunhoferstr. 1, 76131 Karlsruhe, Germany. [3]Signal Processing Laboratory (LTS5), École Polytechnique Fédérale de Lausanne (EPFL), EPFL STI IEL LTS5, ELD 241 (Bâtiment ELD), Station 11, CH-1015 Lausanne, Switzerland. [4]Faculty of Computer and Informatics, Istanbul Technical University (ITU), 34469, Maslak, Istanbul, Turkey.

References
1. C Tomasi, T Kanade, Shape and motion from image streams under orthography: a factorization method. Int. J. Comput. Vis. **9**(2), 137–154 (1992)
2. V Blanz, T Vetter, in *Proceedings of the Annual International Conference on Computer Graphics and Interactive Techniques*. A morphable model for the synthesis of 3D faces, (1999), pp. 187–194
3. S Ullman, Maximizing rigidity: The incremental recovery of 3-D structure structure from rigid and nonrigid motion. Perception. **13**, 255–274 (1983)
4. M Brand, in *Proceedings of the IEEE Computer Society Conference on Computer Vision and Pattern Recognition*. Morphable 3D models from video, vol. 2, (2001), pp. 456–463
5. M Brand, in *Proceedings of the IEEE Computer Society Conference on Computer Vision and Pattern Recognition*. A direct method for 3D factorization of nonrigid motion observed in 2D, vol. 2, (2005), pp. 122–128
6. L Torresani, DB Yang, EJ Alexander, C Bregler, in *Proceedings of the IEEE Computer Society Conference on Computer Vision and Pattern Recognition*. Tracking and modeling non-rigid objects with rank constraints, vol. 1, (2001), pp. 493–500
7. L Torresani, A Hertzmann, in *Proceedings of the European Conference on Computer Vision*. Automatic non-rigid 3D modeling from video, (2004), pp. 299–312
8. J Xiao, J Chai, T Kanade, A closed-form solution to non-rigid shape and motion recovery. Int. J. Comput. Vis. **67**(2), 233–246 (2006)
9. L Torresani, A Hertzmann, C Bregler, Nonrigid structure-from-motion: estimating shape and motion with hierarchical priors. IEEE Trans. Pattern Anal. Mach. Intell. **30**(5), 878–892 (2008)
10. SJD Prince, JH Elder, in *Proceedings of the IEEE International Conference on Computer Vision*. Probabilistic linear discriminant analysis for inferences about identity, (2007), pp. 1–8
11. C Bregler, A Hertzmann, H Biermann, in *Proceedings of the IEEE Computer Society Conference on Computer Vision and Pattern Recognition*. Recovering non-rigid 3D shape from image streams, vol. 2, (2000), pp. 690–696

12. J Xiao, T Kanade, in *Proceedings of the IEEE Computer Society Conference on Computer Vision and Pattern Recognition*. Non-rigid shape and motion recovery: degenerate deformations, vol. 1, (2004), pp. 668–675
13. I Akhter, Y Sheikh, S Khan, in *Proceedings of the IEEE Computer Society Conference on Computer Vision and Pattern Recognition*. In defense of orthonormality constraints for nonrigid structure from motion, (2009), pp. 1534–1541
14. Y Dai, H Li, M He, A simple prior-free method for non-rigid structure-from-motion factorization. Int. J. Comput. Vis. **107**(2), 101–122 (2014)
15. A Shaji, S Chandran, in *Proceedings of the IEEE Computer Society Conference on Computer Vision and Pattern Recognition Workshops*. Riemannian manifold optimisation for non-rigid structure from motion, (2008), pp. 1–6
16. V Rabaud, S Belongie, in *Proceedings of the IEEE Computer Society Conference on Computer Vision and Pattern Recognition*. Linear embeddings in non-rigid structure from motion, (2009), pp. 2427–2434
17. L Tao, BJ Matuszewski, in *Proceedings of the IEEE Computer Society Conference on Computer Vision and Pattern Recognition*. Non-rigid structure from motion with diffusion maps prior, (2013), pp. 1530–1537
18. I Akhter, Y Sheikh, S Khan, T Kanade, in *Advances in Neural Information Processing Systems*. Nonrigid structure from motion in trajectory space, (2008), pp. 41–48
19. PFU Gotardo, AM Martinez, Computing smooth time trajectories for camera and deformable shape in structure from motion with occlusion. IEEE Trans. Pattern Anal. Mach. Intell. **33**(10), 2051–2065 (2011)
20. PFU Gotardo, AM Martinez, in *Proceedings of the IEEE Computer Society Conference on Computer Vision and Pattern Recognition*. Non-rigid structure from motion with complementary rank-3 spaces, (2011), pp. 3065–3072
21. J Valmadre, S Lucey, in *Proceedings of the IEEE Computer Society Conference on Computer Vision and Pattern Recognition*. General trajectory prior for non-rigid reconstruction, (2012), pp. 1394–1401
22. HS Park, T Shiratori, I Matthews, Y Sheikh, 3D trajectory reconstruction under perspective projection. Int. J. Comp. Vis. **115**(2), 115–135 (2015)
23. ME Tipping, CM Bishop, Probabilistic principal component analysis. J. R. Stat. Soc. **61**, 611–622 (1999)
24. DJ Bartholomew, *Latent Variable Models and Factor Analysis*. (Charles Griffin & Co. Ltd., London, 1987)
25. A Edelman, TA Arias, ST Smith, The geometry of algorithms with orthogonality constraints. SIAM J. Matrix Anal. Appl. **20**(2), 303–353 (1999)
26. Y Ma, Košecká, S Sastry, Optimization criteria and geometric algorithms for motion and structure estimation. Int. J. Comput. Vis. **44**(3), 219–249 (1999)
27. RM Murray, Z Li, SS Sastry, *A Mathematical Introduction to Robotic Manipulation*. (CRC Press, Boca Raton, FL, 1994)
28. PN Belhumeur, J Hespanha, DJ Kriegman, Eigenfaces *vs*, Fisherfaces: Recognition using class specific linear projection. IEEE Trans. Pattern Anal. Mach. Intell. **19**(7), 711–720 (1997)
29. AM Martínez, AC Kak, PCA versus LDA. IEEE Trans. Pattern Anal. Mach. Intell. **23**(2), 228–233 (2001)
30. L Yin, X Wei, Y Sun, J Wang, MJ Rosato, in *Proceedings of the International Conference on Automatic Face and Gesture Recognition*. A 3D facial expression database for facial behavior research, (2006), pp. 211–216
31. R Gross, I Matthews, J Cohn, T Kanade, S Baker, in *Proceedings of the International Conference on Automatic Face and Gesture Recognition*. Multi-PIE, (2008), pp. 1–8
32. K Messer, J Matas, J Kittler, J Luettin, G Maître, in *Proceedings of the International Conference on Audio and Video-based Biometric Personal Verification*. XM2VTSDB: The extended M2VTS database, (1999), pp. 72–77

11

Developing a unit selection voice given audio without corresponding text

Tejas Godambe[1], Sai Krishna Rallabandi[1], Suryakanth V. Gangashetty[1], Ashraf Alkhairy[2] and Afshan Jafri[3*]

Abstract

Today, a large amount of audio data is available on the web in the form of audiobooks, podcasts, video lectures, video blogs, news bulletins, etc. In addition, we can effortlessly record and store audio data such as a read, lecture, or impromptu speech on handheld devices. These data are rich in prosody and provide a plethora of voices to choose from, and their availability can significantly reduce the overhead of data preparation and help rapid building of synthetic voices. But, a few problems are associated with readily using this data such as (1) these audio files are generally long, and audio-transcription alignment is memory intensive; (2) precise corresponding transcriptions are unavailable, (3) many times, no transcriptions are available at all; (4) the audio may contain dis-fluencies and non-speech noises, since they are not specifically recorded for building synthetic voices; and (5) if we obtain automatic transcripts, they will not be error free. Earlier works on long audio alignment addressing the first and second issue generally preferred reasonable transcripts and mainly focused on (1) less manual intervention, (2) mispronunciation detection, and (3) segmentation error recovery. In this work, we use a large vocabulary public domain automatic speech recognition (ASR) system to obtain transcripts, followed by confidence measure-based data pruning which together address the five issues with the found data and also ensure the above three points. For proof of concept, we build voices in the English language using an audiobook (read speech) in a female voice from LibriVox and a lecture (spontaneous speech) in a male voice from Coursera, using both reference and hypotheses transcriptions, and evaluate them in terms of intelligibility and naturalness with the help of a perceptual listening test on the Blizzard 2013 corpus.

Keywords: Unit selection synthesis, Found data, Unsupervised TTS, LibriSpeech, Phone recognition, Confidence measures, Data pruning

1 Introduction

1.1 Motivation

Unit selection speech synthesis is one of the techniques for synthesizing speech, where appropriate units from a database of natural speech are selected and concatenated [1–3]. Unit selection synthesis can produce natural-sounding and expressive speech output given a large amount of data containing various prosodic and spectral characteristics. As a result, it is used in several commercial text-to-speech (TTS) applications today.

1.1.1 Overhead of data preparation for building general-purpose synthetic voices

Building a new general-purpose (non-limited domain) unit selection voice in a new language from scratch includes a huge overhead of data preparation, which includes preparing phonetically balanced sentences, recording them from a professional speaker in various speaking styles and emotions in a noise-free environment, and manually segmenting or correcting the automatic segmentation errors. All of it is time consuming, laborious, and expensive, and it restricts rapid building of synthetic voices. A free database such as CMU ARCTIC [4] has largely helped to rapidly build synthetic voices in the English language. But CMU ARCTIC is a small database, contains only a few speakers data, and is not prosodically rich (contains short declarative utterances only). Today, (1) a large amount of audio data has become available on the web in the form of audiobooks, podcasts, video lectures, video blogs, news bulletins, etc, and (2) thanks to technology, we can effortlessly record and store large amounts of high-quality single speaker audio such as lecture, impromptu, or read speech. Unlike

*Correspondence: AfshanJafri@gmail.com
[3]King Saud University, Riyadh, Saudi Arabia
Full list of author information is available at the end of the article

CMU ARCTIC, these data are rich in prosody and provide a plethora of voices to choose from, and their use can significantly ease the overhead of data preparation thus allowing to rapidly build general-purpose natural-sounding synthetic voices.

1.1.2 Problems with using found data for building synthetic voices

Now, the questions to be asked are whether we can readily use such data to build expressive unit selection synthetic voices [5] and will the synthesis be good? In this paper, we try to answer these questions. There are a few problems related to it such as the following: (1) the audio files are generally long and audio-text alignment becomes memory intensive; (2) precise corresponding transcriptions are unavailable; (3) often, no transcriptions are available, and manually transcribing the data from scratch or even correcting the imprecise transcriptions is laborious, time consuming, and expensive; (4) the audio may contain bad acoustic (poorly articulated, dis-fluent, unintelligible, inaudible, clipped, noisy) regions as the audio is not particularly recorded for building TTS systems; and (5) if we obtain automatic transcripts using a speech recognition system, the transcripts will not be error free.

1.1.3 Previous works on long audio alignment

Earlier works have addressed the abovementioned first and second issues of long audio alignment in the following three ways: (1) audio-to-audio alignment, (2) acoustic model-to-audio alignment, and (3) text to text alignment. Each method has its advantages and limitations.

1. *Audio-to-audio alignment*: Here, the text is converted to speech using a TTS system, and the synthesized speech is aligned with the audio [6, 7]. This method requires the existence of a TTS system.
2. *Acoustic model-to-audio alignment*: Here, acoustic models are aligned with the audio. In [8], a modified Viterbi algorithm to segment monologues was used. Their method assumed a good (at least 99 %) correspondence between speech and text, required manual intervention to insert text at the beginnings and endings of monologues, did not handle mispronunciations, and propagated an error in one segment to subsequent segments. In [9], a Java-based GUI to align speech and text was released. They also used acoustic models, assumed good correspondence between audio and text, and required manual intervention.
3. *Text-to-text alignment*: Here, a full-fledged automatic speech recognition (ASR) system including an acoustic and language model is used. Basically, long files are chunked into smaller segments based on silence. Hypothesis transcriptions are obtained for these smaller segments. In [10], they proposed a method where a search is made to see where the sequence of words in the reference and hypothesis transcriptions match. The stretch where they match is aligned with the audio using the Viterbi algorithm. This process is repeated until a forced alignment is done for each audio chunk. The process is practically difficult to implement and relies on correctness of the reference and hypothesis transcriptions. In [11], a finite state transducer-based language model instead of N-grams was used. In [12], they used a phone-level acoustic decoder without any phonotactic or language model and then found the best match within the phonetic transcripts. This approach was inspired by the fact that the data to be aligned could have a mixture of languages. But phonetic alignment is less robust than that at the word level. In [13], they quantified the number of insertions, substitutions, and deletions made by the volunteer who read the book "A Tramp Abroad" by Mark Twain and proposed a lightly supervised approach that accounts for these differences between the audio and text. Their method, unlike the forced-alignment approach in [11] which uses beam pruning to identify erroneous matches, could find also the correct sequence and not only the best match in terms of the state sequence between the text and audio chunk. In [14], a dynamic alignment method to align speech at the sentence level in the presence of imperfect text data was proposed. The drawback of this method is that it cannot handle phrase reordering within the transcripts.

The above works on long audio alignment addressing the first and second problems with found data generally prefer reasonable transcripts and mainly focus on (1) less manual intervention, (2) mispronunciation detection, and (3) segmentation error recovery. In this work, we used a large vocabulary public domain ASR system to obtain transcripts, followed by confidence measure-based data pruning which together address the five issues with the found data and also ensure the above three points. We used posterior probability obtained from the ASR system and unit duration as confidence measures for data pruning. Posterior probability helps detecting mislabeled and bad acoustic regions while unit durational measure helps detect unnaturally short or long units which may have high posterior probability values but they can make words unintelligible or sound hyper-articulated, respectively. Thus, both these confidence features are directly related to the intelligibility and naturalness of speech. For proof of concept, we built voices in English language using an audiobook (read speech) in a female voice from LibriVox and a lecture (spontaneous speech) in a male voice from Coursera, using both the reference and hypotheses transcriptions, and evaluate them in terms of intelligibility and naturalness with the help of a perceptual listening test on the Blizzard 2013 corpus.

1.2 System overview

Figure 1 shows the architecture of the entire system. Each module in Fig. 1 is explained in detail in the succeeding sections. First, the ASR system accepts the audio data and produces corresponding labels. Then, data pruning using confidence measures takes place. The pruned audio and label data form the unit inventory for the TTS system. During synthesis time, the TTS system accepts normalized text, takes into account the duration and phrase break information predicted by a statistical parametric speech synthesizer trained using the same audio data and hypothesized transcriptions, and chooses an appropriate sequence of units that minimizes the total of the target and concatenation costs. The output of the TTS system is an audio file.

1.3 Experiments and evaluation

In the first experiment, we compare the recognition performance of an ASR system trained on LibriSpeech data against those trained on TTS data (Olive and lecture speech). In the second experiment, we check the effectiveness of posterior probability as a confidence measure to prune bad data. In the third experiment, we check the effect of pruning using a combination of posterior probability and unit duration on intelligibility and naturalness of the synthesized voice.

The rest of the paper is organized as follows. Section 2 describes data preparation. Section 3 gives ASR and TTS system development details. Section 4 discusses data pruning using confidence measures. Section 5 explains the experiments and evaluation. We conclude the paper in the Section 6.

2 Data preparation

2.1 Data used for building the ASR system

In this paper, we built three ASR systems, one of which was built using LibriSpeech data [15]. LibriSpeech is a fairly recently made available continuous speech corpus in English language, which is prepared by collating parts of several audiobooks available at the LibriVox website. It contains two parts: 460 h of clean speech and 500 h of speech data containing artificially added noise. We used 460 h of clean speech[1] to build acoustic models, a 3-gram language model[2] pruned with a threshold of 3×10^{-7} to generate the lattices, and a higher order 4-gram language model[3] to rescore the lattices and find the 1-best Viterbi path for the ASR system.

The other two ASR systems (both the acoustic and language models) were built using the TTS data described in the next subsection.

2.2 Data preparation for building the TTS system

Details of the audio used for building voices are given in Tables 1 and 2. One is an audiobook (read speech) in a female voice downloaded from LibriVox, and the other is a lecture (spontaneous speech) in a male voice downloaded from Coursera. The audio files were converted to 16-kHz WAV format and power normalized. Before downloading the audio, we checked that the voice quality and speech intelligibility of the speakers are good and that the audio has not been recorded in a noisy background. For the audiobook, we also made sure that it is not a part of the 460-h clean speech LibriSpeech corpus which was used for training the ASR system so that we can simulate the situation that the found audiobook data is unseen by the ASR system. For the lecture speech, a few lectures contained TED talks and voices from other speakers, both of which were removed. The audiobook and lecture audio files were long and could not be directly used for decoding as memory shortage problems can arise while running

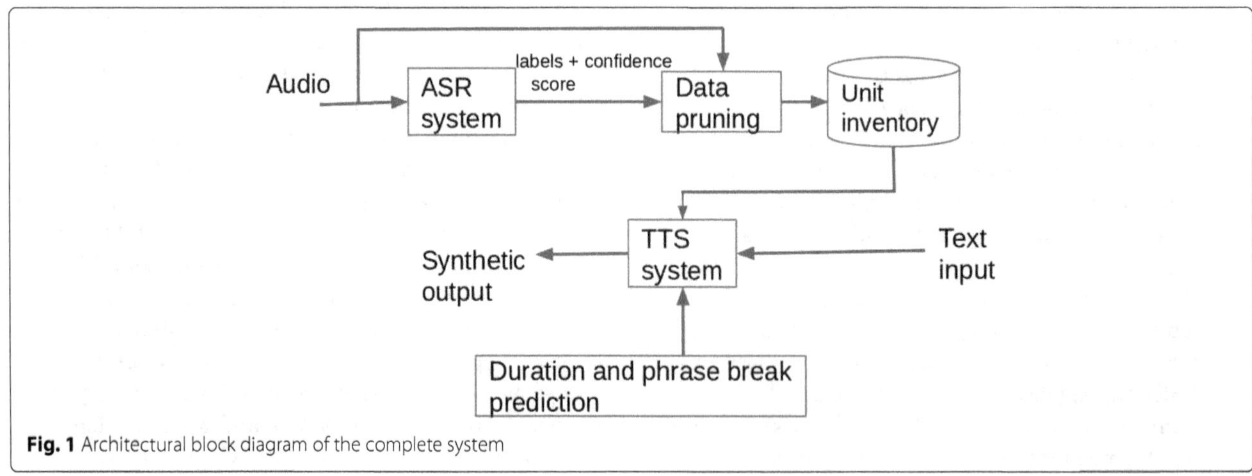

Fig. 1 Architectural block diagram of the complete system

Table 1 Details of the audiobook used for building the unit selection voice

Name of the audiobook	Author	Read by	Running time
Olive (voice 1)	Dinah Maria Mulock CRAIK	Arielle Lipshaw	14:03:13

the Viterbi algorithm. So, silence-based chunking of the long audio files is usually performed. In this work, however, we obtained audio chunks using the open source tool Interslice [8] as we also wanted to obtain chunks of corresponding reference transcripts for building the TTS system using reference transcripts for comparing it with the TTS system built using the hypothesis transcripts. The start and end of each spoken chapter of the audiobooks generally contain metadata such as "This is a LibriVox recording" and the reader's name which are not present in the text chapters downloaded from Project Gutenberg. Since Interslice requires an agreement between the text and speech, we manually checked and added/deleted text at the start and end of each chapter. The same process was carried out even for the lecture speech. Since Interslice does not have a mechanism to prevent the propagation of segmentation error to subsequent segments, we also manually verified the agreement between the start and end of the resulting speech and text chunks before using them for building the TTS systems.

3 ASR and TTS systems

3.1 ASR system development details

Lately, ASR systems have become much more accurate and robust thanks to deep neural networks (DNNs) [16–18]. We used scripts provided with the Kaldi toolkit [19] for training DNN-based ASR systems and the IRSTLM tool [20] for building language models. Kaldi is based upon finite-state transducers, and it is compiled against the OpenFst toolkit [21].

3.1.1 Acoustic modeling

Figure 2 shows the flow of the steps followed for training acoustic models.

1. *Feature extraction*: First, 13 dimensional Mel frequency cepstral coefficients (MFCCs) [22] are extracted. A Hamming window of 25-ms frame size and 10-ms frame shift was used. Then, cepstral mean subtraction is applied on a per-speaker basis. Then, MFCCs are appended with the velocity and acceleration coefficients.
2. *Training monophone system*: A set of context-independent or monophone Gaussian mixture model-hidden Markov model (GMM-HMM) acoustic models were trained on the above features.
3. *Training triphone system with LDA + MLLT*: MFCCs without the deltas and acceleration coefficients were spliced in time taking a context size of seven frames (i.e., ± 3). These features were de-correlated, and their dimensionality was reduced to 40 using linear discriminant analysis (LDA) [23]. Further de-correlation was applied on resulting features using maximum likelihood linear transform (MLLT) [24] which is also known as global semi-tied covariance (STC) [25]. The resulting features were used to train triphone acoustic models.
4. *Training triphone system with LDA + MLLT + SAT*: Then, speaker normalization was applied on above features using feature-space maximum likelihood linear regression (fMLLR), also known as constrained MLLR (CMLLR) [26]. The fMLLR was estimated using the GMM-based system applying speaker-adaptive training (SAT) [26, 27]. A triphone system was again trained with these resulting features.
5. *Training DNN system*: A DNN-HMM system with p-norm non-linearities [28] was trained on top of the SAT features. Here, GMM likelihoods are replaced with the quasi-likelihoods obtained from DNN posteriors by dividing them by the priors of the triphone HMM states.

Table 2 Details of the lecture speech used for building the unit selection voice

Name of the course	Instructor	Running time
Introduction to Public Speaking (voice 2)	Dr. Matt McGarrity	$\approx$ 12 h

3.1.2 Lexicon

A lexicon was prepared from the most frequent 200,000 words in the LibriSpeech corpus. Pronunciations for around one third of them were obtained from CMU-dict. The pronunciations for the remaining words were generated using the Sequitur G2P toolkit [29].

3.1.3 Language modeling

We used the IRSTLM toolkit [20] for training language models. A modified Kneser-Ney smoothing was used [30, 31].

3.1.4 Decoding

First, hypothesis transcriptions were produced using the spliced MFCC features. These transcriptions were then used to estimate the fMLLR transforms as explained above. The accuracy of the hypothesis transcriptions obtained after SAT was much better than that before SAT.

Decoding of the audiobooks was done in two passes. In the first pass, lattices containing competing alternative hypothesis were generated, while in the second pass, Viterbi decoding was applied to find the 1-best hypothesis. While decoding the Olive and lecture data with the ASR

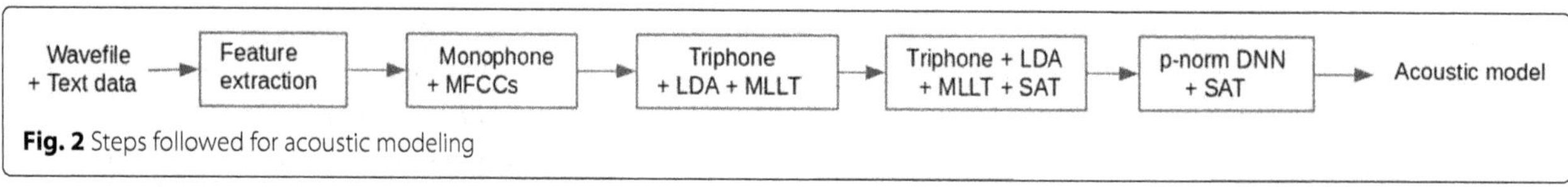

Fig. 2 Steps followed for acoustic modeling

systems trained on themselves, the same 3-gram language model was used for both lattice generation and 1-best Viterbi decoding. In contrast, while decoding the Olive and lecture data with the LibriSpeech ASR system, a 3-gram inexpensive language model pruned with a threshold of 3×10^{-7} was used for lattice generation and a higher order 4-gram language model was used to re-score the language model likelihoods in the lattice, re-rank the set of alternative hypotheses, and find the 1-best hypothesis.

Even though we required phone labels for the audio for building the TTS system, direct phone decoding was not performed as it normally leads to high errors. Rather, word decoding was performed first, and then, word lattices were converted to phone lattices using the lexicon lookup.

3.2 TTS system development details

For synthesis, we made modifications to the TTS system submitted to the Blizzard challenge 2015 [32].

3.2.1 Feature extraction and unit inventory preparation

1. *Unit size*: Units of different sizes ranging from frame-sized units [33, 34], HMM-state sized [35, 36], half-phones [1], and diphones [37] to syllables [38] and to much larger and non-uniform units [39] have been investigated in the literature. We used a quinphone as a context to select appropriate phone-level units. While earlier works mostly used diphone units, where only the previous phone was used as context, the current use of a quinphone is a superset of such selection. Thus, the quinphone context is quite powerful than the regular diphone context. The actual unit used in synthesis is a context-sensitive "phone." This is quite a standard process in unit selection now. Please refer to [2] where the units are called as phones, although sufficient context is used in choosing them. A backoff context (triphone and diphone) is used when a quinphone context is not met. This not only ensures fewer joins and consequently fewer signal processing artifacts, this also leads to good prosody.
2. *Acoustic feature extraction*: Log-energy, 13 dimensional MFCCs, and fundamental frequency (F_0) were extracted for every wave file. A frame size of 20- and 5-ms frame shift was used. The F_0 were extracted using the STRAIGHT tool [40]. The durations and posterior probabilities of the phones which we use as the confidence measure were obtained from the Kaldi decoder.
3. *Preparing the catalog file*: A catalog or dictionary file (which is basically a text file) was prepared which contained the list of all units (including monphone to quinphone) and the attributes of each unit such as duration, start and end times, the duration zscore of each unit type, F_0, log-energy, MFCC values of boundary frames, and posterior probability scores (computed as the minimum of posterior probabilities of phones in that unit). This file was used during synthesis time to compute the target and join costs explained below.
4. *Pre-clustering units*: Pre-clustering is a method that allows the target cost to be effectively pre-calculated. Typically, units of the same type (phones, diphones, etc.) are clustered based on acoustic differences using decision trees [37]. In this work, we clustered units of a type (i.e., units containing the same sequence of phones) on the basis of their positions in words (such as beginning, internal, ending, and singleton), as such clustering implicitly considers acoustic similarity. Units in a cluster are typically called as "candidate" units.

3.2.2 Steps followed at synthesis time

Figure 3 shows the flow of the steps described below.

1. *Text normalization*: A test sentence was first tokenized, punctuations were removed, non-standard words such as time and date were normalized, and abbreviations and acronyms were converted to full forms.
2. *Phonetic analysis*: Each word was broken into a sequence of phones using lexicon. A grapheme-to-phoneme converter [29] was used to convert out-of-vocabulary words and proper names into a phone sequence.
3. *Prediction of phrase-break locations*: Using the audio and automatic transcriptions obtained from the ASR systems, we built statistical parametric voices using the Clustergen synthesizer [41] in the Festival framework [42, 43]. In the current implementation, we took help of the phrase-break locations predicted using classification and regression trees (CART) [44] in Clustergen. For each text input, we first synthesized a statistical parametric voice. A pause unit of appropriate duration was placed at the predicted phrase-break locations during concatenation.

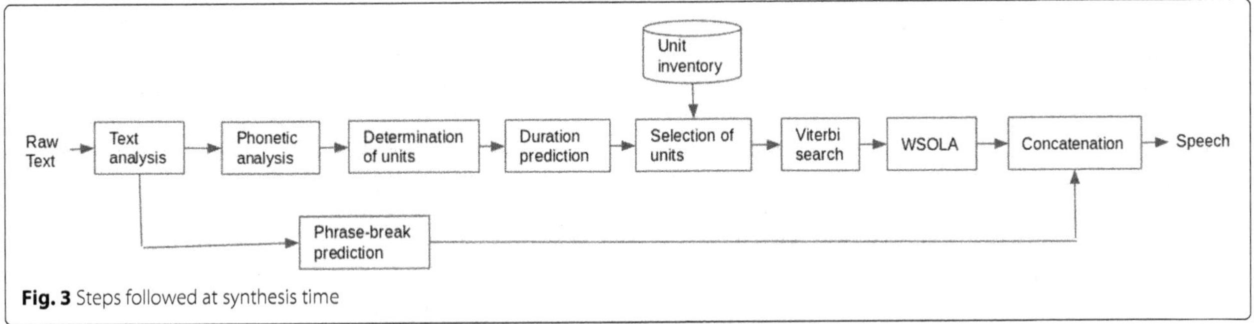

Fig. 3 Steps followed at synthesis time

4. *Determination of units*: For every word in the test sentence, we first searched for units of maximum length which are quinphones or units of length equal to the length of the word if the word is comprised of fewer than five phones. If units of maximum length were not found, then we searched for units of length which is one less than the maximum length, and so on. In short, we joined units of maximum length when they were available in the database or used backoff units of shorter lengths. This approach resulted in fewer joins and a more natural and faster synthesis.
5. *Predicting the durations of units*: In the current implementation, we used the CART-based duration prediction module in Clustergen. For each text input, we first synthesized a statistical parametric voice and used the predicted phone/word durations to select units close to the predicted durations.
6. *Selection of units*:

   - *Target cost computation*: Target cost indicates how close a database unit is to the desired unit. The difference between duration predicted by the CART module in Clustergen [41] and duration of candidate units in the database was used as the target cost.
   - *Join cost computation*: Join cost indicates how well the two adjacently selected units join together. The join cost between two adjacent units u_{i-1} and u_i was calculated using the following equation, which is a linear weighted combination of the distance between log energy, fundamental frequency F_0 (extracted using STRAIGHT tool), and MFCCs of frames near the joining of u_{i-1} and u_i. In the following equation, the symbols α, β, and γ respectively denote the weights for log energy, F_0, and MFCC.

$$\text{Join\_cost} = \alpha C_{F_0}(u_{i-1}, u_i) + \beta C_{\text{log\_energy}}(u_{i-1}, u_i) + \gamma C_{\text{MFCC}}(u_{i-1}, u_i) \tag{1}$$

   Following [32, 45], we used four context frames while computing the distance between the log energies and F_0 of u_{i-1} and u_i, as it helped minimize perceived discontinuities.
   - *Viterbi search*: The Eq. 2 below explains the way the total cost is computed. The term $\mathrm{T}_{\text{dist}}(u_i)$ is the difference between the duration of unit u_i and the predicted duration, and the term $\mathrm{J}_{\text{dist}}(u_i, u_{i-1})$ is the join cost of the optimal coupling point between candidate unit u_i and the previous candidate unit it is to be joined to. W_1 and W_2 denote the weights given to target and join costs, respectively. N denotes the number of units to be concatenated to synthesize the sentence in question. We then used a Viterbi search to find the optimal path through candidate units that minimized the total cost which is the sum total of target and concatenation costs.

$$\text{Total cost} = \sum_{i=1}^{N} W_1 \mathrm{T}_{\text{dist}}(u_i) + W_2 \mathrm{J}_{\text{dist}}(u_i, u_{i-1}) \tag{2}$$

7. *Waveform similarity overlap addition (WSOLA)*: We used an overlap addition-based approach for smoothing the join at the boundaries. Specifically, the cross-correlation formulation of WSOLA [46] was used. The algorithm was reformulated in order to first find a suitable temporal point for concatenating the units at the boundary. This ensured that the concatenation is performed at a point where maximal similarity exists between the units. In different words, this ensured that sufficient signal continuity exists at the concatenation point. For this, cross-correlation between the units was used as a measure of similarity between the units. Next, the units were joined at the point of maximal correlation using a cross-fade technique [33] which further helped remove the phase discontinuities. The number of frames used to calculate the correlation was limited by the duration of the available subword unit. In the current framework, we used the two boundary frames of the individual units to calculate the cross-correlation.

4 Data pruning using confidence measures

In the context of TTS systems, data pruning involves the removal of spurious units (which may be a result of mislabeling or bad acoustics) and units that are redundant in terms of prosodic and phonetic features. Pruning spurious units improves the TTS output [37, 47–51] while pruning redundant units reduces database size thus enabling portability [52–54] and real-time concatenative synthesis [2, 55, 56]. In this work, we focus on removing spurious units and not redundant units.

4.1 Previous works to prune spurious units

In [37], each unit is represented as a sequence of MFCC vectors, and clustering using decision tree proceeds based on questions related to prosodic and phonetic context; each unit is then assessed for its frame-based distance to the cluster center. Units which lie far from their cluster centers are termed as outliers and hence pruned. In [47], this evaluation is done on the basis of an HMM framework: only instances which have the highest HMM scores are retained to represent a cluster of similar units. Confidence features such as log-likelihood ratio [48], transcription confidence ratio [49], and generalized posterior probability [50] have also been used for pruning. We used posterior probability [50, 57] and unit duration [51] obtained from the ASR system as confidence measures.

4.2 Relevance of posterior probability and unit duration as confidence features

Posterior probability helps detecting mislabeled and bad acoustic regions while unit durational measure helps detecting unnaturally short or long units (which may have high posterior probability values), but they can make words unintelligible or sound hyper-articulated, respectively. Thus, both these confidence features are directly related to the intelligibility and naturalness of speech.

4.3 Other advantages of posterior probability

Other motivations to use posterior probability as confidence feature are (1) posterior probability, by definition, tells the correctness or confidence of a classification; (2) it has been shown to work consistently better than the other two formulations of confidence measures, which are the confidence measure as a combination of predictor features and the confidence measures posed as an utterance verification problem [58]; and (3) posterior probability becomes more reliable when robust acoustic [28] and language models are used (as in this case) [59].

4.4 Computation of posterior probability and unit durational zscore

In an ASR system, the posterior probability of a phone or a word hypothesis w given a sequence of acoustic feature vectors $O_1^T = O_1O_2..O_T$ is computed (as given in Eq. 3) as the sum of posterior probabilities of all paths passing through w (in around the same time region) in the lattice. It is computed using a forward-backward algorithm over the lattice. In the equation below, W_s and W_e respectively indicate the sequence of words preceding and succeeding w in a path in the lattice.

$$
\begin{aligned}
p(w|O_1^T) = \\
&= \sum_{W_s}\sum_{W_e} p\left(W_s w W_e | O_1^T\right) \\
&= \frac{\sum_{W_s}\sum_{W_e}\left[p\left(O_1^{t_s}|W_s\right)p\left(O_{t_s}^{t_e}|w\right)p\left(O_{t_e}^{T}|W_e\right)p\left(W'\right)\right]}{p\left(O_1^T\right)} \\
&= \frac{\sum_{W_s}\sum_{W_e}\left[p\left(O_1^{t_s}|W_s\right)p\left(O_{t_s}^{t_e}|w\right)p\left(O_{t_e}^{T}|W_e\right)p\left(W'\right)\right]}{\sum_{W}\left[\sum_{W_s}\sum_{W_e}\left[p\left(O_1^{t_s}|W_s\right)p\left(O_{t_s}^{t_e}|w\right)p\left(O_{t_e}^{T}|W_e\right)p\left(W'\right)\right]\right]}
\end{aligned}
\quad (3)
$$

Normally, the phone hypotheses in the neighborhood of a low confidence phone are also affected. Hence, we discard all units containing even a single phone below a specified threshold. The posterior probability of a unit is calculated as the minimum of posterior probability of phones in that unit.

The unit durational zscore for every unit class is computed as in the following equation.

$$
\text{zscore} = \frac{\text{duration} - \text{mean}}{\text{standard deviation}} \quad (4)
$$

5 Experiments and evaluation

5.1 Experiment 1: checking the performance of ASR systems

We trained a p-norm DNN-HMM acoustic model for all three ASR systems trained on the Olive, lecture, and LibriSpeech data. In the first pass, lattices containing a competing alternative hypothesis were generated, while in the second pass, Viterbi decoding was applied to find the 1-best hypothesis. While decoding the Olive and lecture data with ASR systems trained on themselves, the same 3-gram language model was used for both lattice generation and 1-best Viterbi decoding. In contrast, while decoding the Olive and lecture data with the LibriSpeech ASR system, a 3-gram inexpensive language model pruned with a threshold of 3×10^{-7} was used for lattice generation and a higher order 4-gram language model was used to re-score the language model likelihoods in the lattice, re-rank the set of alternative hypotheses, and find the 1-best hypothesis.

Table 3 shows the word error rates (WERs) and phone error rates (PERs) given by the ASR systems trained on the Olive and lecture data and tested on the Olive and lecture data (TTS data), respectively. It also shows the WER and PER of the ASR system trained with the LibriSpeech data and tested on the Olive and lecture

Table 3 WERs and PERs of ASR systems trained on TTS data and LibriSpeech data

Training data for ASR	Test data to ASR	WER (%)	PER (%)
Olive	Olive	2.54	1.02
LibriSpeech	Olive	3.74	1.57
Lecture	Lecture	5.91	5.19
LibriSpeech	Lecture	21.93	10.20

data. Note, for computing WERs for both the Olive and lecture data, we respectively used the word-level transcriptions available at the Project Gutenberg website and those available at Coursera, as reference transcriptions. These transcriptions are reliable, *but not gold standard.* For computing the PERs, phone-level reference transcriptions were obtained by converting word-level reference transcriptions to phones using a lexicon lookup. A lexicon containing 200,000 words provided with the Kaldi setup was used for that purpose. We can observe the following things in Table 3:

1. As expected, the performance of the ASR system trained with LibriSpeech data is relatively poor, but it is still quite decent.
2. The performance gap between the ASR system trained on lecture data and LibriSpeech is big as compared to that between the ASR system trained on Olive and LibriSpeech data because LibriSpeech is a read speech corpus just like Olive, while lecture data is spontaneous data.
3. The performance of the ASR system trained on lecture data and tested on lecture data is slightly poorer than the ASR system trained on Olive data and tested on Olive data. The reason could be that there is relatively more uniformity in the Olive data (read speech) compared to lecture speech. Lecture speech is more spontaneous and contains a lot of filled pauses, dis-fluencies, fast spoken (at times unintelligible words), and also emphasized words.

5.2 Experiment 2: checking the effectiveness of posterior probability as a confidence measure

We saw that the ASR system trained with LibriSpeech data produces more accurate and reasonably accurate transcripts for Olive and lecture data, respectively. The incorrect phone hypotheses should not be a part of the unit inventory and need to be automatically removed to prevent them from corrupting a synthesized voice. We used the posterior probability given by the ASR system as a confidence measure to prune the erroneous data, where all phones below an optimal posterior probability threshold were pruned. In this experiment, we see to what extent our confidence measure is useful to automatically detect bad acoustics and incorrect hypotheses of the ASR system. Table 4 shows the PER as in Table 3 and its breakup in terms of percentage substitution, insertion, and deletion errors. Note that we can have posterior probabilities only for a hypothesis produced by the ASR system. The phone hypotheses could not be correct phones, substitutions, or insertions. So, deletions could not be detected by the confidence measure, but as their amount was small, we ignored them. The confidence measure is expected to truly reject (TR) as many substitution and insertion errors. Table 4 also shows the percentage of true acceptances (TA), false rejections (FR), true rejections (TR), and false acceptances (FA) obtained at the maximum and optimal posterior probability threshold value equal to 1.0 for all the four cases in Table 3. This threshold value is optimal in the sense that it yields the least number of false acceptances (which are the ASR system's erroneous phone hypotheses termed as correct and hence left unpruned by the confidence measure). We would want the least number of erroneous hypotheses/spurious phones, and hence, we prefer the least number of false acceptances. We observe that the posterior probability does a decent job to harness most of the correct data (as can be seen from the percentage of TAs) leaving just a small amount of erroneous data behind (as can be seen from the amount of false acceptances). Specifically, in the case of the lecture speech recognized by the ASR system trained with Librispeech data, we can see that 10.20 %PER is effectively reduced to $2.77 + 1.63 = 4.4\,\%$ (percentage of FAs) with the use of a confidence measure. There is also a sizeable amount of false rejections that we can see, but we could afford to lose that data since we were using large data for synthesis.

5.3 Experiment 3: checking the effect of pruning based on the posterior probability and unit duration on WER and MOS

There are several examples of fast unintelligible speech (in the case of common and short words such as "to," "the,"

Table 4 Performance of ASR systems and posterior probability confidence measure

Training data for ASR	Test data to ASR	%PER	%sub	%ins	%del	%FA	%TR	%FR	%TA
Olive	Olive	1.02	0.20	0.66	0.16	0.74	0.12	0.54	98.60
LibriSpeech	Olive	1.57	0.58	0.69	0.30	1.00	0.27	4.94	93.79
Lecture	Lecture	5.19	0.96	3.52	0.71	3.09	1.39	3.28	92.24
LibriSpeech	Lecture	10.20	3.27	5.30	1.63	2.77	5.80	19.61	71.82

Table 5 Posterior probability and duration zscore thresholds used to achieve different amounts of data pruning, for all four voices

Percent units used	Posterior probability and duration zscore thresholds			
	Test data = Olive		Test data = Intro. to Public Speaking	
	ASR trained on Olive data	ASR trained on LibriSpeech	ASR trained on lecture data	ASR trained on LibriSpeech
100	–	–	–	–
	1.00, ≈ 97 %	1.00, ≈ 92 %	1.00, ≈ 93 %	1.00, ≈ 65 %
50	1.00, ± 0.51	1.00, ± 0.70	1.00, ± 0.57	1.00, ± 0.98
30	1.00, ± 0.35	1.00, ± 0.45	1.00, ± 0.39	1.00, ± 0.60

"for," and "and" plus other short words) and unnaturally long or emphasized/hyper-articulated words particularly in lecture speech. Several instances of such words have a posterior probability value equal to 1.0, and are left unpruned. Hence, we also prune the units which are much deviant from their mean duration. In addition, pruning units based on duration allows us to prune many more units than is possible using posterior probability alone.

Table 5 shows, for all four voices, the different posterior probability and duration zscore thresholds used to achieve different amounts of unit pruning. The first number in every cell is the posterior probability threshold which is 1.0. The second entry in the second row indicates the percentage of units retained when only posterior probability based pruning is applied. This value is different (97 %, 92 %, 93 %, 65 %) in case of all four systems. No duration threshold was applied in this case. The second entry in third and fourth rows indicates the duration thresholds applied for performing duration based pruning in addition to posterior probability based pruning.

We used the hypotheses and the time stamps given by the ASR systems trained with Olive, lecture, and Librispeech data for synthesis. Even in the case of Olive and lecture, respectively, we used hypotheses of the ASR system instead of force-aligned reference transcriptions from Project Gutenberg and Coursera because the reference transcriptions are reliable, but not gold standard, and the ASR system trained and adapted to a single speaker generally gives better transcriptions and is able to detect the inconsistencies in the reference speech and text.

Table 5 contains 16 different combinations of posterior probability and duration zscore thresholds. We synthesized (for each combination in the table) 10 semantically unpredictable sentences (SUS) [60] and 10 news sentences from the Blizzard 2013 test corpus. So, 20 sentences were synthesized for each combination. In all, 320 sentences were synthesized. A few of the samples used for this experiment can be listened to at https://researchweb.iiit.ac.in/~tejas.godambe/EURASIP/. These sentences were randomly distributed among 16 listeners for perceptual test. So, each listener transcribed 10 SUS (from which we computed the WER indicating the speech intelligibility) and rated the naturalness of the news utterances on a scale of 1 (worst) to 5 (best) from which we calculated the mean opinion score (MOS).

Tables 6 and 7 respectively show the WER and MOS for all four voices synthesized using different amounts of pruned data. We can see that the WERs and MOS are quite good for all the four voices. We can observe the following things in Table 6.

1. The WERs are high for lecture speech than audiobook speech.
2. The WERs are high for unpruned data (as can be seen in the first row). They become slightly better in the second row when we use only units having a posterior probability value equal to 1.0 to synthesize the sentences. The improvement is maximum in the last column where the amount of pruned data having posterior probability less than 1.0 is the highest.
3. Selecting units close to mean duration (as in the third row) decreases WER even further, as short units which are much deviant from the mean duration are pruned.
4. The improvement in WER observed with duration pruning (difference in WERs of the second and third

Table 6 Word error rates for all four voices for different amounts of data pruning

Percentage units used	Word error rate (%)			
	Test data = Olive		Test data = lecture	
	ASR trained on Olive	ASR trained on LibriSpeech	ASR trained on lecture	ASR trained on LibriSpeech
100	14.25	17.21	22.13	28.17
	13.50	15.95	20.56	23.90
50	8.11	9.56	16.25	17.15
30	6.26	6.28	13.25	13.87

Table 7 MOS scores for all four voices for different amounts of data pruning

Percentage units used	Mean opinion score			
	Test data = Olive		Test data = lecture	
	ASR trained on Olive	ASR trained on LibriSpeech	ASR trained on lecture	ASR trained on LibriSpeech
100	3.49	3.52	3.18	2.91
	3.51	3.47	3.21	3.22
50	3.28	3.22	2.99	3.05
30	3.08	3.00	2.90	2.93

rows) is more than the difference in the WERs of first and second rows observed with pruning units having a posterior probability less than 1.0. This difference is more evident in the case of the lecture speech (which contains more units corresponding to fast speech than audiobook contributing to less intelligible speech). The WER further reduces when more units based on duration are pruned (even when only 30 % units are retained).

In the case of naturalness of speech in Table 7, we can observe the following things.

1. Voices built using an audiobook seem to be more natural than those built using lecture speech.
2. The MOS is almost the same for the first and second rows except the case of the last column where a noticeable improvement is observed in MOS.
3. The MOS decreases as we move down rows as it becomes difficult to find units having a duration close to predicted duration and which can also maintain continuity in terms of energy, F_0, and MFCCs.

6 Conclusions

Today, a large amount of audio data has become available to us via the web, and also, we can easily record and store a huge amount of audio data on handheld devices, etc. These data are rich in prosody and provide many voices to choose from, and their availability can help to rapidly build general-purpose unit selection voices. But, there are a few hurdles such as the unavailability of transcriptions or availability of imprecise transcriptions and the presence of speech and non-speech noises. In this paper, we built voices for an audiobook (read speech) in a female voice and a lecture (spontaneous speech) in a male voice using reference transcripts and a combination of automatic transcripts and confidence measure-based data pruning and showed that voices of comparable quality as that using reference transcripts can be rapidly built using found data.

Endnotes

[1]http://www.openslr.org/12/.

[2]http://www.openslr.org/resources/11/3-gram.pruned.3e-7.arpa.gz.

[3]http://www.openslr.org/resources/11/4-gram.arpa.gz.

Competing interests

The authors declare that they have no competing interests.

Acknowledgements

This project was funded by the National Plan for Science, Technology and Innovation (MAARIFAH), King Abdulaziz City for Science and Technology, Kingdom of Saudi Arabia, Award Number 10-INF1038-02.

Author details

[1]International Institute of Information Technology Hyderabad, Hyderabad, India. [2]King Abdulaziz City for Science and Technology, Riyadh, Saudi Arabia. [3]King Saud University, Riyadh, Saudi Arabia.

References

1. M Beutnagel, A Conkie, J Schroeter, Y Stylianou, A Syrdal, in *Joint Meeting of ASA, EAA, and DAGA*. The AT&T next-gen TTS system (Citeseer, Berlin, Germany, 1999), pp. 18–24
2. AJ Hunt, AW Black, in *Proc. of ICASSP*. Unit selection in a concatenative speech synthesis system using a large speech database, vol. 1 (IEEE, Atlanta, Georgia, USA, 1996), pp. 373–376
3. S Ouni, V Colotte, U Musti, A Toutios, B Wrobel-Dautcourt, M-O Berger, C Lavecchia, Acoustic-visual synthesis technique using bimodal unit-selection. EURASIP J. Audio Speech Music Process. **2013**(1), 1–13 (2013)
4. J Kominek, AW Black, in *Proc. of Fifth ISCA Workshop on Speech Synthesis*. The CMU ARCTIC speech databases (ISCA, Pittsburgh, PA, USA, 2003). http://festvox.org/cmuarctic
5. H Harrod, How do you teach a computer to speak like Scarlett Johansson? (2014). http://goo.gl/xn5gBw. Accessed 15 Feb 2014
6. X Anguera, N Perez, A Urruela, N Oliver, in *Proc. of ICME*. Automatic synchronization of electronic and audio books via TTS alignment and silence filtering (IEEE, Barcelona, Spain, 2011), pp. 1–6
7. N Campbell, in *Proc. of ICSLP*. Autolabelling Japanese TOBI, vol. 4 (IEEE, Philadelphia, USA, 1996), pp. 2399–2402
8. K Prahallad, AW Black, Segmentation of monologues in audio books for building synthetic voices. IEEE Trans. Audio Speech Lang. Process. **19**(5), 1444–1449 (2011)
9. C Cerisara, O Mella, D Fohr, in *Proc. of Interspeech*. JTrans, an open-source software for semi-automatic text-to-speech alignment (ISCA, Brighton, UK, 2009), pp. 1823–1826
10. PJ Moreno, CF Joerg, J-M Van Thong, O Glickman, in *Proc. of ICSLP*. A recursive algorithm for the forced alignment of very long audio segments, vol. 98 (ISCA, Sydney, Australia, 1998), pp. 2711–2714
11. PJ Moreno, C Alberti, in *Proc. of ICASSP*. A factor automaton approach for the forced alignment of long speech recordings (IEEE, Taipei, Taiwan, 2009), pp. 4869–4872
12. G Bordel, M Peñagarikano, LJ Rodríguez-Fuentes, A Varona, in *Proc. of Interspeech*. A simple and efficient method to align very long speech

signals to acoustically imperfect transcriptions (ISCA, Portland, USA, 2012), pp. 1840–1843
13. N Braunschweiler, MJ Gales, S Buchholz, in *Proc. of Interspeech*. Lightly supervised recognition for automatic alignment of large coherent speech recordings (ISCA, Makuhari, Chiba, Japan, 2010), pp. 2222–2225
14. Y Tao, L Xueqing, W Bian, A dynamic alignment algorithm for imperfect speech and transcript. Comput. Sci. Inf. Syst. **7**(1), 75–84 (2010)
15. V Panayotov, G Chen, D Povey, S Khudanpur, in *Proc. of ICASSP*. LibriSpeech: an ASR corpus based on public domain audio books (IEEE, Brisbane, Queensland, Australia, 2015), pp. 5206–5210
16. V Peddinti, D Povey, S Khudanpur, in *Proceedings of INTERSPEECH*. A time delay neural network architecture for efficient modeling of long temporal contexts (ISCA, Dresden, Germany, 2015), pp. 2440–2444
17. L Tóth, Phone recognition with hierarchical convolutional deep maxout networks. EURASIP J. Audio Speech Music Process. **2015**(1), 1–13 (2015)
18. P Motlicek, D Imseng, B Potard, PN Garner, I Himawan, Exploiting foreign resources for DNN-based ASR. EURASIP J. Audio Speech Music Process. **2015**(1), 1–10 (2015)
19. D Povey, A Ghoshal, G Boulianne, L Burget, O Glembek, N Goel, M Hannemann, P Motlíček, Y Qian, P Schwarz, et al, in *Proc. of ASRU*. The Kaldi speech recognition toolkit (IEEE, Waikoloa, HI, USA, 2011). EPFL-CONF 192584
20. M Federico, N Bertoldi, M Cettolo, in *Proc. of Interspeech*. Irstlm: an open source toolkit for handling large scale language models (ISCA, Brisbane, Australia, 2008), pp. 1618–1621
21. C Allauzen, M Riley, J Schalkwyk, W Skut, M Mohri, in *Implementation and Application of Automata*. Openfst: a general and efficient weighted finite-state transducer library (Springer, Berlin, Heidelberg, 2007), pp. 11–23
22. SB Davis, P Mermelstein, Comparison of parametric representations for monosyllabic word recognition in continuously spoken sentences. IEEE Trans. Acoust. Speech Signal Process. **28**(4), 357–366 (1980)
23. RO Duda, PE Hart, DG Stork, *Pattern classification*. (Wiley, 2012)
24. RA Gopinath, in *Proc. of ICASSP*. Maximum likelihood modeling with Gaussian distributions for classification, vol. 2 (IEEE, Washington, USA, 1998), pp. 661–664
25. MJ Gales, Semi-tied covariance matrices for hidden Markov models. IEEE Trans. Speech Audio Process. **7**(3), 272–281 (1999)
26. MJ Gales, Maximum likelihood linear transformations for HMM-based speech recognition. Comput. Speech Lang. **12**(2), 75–98 (1998)
27. S Matsoukas, R Schwartz, H Jin, L Nguyen, in *DARPA Speech Recognition Workshop*. Practical implementations of speaker-adaptive training (Citeseer, Chantilly, VA, 1997), pp. 11–14
28. X Zhang, J Trmal, D Povey, S Khudanpur, in *Proc. of ICASSP*. Improving deep neural network acoustic models using generalized maxout networks (IEEE, Florence, Italy, 2014), pp. 215–219
29. M Bisani, H Ney, Joint-sequence models for grapheme-to-phoneme conversion. Speech Commun. **50**(5), 434–451 (2008)
30. R Kneser, H Ney, in *Proc. of ICASSP*. Improved backing-off for m-gram language modeling, vol. 1 (IEEE, Detroit, Michigan, USA, 1995), pp. 181–184
31. SF Chen, J Goodman, in *Proc. of ACL*. An empirical study of smoothing techniques for language modeling (Association for Computational Linguistics, Santa Cruz, California, USA, 1996), pp. 310–318
32. SK Rallabandi, A Vadapalli, S Achanta, S Gangashetty, in *Proc. of Blizzard Challenge 2015*. IIIT Hyderabad's submission to the Blizzard Challenge 2015 (ISCA, Dresden, Germany, 2015)
33. T Hirai, S Tenpaku, in *Fifth ISCA Workshop on Speech Synthesis*. Using 5 ms segments in concatenative speech synthesis (ISCA, Pittsburgh, PA, USA, 2004), pp. 37-42
34. Z-H Ling, R-H Wang, in *Ninth International Conference on Spoken Language Processing*. HMM-based unit selection using frame sized speech segments (ISCA, Pittsburgh, PA, USA, 2006)
35. RE Donovan, PC Woodland, in *Eurospeech Proceedings: 4th European Conference on Speech Communication and Technology*. Improvements in an HMM-based speech synthesiser, vol. 1 (ISCA, Madrid, Spain, 1995), pp. 573–576
36. X Huang, A Acero, J Adcock, H-W Hon, J Goldsmith, J Liu, M Plumpe, in *Spoken Language, 1996. ICSLP 96. Proceedings., Fourth International Conference On*. Whistler: a trainable text-to-speech system, vol. 4 (IEEE, Philadelphia, PA, USA, 1996), pp. 2387–2390
37. AW Black, PA Taylor, in *Proc. of Eurospeech*. Automatically clustering similar units for unit selection in speech synthesis (ISCA, Rhodes, Greece, 1997), pp. 601-604
38. SP Kishore, AW Black, in *Proc. of Interspeech*. Unit size in unit selection speech synthesis (ISCA, Geneva, Switzerland, 2003), pp. 1317–1320
39. H Segi, T Takagi, T Ito, in *Fifth ISCA Workshop on Speech Synthesis*. A concatenative speech synthesis method using context dependent phoneme sequences with variable length as search units (ISCA, Pittsburgh, PA, USA, 2004), pp. 115–120
40. H Kawahara, I Masuda-Katsuse, A De Cheveigne, Restructuring speech representations using a pitch-adaptive time-frequency smoothing and an instantaneous-frequency-based F0 extraction: possible role of a repetitive structure in sounds. Speech Commun. **27**(3), 187–207 (1999)
41. AW Black, in *Proc. of Interspeech*. Clustergen: a statistical parametric synthesizer using trajectory modeling (ISCA, Pittsburgh, PA, USA, 2006), pp. 1394–1397
42. A Black, P Taylor, R Caley, R Clark, K Richmond, S King, V Strom, H Zen, The festival speech synthesis system, version 1.4. 2 (2001). Unpublished document available via http://www.cstr.ed.ac.uk/projects/festival.html
43. P Taylor, AW Black, R Caley, in *Proc. of Eurospeech*. The architecture of the festival speech synthesis system (ISCA, Jenolan Caves, Australia, 1998), pp. 147-151
44. L Breiman, J Friedman, CJ Stone, RA Olshen, *Classification and regression trees*. (CRC press, 1984)
45. BSR Rajaram, KHR Shiva, R A G, in *Blizzard Challenge 2014*. MILE TTS for Tamil for Blizzard Challenge 2014 (ISCA, Singapore, 2014)
46. W Verhelst, M Roelands, in *Proc. of ICASSP*. An overlap-add technique based on waveform similarity (WSOLA) for high quality time-scale modification of speech, vol. 2 (IEEE, Minneapolis, Minnesota, USA, 1993), pp. 554–557
47. H Hon, A Acero, X Huang, J Liu, M Plumpe, in *Proc. of ICASSP*. Automatic generation of synthesis units for trainable text-to-speech systems, vol. 1 (IEEE, Seattle, Washington, USA, 1998), pp. 293–296
48. H Lu, Z-H Ling, S Wei, L-R Dai, R-H Wang, in *Proc. of Interspeech*. Automatic error detection for unit selection speech synthesis using log likelihood ratio based SVM classifier (ISCA, Makuhari, Chiba, Japan, 2010), pp. 162–165
49. J Adell, PD Agüero, A Bonafonte, in *Proc. of ICASSP*. Database pruning for unsupervised building of text-to-speech voices (IEEE, Toulouse, France, 2006), pp. 889–892
50. L Wang, Y Zhao, M Chu, FK Soong, Z Cao, in *Proc. of Interspeech*. Phonetic transcription verification with generalized posterior probability (ISCA, Lisbon, Portugal, 2005), pp. 1949-1952
51. J Kominek, AW Black, in *Fifth ISCA Workshop on Speech Synthesis*. Impact of durational outlier removal from unit selection catalogs (ISCA, Pittsburgh, PA, USA, 2004), pp. 155–160
52. H Lu, W Zhang, X Shao, Q Zhou, W Lei, H Zhou, A Breen, in *Proc. of Interspeech*. Pruning Redundant synthesis units based on static and delta unit appearance frequency (ISCA, Dresden, Germany, 2015), pp. 269–273
53. R Kumar, SP Kishore, in *Proc. of Interspeech*. Automatic pruning of unit selection speech databases for synthesis without loss of naturalness (ISCA, Jeju island, Korea, 2004), pp. 1377-1380
54. V Raghavendra, K Prahallad, in *Proc. of ICON*. Database pruning for Indian language unit selection synthesizers (ACL, Hyderabad, India, 2009), pp. 67–74
55. D Schwarz, G Beller, B Verbrugghe, S Britton, et al, in *Proceedings of the COST-G6 Conference on Digital Audio Effects (DAFx), Montreal, Canada*. Real-time corpus-based concatenative synthesis with catart (Citeseer, 2006), pp. 279–282
56. RE Donovan, in *Proc. of ICASSP*. Segment pre-selection in decision-tree based speech synthesis systems, vol. 2 (IEEE, Istanbul, Turkey, 2000), pp. 11937–11940
57. F Wessel, R Schlüter, K Macherey, H Ney, Confidence measures for large vocabulary continuous speech recognition. IEEE Trans. Speech Audio Process. **9**(3), 288–298 (2001)
58. H Jiang, Confidence measures for speech recognition: a survey. Speech Commun. **45**(4), 455–470 (2005)
59. NT Vu, F Kraus, T Schultz, in *Proc. of Spoken Language Technology Workshop*. Multilingual A-stabil: a new confidence score for multilingual unsupervised training (ISCA, Berkley, CA, USA, 2010), pp. 183–188
60. C Benoît, M Grice, V Hazan, The sus test: a method for the assessment of text-to-speech synthesis intelligibility using semantically unpredictable sentences. Speech Commun. **18**(4), 381–392 (1996)

12

Driver aggressiveness detection via multisensory data fusion

Omurcan Kumtepe[1*], Gozde Bozdagi Akar[1] and Enes Yuncu[2]

Abstract

Detection of driver aggressiveness is a significant method in terms of safe driving. Every year, a vast number of traffic accidents occur due to aggressive driving behaviour. These traffic accidents cause fatalities, severe disorders and huge economical cost. Therefore, detection of driver aggressiveness could help in reducing the number of traffic accidents by warning related authorities to take necessary precautions. In this work, a novel method is introduced in order to detect driver aggressiveness on vehicle. The proposed method is based on the fusion of visual and sensor features to characterize related driving session and to decide whether the session involves aggressive driving behaviour. Visual information is used to detect road lines and vehicle images, whereas sensor information provides data such as vehicle speed and engine speed. Both information is used to obtain feature vectors which represent a driving session. These feature vectors are obtained by modelling time series data by Gaussian distributions. An SVM classifier is utilized to classify the feature vectors in order for aggressiveness decision. The proposed system is tested by real traffic data, and it achieved an aggressive driving detection rate of 93.1 %.

Keywords: Driver aggressiveness, Driving behaviour, Road safety, Road line detection, Lane detection, Vehicle detection

1 Introduction

Traffic accidents has become an important problem in the last few decades due to increasing number of vehicles on the roads. Every year, 1.24 million fatalities occur due to traffic accidents globally [1]. Some of these traffic accidents are caused by physical reasons such as road and vehicle conditions. However, mostly, human factor is effective in the occurrence of traffic accidents. Among the human factors, aggressive driving behaviour constitutes a huge portion of traffic accident reasons. According to a report of the American Automobile Association Foundation for Traffic Safety, published in 2009, 56 % of traffic accidents occur due to aggressive driving behaviour [2]. Moreover, traffic accidents brings about billions of dollars of economical cost for people, governments and companies [1]. For these reasons, reduction of the number of traffic accidents is an important issue. Considering human factors, detection of aggressive driving behaviour could help in reducing the number of traffic accidents by giving necessary warnings to drivers and related authorities.

Aggressive driving behaviour is defined as an action "when individuals commit a combination of moving traffic offences so as to endanger other persons or property" by the National Highway Traffic Safety Administration (NHTSA) [3]. Aggressive driving behaviour is a psychological concept that does not have a quantitative measure. However, there exist some certain behaviours associated with aggressive driving such as excess and dangerous speed, following the vehicle in front too closely, in other words tailgating, erratic or unsafe lane changes, improperly signalling lane changes and failure to obey traffic control devices (stop signs, yield signs, traffic signals, etc.) [3]. Also, in [4], it is stated that lane changing and acceleration are the characteristics of driving behaviours that define driving style. Therefore, detecting these behaviours and constituting features from these information can yield quantitative information about the driving style of the driver.

Although these behaviours are indication of driver aggressiveness, detection of these behaviours in real time is a challenging task. Existing methods in the literature

*Correspondence: omurcan@metu.edu.tr
[1] Department of Electrical and Electronics Engineering, Middle East Technical University, Dumlupınar Blv. No:1, 06800 Ankara, Turkey
Full list of author information is available at the end of the article

mostly based on driving simulator data which do not work for real-time aggressive driving behaviour detection and do not fully reflect the real-world conditions [5]. There also exist sensor platform-based methods in literature; however, these methods do not consider vehicle following distance and lane following pattern which are very significant for indicating driver aggressiveness. The proposed system enables detection of driver aggressiveness in real time by considering a wide range of aggressiveness-associated driving behaviours.

In this paper, we proposed an automated aggressive driving behaviour detection system that works in real time. The system performs robust operation with simple and low complexity algorithm in order to be able to work efficiently in real time. Multisensory information is used by this system in order to extract features that characterize the related driving session. The system collects data about lane following, vehicle following, speed and engine speed patterns which are important for aggressive driving detection since aggressive driving behaviour is associated with sudden lane changes, tailgating and abrupt acceleration/deceleration. Features that are extracted utilizing these data are used to train an SVM classifier. The classifier is trained with annotated data so that aggressiveness decision can be modelled regarding the subjective point of view, that is, aggressive driving behaviour, which is a subjective and psychological phenomenon, can be modelled quantitatively. The system uses different types of features and feature extraction methods that works in real time; therefore, the system can create a decision at the end of each session. Session length is a design parameter which will be discussed in test results.

The organization of this paper is as follows: The next part describes the related work about aggressive driving behaviour detection. It is followed by the proposed method description and its advantages and novelty. Then, the test results are presented and concluding remarks are given.

2 Related work

Aggressive driving behaviour detection has been examined via different approaches in recent years [5]. The simplest method for detecting aggressive driving behaviour is to conduct surveys about the driving experience or psychological mood. In literature, there exist some methods that are based on observing the behaviours of subjects in the simulator environment. In [4], subjects are requested to drive via a simulator with different scenarios which contain events such as traffic light existence, intersection crossing and frustrating environment. Then, the findings are illustrated by probabilistic models. Similarly Danaf et al. [6] use a simulator environment to collect data about driving behaviour and expresses anger (or aggressiveness) as a dynamic variable. Hamdar et al. [7] define and develop a quantitative aggressiveness propensity index in order to model driving behaviour by testing its proposal with a driving simulator. The main drawback of these works is that they are using a synthetic environment to measure the driving behaviour. Therefore, they do not fully reflect the real-world conditions and reactions of driver in real traffic environment.

In order to acquire real world data, Gonzalez et al. [5] propose a sensor platform-based system to detect driver aggressiveness. Their method monitors external driving signals such as lateral and longitudinal accelerations and speed and models aggressiveness as a linear filter operating on these signals [5]. Johnson and Trivedi [8] use sensor data which is obtained by a smart phone in order to characterize the driving style. Kang [9] examines driver drowsiness and distraction by collecting visual information such as eye gaze and yawning and physiological data such as ECG signals.

Satzoda and Trivedi [10] use multisensory information in order to analyse the drive and certain driving events such as lane changes, mean speed, etc. However, no interpretation is given about the aggressiveness of driver. Jian-Qiang and Yi-Ying [11] present a dangerous driving behaviour detection scheme using a CCD camera to acquire visual information about driving behaviour and identifies dangerous driving style. Nevertheless, the system uses only visual information and tries to identify the driving with a few features. The work presented in [12] exploits a sensor and a camera platform to detect independent driving events such as lane departure, acceleration, zig-zag driving, etc. Then, it uses a fuzzy technique to indicate whether the driving is dangerous. Although the system shows good results for identifying different driving events, the presented work focuses on dangerous driving rather than aggressiveness and does not propose any technique to verify aggressiveness with subjective observations.

Besides the systems that are specialized on detecting driver aggressiveness, there also exist advanced driving assistance systems (ADAS) in literature. ADAS are very popular in recent years and used in order to provide assistance to the driver about the current driving conditions such as lane departure or forward collision possibility [13]. They are used for collecting data about the driving and for warning the driver by giving feedback about the driving behaviour. However, ADAS do not interpret the driving data to reach an aggressiveness conclusion.

3 Proposed method

As indicated in [3] and [4], aggressive driving is associated with certain behaviour such as sudden lane changes, tailgating behaviour, speed and acceleration basically. Therefore, aggressive driving behaviour can be identified by observing these events. In order to obtain quantitative measures of these events, lane following, vehicle following, speed and engine speed patterns of a driving session is collected and processed automatically. As a result of the process, four different feature types are obtained to represent the related driving session. Lane deviation and forward car distance are extracted as visual information and vehicle and engine speed as sensor information. Since the operation of the system is in real time, robust and algorithmically simple methods are used for extraction of the related information. These information are collected and feature vectors are retrieved. Obtained feature vectors are given to a pre-trained classifier to detect aggressive driving behaviour. The overall system flow can be seen in Fig. 1.

3.1 Road line detection

In order to find the position of the host vehicle, which is the equipped and examined vehicle, inside the road lane, road line detection is required. Drivers who change lanes suddenly and continuously and do not follow the lane properly may involve in aggressive driving attitude. Therefore, detecting the position of the host vehicle inside the lane by detecting the road lines is an important information. For road line detection problem, non-uniformity of road lines is the major challenge [14]. In order to accomplish road line detection task with a robust operation to non-uniformities in road lines, we used a method based on temporal filtering and inverse perspective mapping which is a robust, simple and low-cost method and proper for the real-time operation of the overall system. However, in order to decrease the computation load and satisfy the real time operation condition, we modelled road lines with straight lines instead of curves which provides sufficient results for our application.

In recent years, many different techniques and studies are conducted on road line detection, mainly caused by the current interest in advanced driving assistance systems and autonomous driving systems. Road line detection algorithms in the literature mostly consist of two stages, preprocessing and detection stages. In preprocessing stage, different image processing techniques are used in order to provide enhanced data for detection task. Preprocessing methods in the literature can be exemplified as follows. Somasundaram et al. [15] use transformation from RGB colour space to HSV colour space for reducing redundancy. Morphological filtering is used in [16]. In [17], canny edge detector is used to indicate and emphasis road lines, and in [18], Gaussian smoothing is used to eliminate noise. Transformation to binary image is used in [19]. Inverse perspective mapping and road segmentation methods are used as preprocessing in [20, 21]. A study conducted by Jung et al. [22] proposed constructing spatiotemporal images which exploits the temporal dependency of the video frames.

The widely used method for line detection after preprocessing stage is Hough transform. Hough transform is a generic line detection algorithm and used to find road lines [23, 24]. Borkar et al. [21] use a gaussian template matching method after Hough transform in order to increase the detection efficiency. Wang et al. [25] proposed using B-snakes to detect and track road lines. Ridge detection is performed in [20] with convolution with a Gaussian kernel. Another study proposed in [26] combines the self-clustering algorithm (SCA), fuzzy C-mean and fuzzy rules to process the spatial information and

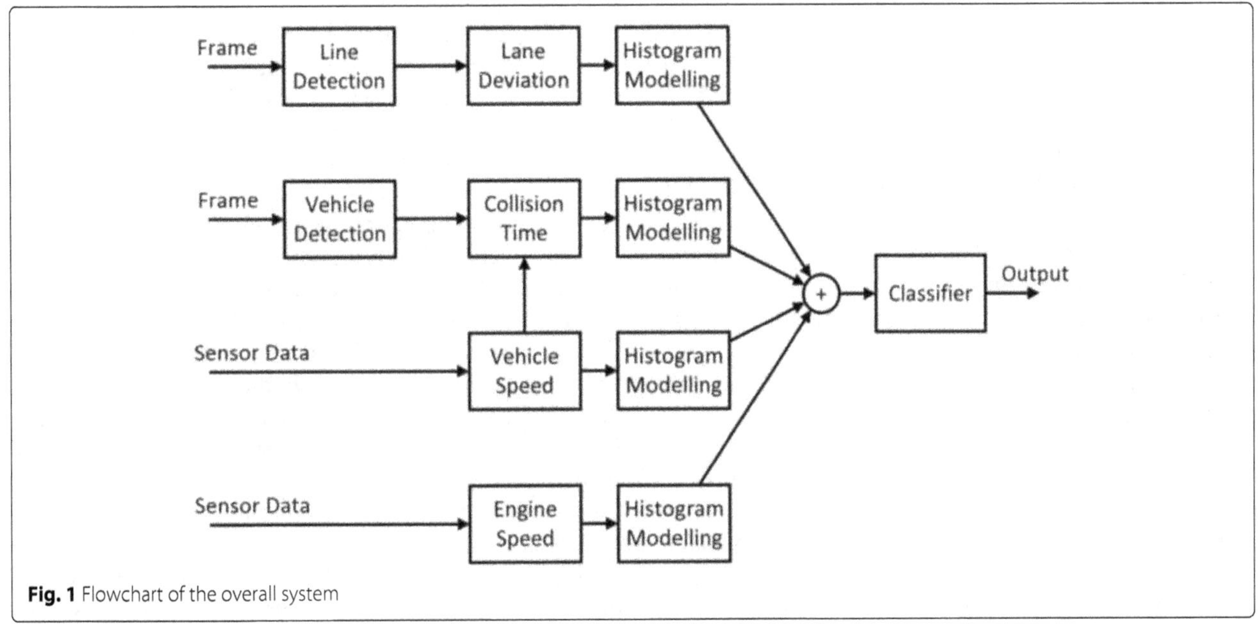

Fig. 1 Flowchart of the overall system

Canny algorithms to get good edge detection. Wu et al. [19] exploits angular relations between lane boundaries. Mu and Ma [27] uses piecewise fitting and object segmentation method to indicate road line positions. In [21], template matching in inverse perspective mapping applied images is used with a tracking scheme.

Although these presented methods perform promising results, they are not fully useful for our application. In our case, the main concern is to use a robust, simple and low complexity algorithm to detect the position of the host vehicle between two road lines correctly and fast enough to work in real time. Although these methods performs well regarding the detection rate, they are not providing low complexity, simplicity and robustness together, that is, a well-performing method may require high computation power and complex implementation which is a big disadvantage for real-time applications. Therefore, the main objective is to implement an algorithmically simple method which performs a robust operation. In order to satisfy this condition, we use temporal filtering and inverse perspective mapping which is a simple method as well as having low complexity and high robustness as explained in [14].

One of the most important problems regarding the robustness of the system is non-uniformity of road conditions [21]. In order to overcome the problems that are caused by shadows, different light conditions and discontinuities on the road line, a method based on temporal filtering is used [14, 21] with inverse perspective mapping which gives robust, fast and simple results.

First, the captured image is temporally filtered in order to eliminate dashed lines and discontinuities according to (1)

$$I'_k(x,y) = \max\{I_k(x,y),\ldots,I_{k-K}(x,y)\} \tag{1}$$

where I_k represents the current frame, I_{k-K} represents the Kth previous frame and (x,y) are pixel coordinates. K is chosen according to the frame rate and dashed line length so that all road lines can be seen as a continuous line as in Fig. 2.

Then, the gradient image of $I'_k(x,y)$ is calculated and the high-gradient pixels are cleared from $I'_k(x,y)$ to obtain $I''_k(x,y)$. This operation gives the low gradient pixels which represent the road plane. Then, the mean and variance values of $I''_k(x,y)$ is calculated so that the mean intensity value of road part can be known. Once these values are obtained, the pixels that are representing road plane are cleared from the image $I'_k(x,y)$. This operation helps to eliminate noise and indicate road lines better. A simple derivative filter $F = [-1\ 0\ 1]$ is used to indicate the lines. After this operation, binary image is obtained using an adaptive threshold according to Otsu's method [28].

Inverse perspective mapping is an efficient method for road line detection. Camera placed at the front of a vehicle gives the road lines as straight lines intersecting at the horizon level. However, inverse perspective mapping enables the road lines to be seen as parallel lines. Moreover, since monocular vision system is used, inverse perspective mapping will be exploited to measure the distance between vehicles. In order to achieve inverse perspective mapping, four points are chosen in the filtered image and they are mapped to four other points in the birds-eye perspective assuming the surfaces are planar as in Fig. 3. This mapping procedure results in a 3×4 H matrix that contains the transformation parameters. This matrix is calculated before the operation and loaded to the system. Then, during the operation, inverse perspective mapping is done by transforming each ith point using H matrix as in (3).

$$p^i_k = HP^i_k \tag{2}$$

Since the aforementioned procedures work well enough to indicate the line positions, a simple procedure is done to locate road lines. Horizontal projection of the image is taken in a limited region so that the line locations appear as peaks in the horizontal projection vector. Nieto et al. [14] solve the line localization problem with a parametric curve fitting. However, since we exploit simple methods for the sake of real time application, we modelled the road lines with simple lines. This procedure is based on the assumption that curved roads are seen as straight up to a certain distance. And the region whose horizontal projection taken is chosen to minimize the noise in peak detection as shown in Fig. 4.

One last step that is used to increase the stability and accuracy of line detection is tracking the detected lines with kalman filter [29]. This tracking scheme includes denoising with Kalman filter as well as keeping the visibility counts of lines and recovering missing detections for a specific frame. This scheme significantly improves the efficiency of the overall process.

The two closest detected lines from the camera center, which is defined beforehand as a pixel value according to horizontal positioning of the camera, are chosen as own lane boundaries. The horizontal position of the camera center from the lane boundary is determined as parameter between -50 and 50 for each frame I_k.

The presented method is tested with real set-up data which includes different road conditions such as shadows, occlusion and road curve. As can be seen in the sample figures (Fig. 5), line detection method show robustness to these environment conditions.

In order to test the accuracy and reliability, the presented method is tested with Borkar's dataset [21]. In Borkar's dataset, there exist video sequences containing driving sessions at urban road, metropolitan highway and

Fig. 2 Raw image and temporal filtered image

isolated highway whose ground truth road line positions are provided. Since the main aim of road line detection module is to extract the position information of the host vehicle inside the lane, these ground truth values are used to determine the ground truth values of lane position. For this task, pixel value of camera center is estimated by visual inspection and position information is calculated accordingly. As can be seen in Fig. 6 for different video sequences, lane position information is determined accurately. In order to quantify the accuracy of the position information over a video sequence, mean absolute error mean absolute error (MAE) values are calculated. As indicated in [30], MAE can be used for measuring estimation accuracy of driving signals such as speed, orientation, etc. Hence, for each presented sequence in Fig. 6, a MAE value is calculated as in Eq. 3 and shown in Table 1. Regarding the mean absolute error values of the sequences, it can be said that for different conditions, the presented method performs lane position detection with a limited error rate. To illustrate MAE value for isolated highway, data is found as 1.51 which means that average error of lane position detection is 1.51 in −50:50 scale for a frame which is a very small error rate.

$$\mathrm{MAE} = \sum_{i=1}^{N} \frac{|\mathrm{LaneDeviation}_{GT}(i) - \mathrm{LaneDeviation}_{\mathrm{measured}}(i)|}{N} \tag{3}$$

We compared our lane deviation detection results with other methods in the literature which is tested for Borkar's dataset. As presented in [22], Jung et al. stated that lane detection rate for their method and Borkar's method are as in Table 2. We tested our method with Borkar's dataset with video sequences containing different conditions and presented the results in Table 2. Our method provides similar results with existing methods, performing better for urban dataset which is more critical in terms of aggressiveness detection. Moreover, the line deviation values over frames will be represented as distributions which is explained in the "Feature extraction and classification" section. This process will further compensate the deteriorating effect of errors regarding aggressiveness detection.

3.2 Vehicle detection

Vehicle detection process is required in order to find the distance between host car and other cars that can be seen from the camera. This distance will be used to build up a feature which characterize tailgating or unsafe following distance behaviour. For vehicle detection task, we used a simple and robust approach for the sake of real-time operation and we employed histogram of oriented gradients (HOG) features with a cascade classifier. We also improved the algorithmic efficiency and accuracy of vehicle detections by exploiting lane detection results since we

Fig. 3 Four corner point selection and perspective transformation. The corners of the *red box* represent the four chosen points, and these points are mapped to the corners of the *trapezoidal region* in the figure at the *right*

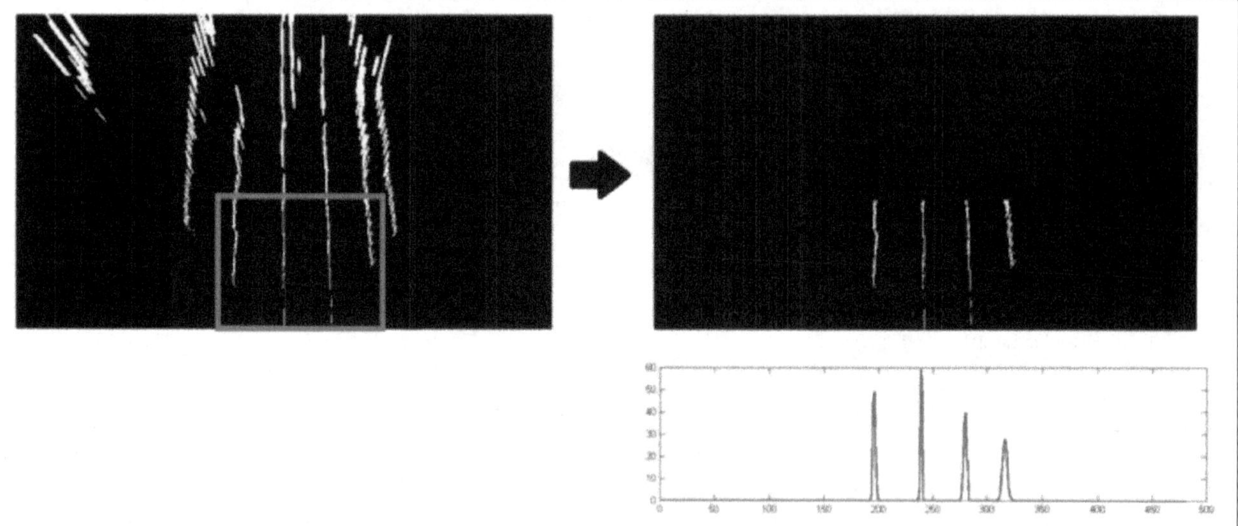

Fig. 4 Line position detection by horizontal projection. The figure at the *top left* represents the processed and transformed image. The *red box* represents the limited interest region. The figure at the *top right* is the masked version according to interest region. The graph at the *bottom right* is the horizontal projection of the image

Fig. 5 Correctly detected lines in different frames

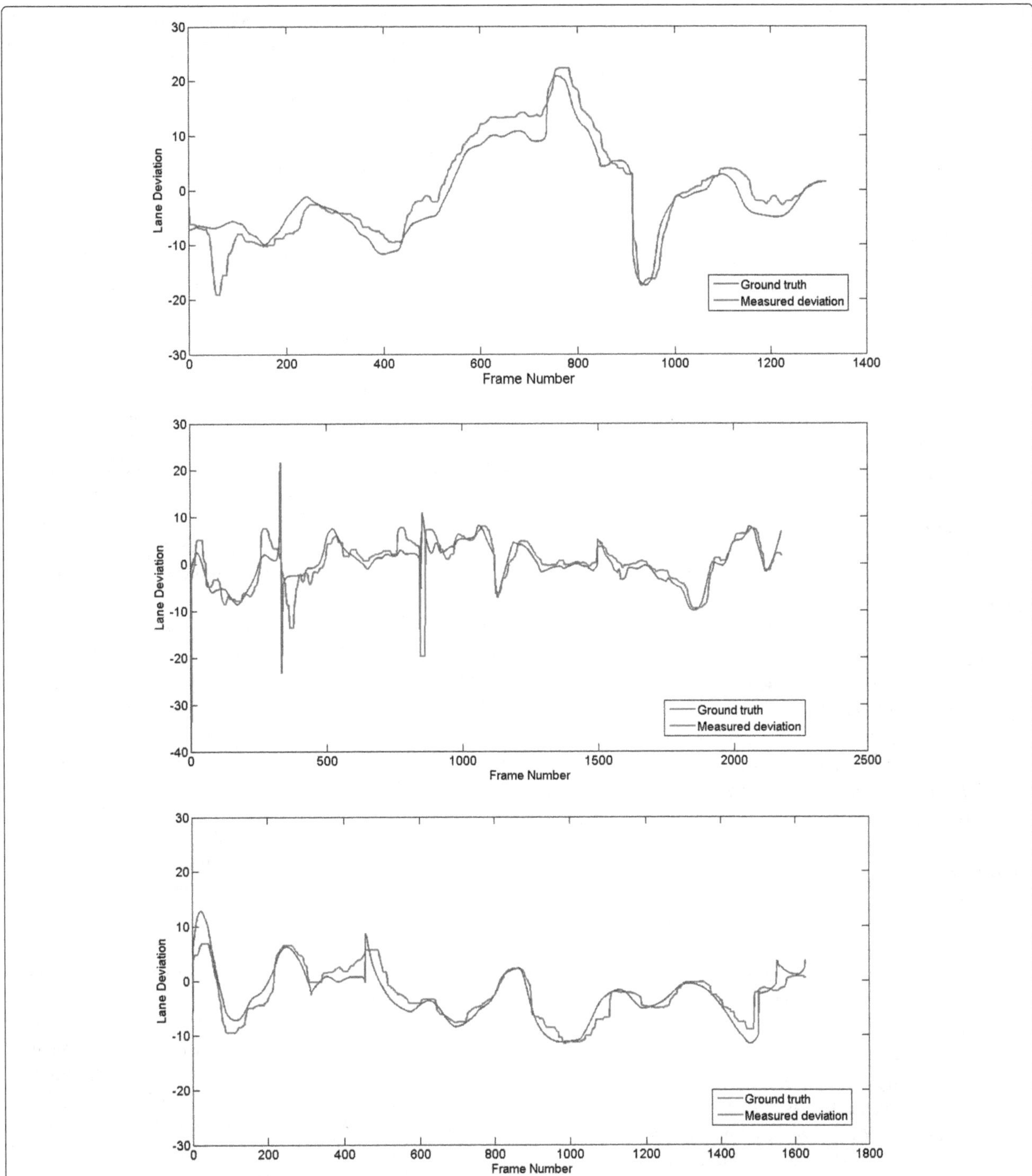

Fig. 6 Comparison of lane position detection with ground truth values in Borkar's dataset. The figure at the *top* belongs to video sequence of urban area in low traffic condition, the figure in the *middle* belongs to video sequence of metro highway in dense traffic condition and the figure at the *bottom* belongs to isolated highway in moderate traffic condition

are interested in only the vehicles which are in the same lane with host vehicle. This condition enabled us to run vehicle detection process in a specific region of interest.

There exist different approaches in previous studies about on-road vehicle detection. In most of the previous studies, vehicle detection is associated with forward

Table 1 Mean absolute error values for different road and traffic conditions

Road condition	Mean absolute error
Urban area, low traffic	2.47
Metro highway, dense traffic	1.96
Isolated highway, moderate traffic	1.51

collision warning systems (FCWS) which is a part of driving assistance systems (DAS). In these systems, vehicle detection and distance estimation can be performed by radars or simple sensors as explained in [31, 32]. Another alternative of radar sensors are lidars. Lidars are also used for this task collaboratively with radars [33]. However, the state-of-the-art forward collision systems are based on camera-based platforms and image processing techniques. In literature, among on-road vehicle detection methods, Kim et al. [34] do vehicle detection by scanning the image so as to find a shadow region by the help of some morphological operations. The work presented in [35] depends on the active training of images represented by Haar-like features. In [10], HOG features are extracted from the frames, then a support vector machine (SVM) classification is utilized to find the vehicles. Considering forward vehicle distance estimation, both [34] and [10] use inverse perspective mapping to find the distance of the target vehicles. Other than these monocular camera-based methods, there exist studies that depend on stereo vision. The method presented in [36] detects objects in both images by motion segmentation and determine the vehicle distance by creating a depth map. Kowsari et al. [37] use Haar-like feature extraction, a feature classification with the power of stereo vision. Similarly, Seo et al. [38] use an omnidirectional camera and stereo vision techniques for vehicle detection and distance estimation. As can be seen in these studies, stereo vision methods give good results for estimating vehicle distance while increasing the hardware and computation complexity.

For our application, we employed HOG feature extraction since it is known to be a robust approach for object detection. And a cascade classifier detection technique is utilized in order to detect vehicle because cascade classifier is a robust and fast method which is proper for real time applications. In order to determine the distances of detected vehicles, inverse perspective mapping is used. Since it involves a training process its performance can be improved by using required number and variety of samples during the training phase.

In order to achieve object detection, first, vehicle images from real traffic data are collected to train a classifier. Image patches that contain a rear images of vehicles are cropped from collected images and tagged as positive. During this phase, different types of vehicles are chosen as samples in order to increase the accuracy. On each of these samples, HOG features are calculated and fed to the classifier.

A cascade classifier is trained according to the process that is described in [39]. During the implementation, each new-coming frame is scanned with a sliding window in different scales; HOG features are calculated over these windows and fed to the classifier to be tested. According to classifier result, detected objects are located by a bounding box. This process may create some false positive that are appearing for a few consecutive frames. Therefore, a Kalman tracking scheme [29] as described in previous section is used to track the detected objects. This process improves the detection rate and eliminates false positives. Some examples of vehicle detection can be seen in Fig. 7. In these figures, it can be seen that different types of vehicles are correctly detected.

In order to find the following distance, a vehicle is chosen as the target vehicle (if there exist a vehicle in the scene). The target vehicle is determined as the nearest vehicle in the own lane of host vehicle. The inverse perspective mapping information, that were found in the previous section, is used to transform the position of the target vehicle to the birds-eye view perspective which enables us to determine the distance between host vehicle target vehicle in pixel units. This difference in pixel units is converted to metric unit with a constant C which is predefined according to the perspective transformation values before the overall process.

The presented vehicle detection method is a well-known and simple scheme, and it gives satisfying results regarding our problem definition. So as to assess the performance of the method, we utilized LISA-Q Front FOV Dataset [35] which contains three different annotated video sequences. In [35], the presented method is tested with LISA dataset and the results are given according to several performance metrics. The details of these metrics can be found in [35].

Since the ultimate aim of the method is to find the distance between host vehicle and target vehicle, we reduced the region of interest in the front view image according to the results of lane detection. In other words, we aimed to detect the vehicles which are in the same lane with the vehicle. To accomplish this, we eliminated the other detections which are in different lanes but ours.

Table 2 Correct detection rate of different methods of Borkar's dataset

Category	Borkar (%)	Jung (%)	Proposed (%)
Isolated highway	98.24	98.31	93.92
Metro highway	98.12	98.33	95.04
Urban	87.12	90.52	93.31

Fig. 7 Examples of detected vehicles

This approach improved the results significantly for each dataset. In Tables 3, 4 and 5, performance results of our method with region of interest selection and comparison with the given method in [35] can be seen for dense, urban and sunny datasets, respectively.

As can be seen in these tables, proposed method performs an average accuracy over 95 % for different conditions. Combining lane detection results with vehicle detection results significantly improved the performance by increasing true positive rate in dense dataset which includes dense traffic images. Furthermore, it decreased the false positive rate in all cases by outperforming the benchmark results in two datasets.

3.3 CAN bus data acquisition

Most of the new cars are equipped with a controller area network (CAN) bus which enables the communication between different microchips and sensors inside the vehicles. It became mandatory in the USA for the cars that are produced after year 1996. CAN bus has a standardized physical connector and a protocol so that the vehicle data can be obtained using the CAN bus port for analysis and diagnosing purposes. In our application, vehicle speed and engine speed are used as the sensor-based information since certain patterns of these information are associated with aggressive driving behaviour. As indicated in [3], abrupt acceleration and deceleration can be an indication of aggressive driving. Therefore, vehicle and engine speed values are exploited for characterizing driver aggressiveness. In order to collect these data, external sensors can be used as performed in [5]. Instead of using external sensors, CAN bus system of the host vehicle can provide this information [10] with a proper adapter as shown in Fig. 8.

In order to read vehicle and engine speed data from the CAN bus of the vehicle, a proper adapter is used and related data is obtained with timestamps during driving in

Table 3 Performance evaluation of different methods for dense dataset

Method	TPR (%)	FDR (%)	FP/frame (%)	TP/frame (%)
Sivaraman's method [35]	95.0	6.4	0.29	4.20
Our method without lane selection	78.4	43.0	2.44	3.23
Our method with lane selection	89.9	9.8	0.09	0.85

Table 4 Performance evaluation of different methods for urban dataset

Method	TPR (%)	FDR (%)	FP/frame (%)	TP/frame (%)
Sivaraman's method [35]	91.7	25.5	0.39	1.14
Our method without lane selection	99.0	36.5	0.57	0.99
Our method with lane selection	99.0	20.6	0.25	0.99

Table 5 Performance evaluation of different methods for sunny dataset

Method	TPR (%)	FDR (%)	FP/frame (%)	TP/frame (%)
Sivaraman's method [35]	99.8	8.5	0.28	3.17
Our method without lane selection	98.7	25.9	1.04	2.97
Our method with lane selection	98.7	4.8	0.05	0.99

order to synchronize the CAN bus data with visual data. Vehicle and engine speed data are collected with a period of 1 s. Therefore, in order to use this data combined with a higher frequency visual data (i.e. 10 fps frame rate), it is up-sampled by a factor of 10.

3.4 Feature extraction and classification

The aforementioned stages are performed to collect information about the behaviour of the driver in the traffic. These collected information is utilized by a feature extraction and classification stage in order to determine whether the related driving session is aggressive or not. For the characterization of the driving session, four different features are chosen considering the aggressive driving indicating behaviours as explained in the "Proposed method" section. These features are as follows:

- Lane deviation
- Collision time
- Vehicle speed
- Engine speed

The line detection results and lane position determination are used to construct lane deviation feature which characterizes the abrupt lane changing and not following the lane properly. The information obtained from the CAN bus, vehicle and engine speed is directly used as the features since drivers who show aggressive driving behaviour tend to drive with high and varying speed, therefore changing engine speed abruptly. The last feature which characterizes the tailgating and unsafe following distance behaviours is the collision time. Collision time feature defines the duration to collision if the vehicle in front would stop suddenly. Therefore, this feature utilizes both speed and target vehicle distance information. Collision time is calculated with a unit of seconds according to (5) where d_k is the distance of the target vehicle in meters and v'_k is the vehicle speed in meters per second at that instant.

$$\text{Collision time}(k) = \frac{d_k}{v'_k} \tag{4}$$

Considering all features that characterize the driving session, their variation pattern in a certain amount of time is more informative for us rather than the time series signal itself in terms of driver aggressiveness. For instance, the frequency that a driver changes lanes is a more important information than the lane position value at a specific time frame. Therefore, we represented time series signals as density functions and modelled them using Gaussian mixture model (GMM) which is a powerful technique for density representation [40]. Since we are handling the collected data by batch process, Gaussian modelling provides an effective representation of driving

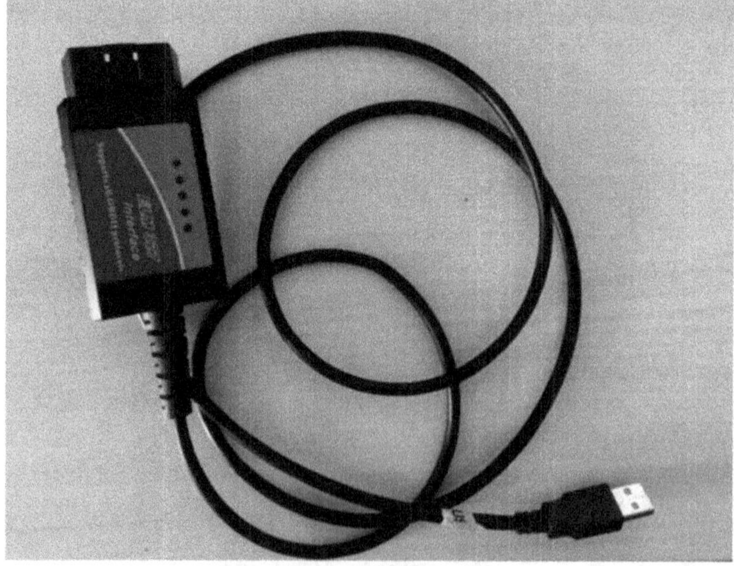

Fig. 8 CAN bus serial port adapter

data. The works presented in [5] and [40] use Gaussian modelling of driving signals for making inferences about driving profiles and present effective results in terms of accuracy.

For our application, each feature is transformed into density functions (i.e. histograms). These histograms are filtered with a median filter in order to eliminate noisy data. Then, they are normalized so that all histograms represent the frequency of the data in the same base. A sample representation of an aggressive and smooth data can be seen in Fig. 9.

During the experiments, we observed that the density functions of driving signals have one dominant Gaussian component. Hence, we modelled histograms using one GMM component which is denoted by a mean μ and a standard deviation σ value which are enough for representing a Gaussian distribution. GMM components of density function are estimated using maximum likelihood estimation. Each driving feature provided one μ and one σ value. Then, these four mean and four standard deviation values are utilized to construct a feature vector consisting eight dimensions. An SVM classifier is employed [41] in order to classify the feature vectors to determine whether a driving session involves aggressive driving behaviour.

Although the presented feature extraction methods are proven to be reliable and comparable with the methods in the literature, the performance of lane deviation detection and collision time estimation modules will effect the result of the aggressiveness classification. Nevertheless, the histogram representation of the features provides robustness to the process and reduces the deteriorating effect of missing detections in line detection and vehicle detection stages. In Fig. 10, histogram modelling of lane deviation and collision time values of an aggressive and a smooth driving session is presented. Mean and standard deviation values of these histograms are presented in Tables 6 and 7 with mean absolute error values between ground truth and measured time series signals. The data presented in Table 6 belong to the sample aggressive session whose histogram is given in Fig. 10, while the data presented in Table 7 belong to the smooth session. As can be seen in these tables, the effect of errors in the detection stage can be eliminated significantly utilizing the histogram modelling.

4 Experimental results

For test purposes, a mobile set up is constructed in order to collect visual and CAN bus data by vehicle. For visual data collection, a portable mini computer (Fig. 11) and a CCD camera (Fig. 12) is used. By this platform, video frames are captured at 10 fps with a resolution of 800 × 600 pixels. For CAN bus data collection, the adapter in (Fig. 8) is connected to CAN bus port of the vehicle and data is acquired through the serial port of the mini computer. The data collected from CAN bus is obtained at each second. Therefore, data is interpolated so that the sensor data exist for each frame. So as

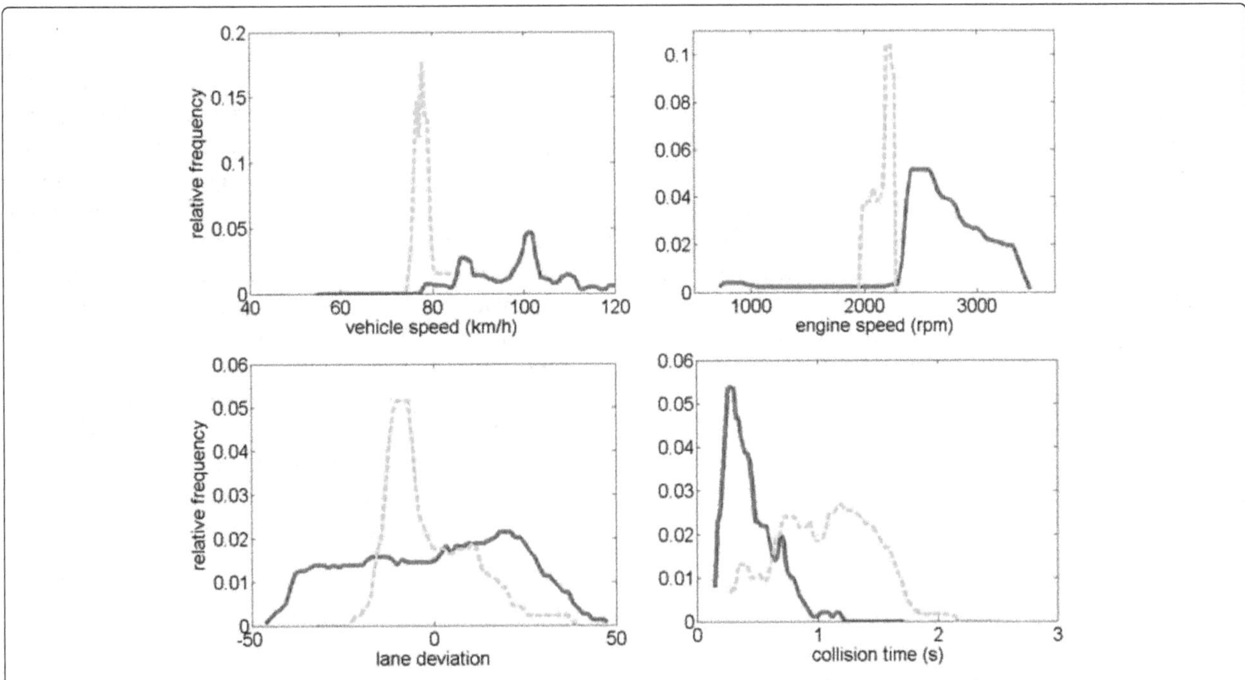

Fig. 9 Examples of histogram comparison of aggressive and smooth driving sessions for different features. *Red solid lines* represent an aggressive driving session while *green dashed lines* represent a smooth driving session in each graph

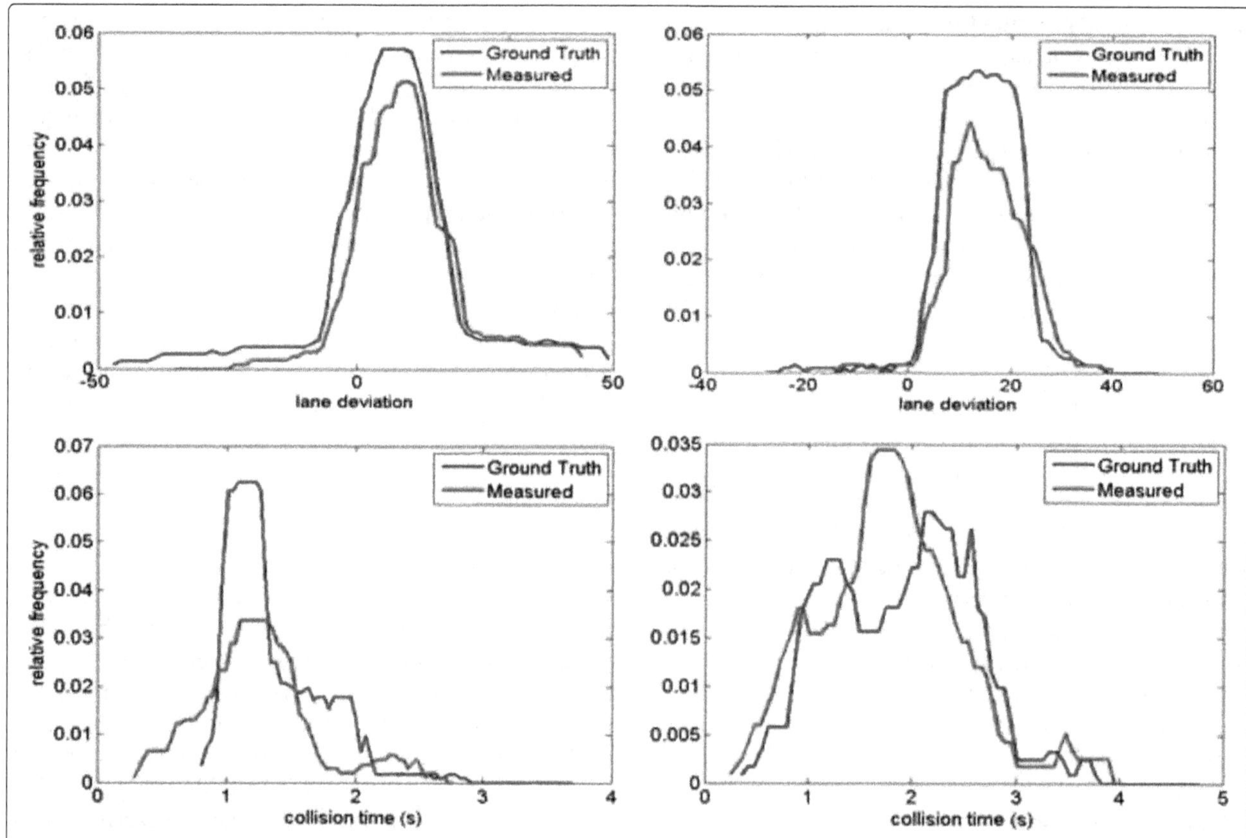

Fig. 10 Comparison of histograms obtained by ground truth values and measured values. The figure at the *top left* belongs to lane deviation values of an aggressive session, the figure at the *top right* presents the histograms of lane deviation of a smooth driving session, the figure at the *bottom left* belongs to collision time distribution of an aggressive driving session and the figure on *bottom right* presents the histograms of collision time of a smooth driving

to synchronize the visual and CAN bus data, the data is timestamped.

Utilizing this set up, real traffic data is collected at different times of the day so that different traffic conditions are included in the dataset. The dataset also includes different road conditions with occlusions, shadows and different illumination. Whole dataset contains driving sessions of six different drivers. During driving, three different observers annotated the last 40 s as aggressive or smooth. The majority voting of the observers are recorded as the ground truth of the related driving session.

One important parameter that effects the performance of the proposed method is the duration of the driving session. In other words, how long multisensory data is required in order to efficiently determine if that driving session is aggressive? In order to answer this question, the collected data is tested with driving sessions with lengths 40, 80, and 120 s. From the whole collected dataset, a total of 83 driving sessions including 41 aggressive and 42 smooth sessions having a duration of 40 s, 51 driving sessions including 22 aggressive 29 smooth sessions having a duration of 80 s and 22 driving sessions including 11 aggressive 11 smooth sessions having a duration of 120 s are tested according to proposed algorithm.

Due to the limited amount of data, k-fold cross validation technique is used for performance assessment. According to this technique, test samples are chosen randomly among the samples; the remaining samples are

Table 6 Comparison of ground truth and measured features of the sample aggressive driving session

	μ_{lane}	σ_{lane}	$\mu_{collision}$	$\sigma_{collision}$	MAE_{lane}	$MAE_{collision}$
Measured	14.88	7.57	2.01	0.81	3.23	0.58
Ground truth	14.07	8.01	1.93	0.89		

Table 7 Comparison of ground truth and measured features of the sample smooth driving session

	μ_{lane}	σ_{lane}	$\mu_{collision}$	$\sigma_{collision}$	MAE_{lane}	$MAE_{collision}$
Measured	9.64	10.73	1.47	0.52	6.78	0.46
Ground Truth	6.70	14.38	1.41	0.48		

Fig. 11 Mini computer used in data collection

used for training the SVM classifier. This process is performed 10 times, and at each run, the classifier results are compared with the ground truth. For the 40-s-long samples, 20 of them; for the 80-s-long samples, 15 of them; and for the 120-s-long samples, 9 of them are chosen randomly as test samples. In Tables 8, 9 and 10, the related confusion matrices of the test results are given for 40-, 80- and 120-s-long samples, respectively.

According to the test results, it is observed that the proposed method achieved 91, 94 and 82.2 % detection rate for 40-, 80- and 120-s-long samples, respectively. As can be inferred from these results, 80-s-long driving sessions are more efficiently representing the driving characteristics while 40-s samples may not allocate enough data or 120-s samples may contain confusing data.

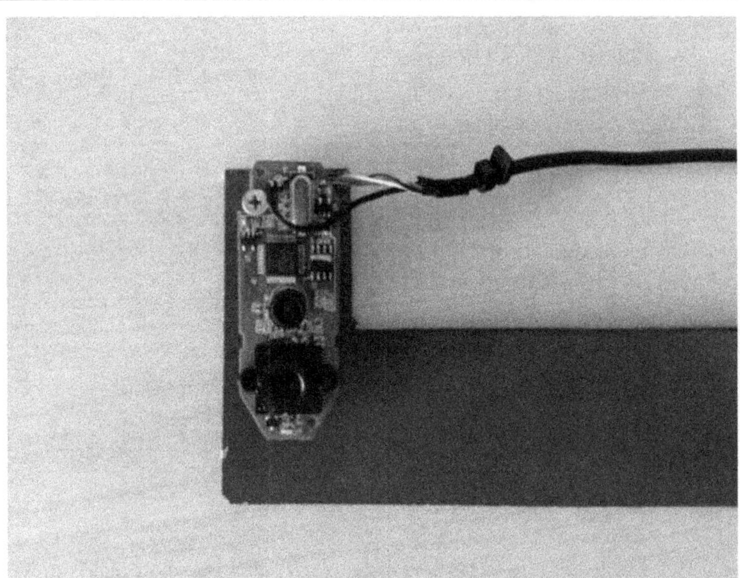

Fig. 12 Camera to capture visual information

Table 8 Confusion matrix of aggressiveness classification for 40-s-long data

		Predicted class	
		Aggressive	Smooth
Real class	Aggressive	90	3
	Smooth	15	92

Table 10 Confusion matrix of aggressiveness classification for 120-s-long data

		Predicted class	
		Aggressive	Smooth
Real class	Aggressive	33	7
	Smooth	9	41

Proposed aggressiveness detection method also tested with real-world data from 100-car dataset [42]. This dataset is the output of a naturalistic driving study and collected via instrumented vehicles in a large scale. In the publicly available part of this dataset, some driving sessions which are approximately 30-s long are given with narratives. These narratives explain the events in the driving session. We investigated these narratives and selected the ones which can be interpreted as an aggressiveness involvement and which cannot. According to narratives, the ones which include aggressive and sharp actions are annotated as "aggressive" and the ones which includes stable actions as "smooth". We selected a total of 76 driving sessions according to narratives and tagged 40 of them as aggressive and 53 of them as smooth. In Table 11, some sample narratives of 100-car data and their interpretation is presented.

The vehicle speed, lane deviation and collision time data are directly present at 100-car dataset. However, instead of engine speed, gas pedal position data is used due to the direct correlation between them. Using these information, the aforementioned feature extraction procedure is applied to the data. In order to validate the reliability of the 100-car data, k-fold cross validation technique is utilized. In each run, 29 of the 93 driving session samples are chosen randomly to train an SVM classifier, and this procedure is repeated 10 times. The classifier achieved a correct detection at an average rate of 93.1 %. Confusion matrix of this process can be seen in Table 12.

5 Conclusions

In this paper, a driver aggressiveness detection method is presented. The proposed method utilizes multisensory information to conceive feature vectors, and using these, feature vectors classify the driving session as aggressive or smooth. The aggressiveness classifier is trained with data annotated by observers and performs classification using data collected in real-world conditions. The paper also studies the required driving session duration that can be efficiently decided if it involves aggressive driving behaviour. According to test results the proposed system performs good results in terms of detecting driver aggressiveness since it considers different driving behaviours in a real time operation. As a future work, the proposed system

Table 9 Confusion matrix of aggressiveness classification for 80-s-long data

		Predicted class	
		Aggressive	Smooth
Real class	Aggressive	67	5
	Smooth	4	74

Table 11 Sample driving sessions with their narratives and aggressiveness interpretation

Sample number	Narrative of the session	Aggressiveness
8354	Subject vehicle is driving relatively fast in the left lane as the traffic is merging into right lane. Lead vehicle is decelerating with right turn signal on, preparing to merge into right lane, and the subject vehicle must brake to avoid hitting lead vehicle in the rear. Subject vehicle is trying to get ahead of right lane traffic before merging.	Aggressive
8392	Subject vehicle is travelling in the rain and almost misses the intended exit. Subject vehicle enters the exit ramp at the last minute, nearly side swiping a vehicle already on the ramp beside it. Subject driver steered slightly left to avoid the crash and the other vehicle went ahead on the exit ramp.	Aggressive
8420	Subject vehicle is preparing to merge onto an exit ramp and a vehicle from the adjacent left lane realizes that they need to get onto the exit ramp also, and the lead vehicle suddenly crosses the subject vehicle's left lane line into the subject vehicle's lane. The subject brakes hard to avoid hitting the lead vehicle in the rear.	Aggressive
8374	Subject driver is talking/singing to herself and stops behind a line of cars at a light. A following vehicle approaches rapidly and almost hits the subject vehicle in the rear.	Smooth
8471	There are 2 left turn lanes with the subject driver in the far left lane. Vehicle 2 at the left turn lane to the right of the subject's vehicle starts to turn left and cuts the subject driver off.	Smooth
9059	Both the subject driver and lead vehicle are decelerating when the subject driver glances out his right side window. When the subject driver glances, the lead vehicle comes to a stop in front of him.	Smooth

Table 12 Confusion matrix of aggressiveness classification for 100-car data

		Predicted class	
		Aggressive	Smooth
Real class	Aggressive	118	12
	Smooth	8	152

will be tested with more data to observe its performance with different classifiers. The system will be improved in order to provide a rate for driver aggressiveness in a granular approach. In other words, the measurement of aggressiveness level will be provided quantitatively.

Competing interests
The authors declare that they have no competing interests.

Acknowledgements
The authors would like to thank to ISSD Informatics and Electronics for providing hardware and equipments to realize the experiments.

Author details
[1]Department of Electrical and Electronics Engineering, Middle East Technical University, Dumlupınar Blv. No:1, 06800 Ankara, Turkey. [2]ISSD Electronics, METU Technopolis, Dumlupınar Blv. No:1, 06800 Ankara, Turkey.

References

1. World Health Organization. Violence and Iznjury Prevention and World Health Organization, *Global Status Report on Road Safety 2013: Supporting a Decade of Action*. (World Health Organization, Geneve, 2013)
2. Aggressive driving: Research update. Technical report, AAA Foundation for Traffic Safety, Washington DC (2009)
3. Aggressive Driving Enforcement: Strategies for Implementing Best Practices. Technical Report, US Dept of Transportation National Highway Traffic Safety Administration, Washington DC (March 2000)
4. T Toledo, HN Koutsopoulos, M Ben-Akiva, Integrated driving behavior modeling. Transp. Res. C Emerg. Technol. **15**(2), 96–112 (2007)
5. ABR Gonzalez, MR Wilby, JJV Diaz, CS Avila, Modeling and detecting aggressiveness from driving signals. Intell. Transp. Syst. **15**(4), 1419–1428 (2014)
6. M Danaf, M Abou-Zeid, I Kaysi, Modeling anger and aggressive driving behavior in a dynamic choice—latent variable model. Accid. Anal. Prev. **75**(0), 105–118 (2015)
7. SH Hamdar, HS Mahmassani, RB Chen, Aggressiveness propensity index for driving behavior at signalized intersections. Accid. Anal. Prev. **40**(1), 315–326 (2008)
8. DA Johnson, MM Trivedi, in *Intelligent Transportation Systems (ITSC), 2011 14th International IEEE Conference On*. Driving Style Recognition Using a Smartphone as a Sensor Platform, (Washington, DC, 2011), pp. 1609–1615
9. H-B Kang, in *Computer Vision Workshops (ICCVW), 2013 IEEE International Conference On*. Various Approaches for Driver and Driving Behavior Monitoring: A Review, (Sydney, NSW, 2013), pp. 616–623
10. RK Satzoda, MM Trivedi, Drive analysis using vehicle dynamics and vision-based lane semantics. IEEE Trans. Intell. Transp. Syst. **16**(1), 9–18 (2015)
11. G Jian-Qiang, W Yi-Ying, in *Intelligent Systems Design and Engineering Applications (ISDEA), 2014 Fifth International Conference On*. Research on Online Identification Algorithm of Dangerous Driving Behavior, (Hunan, 2014), pp. 821–824
12. B-F Wu, Y-H Chen, C-H Yeh, in *ITS Telecommunications (ITST), 2012 12th International Conference On*. Fuzzy Logic Based Driving Behavior Monitoring Using Hidden Markov Models, (Taipei, 2012), pp. 447–451
13. M Rezaei, M Sarshar, MM Sanaatiyan, in *Computer and Automation Engineering (ICCAE), 2010 The 2nd International Conference On*. Toward Next Generation of Driver Assistance Systems: A Multimodal Sensor-Based Platform, vol. 4, (Singapore, 2010), pp. 62–67
14. M Nieto, L Salgado, F Jaureguizar, J Arrospide, in *Image Processing, 2008. ICIP 2008. 15th IEEE International Conference On*. Robust Multiple Lane Road Modeling Based on Perspective Analysis, (San Diego, CA, 2008), pp. 2396–2399
15. G Somasundaram, Kavitha, K. I Ramachandran, in *Computer Science and Information Technologies CSIT, Computer Science Conference Proceedings (CSCP)*, ed. by DC Wyld. Lane Change Detection and Tracking for a Safe Lane Approach in Real Time Vision Based Navigation Systems, (Chennai, India, 2011). July 15 - 17, 2011
16. M-G Chen, C-L Ting, R-I Chang, Safe driving assistance by lane-change detecting and tracking for intelligent transportation system. Int. J. Inform. Process. Manag. **4**(7), 31–38 (2013)
17. M Oussalah, A Zaatri, H Van Brussel, Kalman filter approach for lane extraction and following. J. Intell. Robotics Syst. **34**(2), 195–218 (2002)
18. C Nuthong, T Charoenpong, in *Image and Signal Processing (CISP), 2010 3rd International Congress On*. Lane Detection Using Smoothing Spline, vol. 2, (Yantai, 2010), pp. 989–993
19. C-F Wu, C-J Lin, C-Y Lee, Applying a functional neurofuzzy network to real-time lane detection and front-vehicle distance measurement. IEEE Trans. Syst. Man Cybern. C. **42**(4), 577–589 (2012)
20. M Beyeler, F Mirus, A Verl, in *Robotics and Automation (ICRA), 2014 IEEE International Conference On*. Vision-Based Robust Road Lane Detection in Urban Environments, (Hong Kong, 2014), pp. 4920–4925
21. A Borkar, M Hayes, MT Smith, A novel lane detection system with efficient ground truth generation. IEEE Trans. Intell. Transp. Syst. **13**(1), 365–374 (2012)
22. S Jung, J Youn, S Sull, Efficient lane detection based on spatiotemporal images. IEEE Trans. Intell. Transp. Syst. **PP**(99), 1–7 (2015)
23. V Gaikwad, S Lokhande, Lane departure identification for advanced driver assistance. IEEE Trans. Intell. Transp. Syst. **16**(2), 910–918 (2015)
24. C Tu, BJ van Wyk, Y Hamam, K Djouani, S Du, Vehicle Position Monitoring Using Hough Transform. {IERI} Procedia. **4**(0), 316–322 (2013). 2013 International Conference on Electronic Engineering and Computer Science (EECS 2013)
25. Y Wang, EK Teoh, D Shen, Lane detection and tracking using b-snake. Image Vis. Comput. **22**(4), 269–280 (2004)
26. J-G Wang, C-J Lin, S-M Chen, Applying fuzzy method to vision-based lane detection and departure warning system. Expert Syst. Appl. **37**(1), 113–126 (2010)
27. C Mu, X Ma, Lane detection based on object segmentation and piecewise fitting. TELKOMNIKA Indones. J. Electr. Eng. TELKOMNIKA. **12**(5), 3491–3500 (2014)
28. N Otsu, A threshold selection method from gray-level histograms. IEEE Trans. Syst. Man Cybern. **9**(1), 62–66 (1979)
29. S Thrun, W Burgard, D Fox, *Probabilistic Robotics (Intelligent Robotics and Autonomous Agents)*. (The MIT Press, Cambridge, 2005)
30. R Danescu, F Oniga, S Nedevschi, Modeling and tracking the driving environment with a particle-based occupancy grid. IEEE Trans. Intell. Transp. Syst. **12**(4), 1331–1342 (2011)
31. Y-C Hsieh, F-L Lian, C-M Hsu, in *Intelligent Transportation Systems Conference, 2007. ITSC 2007. IEEE*. Optimal Multi-Sensor Selection for Driver Assistance Systems Under Dynamical Driving Environment, (Seattle, WA, 2007), pp. 696–701
32. M Satake, T Hasegawa, in *Vehicular Electronics and Safety, 2008. ICVES 2008. IEEE International Conference On*. Effects of Measurement Errors on Driving Assistance System Using On-Board Sensors, (Columbus, OH, 2008), pp. 303–308
33. Y Wei, H Meng, H Zhang, X Wang, in *Intelligent Transportation Systems Conference, 2007. ITSC 2007. IEEE*. Vehicle Frontal Collision Warning System Based on Improved Target Tracking and Threat Assessment, (2007), pp. 167–172
34. S Kim, S-y Oh, J Kang, Y Ryu, K Kim, S-C Park, K Park, in *Intelligent Robots and Systems, 2005. (IROS 2005). 2005 IEEE/RSJ International Conference On*. Front and Rear Vehicle Detection and Tracking in the Day and Night Times Using Vision and Sonar Sensor Fusion, (Alberta Canada, 2005), pp. 2173–2178
35. S Sivaraman, MM Trivedi, A general active-learning framework for on-road vehicle recognition and tracking. IEEE Trans. Intell. Transp. Syst. **11**(2), 267–276 (2010)

36. M Miyama, Y Matsuda, in *Signal and Image Processing Applications (ICSIPA), 2011 IEEE International Conference On*. Vehicle Detection and Tracking with Affine Motion Segmentation in Stereo Video, (Kuala Lumpur, 2011), pp. 271–276
37. T Kowsari, SS Beauchemin, J Cho, in *Intelligent Transportation Systems (ITSC), 2011 14th International IEEE Conference On*. Real-Time Vehicle Detection and Tracking Using Stereo Vision and Multi-View AdaBoost, (Washington, DC, 2011), pp. 1255–1260
38. D Seo, H Park, K Jo, K Eom, S Yang, T Kim, in *Industrial Electronics Society, IECON 2013—39th Annual Conference of the IEEE*. Omnidirectional Stereo Vision Based Vehicle Detection and Distance Measurement for Driver Assistance System, (Vienna, 2013), pp. 5507–5511
39. P Viola, M Jones, in *Computer Vision and Pattern Recognition, 2001. CVPR 2001. Proceedings of the 2001 IEEE Computer Society Conference On*. Rapid Object Detection Using a Boosted Cascade of Simple Features, vol. 1, (Kauai, HI, 2001)
40. A Wahab, C Quek, CK Tan, K Takeda, Driving profile modeling and recognition based on soft computing approach. IEEE Trans. Neural Netw. **20**(4), 563–582 (2009)
41. C Cortes, V Vapnik, Support-vector networks. Mach. Learn. **20**(3), 273–297 (1995)
42. VL Neale, TA Dingus, SG Klauer, J Sudweeks, M Goodman, An overview of the 100-car naturalistic study and findings. National Highway Traffic Safety Administration, Paper 05-0400 (2005)

13

On the influence of interpolation method on rotation invariance in texture recognition

Gustaf Kylberg[1] and Ida-Maria Sintorn[1,2*]

Abstract

In this paper, rotation invariance and the influence of rotation interpolation methods on texture recognition using several local binary patterns (LBP) variants are investigated.

We show that the choice of interpolation method when rotating textures greatly influences the recognition capability. Lanczos 3 and B-spline interpolation are comparable to rotating the textures prior to image acquisition, whereas the recognition capability is significantly and increasingly lower for the frequently used third order cubic, linear and nearest neighbour interpolation. We also show that including generated rotations of the texture samples in the training data improves the classification accuracies. For many of the descriptors, this strategy compensates for the shortcomings of the poorer interpolation methods to such a degree that the choice of interpolation method only has a minor impact. To enable an appropriate and fair comparison, a new texture dataset is introduced which contains hardware and interpolated rotations of 25 texture classes. Two new LBP variants are also presented, combining the advantages of local ternary patterns and Fourier features for rotation invariance.

Keywords: Local binary patterns, Texture dataset, Filter banks

1 Introduction

In many computer vision and image analysis applications, the texture of an object is an important property that can be utilized for classification or segmentation procedures. However, texture analysis in digital images is not a trivial task and numerous texture descriptors have been proposed. In some applications, e.g. face recognition as in [1], the orientation of the object is known, while in many other applications the orientation of an object may be arbitrary and hence also the texture. In the latter case, the texture can be rotated to a main orientation or principal direction, see e.g. [2] where the Radon transform is used to accomplish this, or alternatively, a texture descriptor invariant to rotation can be used. A third way to achieve rotation invariance is to add rotated versions of the textures to the training data. This technique of adding a priori information to achieve invariance towards something through adding virtual training samples is explored in [3].

Rotation invariant texture descriptors have been widely studied, as reviewed in [4]. Texture descriptors invariant to viewpoint (adding even more degrees of freedom) have also been studied in for example [5]. However, the problem of arbitrary viewpoints is not addressed in this paper.

Rotation invariance for a texture descriptor can be achieved locally (at each pixel position) or globally (for the region/patch investigated). In [6], clusters of filter bank responses, denoted textons, are computed and the histogram of occurring textons are used as features. They propose a filter bank, denoted MR8, for which global rotation invariance is achieved by using the maximum response over six different orientations. A possible advantage with globally rotation invariant descriptors is that such a descriptor can retain the distribution of local orientations while this information will be lost in a local rotational invariant descriptor. Note however, that in the case of MR8, this information is lost when only keeping the maximum response over orientations. In [7], a globally rotation invariant descriptor retaining distributions over orientations is introduced based on Fourier transformed responses from Gabor filter banks. In [8], a local rotation invariant descriptor is introduced based on the

*Correspondence: ida.sintorn@it.uu.se
[1] Vironova AB, Gävlegatan 22, SE-11330 Stockholm, Sweden
[2] Uppsala University, Department of Information Technology, Lägerhyddsvägen 20, SE-75105 Uppsala, Sweden

local binary pattern (LBP) descriptor and [9, 10] proposed globally rotation invariant descriptors based on Fourier transformed of the LBP.

The LBP descriptor has over the past decades resulted in a whole family of texture descriptors. For this study on rotation invariance, we have selected the classic LBP descriptor together with seven extensions, including two different approaches to rotation invariance. The descriptors are the following: (i) the classic **LBP** descriptor [8], (ii) **LBP^{ri}**—the approach to local rotation invariance [8], (iii) **LBP^{DFT}**—an approach to rotation invariance where globally rotation invariant Fourier features are extracted [10], (iv) **ILBP** (improved LBP) which is a more noise robust extension of the LBP [11], (v) **ILBP^{ri}** using the approach to local rotation invariance, (vi) **ILBP^{DFT}** using the approach to global rotation invariance, (vii) **LTP^{DFT}** (local ternary patterns) where three rather than two states are considered in the local neighbourhood ([12]) and using the approach to global rotation invariance, (viii) **ILTP^{DFT}** combining ILBP, LTP and the approach to global rotation invariance. By studying these descriptors, we can get a baseline performance from the classic LBP and see how two well used extensions to the LBP perform. In addition, we can see how invariant the two different approaches to rotation invariance are.

In this paper, we investigate and compare the following: local rotation invariance, global rotation invariance, including rotations in the training data, and the effect of the different interpolation methods when rotating textures, in the setting of retaining discriminant texture information. In order to do this, we introduce a new texture dataset, the Kylberg Sintorn Rotation Dataset. The dataset includes images of hardware-rotated textures as well as texture images rotated by the interpolation kernels: nearest neighbour, linear, third order cubic, cubic B-spline and Lanczos 3. The dataset has 25 classes of different types of textured surfaces and is publicly available [13]. The images in the dataset are acquired in raw format avoiding compression artefacts.

2 The Kylberg Sintorn rotation dataset

There are, to our knowledge, two texture datasets available which contain images of rotated textures; the Outex dataset [14] and the Mondial Marmi dataset [10].

The Mondial Marmi dataset contains images of slabs of Italian marble. The textured surfaces are rotated prior to image acquisition to allow for studying rotation invariance in texture analysis. The imaging is done with a compact camera storing the images in JPEG format with notable compression artefacts.

The Outex dataset is a large dataset with 320 texture classes imaged at different resolutions, orientations and illuminations. Unfortunately, there are periodic stripe-like artefacts in the images. In addition, for a given class, orientation, resolution and scale, the image data to generate samples from is rather limited.

Due to, for our purposes, undesirable artefacts in these rotation texture datasets (periodic stripes in Outex and JPEG compression in Mondial Marmi), a new dataset with rotations of textures was acquired. The Kylberg Sintorn Rotation Dataset is a generic texture dataset with similar types of textured surfaces as in the Outex dataset but with the same general dataset structure (many samples of each texture-rotation combination) as the Mondial Marmi dataset. Furthermore, the acquisition setup used in the Kylberg Sintorn Rotation Dataset avoids the aforementioned limitations and artefacts.

The new dataset contains 25 texture classes, mainly consisting of fabrics and arranged textured surfaces using small articles. Figure 1 shows an example patch from each class. The images were acquired using a Canon EOS 550D DSLR camera with a Sigma 17–70-mm zoom lens. The camera was mounted above the textured surfaces, as shown in Fig. 2. Fluorescent lights were placed on two sides of the camera, just above the lens opening. Focus and exposure settings were manually set once for each texture class. The 5184 × 3456 images were acquired as lossless compressed raw files (CR2). The raw files were corrected for lens distortion, chromatic aberration and vignetting formed by the Sigma lens. The corrections were performed according to the settings in the "Adobe (SIGMA 17-70 mm F2.8-4 DC Macro OS HSM, Canon)" lens profile in Adobe Photoshop CS5 and then saved as lossless PNG files. The images were next converted to grey scale as 0.2989 R + 0.5870 G + 0.1140 B, where R, G and B are the red, green and blue intensities, respectively. The selected imaging conditions led to spatially over-sampled images, and to get the textures in a more suitable scale for the relatively small local neighbourhoods used in many texture descriptors, the images were sub-sampled to half the original size (2 592 × 1 728 pixels) using a Lanczos 3 kernel. By this, the influence of sensor noise was also reduced.

2.1 Hardware rotation

The acquisition setup allows for rotation of the camera around the central axis of the camera lens. By rotating the camera rather than the textured surface, the same lighting conditions are kept throughout the image acquisition. For each texture class, one image is acquired for each orientation. The textures are imaged in the nine orientations $\theta \in \{0°, 40°, 80°, \ldots, 320°\}$ chosen not to be even multiples of 90° for which the choice of interpolation methods would not make a difference.

2.2 Rotation by interpolation

Five interpolation methods were used: nearest neighbour, linear, third order cubic, cubic B-spline and Lanczos. The images from the zero orientation were used to interpolate

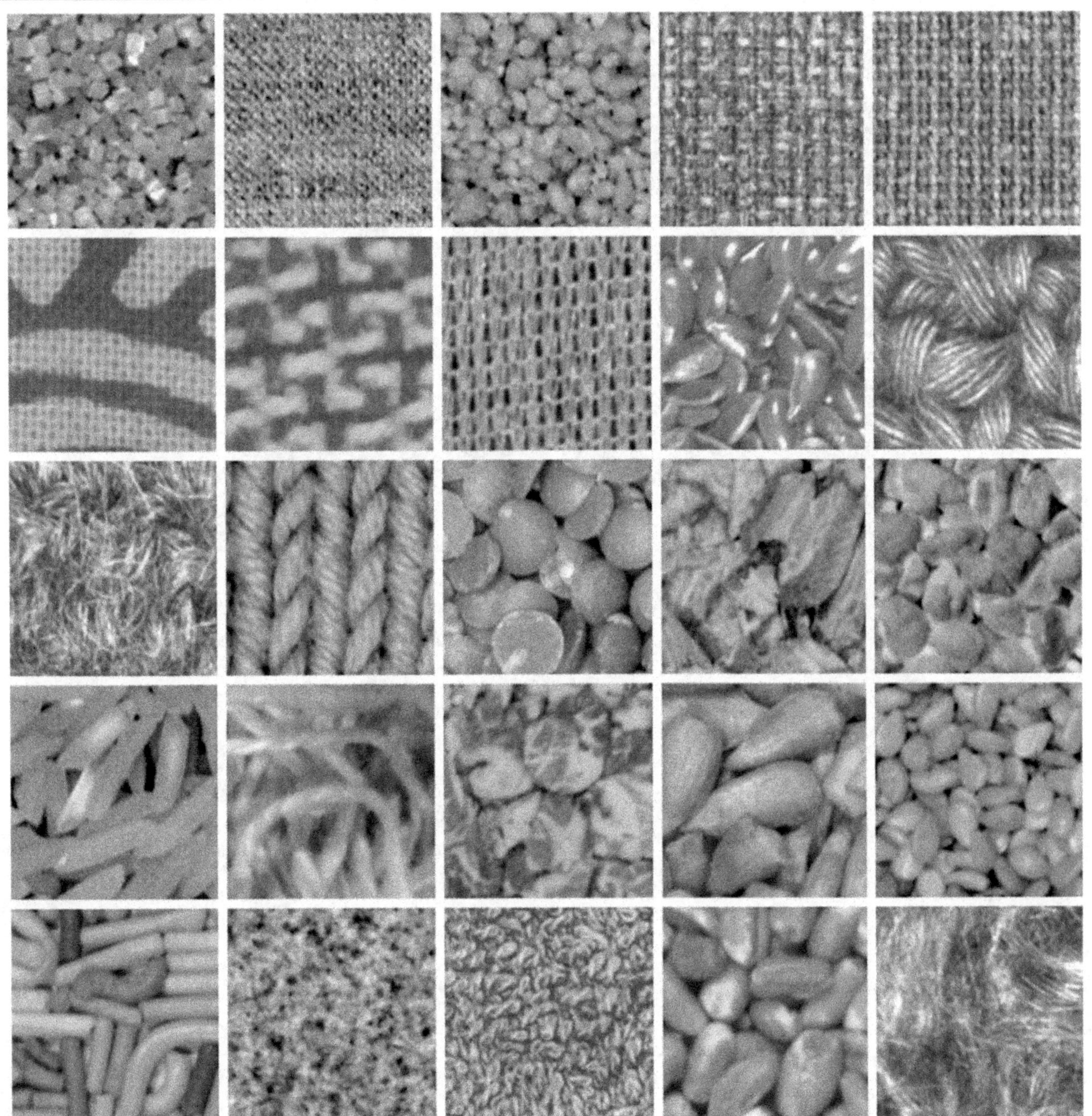

Fig. 1 Examples from the 25 texture classes in the Kylberg Sintorn Rotation Dataset. Row 1: cane sugar, canvas, couscous, fabric 1, fabric 2. Row 2: fabric 3, fabric 4, fabric 5, flax seeds, knitwear 1. Row 3: knitwear 2, knitwear 3, lentils, oatmeal, pearl sugar. Row 4: rice, rug, rye flakes, seeds 1, seeds 2. Row 5: sprinkles, floor tile, towel, wheat, wool fur

the eight other orientations using each of the interpolation approaches. For the interpolation methods based on convolution, Fig. 3 shows 1-D versions of the kernels in spatial as well as in the Fourier domain. The ideal sinc function is also shown for reference. The 1-D definitions of the interpolation methods are as follows:

2.2.1 Nearest neighbour

The interpolation points are assigned the value of the closest pixel:

$$k_{\text{nearest}}(x) = \begin{cases} 1, & |x| \in [0, 0.5] \\ 0, & \text{otherwise} \end{cases} .$$

2.2.2 Linear

Assuming a linear transition between pixel intensities gives

$$k_{\text{linear}}(x) = \begin{cases} 1 - |x|, & |x| \in [0, 1] \\ 0, & \text{otherwise} \end{cases} .$$

Fig. 2 The texture dataset image acquisition setup. From *left* to *right*: overview, view from the camera towards the texture surface beneath, view from the texture surface towards the camera and lights

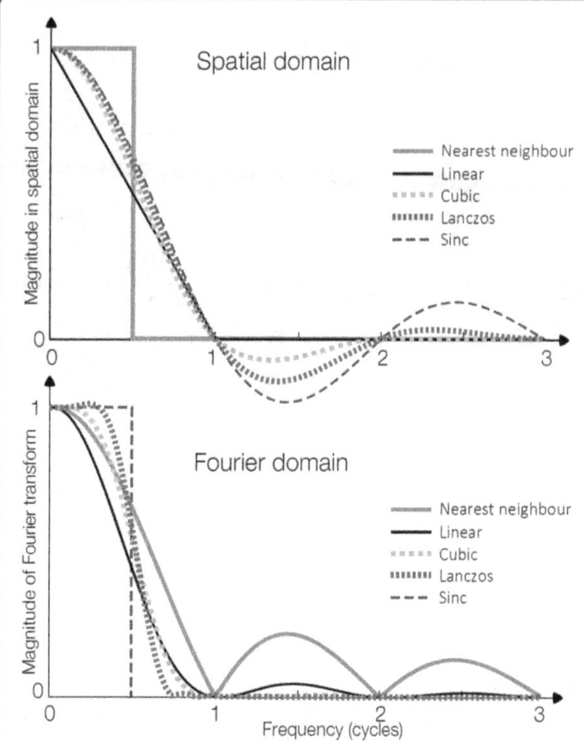

Fig. 3 Interpolation kernels. 1-D plots of the interpolation kernels and the sinc function in the spatial domain (*top*) and Fourier domain (*bottom*)

2.2.3 Cubic

The third order cubic interpolation kernel is a third order polynomial defined as in [15]:

$$k_{\text{cubic}}(x) = \begin{cases} \frac{3}{2}|x|^3 - \frac{5}{2}|x|^2 + 1, & |x| \in [0,1] \\ -\frac{1}{2}|x|^3 + \frac{5}{2}^2 - 4|x| + 2, & |x| \in (1,2] \\ 0 & \text{otherwise} \end{cases} .$$

2.2.4 Lanczos

The Lanczos kernel is introduced in [16] and is a windowed approximation to the sinc function:

$$k_{\text{Lanczos}}(x) = \begin{cases} \text{sinc}(x)\text{sinc}(x/a), & x \in (-a,a) \\ 0 & \text{otherwise} \end{cases} ,$$

where

$$\text{sinc}(x) = \frac{\sin(x)}{x},$$

and $a \in \mathbb{N}$ sets the number of lobes to include of the sinc-function. In this paper, $a = 3$ is used, resulting in the Lanczos kernel shown in Fig. 3. Interpolation of 2-D signals using Lanczos kernels was introduced in [17].

2.2.5 Spline interpolation

In contrast to the interpolation methods based on convolution, spline interpolation fits piecewise polynomials between the data points. The interpolated values are then the values the polynomial assumes at the new sample positions. The cubic B-spline interpolation used here, introduced in [18], fits a third order polynomial of the following form:

$$k_{\text{Bspline}}(x) = \begin{cases} \frac{1}{2}|x|^3 - |x|^2 + \frac{4}{6}, & |x| \in [0,1] \\ -\frac{1}{6}|x|^3 + |x|^2 - 2|x| + \frac{8}{6}, & |x| \in (1,2] \\ 0 & \text{otherwise} \end{cases} .$$

The pieces fit smoothly together, forming a continuous function, going through the original data points.

MATLAB R2012b and the toolbox DIPimage [19] were used for all the interpolations.

2.3 Texture sample generation

All the images, rotated by hardware and software, are divided into smaller texture samples. The region of an image that is common among the different orientations is a disk centred in the image with the diameter that equals the height of the images (1728 pixels). The largest square within this disk (with the side $\lfloor 1728/\sqrt{2} \rfloor = 1221$) is divided into 100 sub-squares with a size of 122×122 pixels (since $\lfloor 1222/10 \rfloor = 122$). The partitioning scheme is illustrated in Fig. 4. Each texture sample is intensity normalized to have a mean value of 127 and a standard deviation of 40. The mean is set to be centred in the interval $[0, 255]$. The standard deviation is set empirically so that intensity values in the dataset generally do not get mapped to integers $\notin [0, 255]$ while retaining most of the dynamic range of the dataset.

3 LBP-based descriptors

In the original definition of LBP [20], the eight connected neighbours of a centre pixel are considered. The neighbours are thresholded by the intensity of the centre pixel and placed in a clockwise order producing an 8-bit binary code. The feature vector consists of the occurrences (histogram) of the different binary codes in a region/patch. The neighbourhood has been generalized to N samples on a radius R from the centre pixel in [8], here denoted $\mathrm{LBP}_{N,R}$. The local binary code for the pixel position (x_c, y_c) with grey value g_c is defined as:

$$\mathrm{LBP}_{N,R}(x, y) = \sum_{p=0}^{N-1} s(g_p - g_c) 2^p, \quad (1)$$

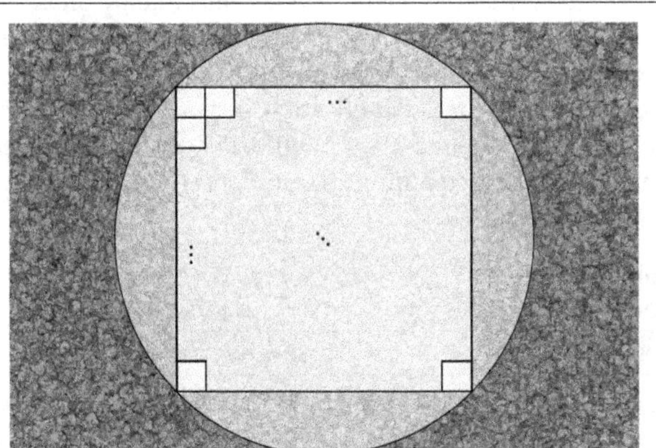

Fig. 4 Partitioning scheme. Illustration of the partitioning scheme used to extract texture samples

where

$$s(x) = \begin{cases} 1, & x \geq 0 \\ 0, & \text{otherwise} \end{cases}. \quad (2)$$

If a point p does not coincide with a pixel centre, linear interpolation is used to compute the grey value g_p. Finally, the histogram of occurring binary codes in a region is the feature vector of this region.

One way of making LBP more robust to noise is to threshold the value of a point, g_p with the mean value of the neighbourhood (including the centre pixel), g_{mean}, rather than with the value of the centre pixel g_c. The centre pixel is also thresholded with the mean value and included in the binary code. This descriptor is called improved local binary patterns (ILBP) and was introduced in [11]. It is defined as

$$\mathrm{ILBP}_{N,R}(x, y) = \sum_{p=0}^{N-1} s(g_p - g_{\mathrm{mean}}) 2^p + \\ + s(g_c - g_{\mathrm{mean}}) 2^N, \quad (3)$$

where

$$g_{\mathrm{mean}} = \frac{1}{N+1} \left(\sum_{p=0}^{N-1} g_p + g_c \right), \quad (4)$$

and the function s is defined as in Eq. 2. Note that p_c is part of the binary code making it $N+1$ bits long.

In [21], local ternary pattern (LTP) descriptor is proposed. The difference between neighbouring values g_p and the centre pixel value g_c are encoded with three values using a threshold value t:

$$s_3(g_p, g_c, t) = \begin{cases} 1, & g_p \geq g_c + t \\ 0, & g_c - t \leq g_p < g_c + t \\ -1, & \text{otherwise} \end{cases}. \quad (5)$$

In our implementation, as described above, the interval coded with zero is half-bound while in [21] it is open. Instead of using a code with base 3 to encode the three states in Eq. 5, LTP uses two binary codes representing the positive and the negative components of the ternary code, i.e., two binary codes coding for the two states $\{-1, 1\}$. These binary codes are collected in two separate histograms, and as a last step, the histograms are concatenated to form the LTP feature vector [21].

In analogy with the extension of LBP to ILBP, where the neighbourhood mean value (g_{mean}) is used as the local threshold and the centre pixel (g_c) is included in the code, LTP can be extended to ILTP. This was done in [22]. For ILTP, the same scheme of using two concatenated binary codes as the final feature vector as for LTP is employed [22].

3.1 Rotation invariance

There are different approaches to making the classical LBP descriptor rotation invariant. One way is to group the binary codes that are rotations of one another (i.e. circular shifts of the binary code). Next, the occurrences of each group are computed and used as feature values. This approach is introduced in [8], and the descriptor is denoted $\text{LBP}^{\text{ri}}_{N,R}$. For example, when $N = 8$ is used, there are 36 such rotation invariant groups. Since occurrences of rotation groups are considered, the relative distribution within rotation groups is lost. It also means that the $\text{LBP}^{\text{ri}}_{N,R}$ descriptor achieves rotation invariance by normalizing rotation locally, as described in [23].

Another way of making LBP rotation invariant is introduced in [9, 23]. The occurrences of rotation codes within the rotation groups are not summed, as in $\text{LBP}^{\text{ri}}_{N,R}$, but Fourier transformed and the resulting power spectrum is used as the feature vector. The descriptor is called LBP histogram Fourier features ($\text{LBP}^{\text{HF}}_{N,R}$). Since the Fourier features are computed on the global histogram of binary codes in the region/patch investigated, $\text{LBP}^{\text{HF}}_{N,R}$ achieves rotation invariance globally, and hence, retains the relative distribution within rotation groups [23]. However, the $\text{LBP}^{\text{HF}}_{N,R}$ descriptor in [9] only considers uniform binary codes (binary codes with at the most two transitions between 0 and 1). It was generalized in [10] to include all binary codes, uniform and non-uniform, and called LBP^{DFT}. Together with the LBP^{DFT} descriptor, the corresponding ILBP^{DFT} descriptor was also introduced in [10].

When having the aforementioned methods at hand, two interesting additional descriptors can be compiled; LTP^{DFT} and ILTP^{DFT}. Thus combining the generalized version of the Fourier features from [10], achieving global rotation invariance, with the promising descriptors LTP from [21] and ILTP from [22].

In the tests reported on here, eight descriptors were used, all with $N = 8$ samples on a radius $R = 1$; $\text{LBP}_{8,1}$, $\text{LBP}^{\text{ri}}_{8,1}$, $\text{LBP}^{\text{DFT}}_{8,1}$, $\text{LTP}^{\text{DFT}}_{8,1}$, $\text{ILBP}_{8,1}$, $\text{ILBP}^{\text{ri}}_{8,1}$, $\text{ILBP}^{\text{DFT}}_{8,1}$, $\text{ILTP}^{\text{DFT}}_{8,1}$. The descriptors are listed in Table 1 together with the length of their respective feature vector. The parameters are selected to describe the textures in their highest scale since the effects of interpolation methods are expected to be most prominent here. $N \neq 9$ is desirable since the dataset has nine equally spaced orientations and they would otherwise line up perfectly. $N \gg 8$ results in feature spaces of very high dimensionality (e.g. $\text{LBP}^{\text{DFT}}_{16,1} \in \mathbb{R}^{36\,883}$ and $\text{ILTP}^{\text{DFT}}_{16,1} \in \mathbb{R}^{147\,532}$) and $N = 8$ samples the local neighbourhood at $R = 1$ well.

Table 1 Texture descriptor dimensionality. Length of feature vector using $N = 8$

Descriptor	Dim.	Descriptor	Dim.
$\text{LBP}_{8,1}$	256	$\text{ILBP}_{8,1}$	511
$\text{LBP}^{\text{ri}}_{8,1}$	36	$\text{ILBP}^{\text{ri}}_{8,1}$	71
$\text{LBP}^{\text{DFT}}_{8,1}$	163	$\text{ILBP}^{\text{DFT}}_{8,1}$	325
$\text{LTP}^{\text{DFT}}_{8,1}$	326	$\text{ILTP}^{\text{DFT}}_{8,1}$	651

4 Classification procedure

The interpolation methods and texture descriptors are evaluated by comparing the obtained classification accuracies. A first nearest neighbour (1-NN) classifier with Euclidean metric is used. The 1-NN classifier is used to be able to compare classification results obtained on the same and fair basis. To validate the trained classifier 10-folded cross-validation is performed by randomly assigning each texture sample an index $n \in \{1, 2, \ldots, 10\}$, creating 10 disjoint subsets with equal number of samples (stratified samples). In the first cross-validation fold, samples with $n \in \{2, 3, \ldots, 10\}$ will be the training data and samples with $n = 1$ will serve as test data. In the second fold, samples with $n = 2$ will be the test data and the rest is used for training, and so on.

The indices for the cross-validation folds are created once and then kept fixed throughout the experiments. The classification results from the 10 folds are combined into a single confusion matrix estimation and the mean and standard deviation of the classification accuracy is computed.

5 Evaluating interpolation methods and rotation invariance

Before the evaluation a baseline is established for each descriptor by training and testing on the $\theta = 0°$ orientation. Table 2 lists the means and standard deviations of the classification accuracies of these baseline tests.

The descriptors are applied to all the texture samples, rotated by hardware and by each of the five interpolation methods. For the evaluation, the 1-NN classifier is trained on features from the $\theta = 0°$ orientation followed by testing on the remaining eight orientations, one by one. The results are shown in Fig. 5. The established baselines are shown as dashed red lines in Fig. 5. One dot in Fig. 5 corresponds to one classification test where the classifier has been trained on $\theta = 0°$ orientation and tested on one of the remaining orientations (using 10-folded cross-validation).

Table 2 Mean classification accuracy at $\theta = 0°$. Standard deviations in brackets

Descriptor	Mean	Std.	Descriptor	Mean	Std.
$\text{LBP}_{8,1}$	96.4	(1.2)	$\text{ILBP}_{8,1}$	95.4	(0.9)
$\text{LBP}^{\text{ri}}_{8,1}$	97.2	(0.9)	$\text{ILBP}^{\text{ri}}_{8,1}$	96.9	(0.9)
$\text{LBP}^{\text{DFT}}_{8,1}$	98.7	(0.7)	$\text{ILBP}^{\text{DFT}}_{8,1}$	97.5	(0.7)
$\text{LTP}^{\text{DFT}}_{8,1}$	98.9	(0.5)	$\text{ILTP}^{\text{DFT}}_{8,1}$	99.6	(0.4)

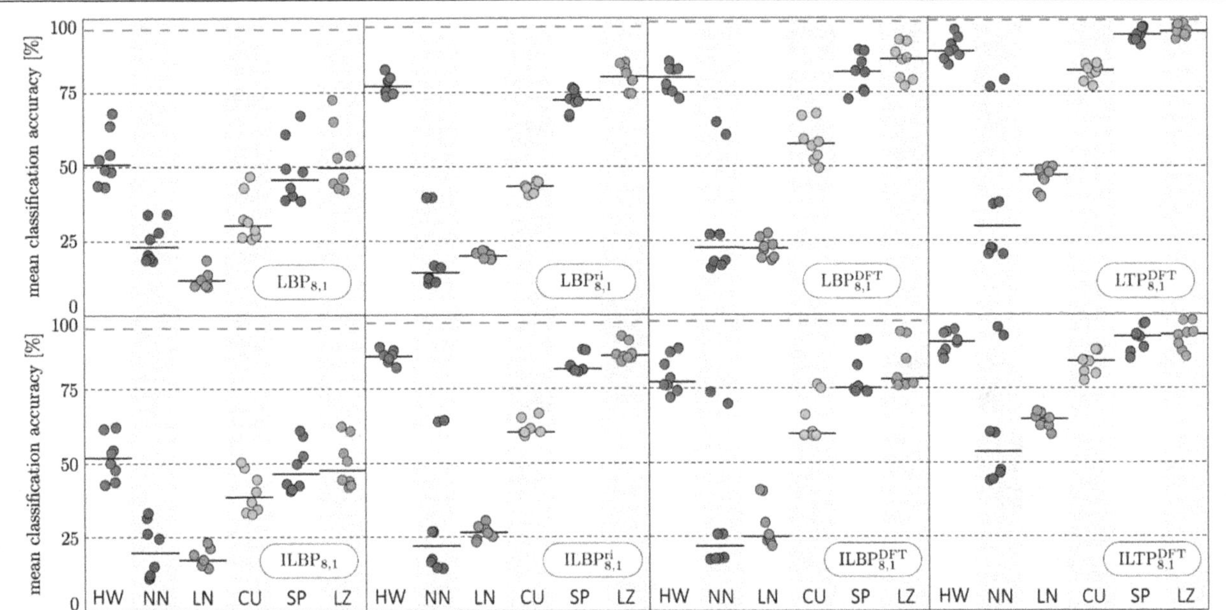

Fig. 5 Rotation invariance test. Plot of mean classification accuracies (in %) when training on the $\theta = 0°$ orientation and testing on $\theta \in \{40°, 80°, \ldots 320°\}$ for each descriptor applied on textures rotated by hardware (HW) or nearest neighbour (NN), linear (LN), cubic (CU), B-spline (SP) and Lanczos 3 (LZ) interpolation. The *horizontal red dashed lines* are the mean accuracies when training and testing on $\theta = 0°$. Each dot corresponds to the mean classification accuracy from one rotation test. The *black horizontal line segments* in each column mark the median in each distribution

5.1 Interpolation method

Figure 5 shows that the obtained classification accuracies differ greatly for the different interpolation methods (Additional file 1). This indicates that the characteristic properties of the textures are retained to a widely varying degree under rotation. Lanczos and B-spline interpolation results in accuracies similar to, or even better than, that of the hardware rotated textures. Nearest neighbour, linear and cubic interpolation all show lower accuracies for all descriptors compared.

For the nearest neighbour rotated textures, some rotation tests (dots in Fig. 5) show relatively high accuracies for several descriptors. To investigate this, Fig. 6 shows the result for the $\text{LBP}^{\text{ri}}_{8,1}$ descriptor in more detail. Here, each dot is a cross-validation fold in a rotation test. It can be seen that the nearest neighbour has noticeably higher accuracies at $\theta = 80°$ and $\theta = 280°$. These two angles correspond to the two dots that often have higher accuracies than the rest for the nearest neighbour interpolation method in Fig. 5. Rotations of a digital image with even multiples of 90° are ideal/lossless. The two orientations of the textures $\theta \in \{80°, 280°\}$ in the dataset are the two that are closest to even multiples of 90° and, hence, closest to ideal rotations for which nearest neighbour is relatively successful.

In a few cases, the result achieved using Lanczos and B-spline interpolation exceeds the result obtained using the hardware rotated textures, especially at $\theta \in \{80°, 280°\}$, see Fig. 6. This can, to a certain degree, be explained by that, in the case of hardware rotations, the sensor noise is sampled again and again for the different orientations, while the interpolated images all originate from one image with the sensor noise sampled once. The set of images which are rotated by interpolation is hence more homogeneous, and in addition, these two orientations are closest to ideal orientations of the original image, as was discussed above with respect to the higher performance for nearest neighbour interpolation for those two angles. Another explanation can be found by studying the per-class accuracies (data not shown). The classifier runs into problems with class number 12 and 20 for the hardware rotated textures while they are easier to classify in the Lanczos and B-spline interpolated data.

The tests were repeated using $N = 8$ and $R = 2$ for the eight LBP descriptors (data not shown). The interpolation methods appeared in the same order, accuracy-wise, with one exception. The nearest neighbour interpolation was found to perform on a level between linear and cubic interpolation. The descriptors generally achieved higher accuracies in the same order indicating that the selected parameters for the comparison is not optimal if the best overall classification result of the dataset is the objective. However, at $R = 1$, the effects of interpolation methods and difference in the descriptor rotation invariance are more prominent and this is the main focus of this study.

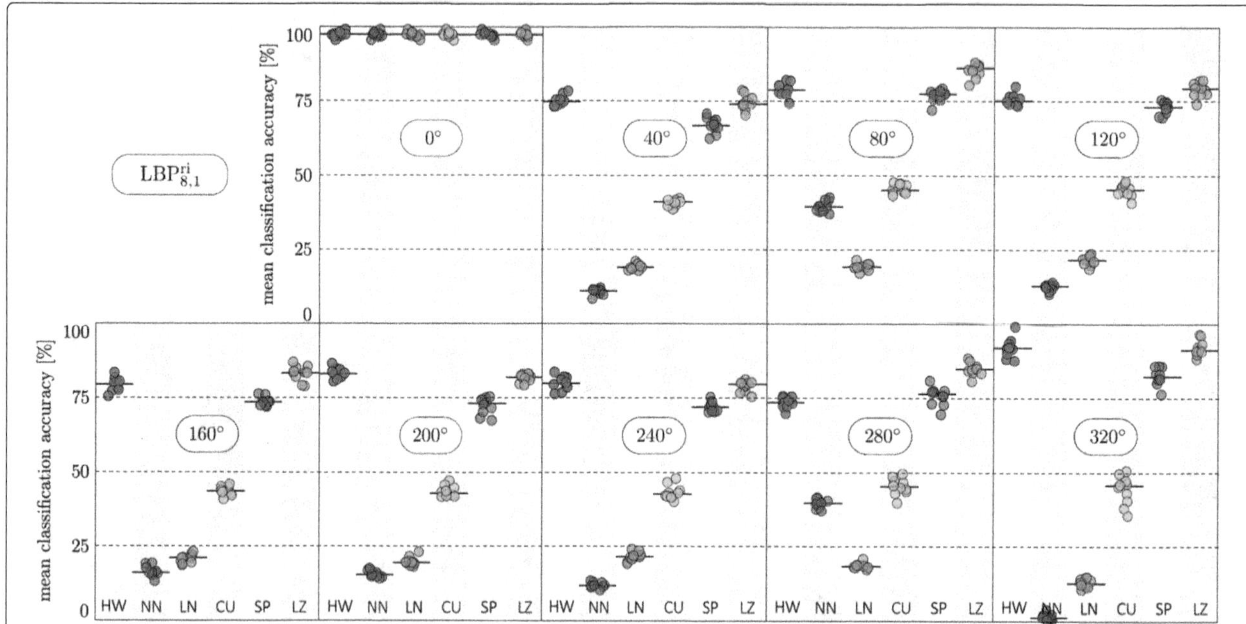

Fig. 6 Rotation invariance test for $LBP^{ri}_{8,1}$ in detail. Plot of mean classification accuracies within each cross-validation fold for each orientation for the $LBP^{ri}_{8,1}$ descriptor applied on textures rotated by hardware (HW) or nearest neighbour (NN), linear (LN), cubic (CU), B-spline (SP) and Lanczos 3 (LZ) interpolation. The *black horizontal line segments* in each column mark the median in each distribution

5.2 Rotation invariance

The rotation invariance of the descriptors can be evaluated in Fig. 5. Mean and standard deviation is also reported in Table 3. A truly rotation invariant texture descriptor should only show subtle variations over different orientations of a texture. The classic LBP has obvious problems in the rotation tests and fall far below the baseline. The LBP-versions design to be rotation invariant do achieve higher accuracies than the classic LBP. However, none of the descriptors reach their baseline accuracy.

Figure 6 shows that the cross-validation folds give very similar results. This means that the standard deviations over orientations, see Fig. 5, are not due to inadequate validation of the classifier but a genuine variation in descriptor performance.

5.2.1 *Reference methods*

To put the rotation invariance of the LBP-based descriptor into perspective, three additional descriptors were evaluated in the same way. The rotation invariant Gabor filter bank achieved a mean accuracy over rotation tests of 90.4 %, Haralick features 75.3 % and MR8 56.6 %. These values should be compared to the mean classification accuracies reported in Table 3 (also showed as black lines in Fig. 5). The Gabor filter bank achieves as good results as the best descriptor tested in the LBP family $\left(ILTP^{DFT}_{8,1}\right)$ while Haralick achieves just below the local rotation invariant $LBP^{r}_{8,1}i$ descriptor. Below follows condensed details of how the reference methods were used.

Table 3 Mean classification accuracy (and standard deviations in brackets) across the eight rotation tests. A rotation test refers to training on $\theta = 0°$ and testing on one of the other rotations $\theta \neq 0°$

Descriptor	Mean	Std.	Descriptor	Mean	Std.
$LBP_{8,1}$	52.9	(8.9)	$ILBP_{8,1}$	52.0	(7.3)
$LBP^{ri}_{8,1}$	77.4	(3.2)	$ILBP^{ri}_{8,1}$	85.9	(2.2)
$LBP^{DFT}_{8,1}$	79.3	(4.6)	$ILBP^{DFT}_{8,1}$	79.4	(6.1)
$LTP^{DFT}_{8,1}$	89.5	(3.9)	$ILTP^{DFT}_{8,1}$	90.4	(3.7)

For the Gabor filter bank, definitions and guidelines as described in [7, 24] were used. The frequency ratio was set to $\sqrt{2}$, the number of orientations was set to eight, and the number of frequencies to six. The Gaussian envelope was set to $(\gamma, \eta) = (3, 2)$, and the kernel size was set to 19×19.

The Haralick features were defined and used as described in [25]. The contrast, correlation, energy and homogeneity measures were computed for four grey-level co-occurrence matrices (four directions) using quantization into 16 grey levels and a distance of two pixels.

MR8 was defined and used as in [26].

6 Computational cost

To assess the computational cost of the different interpolation methods and texture descriptors, the relative computation time was measured.

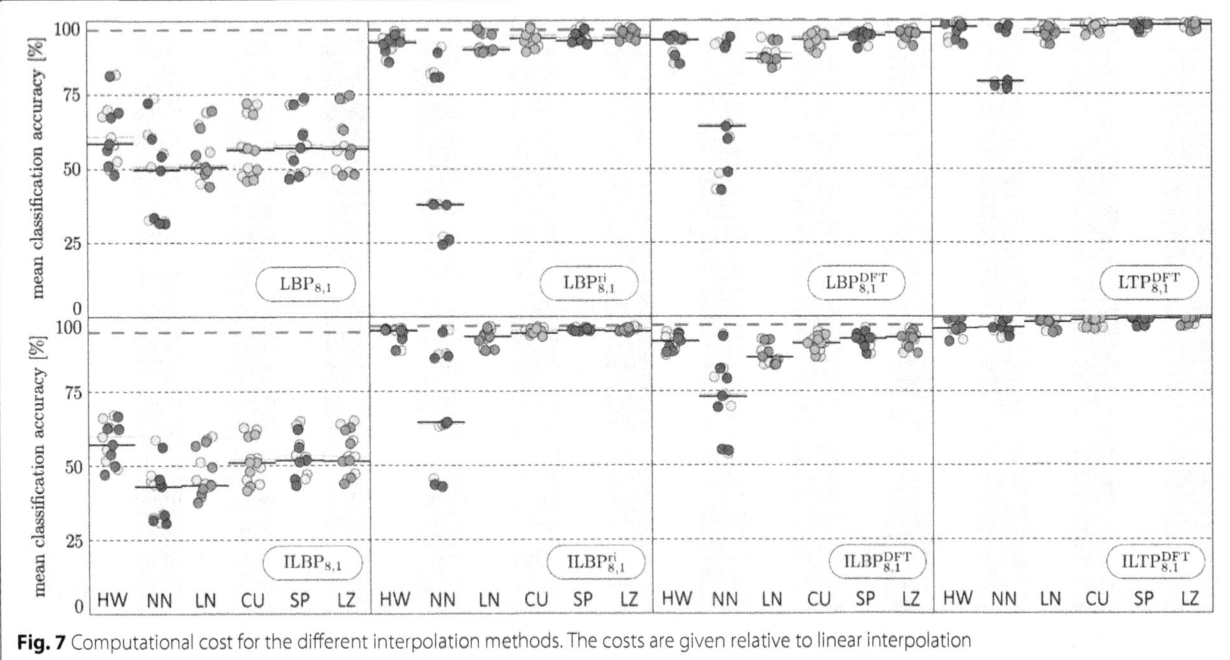

Fig. 7 Computational cost for the different interpolation methods. The costs are given relative to linear interpolation

6.1 Interpolation methods

The computational time was measured while rotating the 2500 texture samples to the eight orientations. Figure 7 shows the computation time relative to rotating the texture sample using linear interpolation. The two best performing interpolation methods (in terms of retaining texture information) are slightly more computationally expensive; a factor of 1.27 for Lanczos and 1.45 for Spline.

6.2 Texture descriptors

The computational time for the different descriptors was measured while applying them to the 2500 texture samples of the 0° orientation. Figure 8 shows the computation time relative to the LBP descriptor. The slowest descriptor is the ILTP^{DFT} being about 2.5 times slower than LBP. However, all the descriptors in the LBP family evaluated in this manuscript are computationally cheap compared to the filter-based texture descriptors used here as reference methods. The Gabor filter bank takes about 37 times as long to compute as the basic LBP. The numbers may vary with implementations and systems but convolving the texture sample with a whole bank of filter kernels is more computationally expensive than computing these LBP-variants.

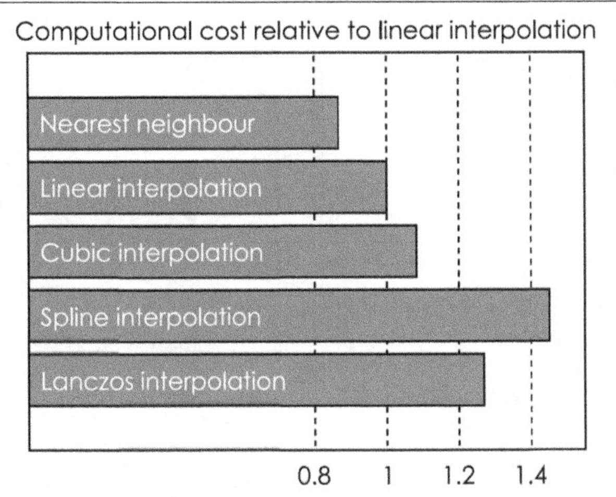

Fig. 8 Computational cost for the texture descriptors. The costs are given relative to the LBP descriptor

7 Rotation representation in classification

To investigate the approach of adding virtual training samples by rotating existing training samples, an increasing number of orientations are used in the training data, from one to eight. The classifier is then tested on the remaining orientations. Consequently, the training and test sets are still disjoint, sample and orientation-wise. The results are shown in Fig. 9. Using two orientations in the training data generally results in greatly improved classification accuracies and lower standard deviations. In fact, several of the texture measures approach their baseline. Another striking result is that the choice of interpolation methods is less important. The clear result seen in Fig. 5, that Lanczos 3 and B-spline outperform the other interpolation methods, is drastically reduced when the training data is somewhat more heterogeneous in terms of represented orientations.

In the first test, using training samples $\in \{0°, 40°\}$, the training set becomes twice as large. To make sure that the observed improvement originate from better

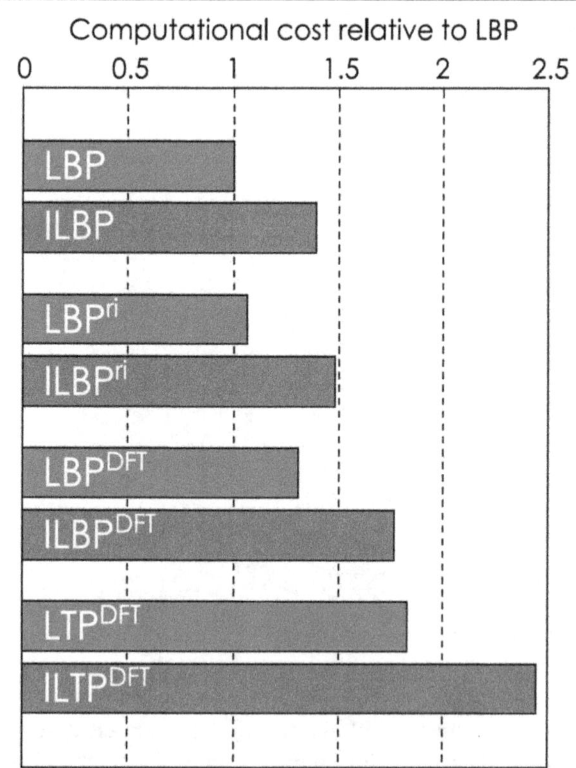

Fig. 9 Rotation invariance when training on two orientations. Plot of mean classification accuracies (in %) when training on the two orientations $\theta \in \{0°, 40°\}$ and testing on $\theta \in \{80°, 120°, \ldots 320°\}$ for each descriptor applied on textures rotated by hardware (HW) or nearest neighbour (NN), linear (LN), cubic (CU), B-spline (SP) and Lanczos 3 (LZ) interpolation. The horizontal *red dashed lines* are the mean accuracies when training and testing on $\theta = 0°$. The *grey dots* in the background correspond to the mean classification accuracies achieved when the size of the training data is halved. The *horizontal line segments* mark the median in each distribution

representation of the rotated textures rather than a larger training set, we carried out a complementary test where every other sample was removed from the training set, making it the same size as in the previous tests. These results are shown as grey dots in the background in Fig. 9 and prove that the improvement is not from an increase in the number of training samples.

Figure 10 shows how the classification accuracies change when even more orientations are used in the training data, ranging from one (as in the evaluation reported on in Section 4) to eight orientations. The additional orientations are the Lanczos interpolated data, following the scenario of adding "artificial" samples to the training data. It shows that the introduction of a second orientation in the training data is very beneficial in terms of texture recognition. For the less rotation invariant texture descriptors, additional orientations (more than two) further improve the classification accuracies even though the achieved accuracies level out and do not reach the levels of the rotation invariant descriptors. This trend is not as clear for the more rotation invariant descriptors, for which a third and fourth orientation only improves the classification accuracies slightly. Overall, $\mathrm{ILTP}_{8,1}^{\mathrm{DFT}}$ and $\mathrm{LTP}_{8,1}^{\mathrm{DFT}}$ are found at the top with mean accuracies between 99 and 100 % when at least four orientations are included in the training data.

8 Conclusions

Based on the performed experiments we conclude that

- Lanczos 3 interpolation, closely followed by B-splines outperform the other interpolation strategies in all tests. The same levels of texture recognition are achieved using these two interpolation methods as those using hardware-rotated textures.

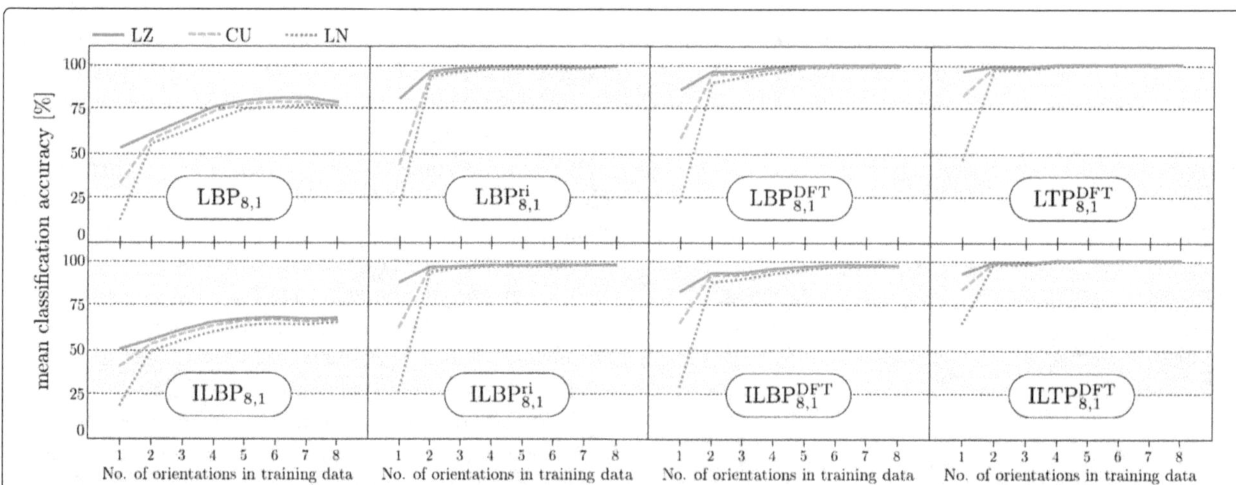

Fig. 10 Increasing the number of rotations in training. Plot of mean classification accuracies with increasing number of Lanczos interpolated orientations in the training data. The additional orientations included in the training data are excluded from the test data

- As expected, the interpolation methods closer to the ideal sinc function retain the texture information best. The commonly used linear and cubic interpolation clearly have shortcomings in preserving texture information in the setting of texture recognition.
- The best performing interpolation methods are only slightly more computationally expensive. Lanczos stands out as it is both the best interpolation method in terms of preserving texture information and faster to compute than the second best interpolation method B-spline.
- Both the local rotation invariant versions (by counting occurrences of rotation groups), and the global rotation invariant versions (by computing Fourier descriptors of the groups) of the LBP-based texture descriptors are less rotation variant than the classic LBP descriptor but they still suffer from rotation variance.
- The LTP^{DFT} and ILTP^{DFT} descriptors perform better than the other tested descriptors. They both show high classification accuracies and low standard deviations over different orientations.
- The ILTP^{DFT} achieves the same high classification accuracy as an optimized and rotation invariant Gabor filter bank. The other reference descriptors are inferior to all the tested rotation invariant forms of the LBP descriptors.
- Even though the more advanced descriptors in the LBP family are more computationally expensive than the basic LBP, they are all computationally inexpensive compared to common filter bank-based texture descriptors such as Gabor filter banks and MR8.
- Including several different orientations of the textures in the training data has great positive impact on the classification accuracies. This also compensates to a large extent the shortcomings of choosing a simple interpolation method. Hence, a simple strategy that should be generally considered!

Linear and cubic interpolation is still commonly used even though they are easily outperformed in terms of retaining texture information. For example, in the image processing software ImageJ, linear interpolation is the default, in GIMP cubic is default (although Lanzos 3 is offered as an option) and MATLAB only supports nearest neighbour, linear and cubic interpolation when rotating images. The software library OpenCV has linear interpolation as default while cubic is optional.

Furthermore, these tests show that the use of rotation invariant texture descriptors may not be enough to achieve rotation invariant texture recognition; representing different orientations of the textures in the training data shows to be very important, even if the extra orientations are artificially generated by rotating already existing samples by means of interpolation.

In the light of these findings, the use of linear interpolation to compute the intensity samples in the local neighbourhoods in all the $\text{LBP}_{N,R}$-based descriptors should perhaps be revised.

Additional file

Additional file 1: Tables for the data illustrated in Figs. 5 and 7. (PDF 108 kb)

Competing interests
The authors do not have any competing interests.

Authors' contributions
Both authors have contributed equally to the text while GK have implemented the texture descriptors and performed most of the tests.

Acknowledgements
The authors would like to thank Cris Luengo for expanding the set of interpolation kernels implemented in DIPimage. This work has been part of the MiniTEM E!6143 project funded by EU and EUREKA through the Eurostars Programme.

References

1. X Tan, B Triggs, Enhanced local texture feature sets for face recognition under difficult lighting conditions. IEEE Trans. Image Process. **19**(6), 1635–1650 (2010)
2. K Jafari-Khouzani, H Soltanian-Zadeh, Radon transform orientation estimation for rotation invariant texture analysis. IEEE Trans. Pattern Anal. Mach. Intell. **27**(6), 1004–1008 (2005)
3. B Schölkopf, *Support vector learning*. (PhD thesis, Technische Universität Berlin, 1997)
4. J Zhang, T Tan, Brief review of invariant texture analysis methods. Pattern Recogn. **35**(3), 735–747 (2002)
5. Y Xu, H Ji, C Fermüller, Viewpoint invariant texture description using fractal analysis. Int. J. Comput. Vis. **83**(1), 85–100 (2009)
6. M Varma, A Zisserman, A statistical approach to texture classification from single images. Int. J. Comput. Vision. **62**(1-2), 61–81 (2005)
7. F Bianconi, A Fernández, A Mancini, in *Proceedings of the 20th International Congress on Graphical Engineering*. Assessment of rotation-invariant texture classification through gabor filters and discrete Fourier transform (ACM, Valencia, Spain, 2008)
8. T Ojala, T Mäenpää, Multiresolution Gray-Scale and Rotation Invariant Texture Classification with Local Binary Patterns. IEEE Trans. Pattern Anal. Mach. Intell. **24**(7), 971–987 (2002)
9. T Ahonen, J Matas, C He, M Pietikäinen, in *Proceedings of the 16th Scandinavian Conference on Image Analysis. LNCS*. Rotation invariant image description with local binary pattern histogram Fourier features, vol. 5575 (Springer, Berlin Heidelberg, 2009), pp. 61–70
10. A Fernández, O Ghita, E González, F Bianconi, PF Whelan, Evaluation of robustness against rotation of LBP, CCR and ILBP features in granite texture classification. Mach. Vis. Appl. **22**(6), 913–926 (2011)
11. H Jin, Q Liu, H Lu, X Tong, in *Proceedings of the Third International Conference on Image and Graphics*. Face detection using improved LBP under Bayesian framework, (IEEE Computer Society, Washington, DC, USA, 2004), pp. 306–309. doi:10.1109/ICIG.2004.62
12. A Fernández, M Álvarez, F Bianconi, Texture description through histograms of equivalent patterns. J. Math. Imaging Vis. **45**, 76–102 (2013)
13. G Kylberg, I-M Sintorn, Kylberg Sintorn Rotation dataset (2015). http://www.cb.uu.se/~gustaf/KylbergSintornRotation/. Accessed 1 Jan 2016

14. T Ojala, T Mäenpää, M Pietikäinen, J Viertola, J Kyllönen, S Huovinen, in *Proceedings of the 16th International Conference on Pattern Recognition, Quebec, Canada*. Outex - New Framework for Empirical Evaluation of Texture Analysis Algorithms (IEEE Computer Society, Washington, DC, USA, 2002), pp. 701–706
15. R Keys, Cubic convolution interpolation for digital image processing. IEEE Trans. Acoust. Speech Signal Process. **29**(6), 1153–1160 (1981)
16. C Lanczos, *Applied Analysis. Prentice-Hall mathematics series*. (Prentice-Hall, New Jersey, 1956)
17. CE Duchon, Lanczos Filtering in One and Two Dimensions. J. Appl. Meteor. **18**(8), 1016–1022 (1979)
18. H Hou, H Andrews, Cubic splines for image interpolation and digital filtering. IEEE Trans. Acoust. Speech Signal Process. **26**(6), 508–517 (1978)
19. C Luengo, DIPimage, a Matlab toolbox for scientific image processing and analysis. http://www.diplib.org/. Accessed 1 Jan 2016
20. T Ojala, M Pietikäinen, D Harwood, A comparative study of texture measures with classification based on featured distributions. Pattern Recogn. **29**(1), 51–59 (1996)
21. X Tan, B Triggs, in *Analysis and Modeling of Faces and Gestures. Lecture Notes in Computer Science*, ed. by SK Zhou, W Zhao, X Tang, and S Gong. Enhanced local texture feature sets for face recognition under difficult lighting conditions, vol. 4778 (Springer, Springer-Verlag Berlin Heidelberg, 2007), pp. 168–182
22. L Nanni, S Brahnam, A Lumini, A local approach based on a Local Binary Patterns variant texture descriptor for classifying pain states. Expert Syst. Appl. **37**(12), 7888–7894 (2010)
23. G Zhao, T Ahonen, J Matas, M Pietikainen, Rotation-invariant image and video description with local binary pattern features. IEEE Trans. Image Process. **21**(4), 1465–1477 (2012)
24. F Bianconi, A Fernández, Evaluation of the effects of Gabor filter parameters on texture classification. Pattern Recog. **40**(12), 3325–3335 (2007)
25. RM Haralick, K Shanmugam, I Dinstein, Textural Features for Image Classification. IEEE Trans. Syst. Man Cybern. **3**(6), 610–621 (1973)
26. M Varma, A Zisserman, in *Proceedings of the 7th European Conference on Computer Vision, Copenhagen, Denmark. Computer Vision*. Classifying images of materials: Achieving viewpoint and illumination independence, vol. 3 (Springer, Berlin, 2002), pp. 255–271

14

Fast colorization based image coding algorithm using multiple resolution images

Kazunori Uruma[1*†], Katsumi Konishi[2†], Tomohiro Takahashi[1] and Toshihiro Furukawa[1]

Abstract

This paper proposes a representative pixel (RP) extraction algorithm and chrominance image recovery algorithm for the colorization-based digital image coding. The colorization-based coding methods reduce the color information of an image and achieve higher compression ratio than JPEG coding; however, they take much more computing time. In order to achieve low computational cost, this paper proposes the algorithm using the set of multiple-resolution images obtained by colorization error minimizing method. This algorithm extracts RPs from each resolution image and colorizes each resolution image utilizing a lower resolution color image, which leads to the reduction of the number of RPs and computing time. Numerical examples show that the proposed algorithm extracts the RPs and recovers the color image fast and effectively.

Keywords: Image colorization, Image coding, Representative pixel, Multiple-resolution images

1 Introduction

The image colorization technique can recover a full-color image from a luminance image and several representative pixels (RPs) which have the chrominance values and their positions [1–4]. Levin et al. proposes a well known colorization algorithm, which achieves a high colorization performance using appropriately given RPs. However, this algorithm requires high computing cost and fails to colorize a whole image if only a few RPs are given. The authors have proposed the colorization algorithm, which can restore a color image from only a few RPs and takes a low computational cost [4].

The development of the colorization technique has led to the generation of a new color image coding method called colorization-based color image coding [5–12]. Because the luminance image can be compressed by standard coding methods such as JPEG coding and because the chrominance information is represented by a small number of RPs, the colorization-based image coding can compress full-color images more than these methods. The colorization-based image coding algorithms proposed in [10, 12] have achieved higher coding performance than JPEG2000, which is a highly optimized standard using many entropy coding techniques together. Therefore, the colorization-based image coding has potential to become standard image coding for the next generation. The performance of the colorization based coding is evaluated based on the number of RPs and the quality of a restore image, and therefore, the objective of this paper is to propose a new coding method which requires less RPs and gives more accurate chrominance information. In order to reduce the amount of information to represent the RPs, the method [5] assumes that all pixels in a line segment have only one color and proposes the algorithm where the RPs are described as a set of color line segments. In [9], the algorithm finds a set of two RPs which give the similar effect for colorization result, and the redundant RPs are deleted. In [12], under the assumption that the chrominance image is given as a sparse linear combination of the basis constructed from the luminance image, the algorithm achieves a high compression performance using the theory of compressed sensing [13, 14]. However, these algorithms require a high computing time depending on the image size. To reduce computing time, the method [5] proposes an algorithm to extract RPs from a single low-resolution image. This paper takes the same approach to reduce the computational costs and

*Correspondence: uru-kaz@ms.kagu.tus.ac.jp
†Equal contributors
[1]Tokyo University of Science, 162-8677 Tokyo, Japan
Full list of author information is available at the end of the article

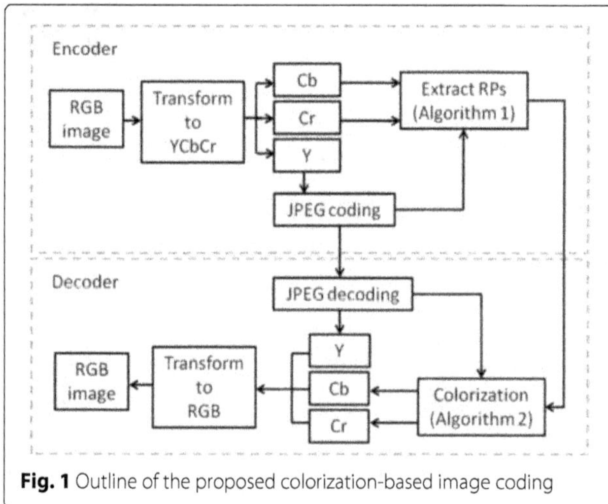

Fig. 1 Outline of the proposed colorization-based image coding

provides an RPs extraction algorithm using multiple low-resolution images. Though a low resolution image used in [5] is obtained by a simple downsampling, this paper gives low resolution images which are optimized such that a good color image is recovered in the colorization phase.

This paper proposes a new colorization-based image coding algorithm consisting of the extraction RPs phase (coding) and the colorization phase (decoding). In the phase of extracting RPs, the algorithm extracts RPs using error feedback between the original image and the colorized image similar to [15]. In order to achieve a low computational cost and to reduce the amount of information to represent RPs, the algorithm extracts RPs from a set of multiple resolution images obtained by multiple downsampling and colorization error minimizing. In the phase of colorization, the colorization algorithm colorizes multiple-resolution images. Performance of the colorization algorithm affects the performance of the colorization-based image coding, that is, the colorization using a few color pixels leads to a high coding performance. To propose a fast and precise coding algorithm, this paper modifies the colorization algorithm proposed in [4] and introduces a multiple resolution scheme to reduce the computational cost. Figures 1 and 2 show the outline and concept of the proposed algorithm, respectively. The major contribution of this paper is to propose a colorization-based image coding algorithm which requires less computational time and achieves a high coding performance.

This paper is organized as follows. In Section 2.2, we give a colorization algorithm based on the algorithm proposed in [4]. Section 2.3 proposes the RPs extraction algorithm which uses recovery error feedback, and it is applied to multiple resolution images. In Section 2.4, a multi-resolution color image recovery algorithm is proposed. We consider the implementation of saving images with RPs in Section 2.5. Numerical

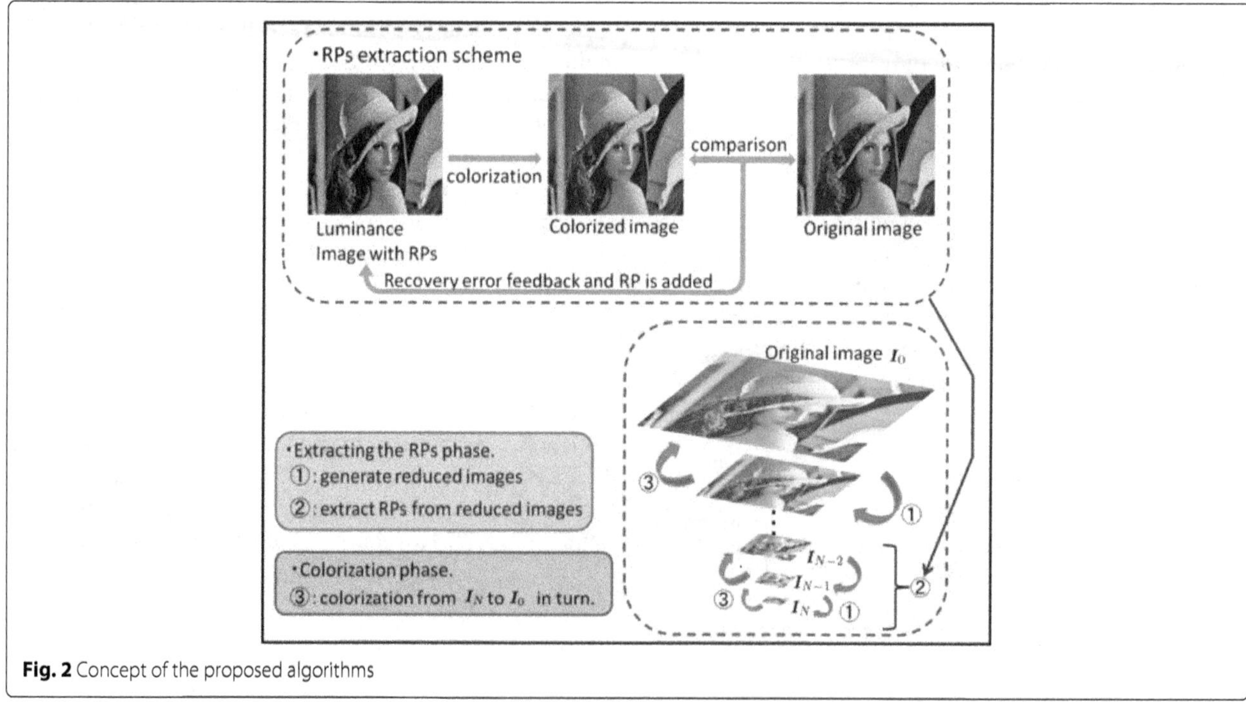

Fig. 2 Concept of the proposed algorithms

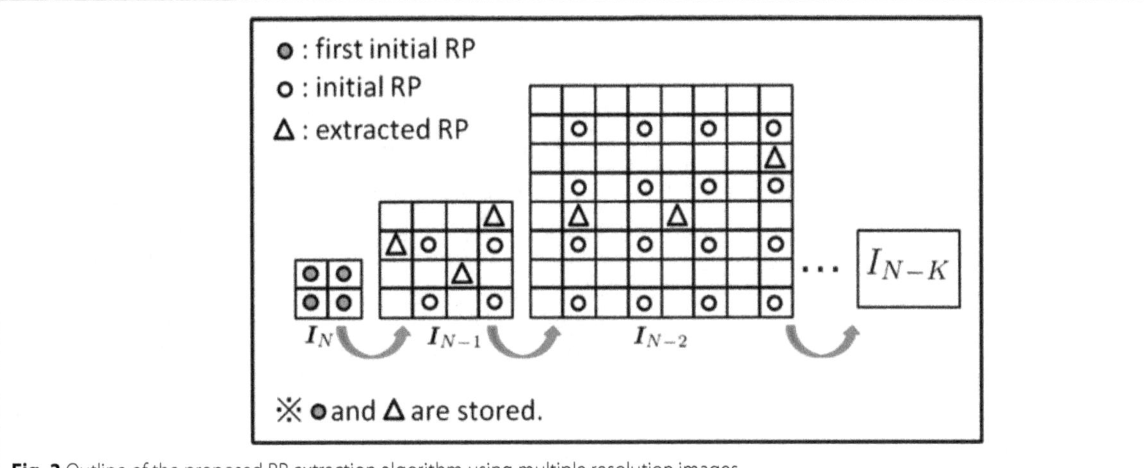

Fig. 3 Outline of the proposed RP extraction algorithm using multiple resolution images

examples show that the proposed algorithm extracts the RPs and recovers the color image fast and effectively in Section 3.

2 Main works

2.1 Notation

We provide here a brief summary of the notations used throughout the paper. $(A)_{i,j}$ and $(\boldsymbol{a})_i$ denote the (i,j)-element of a matrix A and the ith element of a vector $\boldsymbol{a}$. $\|\cdot\|_2$ and $\|\cdot\|_F$ denote the l_2 norm of a vector and the Frobenius norm of a matrix, respectively. $\boldsymbol{0}_k \in R^k$ is denoted as a k-dimension zero vector. We use $\mathrm{diag}(A_1,\ldots,A_m)$ to denote a block diagonal matrix consisting of $A_1, \ldots, A_m$ and $|S|$ to denote the number of elements in a set S. We denote the ceiling function and the floor function by $\lceil a \rceil$ and $\lfloor a \rfloor$, respectively.

2.2 Colorization algorithm

In order to provide a colorization-based image coding, this paper applies the colorization algorithm proposed in [4] because it can recover a full color image from a luminance image using only a few color pixels. The algorithm restores chrominance images C_b and C_r independently to obtain a color image, and this subsection gives a chrominance image restoration method.

Let vectors $\boldsymbol{x} \in \boldsymbol{R}^{mn}$ and $\boldsymbol{y} \in \boldsymbol{R}^{mn}$ denote a desired chrominance image and a rasterized-given luminance image, respectively, where m and n denote height and width of the image. Define $U \in \boldsymbol{R}^{(m-1)\times m}$, $V \in \boldsymbol{R}^{m(n-1)\times mn}$, $\bar{U} \in \boldsymbol{R}^{(m-1)n\times mn}$ and $D \in \boldsymbol{R}^{(2mn-m-n)\times mn}$ as

$$U_{i,j} = \begin{cases} 1, & \text{if } i = j \\ -1, & \text{if } i+1 = j \\ 0, & \text{otherwise} \end{cases},$$

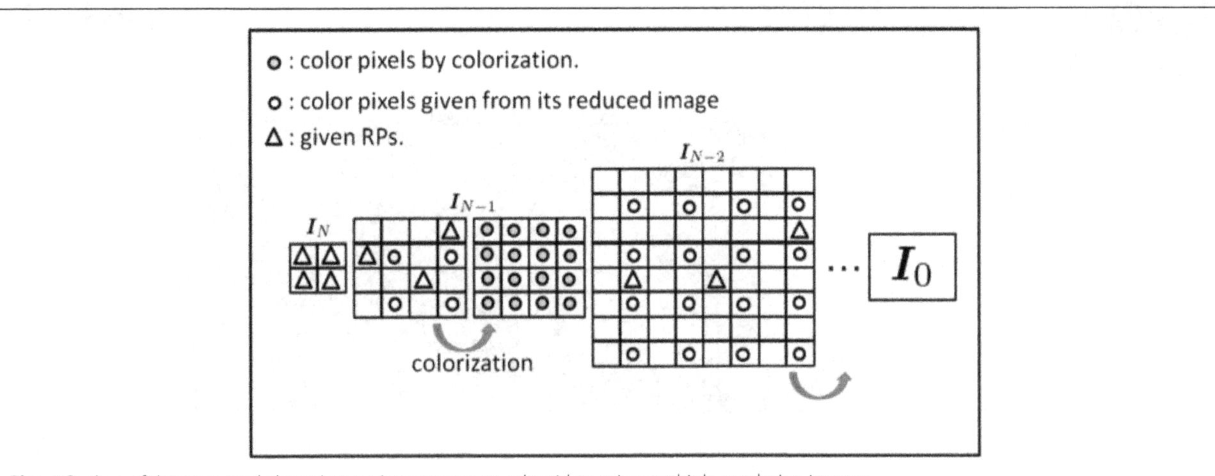

Fig. 4 Outline of the proposed chrominance image recovery algorithm using multiple resolution images

Table 1 Correspondence table (bit unit is 3)

Length	Sign
9	111001
8	111000
7	110
...	...
2	010
1	001
0	000

$$V_{i,j} = \begin{cases} 1, \text{ if } i = j \\ -1, \text{ if } i + M = j \\ 0, \text{ otherwise} \end{cases},$$

$$\bar{U} = \text{diag}(U, \ldots, U),$$

and

$$D = \left[\bar{U}^T \; V^T\right]^T, \tag{1}$$

respectively. The matrices $\bar{U}$ and V denote vertical and horizontal difference operators. Then $D\boldsymbol{x}$ denotes the differences between the neighbor pixels of a whole image. Let us consider the following ℓ_2 norm minimizing colorization problem,

$$\begin{array}{ll} \underset{\boldsymbol{x}}{\text{Minimize}} & \|D\boldsymbol{x}\|_2^2 \\ \text{Subject to} & (\boldsymbol{x})_i = c_i, \forall i \in \mathcal{C}, \end{array} \tag{2}$$

where c_i is a constant corresponding to a given color value, and the index set $\mathcal{C}$ denotes a given set of vector indices of given color pixels. This problem recovers a color image by minimizing the sum of differences between neighbor pixels. In order to colorize an image more appropriately, [4] assumes that the differences between neighbor pixels of a chrominance image have relationship with those of a color image, which is defined by the following equality,

$$(D\boldsymbol{x})_i = \alpha f((D\boldsymbol{y})_i), \tag{3}$$

where α is an unknown constant, and $f : R \to R$ is a given function. Then, we obtain the following problem from (2) under the assumption (3),

$$\begin{array}{ll} \underset{\boldsymbol{x}}{\text{Minimize}} & \|FD\boldsymbol{x}\|_2^2 \\ \text{Subject to} & (\boldsymbol{x})_i = c_i, \forall i \in \mathcal{C}, \end{array} \tag{4}$$

where F is a diagonal matrix whose ith diagonal element is $1/f((D\boldsymbol{y})_i)$. The details are described in [4]. While the problem (4) is a convex quadratic programming and can

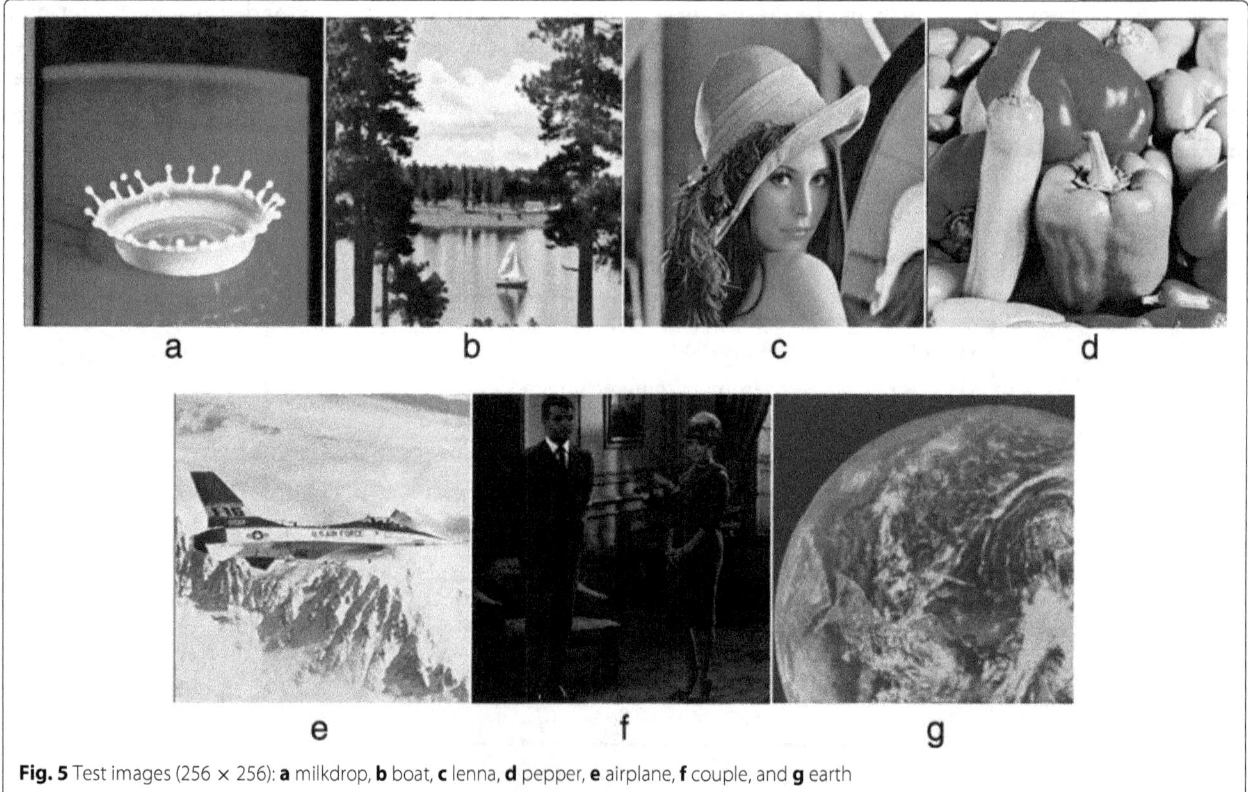

Fig. 5 Test images (256 × 256): **a** milkdrop, **b** boat, **c** lenna, **d** pepper, **e** airplane, **f** couple, and **g** earth

be solved exactly, [4] applies Lagrangian relaxation to reduce the computational cost. This paper provides an exact analytical solution of (4) which can be implemented simply.

Let us define $H \in R^{(2mn-m-n)\times mn}$ by FD. Then, we define $H^* \in R^{(2mn-m-n)\times(mn-p)}$ and $\acute{H}^* \in R^{(2mn-m-n)\times p}$ as submatrices of H composed of columns corresponding to unknown color pixels and given color pixels, respectively. Let $\boldsymbol{x}^* \in R^{mn-p}$ and $\acute{\boldsymbol{x}}^* \in R^p$ denote unknown color vectors and given color vectors, respectively, where p denotes the number of given color pixels. Then the objective function of (4) is rewritten as follows,

$$FD\boldsymbol{x} = H\boldsymbol{x} = H^*\boldsymbol{x}^* + \acute{H}^*\acute{\boldsymbol{x}}^*. \tag{5}$$

Because all entries of $\acute{\boldsymbol{x}}^*$ are given by c_i, that is, they satisfy the constraint of (4), the problem (4) is equal to the following nonconstrained convex quadratic programming,

$$\underset{\boldsymbol{x}^*}{\text{Minimize}} \quad \|H^*\boldsymbol{x}^* + \acute{H}^*\acute{\boldsymbol{c}}^*\|_2^2, \tag{6}$$

where vector $\acute{\boldsymbol{c}}^* \in R^p$ is a given vector consisting of c_i. If $H^{*T}H^*$ is nonsingular, we obtain the solution of (6) as follows,

$$\boldsymbol{x}^* = -\left(H^{*T}H^*\right)^{-1} H^{*T}\acute{H}^*\acute{\boldsymbol{c}}^*. \tag{7}$$

Now we move onto discuss the singularity of $H^{*T}H^*$. For an image $\boldsymbol{a} = [a_1, a_2, \ldots, a_{mn}]^T \in \boldsymbol{R}^{mn}$, the following lemma is provided.

Lemma 1. *It holds that* $H\mathbf{a} = \mathbf{0}_{2mn-m-n}$ *if and only if* $a_1 = a_2 = \cdots = a_{mn}$.

Proof. Because F is an invertible matrix, $H\boldsymbol{a} = \mathbf{0}_{2mn-m-n}$ if and only if $D\boldsymbol{a} = \mathbf{0}_{2mn-m-n}$. Since D is the difference operator defined by (1), $D\boldsymbol{a}$ is a zero vector if and only if all pixels have the same values, that is, $a_1 = a_2 = \cdots = a_{mn}$. □

Let $\boldsymbol{a}^{(i)} \in R^{mn-1}$ denotes a subvector of $\boldsymbol{a}$ and be composed by deleting an ith element of $\boldsymbol{a}$, and $H^{(i)} \in R^{(2mn-m-n)\times(mn-1)}$ is denoted as a submatrix of H and is composed by deleting the ith column of H. Then the Lemma 1 leads to the following lemma:

Lemma 2. *For all* $i \in \{1, 2, \ldots, mn\}$, *it holds that* $H^{(i)}\boldsymbol{a}^{(i)} = \mathbf{0}_{2mn-m-n}$ *if and only if* $\boldsymbol{a}^{(i)} = \mathbf{0}_{mn-1}$.

Proof. We have that

$$H^{(i)}\boldsymbol{a}^{(i)} = H\boldsymbol{a} - H\acute{\boldsymbol{a}} = H(\boldsymbol{a} - \acute{\boldsymbol{a}}), \tag{8}$$

where $\acute{\boldsymbol{a}} = [0, \ldots, 0, a_i, 0, \ldots, 0]^T \in \boldsymbol{R}^{mn}$. Therefore, we obtain followings from Lemma 1,

$$\begin{aligned} H^{(i)}\boldsymbol{a}^{(i)} = \mathbf{0}_{2mn-m-n} &\Leftrightarrow H(\boldsymbol{a} - \acute{\boldsymbol{a}}) = \mathbf{0}_{2mn-m-n} \\ &\Leftrightarrow a_1 = a_2 = \ldots = a_i - a_i = \ldots = a_{mn} \\ &\Leftrightarrow a_1 = a_2 = \cdots = 0 = \ldots = a_{mn} \\ &\Leftrightarrow \boldsymbol{a}^{(i)} = \mathbf{0}_{mn-1}. \end{aligned} \tag{9}$$

□

Theorem 1. $H^{*T}H^*$ *is a nonsingular matrix*

Proof. Lemma 2 guarantees that the subspace of the column space of H is linear independent, that is, the column space of H^* is linear independent. □

From this theorem, there exists the inverse of $H^{*T}H^*$, and therefore we can obtain the exact solution of (4) by (7).

2.3 RPs extraction algorithm

This subsection proposes an algorithm to extract the RPs from an original image utilizing the error feedback between the original image and the colorized image. Let $Y \in \boldsymbol{R}^{m\times n}$, $Cb \in \boldsymbol{R}^{m\times n}$ and $Cr \in \boldsymbol{R}^{m\times n}$ denote a given luminance image and two given chrominance images, respectively. First we provide the following RPs extraction scheme for a given color image, its luminance image, a given initial set of RPs and a small constant $\varepsilon > 0$,

Step 1. Recover two chrominance images $\bar{Cb}$ and $\bar{Cr}$ using (7) with the luminance image and current RPs.

Step 2. Calculate the error between the original image and the colorized image obtained in Step 1 at each pixel as follows,

$$(E)_{i,j} = |(Cb - \bar{Cb})_{i,j}| + |(Cr - \bar{Cr})_{i,j}|. \tag{10}$$

Step 3. Terminate if $(E)_{i,j} \leq \varepsilon$ for all (i, j), else go to Step 4.

Step 4. Add the pixel with the largest error $E_{i,j}$ to the set of RPs.

Step 5. Terminate if the number of iterations is larger than a given threshold, else go to Step 1.

Since this scheme extracts one pixel in one iteration, some pixels are selected appropriately as RPs by repeating Step 1–5. Although this scheme can steadily reduce the recovery error, it requires a lot of iterations to obtain an adequate amount of RPs and takes a lot of computing time for a large size image. In order to reduce the computing cost, this paper proposes an RP extraction algorithm using multiple resolution images.

Table 2 Numerical results for several (N, K)s

Milkdrop	N	N-K 0	1	2	3	Airplane	N	N-K 0	1	2	3
PSNR [dB]	4	38.0759	37.5711	37.1943	34.3947	PSNR [dB]	4	38.2035	33.2393	32.3945	32.0054
	5	37.6920	37.2186	36.8684	34.2402		5	38.1185	33.2027	32.3641	31.9780
	6	37.6314	37.1643	36.8181	34.2058		6	37.7937	33.1094	32.2888	31.9157
	7	37.6244	37.1576	36.8116	34.2006		7	37.7754	33.1061	32.2874	31.9159
Encoding Time [s]	4	14.1671	2.3608	0.6417	0.2239	Encoding Time [s]	4	14.6127	2.3328	0.6553	0.2307
	5	13.9903	2.4329	0.7173	0.2841		5	14.3815	2.3910	0.7077	0.2871
	6	13.9673	2.4428	0.7311	0.3021		6	14.6817	2.4039	0.7209	0.3002
	7	14.0246	2.4431	0.7268	0.3098		7	14.7841	2.4064	0.7275	0.3042
Decoding Time [s]	4	0.2144	0.2189	0.2122	0.2170	Decoding Time [s]	4	0.2099	0.2111	0.2033	0.1969
	5	0.2137	0.1979	0.2156	0.1996		5	0.2240	0.1994	0.2011	0.1988
	6	0.2382	0.2101	0.2075	0.1974		6	0.2193	0.2040	0.2087	0.2146
	7	0.2242	0.2021	0.2336	0.2136		7	0.2181	0.1986	0.2042	0.2231
Volume [byte]	4	1407	1146	913	702	Volume [byte]	4	1411	1148	915	706
	5	1193	933	699	488		5	1197	935	702	492
	6	1195	933	700	489		6	1154	892	659	449
	7	1193	932	699	488		7	1149	886	653	444
Boat	**N**	**N-K 0**	**1**	**2**	**3**	**Couple**	**N**	**N-K 0**	**1**	**2**	**3**
PSNR [dB]	4	34.2850	33.9851	33.6053	32.5525	PSNR [dB]	4	38.5062	38.2618	37.9755	37.0003
	5	34.0394	33.7643	33.4117	32.4187		5	38.3833	38.1462	37.8793	36.9063
	6	34.0133	33.7336	33.3841	32.3973		6	38.3118	38.0789	37.8241	36.8307
	7	34.0132	33.7335	33.3840	32.3969		7	38.3085	38.0757	37.8214	36.8280
Encoding Time [s]	4	14.117	2.3290	0.6535	0.2281	Encoding Time [s]	4	14.7000	2.3234	0.6541	0.2365
	5	14.1622	2.4299	0.7107	0.2882		5	14.3138	2.3749	0.7144	0.2868
	6	14.2034	2.4091	0.7348	0.3129		6	14.3077	2.3927	0.7312	0.3089
	7	14.0931	2.4581	0.7512	0.3148		7	14.4525	2.4048	0.7412	0.3151
Decoding Time [s]	4	0.2119	0.2098	0.2359	0.1954	Decoding Time [s]	4	0.2105	0.1967	0.1976	0.2081
	5	0.2111	0.2058	0.2159	0.2064		5	0.2072	0.1983	0.1980	0.2156
	6	0.2162	0.2037	0.2016	0.1980		6	0.2075	0.2159	0.2026	0.2246
	7	0.2116	0.2069	0.2099	0.2536		7	0.2068	0.2166	0.2120	0.2006
Volume [byte]	4	1413	1150	912	702	Volume [byte]	4	1414	1154	917	705
	5	1198	936	698	487		5	1200	940	703	491
	6	1195	933	695	484		6	1194	934	697	485
	7	1198	936	698	487		7	1191	931	694	482
Lenna	**N**	**N-K 0**	**1**	**2**	**3**	**Earth**	**N**	**N-K 0**	**1**	**2**	**3**
PSNR [dB]	4	36.3927	36.2937	36.0081	35.3116	PSNR [dB]	4	38.5740	38.4291	38.1530	37.4579
	5	36.1112	36.0174	35.7481	35.1009		5	38.1631	38.0269	37.7843	37.0972
	6	36.0689	35.9747	35.7079	35.0432		6	38.0894	37.9473	37.7087	37.0341
	7	36.0716	35.9774	35.7105	35.0497		7	38.0962	37.9539	37.7153	37.0421

Table 2 Numerical results for several (N, K)s *(Continued)*

Encoding Time [s]	4	14.0836	2.2984	0.6657	0.2247	Encoding Time [s]	4	14.5969	2.2922	0.6481	0.2284
	5	14.1050	2.3610	0.7059	0.2838		5	14.2419	2.3374	0.7004	0.2842
	6	14.4354	2.3993	0.7309	0.3194		6	14.3422	2.3709	0.7293	0.3078
	7	14.5874	2.3534	0.7317	0.3150		7	14.1882	2.3787	0.7252	0.3131
Decoding Time [s]	4	0.2023	0.2110	0.2272	0.1950	Decoding Time [s]	4	0.2106	0.2018	0.2078	0.1997
	5	0.2278	0.2029	0.2078	0.2096		5	0.2291	0.2076	0.1970	0.2152
	6	0.2624	0.2173	0.2230	0.2273		6	0.2057	0.2124	0.1964	0.2003
	7	0.2388	0.2115	0.2162	0.1970		7	0.2082	0.2053	0.1983	0.2041
Volume [byte]	4	1420	1153	915	704	Volume [byte]	4	1402	1143	909	702
	5	1206	939	701	490		5	1188	929	695	488
	6	1200	934	696	485		6	1184	926	692	485
	7	1204	937	699	488		7	1183	924	690	484

Pepper	N	N-K			
		0	1	2	3
PSNR [dB]	4	32.6169	30.3128	30.0602	29.1197
	5	32.0963	29.9893	29.7612	28.8955
	6	32.0913	29.9865	29.7585	28.8929
	7	32.0914	29.9867	29.7588	28.8932
Encoding Time [s]	4	14.7880	2.3244	0.6564	0.2304
	5	14.3656	2.3847	0.7095	0.2876
	6	14.4174	2.4086	0.7376	0.3121
	7	14.5826	2.4150	0.7427	0.3207
Decoding Time [sec]	4	0.2103	0.2026	0.1971	0.2081
	5	0.2246	0.2147	0.1985	0.2220
	6	0.2037	0.1981	0.1977	0.1986
	7	0.2061	0.2201	0.2230	0.1985
Volume [byte]	4	1412	1143	911	702
	5	1198	929	697	488
	6	1206	937	705	496
	7	1209	940	708	499

Let $I_0 = [Y_0\ Cb_0\ Cr_0] \in R^{m \times 3n}$ and $I_k = [Y_k\ Cb_k\ Cr_k] \in R^{m_k \times 3n_k}$ $(k = 1, 2, \cdots, N)$ denote the original image and kth reduced image, respectively, and $N \geq 1$ is a given constant. Y_k is obtained by simply multiple downsampling as follows,

$$(Y_k)_{i,j} = (Y_{k-1})_{2i,2j}, \quad i = 1, \cdots, \lfloor \frac{m_{k-1}}{2} \rfloor, j = 1, \cdots, \lfloor \frac{n_{k-1}}{2} \rfloor, \tag{11}$$

where m_k and n_k denote horizontal and vertical sizes of the kth reduced image. Next, we consider a way to generate Cb_k and Cr_k. Since Cb_k and Cr_k are composed by the same method, this paper gives a way to make the reduced image Cb_k. The chrominance images of the kth reduced image are generated so that the $(k-1)$th chrominance image Cb_{k-1} can be restored preciously using Y_{k-1} and Cb_k. Let $\acute{\mathbf{c}}^*_{k-1} \in R^{m_k n_k}$ and $\mathbf{c}^*_{k-1} \in R^{m_{k-1}n_{k-1} - m_k n_k}$ denote vectors formed by pixels of even numbered columns and rows, that is, they do not include pixels of odd numbered columns nor rows. Then, this paper proposes the chrominance image recovery problem of the kth reduced image as follows,

$$\underset{\boldsymbol{x}_k}{\text{Minimize}}\ \|\boldsymbol{c}^*_{k-1} + \left(H^{*T} H^*\right)^{-1} H^{*T} \acute{H}^* \boldsymbol{x}_k\|_2^2 + \lambda \|\acute{\boldsymbol{c}}^*_{k-1} - \boldsymbol{x}_k\|_2^2, \tag{12}$$

Algorithm 1 RP extraction algorithm.

Require: Y_0, Cb_0, Cr_0, N, K, ε, t_{max}
Generate Y_k $k = 1, 2, \ldots, N$ according to (11).
Calculate Cb_k, Cr_k for $k = 1, 2, \ldots, N$ according to (14).
Set $\Omega_N^{RP} \leftarrow \{(i, j, c_b, c_r) | \ i \in \{1, 2, \ldots, m_N\}, \ j \in \{1, 2, \ldots, n_N\}, c_b = (Cb_N)_{i,j}, c_r = (Cr_N)_{i,j}\}$.
for k = N-1,N-2,..., N-K **do**
$\Omega_k^{ini} \leftarrow \{(i, j) | \ i \in \{2, 4, \ldots, 2m_{k+1}\}, \ j \in \{2, 4, \ldots, 2n_{k+1}\}\}$.
$\Omega_k \leftarrow \Omega_k^{ini}$.
$t \leftarrow 1$.
repeat
Calculate $\bar{Cb}_k$ and $\bar{Cr}_k$ from Y_k and (i, j)-elements of Cb_k and Cr_k for all $(i, j) \in \Omega_k$ using (7).
for all (i,j) **do**
$(E)_{i,j} \leftarrow |(Cb_k - \bar{Cb}_k)_{i,j}| + |(Cr_k - \bar{Cr}_k)_{i,j}|$.
end for
$(i^{RP}, j^{RP}) \leftarrow \arg\max_{(i,j)} (E_{i,j})$.
$\Omega_k \leftarrow \Omega_k \cup \{(i^{RP}, j^{RP})\}$.
$t \leftarrow t + 1$.
until $t = t_{max}$ or $\max(E_{i,j}) \leq \varepsilon$
$\Omega_k \leftarrow \Omega_k \setminus \Omega_k^{ini}$.
$\Omega_k^{RP} = \{(i, j, c_b, c_r) | (i, j) \in \Omega_k, c_b = (Cb_k)_{i,j}, c_r = (Cr_k)_{i,j}\}$.
end for
Ensure: Ω_k^{RP} $k = N, N-1, \ldots, N-K$

where $\lambda > 0$ is a given constant. Because the chrominance images Cb_{k-1} and Cr_{k-1} are recovered using (7) with Y_{k-1}, Cb_k and Cr_k in the colorization phase as written Section 2.4, this problem gives Cb_k to achieve a good colorization of the $(k-1)$th image. However, problem (12) requires high computational cost to obtain the optimal solution since it takes high cost to calculate $\left(H^{*T}H^*\right)^{-1}$. Therefore this paper provides its relaxed problem to obtain an approximate solution with low computational cost. Let J denote the first term of the objective function in (12). Then, we have that

$$\begin{aligned} J &= \|\boldsymbol{c}_{k-1}^* + \left(H^{*T}H^*\right)^{-1} H^{*T} \acute{H}^* \boldsymbol{x}_k\|_2^2 \\ &= \| \left(H^{*T}H^*\right)^{-1} H^{*T} \left(H^* \boldsymbol{c}_{k-1}^* + \acute{H}^* \boldsymbol{x}_k\right) \|_2^2 \qquad (13) \\ &\leq \| \left(H^{*T}H^*\right)^{-1} H^{*T} \|_F^2 \ \|H^* \boldsymbol{c}_{k-1}^* + \acute{H}^* \boldsymbol{x}_k\|_2^2. \end{aligned}$$

Since J is bounded by $\|H^* \boldsymbol{c}_{k-1}^* + \acute{H}^* \boldsymbol{x}_k\|_2^2$, we consider the following problem to reduce J instead of (12),

$$\underset{\boldsymbol{x}_k}{\text{Minimize}} \quad \|H^* \boldsymbol{c}_{k-1}^* + \acute{H}^* \boldsymbol{x}_k\|_2^2 + \lambda \|\acute{\boldsymbol{c}}_{k-1}^* - \boldsymbol{x}_k\|_2^2. \quad (14)$$

Utilizing (11) and (14) to obtain a set of multiple resolution images, this paper provides the RPs extraction method as shown in Algorithm 1, where $\bar{Cb}_k$ and $\bar{Cr}_k$ denote the colorized chrominance images of I_k. This algorithm extracts RPs from I_N to I_{N-K} in turn for given constant $K \geq 1$, and the even numbered pixels of I_k are used as initial RPs at each iteration. The performance of Algorithm 1 depends heavily on the value of K. If the value of K is large, lots of RPs are extracted, that is, the K determines the number of RPs, the calculation time and the volume of information to store RPs. The numerical examples in Section 3 show the effects of K.

Note that the chrominance values of these even numbered pixels are deleted in Ω_k^{RP} at the end of each iteration. They are restored as Cb_{k+1} and Cr_{k+1} in the previous iteration and are reused as RPs for recovering I_k. Therefore Algorithm 1 does not store them in Ω_k^{RP} to reduce the amount of information about RPs. All pixels of the Nth reduced image are members of Ω_N^{RP}. Figure 3 shows outline of the algorithm.

Algorithm 2 Colorization algorithm.

Require: Y_0, Ω_k^{RP} $(k = N, N-1, \ldots, N-K)$
Generate Y_k for $k = 1, 2, \ldots, N$ by (11).
$\Omega_k^{RP} \leftarrow \{\phi\}$ for $k = N-K-1, \ldots, 0$.
Generate Cb_N and Cr_N from Ω_N^{RP}.
for k= $N-1$,N-2,...,0 **do**
$\Omega_k^{RP_{even}} = \{(2i, 2j, c_b, c_r) | i \in \{1, 2, \ldots, m_{k+1}\}, \ j \in \{1, 2, \ldots, n_{k+1}\}, c_b = (Cb_{k+1})_{i,j}, c_r = (Cr_{k+1})_{i,j}\}$.
$\Omega_k^{RP} \leftarrow \Omega_k^{RP} \cup \Omega_k^{RP_{even}}$.
Recover the chrominance images Cb_k and Cr_k with Ω_k^{RP} by (7).
end for
Ensure: colorized image.

2.4 Color image recovery

This subsection provides the chrominance image recovery algorithm. Now we consider a way to recover an image $I_0 = [Y_0 \ Cb_0 \ Cr_0] \in R^{m \times 3n}$ from the stored RPs Ω_k^{RP} $(k = N, \ldots N-K)$. First, I_N is completely obtained by Ω_N^{RP}. Next, we focus onto recover the image I_k $(k = N-1, N-2, \ldots, 0)$. Since Ω_k^{RP} includes RPs without the even numbered pixels, image I_{k+1} is used as Ω_k^{RP} corresponding to the even numbered pixels. Then the color images I_k is restored by (7) using Ω_k^{RP} and I_{k+1}. Finally we obtain the colorization algorithm with multiple resolution images as shown in Algorithm 2, and Fig. 4 shows its outline.

The calculation cost of the colorization proposed in subsection 2.2 mostly depends on the calculating of inverse matrix $\left(H^{*T}H^*\right)^{-1}$, and the size of $\left(H^{*T}H^*\right)^{-1}$ depends on the number of unknown color pixels. The number of unknown color pixels in Algorithm 2 is equal to the following,

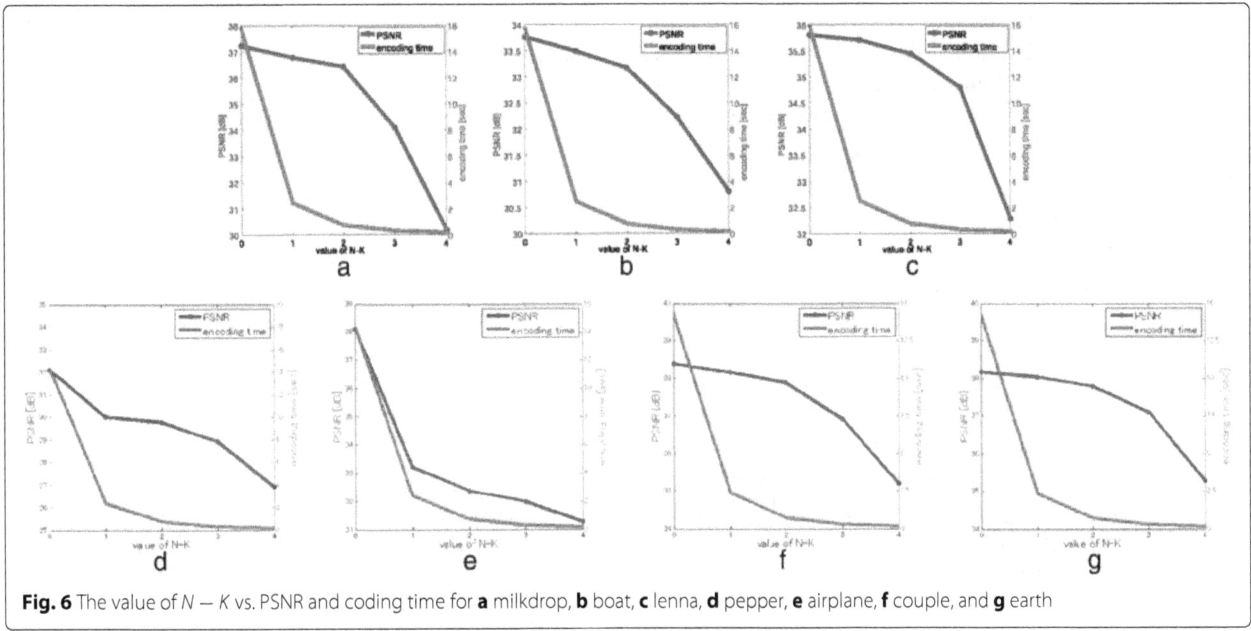

Fig. 6 The value of $N - K$ vs. PSNR and coding time for **a** milkdrop, **b** boat, **c** lenna, **d** pepper, **e** airplane, **f** couple, and **g** earth

$$\begin{aligned}
&\sum_{k=0}^{N-1}\left(m_k \times n_k - m_{k+1} \times n_{k+1} - |\Omega_k^{RP}|\right) \\
&= m_0 \times n_0 - m_N \times n_N - \sum_{k=0}^{N-1} |\Omega_k^{RP}| \qquad (15) \\
&= m_0 \times n_0 - \sum_{k=0}^{N} |\Omega_k^{RP}| .
\end{aligned}$$

This implies that the calculation cost does not depend on N. Numerical examples in Section 3 show that the computing time is independent of N.

2.5 Volume of information of RPs

This subsection gives an encoding method of RPs.

Let us consider the case that an $m \times n$ 24-bit image is compressed with p RPs. If the information of RPs is composed of their coodinates and two chrominance values,

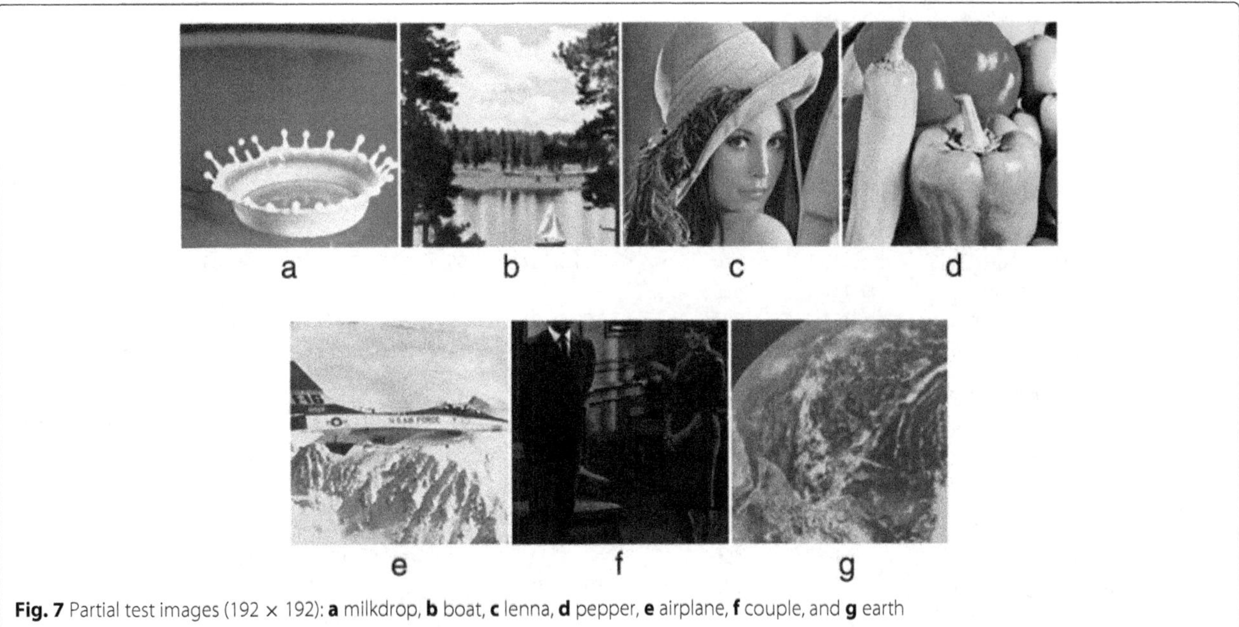

Fig. 7 Partial test images (192 × 192): **a** milkdrop, **b** boat, **c** lenna, **d** pepper, **e** airplane, **f** couple, and **g** earth

$p(2 \times 8 + \lceil \log_2 mn \rceil)$ [bits] are required to represent all RPs because 2×8 [bits] and $\lceil \log_2 mn \rceil$ [bits] are required to represent their chrominance values and coordinates, respectively. The proposed algorithm extracts RPs based on multiple resolution images and requires less information to represent them.

We consider encoding of the coordinates of RPs Ω_k^{RP} $(k = N, N-1, \ldots N-K)$. In Algorithm 1, Ω_N^{RP} is given as the reduced image I_N, the Ω_N^{RP} are given as the reduced image I_N, that is, all pixels of I_N are RPs. Therefore, the coordinates of Ω_N^{RP} are not required, and only the value of N is coded to represent Ω_N^{RP}. Next, we consider coding of the coordinates of RPs in Ω_k^{RP} $(k = N-1, \ldots, N-K)$. For $m \times n$ image, we store the coordinates of RPs as a rasterized vector of a binary matrix of size $m \times n$ whose entries are 1 if corresponding pixels are RPs and 0 if otherwise. Hence, this vector is sparse, and it is coded by run length encoding (RLE), which stores the length of consecutive 0s. Empirical results show that the length of consecutive 0s on the image Ω_{N-k}^{RP} is less than z_k defined by

$$z_{N-k} = \frac{m_{N-k}n_{N-k} - m_{N-k+1}n_{N-k+1}}{t_{max}}. \tag{16}$$

Therefore, this paper proposes to encode the coordinate by RLE per following bit,

$$bitunit_{N-k} = \max\left(\left\lceil \log_2(z_{N-k}) \right\rceil, 2\right). \tag{17}$$

Table 1 shows the coding table of $bitunit_{N-k} = 3$. This decoding method requires the run length data of Ω_k^{RP} $(k = N-1, \ldots, N-K)$ and the values of N and t_{max} to decode the coordinates of RPs.

3 Numerical examples

This section presents numerical examples to show the efficiency of the proposed algorithm. We use the test

Table 3 Numerical results of comparison with other colorization based image coding algorithms

Image	Algorithm	Number of RPs [pixels]	Volume of information [byte]	Time [sec]		PSNR [dB]		SSIM	
				encoding	decoding	Cb	Cr	Cb	Cr
Milkdrop	algorithm proposed in [9]	109	436	1.31×10^3	3.51	35.0257	27.1737	0.9232	0.8424
	random algorithm	108	432	—	0.17	33.2654	27.5243	0.9298	0.8818
	proposed algorithm with (12)	108	273	4.17×10^3	0.11	38.2353	33.7145	0.9574	0.9332
	proposed algorithm with (14)	108	273	0.23	0.11	37.9307	34.5197	0.9588	0.9371
Boat	algorithm proposed in [9]	117	468	1.19×10^3	3.53	33.9007	30.2299	0.8798	0.8216
	random algorithm	117	468	—	0.17	32.5359	28.6484	0.8766	0.8173
	proposed algorithm with (12)	117	308	4.02×10^3	0.12	34.4766	32.3504	0.8962	0.8774
	proposed algorithm with (14)	117	308	0.25	0.12	33.9637	32.1755	0.8946	0.8802
Lenna	algorithm proposed in [9]	117	468	1.20×10^3	3.57	33.6590	32.3297	0.8665	0.8335
	random algorithm	117	468	—	0.18	31.1721	30.8207	0.8546	0.8344
	proposed algorithm with (12)	117	310	4.02×10^3	0.12	34.3554	33.1450	0.8860	0.8660
	proposed algorithm with (14)	117	310	0.24	0.12	33.9673	32.8775	0.8846	0.8662
Pepper	algorithm proposed in [9]	113	452	1.25×10^3	3.50	31.8969	26.8608	0.8555	0.8260
	random algorithm	111	444	—	0.1716	29.9528	24.5312	0.8536	0.8209
	proposed algorithm with (12)	111	285	3.90×10^3	0.12	33.2907	29.4019	0.8950	0.8809
	proposed algorithm with (14)	111	284	0.24	0.12	32.9208	30.2535	0.8930	0.8852
Airplane	algorithm proposed in [9]	119	476	1.21×10^3	3.55	31.0753	34.4833	0.8166	0.9014
	random algorithm	117	468	—	0.17	29.8990	32.9087	0.8169	0.9035
	proposed algorithm with (12)	117	307	3.90×10^3	0.12	33.7608	36.1246	0.8879	0.9154
	proposed algorithm with (14)	117	307	0.27	0.12	33.8210	35.7528	0.8883	0.9155
Couple	algorithm proposed in [9]	118	472	1.26×10^3	3.47	35.9439	33.0840	0.9001	0.8795
	random algorithm	117	468	—	0.17	34.2172	31.2331	0.8992	0.8804
	proposed algorithm with (12)	117	311	3.90×10^3	0.14	37.4989	35.2208	0.9210	0.9120
	proposed algorithm with (14)	117	309	0.25	0.13	37.1587	35.2879	0.9188	0.9115
Earth	algorithm proposed in [9]	115	460	1.19×10^3	3.56	36.1886	33.8077	0.9133	0.8290
	random algorithm	114	456	—	0.17	34.1523	32.4901	0.9068	0.8277
	proposed algorithm with (12)	114	288	3.90×10^3	0.12	37.6137	34.4186	0.9386	0.8598
	proposed algorithm with (14)	114	289	0.26	0.12	37.9318	34.1140	0.9390	0.8602

images as shown in Fig. 5. In order to evaluate the quality of image compression, we measure the differences between the original image and the recovered image using the peak signal to noise ratio (PNSR) and the structural similarity (SSIM) [16]. In order to calculate these evaluates, we use the MATLAB functions `psnr` and `ssim`. In all experiments we use $\varepsilon = 2$ and $\lambda = 0.3$, which are selected empirically from $\varepsilon \in [0, 10]$ and $\lambda \in [0.1, 1]$ to achieve the best performance. We use $t_{max} = 70$ except for the second experiment.

First we examine the effect of the parameters N and K using uncompressed luminance images. Table 2 shows PSNR, the computing time of Algorithm 1, the computing time of Algorithm 2 and the volume of information to store RPs. PSNR is obtained by averaging those of Cb and Cr. As can be seen, the results of $N \geq 5$ are almost the same. Figure 6 shows the PSNR and the computing time of Algorithm 1 with $N = 5$ for $N - K \in \{0, 1, 2, 3, 4\}$. We can see that the algorithm with $(N, K) = (5, 3)$ achieves the best tradeoff between calculation time and PSNR, and therefore we use $(N, K) = (5, 3)$ in the rest of this section.

Next this paper shows the coding performance of the proposed algorithm comparing with the algorithm proposed in [9] and the random algorithm which selects RPs randomly. We chose $\lambda = 5$ in (12), which is selected empirically from $\lambda \in [1, 10]$ to achieve the best performance. To evaluate the performance of the relaxed problem (14) in RPs extraction scheme, we also examine Algorithm 1 with (12) instead of (14). Because we cannot fix the number of RPs exactly in the algorithm proposed in [9], we adjust the parameter t_{max} of Algorithm 1 such that the number of RPs is nearly equal to that of [9]. The random algorithm selects RPs randomly and gives the same number of RPs as the proposed algorithm, and we use colorization algorithm (7). In order to see the performance of colorization based coding method, these algorithms use uncompressed luminance images. Because the algorithm proposed in [9] and the proposed algorithm with (12) use a large amount of memory and take a high computational cost, they cannot be applied to a whole image of the test images. Therefore, we use their partial images as shown in Fig. 7. Table 3 shows the number of RPs, the volume of information to restore the RPs, SSIM, PSNR, the computing time to extract the RPs (encoding

Table 4 Numerical results of comparison with Lee's algorithm

Image	Algorithm	Number of RPs [pixels]	Volume of information [byte]	Time [sec]		PSNR [dB]		SSIM	
				encoding	decoding	Cb	Cr	Cb	Cr
Milkdrop	proposed algorithm	274	699	0.72	0.22	37.8741	35.8627	0.9635	0.9377
	algorithm proposed in [12]	274	959	81.78	31.01	37.1474	32.4973	0.9432	0.9049
	algorithm proposed in [12]	200	700	53.41	29.27	36.8288	32.4900	0.9397	0.8978
Boat	proposed algorithm	274	698	0.71	0.22	34.7794	32.0441	0.8928	0.8734
	algorithm proposed in [12]	274	959	79.87	28.77	34.5627	32.8473	0.8815	0.8744
	algorithm proposed in [12]	200	700	52.57	28.66	34.1386	32.5993	0.8753	0.8707
Lenna	proposed algorithm	274	701	0.71	0.21	36.1906	35.3057	0.9231	0.9018
	algorithm proposed in [12]	274	959	80.88	29.55	36.2129	34.5499	0.9086	0.8925
	algorithm proposed in [12]	200	700	53.39	29.51	35.9824	34.5354	0.9059	0.8903
Pepper	proposed algorithm	274	697	0.71	0.20	29.4435	30.0789	0.8697	0.8762
	algorithm proposed in [12]	274	959	88.56	38.37	32.6670	29.7992	0.8702	0.8510
	algorithm proposed in [12]	199	697	59.50	37.39	30.2358	28.8093	0.8458	0.8256
Airplane	proposed algorithm	274	702	0.71	0.22	33.0319	31.6955	0.9117	0.9149
	algorithm proposed in [12]	274	959	79.14	29.15	34.9049	33.9563	0.9124	0.9185
	algorithm proposed in [12]	200	700	52.83	29.07	34.7446	33.8131	0.9100	0.9165
Couple	proposed algorithm	274	703	0.70	0.20	38.7016	37.0570	0.9216	0.9172
	algorithm proposed in [12]	274	959	84.51	34.82	38.4629	35.9642	0.9143	0.9013
	algorithm proposed in [12]	201	704	56.13	32.76	38.3394	35.8256	0.9134	0.9012
Earth	proposed algorithm	274	695	0.71	0.21	39.4627	36.1058	0.9500	0.9049
	algorithm proposed in [12]	274	959	85.51	35.81	40.2759	37.7125	0.9536	0.9166
	algorithm proposed in [12]	199	697	58.08	35.84	40.0729	37.6271	0.9525	0.9156

time), and to colorize the images (decoding time). As can be seen, the proposed algorithm with (14) works faster and achieves better performance than the other algorithms.

Thirdly, the proposed algorithm is compared with the algorithm proposed in [12] using the test images as shown in Fig. 5. In [12], Lee et al. have proposed the colorization-based image coding algorithm, which achieves a high coding performance. In order to see the performance of colorization-based coding method, these algorithms use uncompressed luminance images. Since an arbitrary number of RPs can be used in Lee's algorithm, we examine the algorithm in two ways by using (1) the same number of RPs as those of the proposed algorithm and (2) RPs whose volume of information to be represented is equal to that of the proposed algorithm.

Table 4 shows the number of RPs, the volume of information to restore the RPs, SSIM, PSNR, the computing time to extract the RPs (encoding time) and to colorize the images (decoding time). We can see that the proposed algorithm achieves similar performance to Lee's algorithm and requires less computational.

Finally, we compare the proposed algorithm with JPEG and JPEG2000 coding method. Luminance images are compressed using JPEG/JPEG2000 coding where the quality parameter (QP) is selected such that the volume of information is equal to about 4000 bytes.

Table 5 shows the volume of information [byte] to represent each color image, SSIM, PSNR and quality parameter (QP) for JPEG/JPEG2000, Figs. 8, 9 and 10 show the recovery images, and Fig. 11 shows zoomed images of Fig. 8. SSIM and PSNR are obtained by averaging those of red, green, and blue images. As can be seen, the proposed algorithm with JPEG2000 can compress color images the best of all while the qualities of compressed images are equal or better than other coding methods. The proposed

Table 5 Numerical results of comparison with JPEG and JPEG2000

Image	Algorithm	Volume of information [byte]	PSNR[dB]	SSIM	QP for JPEG/JPEG2000
Milkdrop	JPEG	4206	28.6925	0.7988	20
	JPEG2000	4094	32.4263	0.8719	46
	proposed algorithm with JPEG	4138	30.7470	0.8540	22
	proposed algorithm with JPEG2000	4036	32.6914	0.8891	20
Boat	JPEG	4139	24.7191	0.7488	10
	JPEG2000	4068	26.9982	0.8219	46
	proposed algorithm with JPEG	4125	25.8622	0.7849	10
	proposed algorithm with JPEG2000	4039	27.2043	0.8374	20
Lenna	JPEG	4136	26.794	0.7864	13
	JPEG2000	4070	29.2353	0.8398	46
	proposed algorithm with JPEG	4159	27.6452	0.8116	13
	proposed algorithm with JPEG2000	4061	29.2051	0.8568	20
Pepper	JPEG	4137	24.0496	0.7078	11
	JPEG2000	4071	27.1189	0.7969	46
	proposed algorithm with JPEG	4147	24.8098	0.7632	12
	proposed algorithm with JPEG2000	4051	25.9898	0.8127	20
Airplane	JPEG	4176	25.1615	0.7894	12
	JPEG2000	4062	28.0905	0.8443	46
	proposed algorithm with JPEG	4192	25.6759	0.8108	12
	proposed algorithm with JPEG2000	4044	26.7533	0.8471	20
Couple	JPEG	4158	30.5207	0.8112	21
	JPEG2000	4018	32.0973	0.8373	46
	proposed algorithm with JPEG	4184	31.0459	0.8314	21
	proposed algorithm with JPEG2000	4064	32.1436	0.8527	20
Earth	JPEG	4277	27.5088	0.7952	13
	JPEG2000	4164	30.0662	0.8527	45
	proposed algorithm with JPEG	4146	28.4777	0.8088	12
	proposed algorithm with JPEG2000	4017	30.2628	0.8604	20

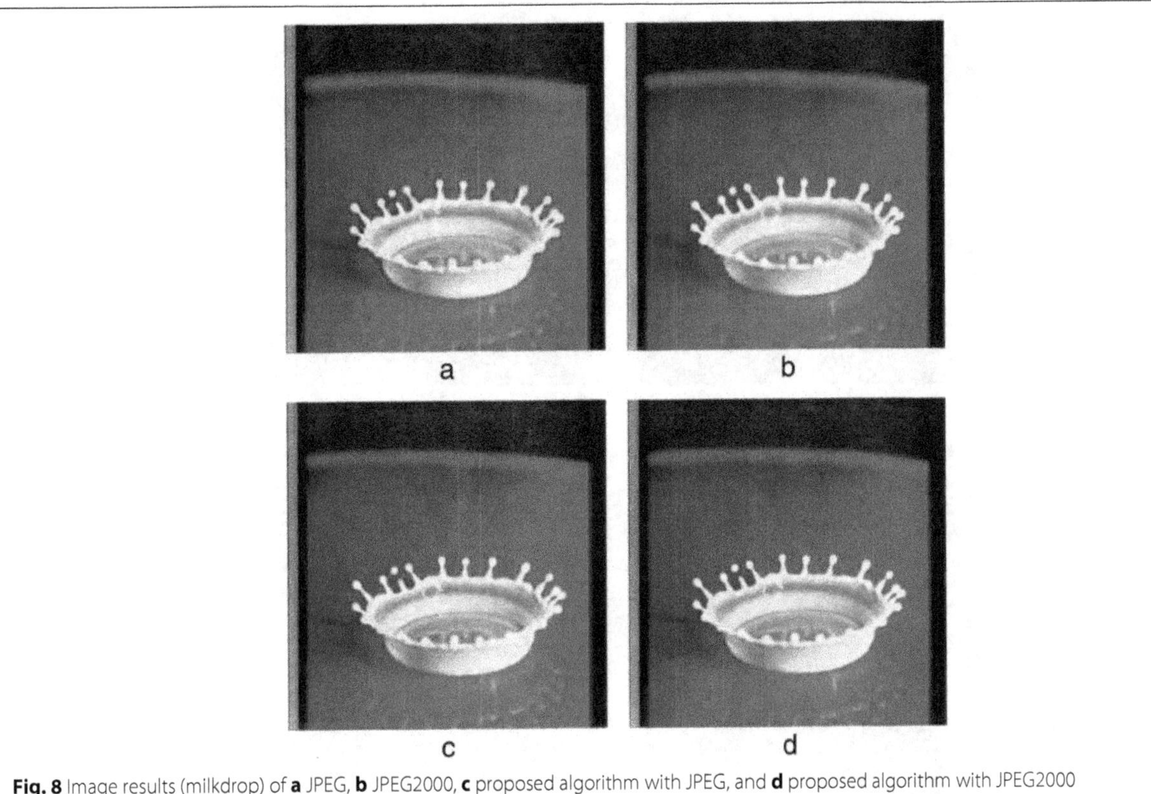

Fig. 8 Image results (milkdrop) of **a** JPEG, **b** JPEG2000, **c** proposed algorithm with JPEG, and **d** proposed algorithm with JPEG2000

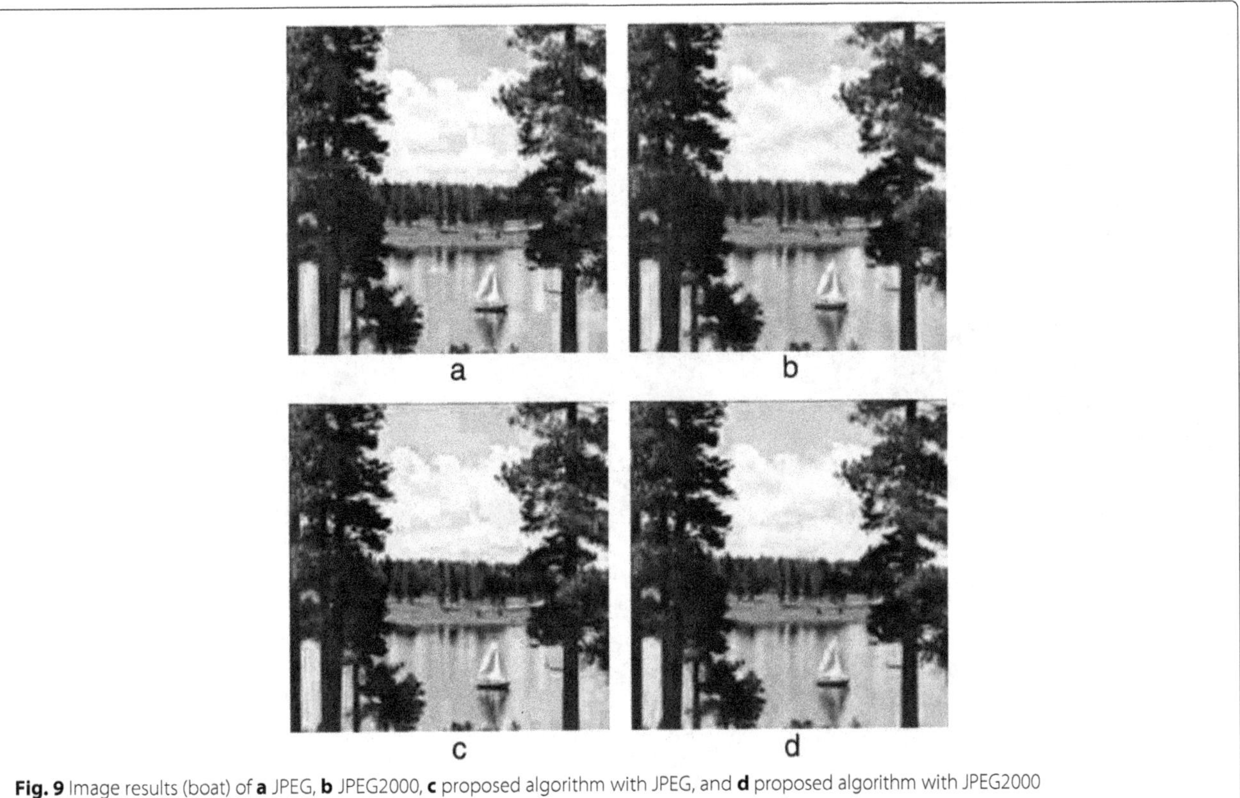

Fig. 9 Image results (boat) of **a** JPEG, **b** JPEG2000, **c** proposed algorithm with JPEG, and **d** proposed algorithm with JPEG2000

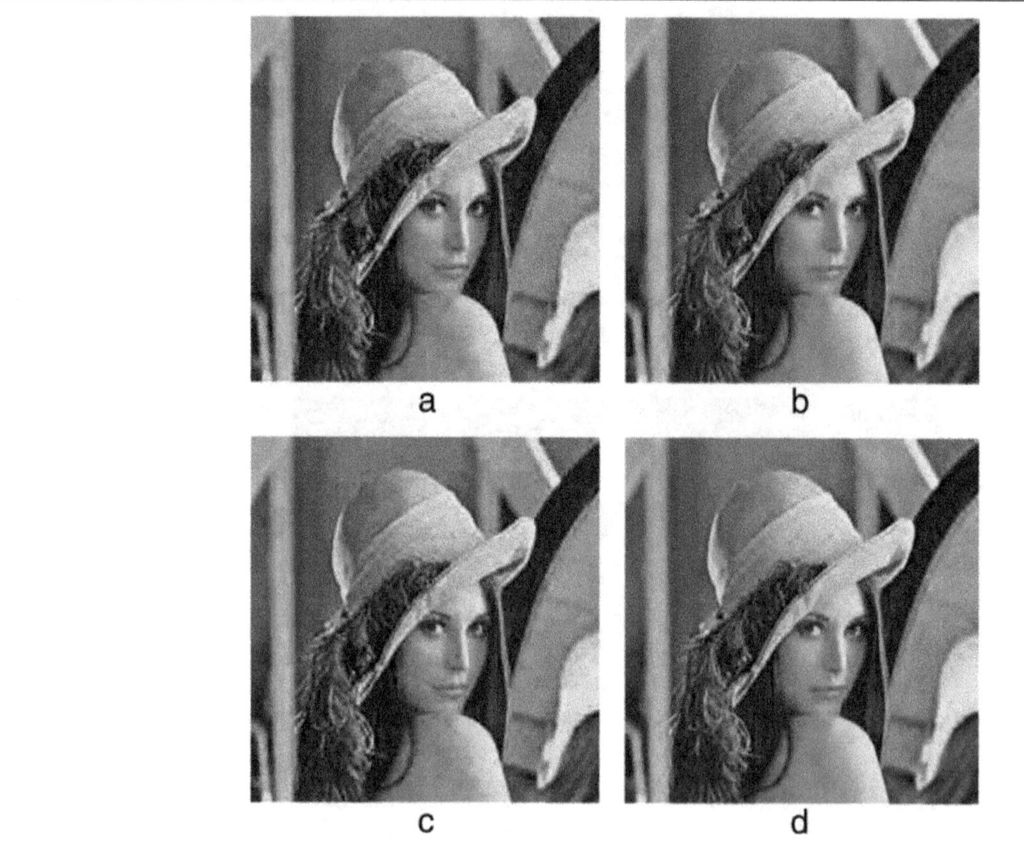

Fig. 10 Image results (lenna) of **a** JPEG, **b** JPEG2000, **c** proposed algorithm with JPEG, and **d** proposed algorithm with JPEG2000

algorithm requires 0.7–0.8 [s] for encoding and 0.2–0.3 [s] for decoding in MATLAB 2014a on a PC with an Intel Core i7 3.4 GHz CPU, 8 GB of RAM memory.

4 Conclusions

This paper proposes a representative pixel (RP) extraction and colorization algorithm for the colorization-based digital image coding. In order to achieve low computing cost and high image coding performance, multiple reduced images are generated by colorization error minimizing method, and the RPs are extracted from these reduced images. Numerical examples show that the proposed algorithm can extract RPs and recover a color image fast and effectively

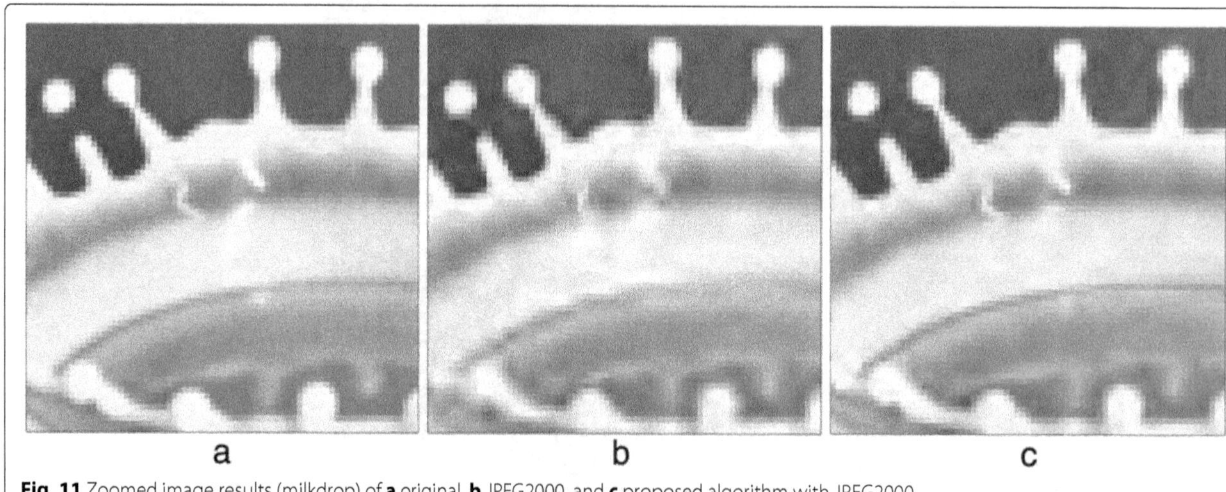

Fig. 11 Zoomed image results (milkdrop) of **a** original, **b** JPEG2000, and **c** proposed algorithm with JPEG2000

comparing with other colorization-based algorithms, and numerical results show that the proposed algorithm achieves higher coding performance than JPEG/JPEG2000.

Competing interests
The authors declare that they have no competing interests.

Acknowledgements
We are grateful to Sukho Lee and his co-authors for providing the source code of the colorization-based coding proposed in [12].
This work was supported by JSPS KAKENHI Grant Numbers 26·6546.

Author details
[1]Tokyo University of Science, 162-8677 Tokyo, Japan. [2]Kogakuin University, Tokyo, Japan.

References
1. A Levin, D Lischinski, Y Weiss, Colorization using optimization. ACM Trans. Graph. **23**(3), 689–694 (2004)
2. L Yatziv, G Sapiro, Fast image and video colorization using chrominance blending. IEEE Trans. Image Process. **15**(5), 1120–1129 (2006)
3. J Pang, OC Au, K Tang, Y Guo, in *Proc. IEEE Int. Conf. Acoust., Speech, Signal Process. (ICASSP)*. Image colorization using sparse representation, (2013), pp. 1578–1582
4. K Uruma, K Konishi, T Takahashi, T Furukawa, in *Proc. IEEE Int. Conf. Acoust., Speech, Signal Process. (ICASSP)*. Image colorization algorithm using series approximated sparse function, (2013), pp. 1215–1219
5. T Miyata, Y Komiyama, Y Inazumi, Y Sakai, in *Proc. Picture Coding Symp.* Novel inverse colorization for image compression, (2009), pp. 1–4
6. H Noda, N Takao, M Niimi, in *Proc. IEEE Int. Conf. Image Process. (ICIP)*. Colorization in YCbCr space and its application to improve quality of JPEG Color Images, (2007), pp. 385–388
7. L Cheng, S Vishwanathan, in *Proc. Int. Conf. Mach. Learn. (ICML)*. Learning to compress images and videos, (2007), pp. 161–168
8. X He, M Ji, H Bao, in *Proc. IEEE Int. Conf. Comput. Vis. Pattern Recognit. (CVPR)*. A unified active and semi-supervised learning framework for image Compression, (2009), pp. 65–72
9. S Ono, T Miyata, Y Sakai, in *Proc. Picture Coding Symp.* Colorization-based coding by focusing on characteristics of colorization bases, (2010), pp. 230–233
10. Y Inoue, T Miyata, Y Sakai, Colorization based image coding by using local correlation between luminance and chrominance. IEICE Trans. Inf. Syst. **95**(1), 247–255 (2012)
11. T Ueno, T Yoshida, M Ikehara, *Color image coding based on the colorization*, Asia-Pacific Signal and Information Processing Association Annual Summit and Conference (APSIPA ASC), (2012), pp. 1–4
12. S Lee, S Park, P Oh, M Kang, Colorization-based compression using optimization. IEEE Trans. Image Proc. **22**(7), 2627–2636 (2013)
13. EJ Candes, J Romberg, T Tao, Stable signal recovery for incomplete and inaccurate measurements. Commun. Pure Appl. Math. **59**(3), 1207–1223 (2013)
14. DL Donoho, Compressed sensing. IEEE Trans. Inf. Theory. **52**, 1289–1306 (2006)
15. C Rusu, SA Tsaftaris, in *Proc. Int. Symp. Image and Signal Processing and Analysis (ISPA)*. Estimation of scribble placement for painting colorization, (2013), pp. 564–569
16. Z Wang, AC Bovik, HR Sheikh, EP Simoncelli, Image quality assessment: from error visibility to structural similarity. IEEE Trans. Image Proc. **13**(4), 600–612 (2004)

15

Comparison of ALBAYZIN query-by-example spoken term detection 2012 and 2014 evaluations

Javier Tejedor[1*], Doroteo T. Toledano[2], Paula Lopez-Otero[3], Laura Docio-Fernandez[3] and Carmen Garcia-Mateo[3]

Abstract

Query-by-example spoken term detection (QbE STD) aims at retrieving data from a speech repository given an acoustic query containing the term of interest as input. Nowadays, it is receiving much interest due to the large volume of multimedia information. This paper presents the systems submitted to the ALBAYZIN QbE STD 2014 evaluation held as a part of the ALBAYZIN 2014 Evaluation campaign within the context of the IberSPEECH 2014 conference. This is the second QbE STD evaluation in Spanish, which allows us to evaluate the progress in this technology for this language. The evaluation consists in retrieving the speech files that contain the input queries, indicating the start and end times where the input queries were found, along with a score value that reflects the confidence given to the detection of the query. Evaluation is conducted on a Spanish spontaneous speech database containing a set of talks from workshops, which amount to about 7 h of speech. We present the database, the evaluation metric, the systems submitted to the evaluation, the results, and compare this second evaluation with the first ALBAYZIN QbE STD evaluation held in 2012. Four different research groups took part in the evaluations held in 2012 and 2014. In 2014, new multi-word and foreign queries were added to the single-word and in-language queries used in 2012. Systems submitted to the second evaluation are hybrid systems which integrate letter transcription- and template matching-based systems. Despite the significant improvement obtained by the systems submitted to this second evaluation compared to those of the first evaluation, results still show the difficulty of this task and indicate that there is still room for improvement.

Keywords: Query-by-example spoken term detection, International evaluation, Search on spontaneous speech

Introduction

The ever-increasing volume of heterogeneous speech data stored in audio and audiovisual repositories promotes the development of efficient methods for retrieving such information. Significant research has been conducted on spoken document retrieval (SDR), keyword spotting (KWS), spoken term detection (STD), query-by-example (QbE), or spoken query approaches to address this issue. STD aims at finding individual words or sequences of words within audio archives. It usually relies on a text-based input, commonly the word/phone transcription of the search term. For this reason, STD is also called text-based STD. STD systems are typically composed of three different stages: (1) the audio is decoded in terms of word/subword lattices using an automatic speech recognition (ASR) subsystem, (2) a term detection subsystem searches the terms within those word/subword lattices and hypothesizes detections, and (3) confidence measures are applied to output reliable detections.

QbE can be defined as 'a method of searching for an example of an object or a part of it in other objects'. This has been widely used in audio applications such as sound classification [1–3], music information retrieval [4, 5], and spoken document retrieval [6]. In QbE STD, we consider the scenario in which the user has found some interesting data within a speech data repository and his/her purpose is to find similar data within that repository. The interesting data consist of one or several speech segments containing the term of interest (henceforth,

*Correspondence: javier.tejedor@depeca.uah.es
[1] GEINTRA, Universidad de Alcalá, Campus Universitario. Ctra. Madrid-Barcelona, km.33,600, Alcalá de Henares, Madrid, Spain
Full list of author information is available at the end of the article

query) and the system outputs other putative hits from the repository (henceforth, utterances). Alternatively, the term of interest can be uttered by the user. Using speech queries offers a big advantage for devices with limited text-based capabilities, which can be effectively used under the QbE STD paradigm. Other advantage is that QbE STD can be employed for building language-independent STD systems [7–10], since prior knowledge of the language involved in the speech data is not necessary.

QbE STD has been mainly addressed in the literature from two perspectives:

- Methods based on word/subword transcription of the query [11–17], so that the text-based STD technology can be applied.
- Methods based on template matching of features extracted from the query and speech repository [9–11, 15, 18–27]. These usually borrow the idea from dynamic time warping (DTW)-based speech recognition and were found to outperform subword transcription-based techniques in QbE STD [28].

Recently, hybrid systems combining both methods have also been proposed [29–33].

Unsupervised spoken term detection techniques, which aim at automatically discovering acoustic patterns (e.g., for training acoustic models) for languages for which manual transcriptions and linguistic knowledge are scarce, have been also investigated [34, 35]. These techniques can also be employed for building language-independent QbE STD systems, since prior knowledge of the language is not necessary.

Recently, several evaluations including SDR, STD, and QbE STD have been held [36–40]. We organized the second international ALBAYZIN QbE STD evaluation in the context of the ALBAYZIN 2014 Evaluation campaign. These campaigns are internationally open sets of evaluations supported by the Spanish Network of Speech Technologies (RTTH)[1] and the ISCA Special Interest Group on Iberian Languages (SIG-IL)[2] held every 2 years from 2006. The evaluation campaigns provide an objective mechanism to compare different systems and promote research in different speech technologies such as audio segmentation [41], speaker diarization [42], language recognition [43], query-by-example spoken term detection [44], and speech synthesis [45].

Spanish is a major language in the world and significant research has been conducted on it for ASR, KWS, and STD tasks [46–52]. In 2012, the first QbE STD evaluation dealing with Spanish was organized in the context of the ALBAYZIN 2012 Evaluation campaign. The success of this first evaluation [44] encouraged us to organize a new QbE STD evaluation for the ALBAYZIN 2014 Evaluation campaign aiming at evaluating the progress in this technology for Spanish. The second ALBAYZIN QbE STD evaluation incorporates new and more difficult queries (i.e., multi-word and foreign queries). In addition, all the queries from the first evaluation are kept so that a comparison between the systems submitted to both evaluations is possible. This paper presents the systems submitted to the ALBAYZIN QbE STD 2014 evaluation and makes a comparison with the systems submitted to the ALBAYZIN QbE STD 2012 evaluation.

The rest of the paper is organized as follows: The next section presents a description of the QbE STD evaluations held in 2012 and 2014. Section 3 presents the different systems submitted to the evaluations. Results along with discussion are presented next, and the work is concluded in the last section.

ALBAYZIN QbE STD evaluations

The ALBAYZIN QbE STD 2012 and 2014 evaluations involve searching for audio content within audio content using an audio content query. These evaluations are suitable for research groups working on speech indexing and retrieval and on speech recognition. The input to the system is an acoustic example per query, and hence prior knowledge of the correct word/subword transcription corresponding to each query is not available.

The evaluations consist in searching for a test query list within test speech data. Participants were provided with a training/development (train/dev) query list and speech data that can be used for system training and tuning, though any additional data can also be employed, as long as it is properly described in the system description.

Participants could submit a primary system and several contrastive systems. No manual intervention is allowed to generate the final output file, and hence all systems must be fully automatic. Listening to the test data, or any other human interaction with the test data is forbidden before all the evaluation results on test data have been sent to the participants. The standard XML-based format corresponding to the NIST STD 2006 evaluation [53] has been used for building the system output file. For both evaluations, about 3 months were given to the participants for system design. Train/dev data (i.e., train/dev speech data, train/dev query list, train/dev ground-truth labels, orthographic transcription and timestamps for phrase boundaries in the train/dev speech data, and evaluation tools) were released at the end of June 2012/2014, test data (i.e., test speech data and test query list) were released at the beginning of September 2012/2014, and the final system submission was due at the end of September 2012/2014. Final results were presented and discussed at IberSPEECH 2012 and IberSPEECH 2014 conferences at the end of November 2012[3]/2014[4].

Evaluation metric

In QbE STD, a hypothesized occurrence is called a *detection*; if the detection corresponds to an actual occurrence, it is called a *hit*, otherwise it is a *false alarm* (FA). An actual occurrence that is not detected is called a *miss*. The Actual Term Weighted Value (ATWV) proposed by NIST [53] for STD has been used as the main metric for the evaluations. This metric integrates the hit rate and false alarm rate of each query into a single metric and then averages over all the queries:

$$\text{ATWV} = \frac{1}{|\Delta|} \sum_{K \in \Delta} \left(\frac{N_{\text{hit}}^{K}}{N_{\text{true}}^{K}} - \beta \frac{N_{FA}^{K}}{T - N_{\text{true}}^{K}} \right) \quad (1)$$

where Δ denotes the set of queries and $|\Delta|$ is the number of queries in this set. N_{hit}^{K} and N_{FA}^{K} represent the numbers of hits and false alarms of query K, respectively, and N_{true}^{K} is the number of actual occurrences of K in the audio. T denotes the audio length in seconds, and β is a weight factor set to 999.9, as in the ATWV proposed by NIST [54]. This weight factor causes an emphasis placed on recall compared to precision in the ratio 10:1.

ATWV represents the TWV for the threshold set by the system (usually tuned on development data). An additional metric, called Maximum Term Weighted Value (MTWV) [53] has also been considered. It is the maximum TWV achieved by the system for all possible thresholds and hence does not depend on the tuned threshold. This MTWV represents the performance of the system if the threshold was perfectly set. Results based on this metric are presented to evaluate the goodness of threshold selection.

In addition to ATWV and MTWV, NIST also proposed a detection error tradeoff (DET) curve [55] to evaluate the performance of a QbE STD system working at various miss/FA ratios. DET curves are also presented in this paper for system comparison.

Database

The database used for ALBAYZIN QbE STD 2012 and 2014 evaluations consists of a set of talks extracted from the MAVIR workshops[5] held in 2006, 2007, and 2008 (corpus MAVIR 2006, 2007, and 2008) that contain speakers from Spain and Latin America (henceforth MAVIR corpus or database). The MAVIR corpus contains 3 recordings in English and 10 recordings in Spanish, but only the recordings in Spanish were used for the evaluations. The MAVIR Spanish data consist of spontaneous speech files, each containing different speakers, which amount to about 7 h of speech, and are further divided for the purpose of these evaluations into train/dev and test sets. There are 20 male and 3 female speakers in the MAVIR Spanish database.

The speech data were originally recorded in several audio formats (pulse-code modulation (PCM) mono and stereo, MP3, 22.05 KHz., 48 KHz., etc.). All data were converted to PCM, 16 KHz., single channel, 16 bits per sample using SoX tool[6]. Recordings were made with the same equipment, a Digital TASCAM DAT model DA-P1, except for one recording. Different microphones were used for the different recordings. They mainly consisted of tabletop or floor standing microphones, but in one case a lavalier microphone was used. The distance from the mouth of the speaker to the microphone varies and was not particularly controlled, but in most cases the distance was smaller than 50 cm. All the speech contain real and spontaneous speech of MAVIR workshops in a real setting. Thus, the recordings were made in large conference rooms with capacity for over a hundred people and a large amount of people in the conference room. This poses additional challenges including background noise (particularly babble noise) and reverberation. The realistic settings and the different nature of the spontaneous speech in this database make it appealing and challenging enough for ALBAYZIN QbE STD evaluations and definitely for further work. The speech data were manually annotated in an orthographic form, but timestamps were only set for phrase boundaries. To prepare the data used for the evaluations, organizers manually added the timestamps for all the occurrences of the train/dev and test search queries. Table 1 includes some database features such as the number of word occurrences, duration, and signal-to-noise ratio (SNR) [56] of each speech file in the MAVIR Spanish database.

Given that the database used in these evaluations consists of spontaneous speech, there is an inherent difficulty for query detection. In addition, QbE STD and, in general, ASR system performance significantly degrades when training data belong to a different domain or pose different acoustic conditions to those of the test data. To alleviate this problem in the ALBAYZIN QbE STD evaluations, organizers provided limited train/dev data from the same domain and with *similar* acoustic conditions (microphone speech from workshops) as the test data. However, it must be noted that different microphones and even different rooms were used for each recorded file in the MAVIR database, and hence, the acoustic conditions can vary significantly from one file to another.

Train/dev data amount to about 5 h of speech extracted from 7 out of the 10 speech files of the MAVIR Spanish database and contain 15 male and 2 female speakers. For the evaluation held in 2012, the train/dev query list consisted of 60 queries. All of them were single words with lengths between 7 and 16 graphemes. There are 1027 occurrences of those queries in the train/dev speech data. For the evaluation held in 2014, the train/dev

Table 1 MAVIR database characteristics

File ID	Dataset	# word occ.	Duration (min)	# speakers	SNR (dB)
Mavir-02	Train/dev	13,432	74.51	7 male	2.1
Mavir-03	Train/dev	6681	38.18	1 male, 1 fem.	15.8
Mavir-06	Train/dev	4332	29.15	2 male, 1 fem.	12.0
Mavir-07	Train/dev	3831	21.78	2 male	10.6
Mavir-08	Train/dev	3356	18.90	1 male	7.5
Mavir-09	Train/dev	11,179	70.05	1 male	12.3
Mavir-12	Train/dev	11,168	67.66	1 male	11.1
Mavir-04	Test	9310	57.36	3 male, 1 fem.	10.2
Mavir-11	Test	3130	20.33	1 male	9.2
Mavir-13	Test	7837	43.61	1 male	11.1
All	Train/dev	53,979	320.23	15 male and 2 fem.	–
All	Test	20,277	121.3	5 male and 1 fem.	–

occ. occurrences, *min* minutes, *fem.* female, *SNR* signal-to-noise ratio, *dB* decibels

query list consists of 94 queries. Each query is composed of one or more words containing between 5 and 18 graphemes. There are 1415 occurrences of those queries in the train/dev speech data. Tables 2 and 3 include information related to the train/dev queries used in both 2012 and 2014 evaluations and the train/dev queries used only in the 2014 evaluation, respectively.

Test data amount to about 2 h of speech extracted from the other 3 speech files not included in train/dev data and contain 5 male and 1 female speakers. For the evaluation held in 2012, the test query list consisted of 60 queries. All of them were single words containing between 7 and 16 graphemes. There are 892 occurrences of those queries in the test speech data. For the evaluation held in 2014, the

Table 2 Training/development queries used in both 2012 and 2014 evaluations. Each cell indicates query text, time length per query (in milliseconds), and number of occurrences per query

Query (time)–(# occ.)	Query (time)–(# occ.)	Query (time)–(# occ.)
Académico (500)–(10)	Gallego (300)–(7)	Cuestión (260)–(8)
Acceder (350)–(7)	General (350)–(43)	Cultural (790)–(10)
Administración (550)–(27)	Indexación (640)–(10)	Desarrollo (750)–(15)
Arquitectura (610)–(8)	Industria (390)–(6)	Después (280)–(38)
Barcelona (670)–(8)	Información (570)–(153)	Directamente (450)–(16)
Cálculo (440)–(6)	Instituto (370)–(22)	Establecer (550)–(8)
Calidad (550)–(33)	Investigación (740)–(52)	Estructura (540)–(13)
Capacidad (670)–(12)	Latinoamérica (690)–(8)	Euskera (530)–(10)
Capital (500)–(11)	Máquina (510)–(8)	Formato (430)–(7)
Castellano (670)–(21)	Ministerio (310)–(9)	Francia (560)–(6)
Catalogación (750)–(6)	Momento (370)–(50)	Sentido (380)–(24)
Cataluña (440)–(11)	Nacional (770)–(7)	Situación (690)–(24)
Cervantes (420)–(25)	Negocio (490)–(18)	Soporte (330)–(6)
Clasificación (620)–(13)	Patrimonio (670)–(7)	Telefónica (540)–(21)
Comentario (540)–(14)	Pequeño (320)–(8)	Todavía (330)–(16)
Compañía (360)–(6)	Validación (520)–(7)	Publicidad (650)–(13)
Picasso (270)–(21)	Conjunto (340)–(16)	Visibilidad (730)–(8)
Trabajo (320)–(36)	Proceso (420)–(13)	Contabilidad (1090)–(7)
Computadora (740)–(12)	Virtual (570)–(12)	Referencia (530)–(9)
Potencial (470)–(13)	Conocimiento (560)–(6)	Volumen (300)–(6)

Table 3 New training/development queries used in the 2014 evaluation and not in the 2012 evaluation, time length per query (in milliseconds), and number of occurrences per query

Query (time)–(# occ.)	Query (time)–(# occ.)
Presentación (760)–(17)	Portugal (340)–(4)
Vosotros (360)–(6)	Parlamento (360)–(3)
Etcétera (490)–(28)	Microsoft (580)–(4)
Empresas (820)–(71)	Mavir (420)–(2)
Porcentaje (490)–(6)	Málaga (450)–(2)
Experimentos (400)–(10)	Isabel (310)–(4)
Noventa (630)–(39)	Garner (320)–(3)
Atenclón (280)–(8)	Galicia (520)–(4)
Mercado (510)–(111)	Erasmus (430)–(2)
Resolver (500)–(8)	Dilbert (640)–(2)
Probablemente (490)–(6)	Complutense (460)–(4)
Dominios (370)–(17)	Cristian (510)–(2)
Wikipedia (670)–(3)	Berrilan (430)–(2)
Webmaster (460)–(2)	Aguilera (480)–(3)
Valladolid (530)–(2)	Premios nobel (650)–(2)
Sevilla (450)–(2)	Universidad de Chile (840)–(3)
Profit (330)–(3)	Nick cohn (410)–(3)

test query list consists of 99 queries. Each query is composed of one or more words containing between 6 and 16 graphemes. There are 1162 occurrences of those queries in the test speech data. Tables 4 and 5 include information related to the test queries used in both 2012 and 2014 evaluations and the test queries used only in the 2014 evaluation, respectively.

Each train/dev query has one or more occurrences in the train/dev speech data and each test query has one or more occurrences in the test speech data. All these queries were extracted from the MAVIR database.

Comparison with other QbE STD evaluations

The most similar evaluations to ALBAYZIN QbE STD evaluations are the MediaEval 2011, 2012, and 2013 Spoken Web Search [38, 57, 58]. The task to be performed in MediaEval and ALBAYZIN evaluations is the same, but these differ in several aspects. This makes it difficult to compare the results obtained in ALBAYZIN evaluations to previous MediaEval evaluations.

The most important difference is the nature of the audio content used for the evaluations. In MediaEval evaluations, the speech is typically telephone speech, either conversational or read and elicited speech, or speech recorded with in-room microphones. In ALBAYZIN evaluations, the audio contains microphone recordings of real talks in real workshops, in large conference rooms with the public. Microphones, conference rooms, and even recording conditions change from one recording to another. Microphones are not close talking microphones but table top and floor standing microphones mainly.

In addition, the MediaEval evaluations deal with Indian and African-derived languages, along with Albanian, Basque, Czech, non-native English, Romanian, and Slovak languages, while ALBAYZIN evaluations deal with Spanish.

Besides MediaEval evaluations, a new QbE STD evaluation has been organized within NTCIR-11 conference [59]. Data used in this evaluation contained spontaneous speech in Japanese provided by the National Institute for Japanese language as well as spontaneous speech recorded during seven editions of the Spoken Document Processing Workshop. As additional information, this evaluation provides participants with the results of a voice activity detection system on the input speech data, the manual transcription of the speech data, and the output of a Large Vocabulary Continuous Speech Recognition (LVCSR) system. Although ALBAYZIN evaluations could be similar in terms of speech nature to this NTCIR QbE STD evaluation (speech recorded in real workshops), ALBAYZIN evaluations do not provide any kind of information apart from the speech content, the list of queries, and the train/dev ground-truth files to participants. In addition, ALBAYZIN evaluations make use of other language and define disjoint train/dev and test query lists to measure the generalization capability of the systems.

Table 6 summarizes the main characteristics of the MediaEval QbE STD evaluations, the NTCIR-11 QbE STD evaluation, and the ALBAYZIN QbE STD evaluations held in 2012 and 2014.

Systems

Four research groups, listed in Table 7, took part in the ALBAYZIN QbE STD evaluations held in 2012 and 2014. Four systems were submitted to the ALBAYZIN QbE STD 2014 evaluation. Two of them (those based on deep neural networks (DNN)) were post-evaluation submissions. In addition, two text-based STD systems (the DNN-based system was a post-evaluation submission) were also submitted to compare the QbE STD systems with other technology that also aims at searching for terms within speech data. For the ALBAYZIN QbE STD 2012 evaluation, four different systems were submitted. Table 8 summarizes the main characteristics of these QbE STD systems, which are further described next along with the text-based STD system.

ALBAYZIN QbE STD 2014 evaluation: Fusion (SGMM)+Posteriorgram system

This system consists of the fusion of three different subsystems, as shown in Fig. 1: a large vocabulary continuous speech recognition system, a dynamic time warping

Table 4 Test queries used in both 2012 and 2014 evaluations. Each cell indicates query text, time length per query (in milliseconds), and number of occurrences per query

Query (tme)–(# occ.)	Query (time)–(# occ.)	Query (time)–(# occ.)
Acuerdo (290)–(7)	Lenguaje (390)–(6)	Referencia (470)–(13)
Análisis (370)–(18)	Mecanismo (470)–(7)	Fuenlabrada (570)–(15)
Aproximación (850)–(7)	Metodología (810)–(10)	General (420)–(11)
Buscador (580)–(7)	Motores (340)–(6)	Gracias (400)–(13)
Cangrejo (490)–(7)	Necesario (650)–(6)	Idiomas (290)–(27)
Castellano (570)–(9)	Normalmente (320)–(6)	Implicación (600)–(31)
Conjunto (490)–(7)	Obtener (380)–(9)	Importante (680)–(19)
Conocimiento (490)–(6)	Orientación (600)–(6)	Incluso (410)–(12)
Desarrollo (460)–(6)	Parecido (400)–(6)	Información (560)–(92)
Detalle (280)–(7)	Personas (540)–(6)	Intentar (420)–(13)
Difícil (410)–(12)	Perspectiva (490)–(7)	Interfaz (480)–(10)
Distintos (450)–(21)	Porcentaje (660)–(8)	Resolver (420)–(6)
Documentos (750)–(7)	Precisamente (680)–(6)	Segunda (520)–(8)
Efectivamente (290)–(10)	Presentación (580)–(15)	Seguridad (350)–(6)
Ejemplo (550)–(54)	Primera (290)–(19)	Siguiente (370)–(11)
Empezar (340)–(7)	También (240)–(93)	Reconocimiento (660)–(6)
Principio (480)–(9)	Entidades (670)–(28)	Trabajar (380)–(39)
Simplemente (650)–(8)	Realidad (270)–(10)	Evaluación (480)–(15)
Encontrar (350)–(19)	Textual (590)–(15)	Recurso (520)–(7)
Propuesta (440)–(19)	Estudiar (500)–(7)	Utilizar (500)–(15)

search-based system with a fingerprint representation of the queries and the utterances, and a DTW search-based system with phoneme posterior probabilities for query and utterance representation. The three subsystems are described in the following sections.

Kaldi LVCSR-based QbE STD system

The architecture of the Kaldi LVCSR-based QbE STD system is shown in Fig. 2. First, an LVCSR system was built using the Kaldi open-source toolkit [60]. Thirteen-dimensional perceptual linear prediction (PLP)

Table 5 New test queries used in the 2014 evaluation and not in the 2012 evaluation, time length per query (in milliseconds), and number of occurrences per query

Query (time)–(# occ.)	Query (time)–(# occ.)	Query (time)–(# occ.)
Académico (690)–(6)	Investigación (520)–(15)	Formularios (520)–(19)
Anselmo (440)–(2)	Madrid (410)–(6)	Transparencia (500)–(6)
Antonio moreno (480)–(2)	Manuel (240)–(6)	Francisco garcía (930)–(2)
Autónoma (690)–(8)	Mencionado (500)–(6)	Unesco (390)–(5)
Bastante (590)–(22)	Objetivos (650)–(9)	Fundamentalmente (900)–(7)
Cindoc (690)–(2)	Obviamente (550)–(10)	Universidades (790)–(13)
Coca cola (470)–(2)	Paloma (310)–(3)	Harvard (540)–(2)
Completamente (450)–(13)	Programa (420)–(11)	Validación (500)–(18)
Consorcio mavir (820)–(2)	Scholar (390)–(2)	Huelva (250)–(2)
Daedalus (660)–(6)	Setenta (430)–(6)	Vicente fox (730)–(2)
Embargo (340)–(7)	Solamente (620)–(10)	Inicial (590)–(8)
Evidentemente (660)–(7)	Solución (560)–(7)	Zagreb (460)–(2)
Felisa (310)–(2)	Soporte (600)–(89)	Internet (410)–(4)

Table 6 QbE STD evaluation characteristics and languages: Albanian ('ALB'), Basque ('BAS'), Czech ('CZE'), non-native English ('NN-ENG'), Isixhosa ('ISIX'), Isizulu ('ISIZ'), Romanian ('ROM'), Sepedi ('SEP'), Setswana ('SET'), and Slovak ('SLO')

Evaluation	Language/s	Type of speech	# queries dev./test	Metric
MediaEval 2011	English, Hindi, Gujarati, and Telugu	Tel.	64/36	ATWV
MediaEval 2012	2011 + isiNdebele, Siswati, Tshivenda, and Xitsonga	Tel.	164/136	ATWV
MediaEval 2013	ALB, BAS, CZE, NN-ENG, ISIX, ISIZ, ROM, SEP, and SET	Tel. and mic.	>600/>600	ATWV
MediaEval 2014	ALB, BAS, CZE, NN-ENG, ROM, and SLO	Tel. and mic.	560/555	C_{nxe}
NTCIR-11 2014	Japanese	mic. workshop	63/203	F-measure
ALBAYZIN 2012	Spanish	mic. workshop	60/60	ATWV
ALBAYZIN 2014	Spanish	mic. workshop	94/99	ATWV

Tel. telephone, *mic.* microphone, *dev.* development, C_{nxe} normalized cross entropy cost

coefficients augmented with delta and double delta coefficients were used to build 39-dimensional feature vectors used as acoustic features. Next, a state-of-the-art maximum likelihood (ML) acoustic model training strategy was employed. This training starts with a flat-start initialization of context-independent phonetic Hidden Markov Models (HMMs) and ends with speaker adaptive training (SAT) of state-clustered triphone HMMs with Gaussian mixture model (GMM) output densities. After the ML-based acoustic model training stage, a universal background model (UBM) is built from speaker-transformed

Table 7 Participants in the ALBAYZIN QbE STD 2012 and 2014 evaluations along with their submitted systems

Team ID	Research institution	Year	System/s
TID	Telefonica Research, Barcelona, Spain	2012	DTW-Zero
GTTS	University of the Basque Country, Bilbao, Spain	2012	P1B-STD
ELiRF	Politechnical University of Valencia, Valencia, Spain	2012	P1L-STD DTW-Spanish
GTM	AtlantTIC Research Center, University of Vigo, Vigo, Spain	2014	Fusion+Post. Fusion

Post. posteriorgram

Table 8 Main characteristics of the ALBAYZIN QbE STD 2012 and 2014 evaluation systems

System ID	System type	Language-dependent
Fusion (SGMM)+Posteriorgram	Fusion: 1 SGMM-based LVCSR and 2 template matching	YES
Fusion (SGMM)	Fusion: 1 SGMM-based LVCSR and 1 template matching	YES
Fusion (DNN)+Posteriorgram	Fusion: 1 DNN-based LVCSR and 2 template matching	YES
Fusion (DNN)	Fusion: 1 DNN-based LVCSR and 1 template matching	YES
DTW-Zero	Template matching	NO
P1B-STD	Phone transcription	NO
P1L-STD	Phone transcription	YES
DTW-Spanish	Template matching	YES

LVCSR Large Vocabulary Continuous Speech Recognition

training data, which is next used to train a Subspace GMM (SGMM) employed in the decoding stage to generate word lattices and word sequences.

The acoustic models were trained using the Spanish data from 2006 TC-STAR ASR evaluation campaign[7]. Specifically, the training data from the European Parliamentary plenary sessions and the Spanish Parliament sessions, which were manually transcribed, were used for acoustic model training [61]. All the non-speech parts, the speech parts corresponding to transcriptions with pronunciation errors, incomplete sentences, and short speech utterances from the speech data were discarded. After this, the training data amount to about 79 h of speech.

The language model (LM) was trained using a text database of 160 million words extracted from several sources: transcriptions of European and Spanish Parliaments of the TC-STAR database, subtitles, books, newspapers, online courses, and the transcriptions of the MAVIR sessions included in the train/dev data provided by the organizers[8]. For development experiments, a different LM was created for each MAVIR session, using the transcription of the session to obtain the optimum mixture of the partial LMs. The LM and the corresponding vocabulary created from all the train/dev data files except one were then used to compute the query detections of that file in a leave-one-out strategy. For the test data, the LM was generated using a normalized average of the weights obtained from the development sessions. It must be noted that the vocabulary was selected at the last stage of the LM training, once the partial LMs and their weights were computed. A trigram word LM trained with a vocabulary of 60,000 words and a Kneser-Ney discount strategy was

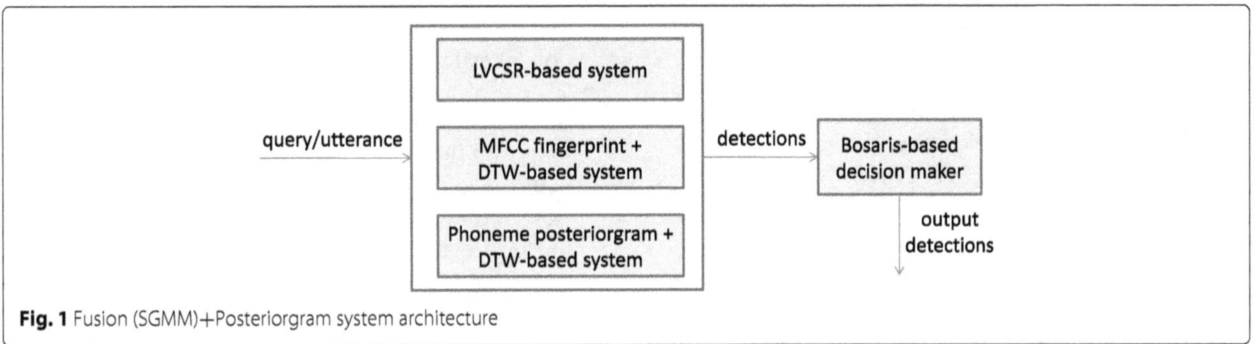

Fig. 1 Fusion (SGMM)+Posteriorgram system architecture

used for decoding. LMs have been built using the SRILM toolkit [62].

The Kaldi-based LVCSR system generates word lattices [63] for the utterances and a word sequence for each query using the SGMM acoustic models and the word-based LM. The word sequence of the query is then used as the word transcription of the query term so that a text-based STD system can be effectively used to hypothesize query detections. To do so, the system integrates the Kaldi term detector [60, 64, 65], which searches for the input queries within those word lattices. The lattice indexing technique, described in [66], first converts the word lattices of all the utterances from individual weighted finite state transducers (WFST) to a single generalized factor transducer structure that stores the start-time, end-time, and the lattice posterior probability of each word token as a 3-dimensional cost. This factor transducer represents an inverted index of all the word sequences contained in the lattices. Thus, given a search query, a simple finite state machine that accepts the input query (from its word sequence) is created and composed with the factor transducer in order to obtain all the occurrences of the query in the utterances. The posterior probabilities of the lattice corresponding to all the words of the input query are accumulated, assigning a confidence score to each detection.

MFCC fingerprint-based QbE STD system

This system employs a Mel-frequency cepstral coefficient (MFCC) fingerprint representation of the queries and utterances [67], and search is performed using a dynamic time warping approach.

An audio fingerprint is a compact representation of an audio signal that extracts the most meaningful information of an audio excerpt. This representation, which is often restricted to binary values, has been used in different tasks [68–70] for several reasons: (1) the storage requirements of fingerprints are relatively small since binary values are employed, (2) the comparison of different fingerprints is efficient since perceptual irrelevancies have been removed, and (3) searching on fingerprint databases is efficient because the searching space is small [71] and almost all the operations are performed in a binary domain.

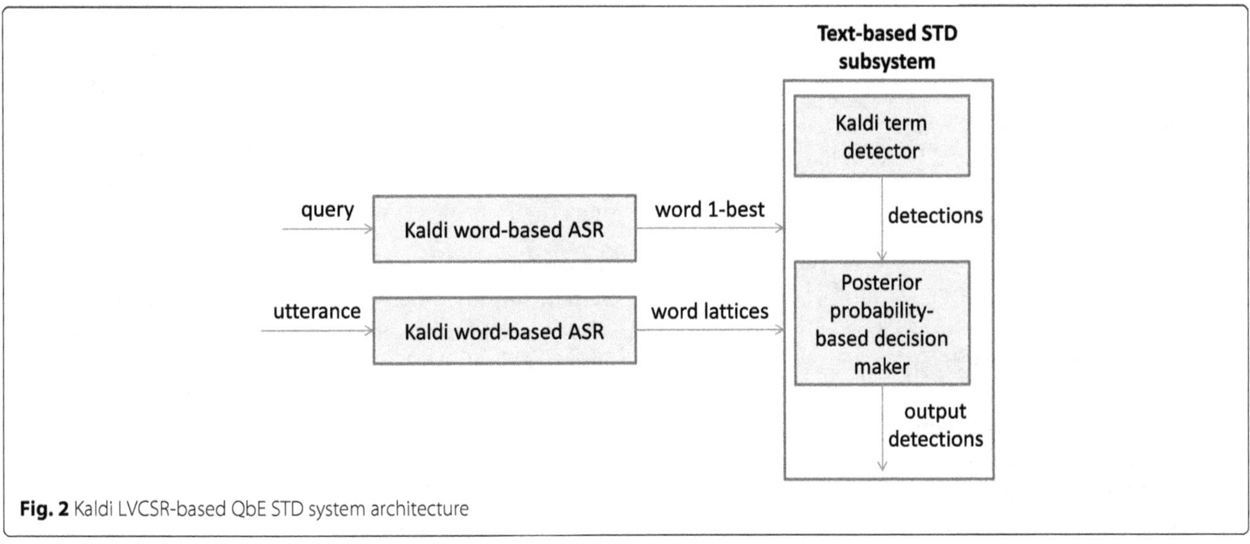

Fig. 2 Kaldi LVCSR-based QbE STD system architecture

The fingerprint extraction comprises two different steps:

- Feature extraction. First, acoustic features are obtained from the audio signal that represents the queries and the utterances. Twelve-dimensional MFCCs augmented with C0 and first and second order derivatives were employed to build a 39-dimensional feature vector. These MFCCs were extracted every 10 ms using a 20-ms sliding window.
- Frame-level fingerprints. The fingerprints corresponding to the acoustic features of each query and utterance frame were obtained from the MFCCs as described in [68]. A convolution mask is used to binarize the acoustic features. Specifically, a mask for finding negative slopes on the spectrogram in two consecutive frames is applied. Given a set of acoustic features $S \in \Re^{I \times J}$, where $S_{i,j}$ is the feature corresponding to energy band i and frame j, the value $F_{i,j}$ of the frame-level fingerprint corresponding to frame j is obtained after applying the convolution mask as follows:

$$F_{i,j} = \begin{cases} 1 \text{ if } \phi > 0 \\ 0 \text{ if } \phi \leq 0 \end{cases} \tag{2}$$

where $\phi = S_{i,j} - S_{i,j+1} + S_{i-1,j} - S_{i-1,j+1}$.

The DTW search was adopted from [29] and comprises four different steps:

- Similarity measure calculation. Euclidean distance between each pair of query frame and utterance frame has been used to compute a matrix that stores the similarity between query and utterance frames. Given a query $Q = \{q_1, \ldots, q_N\}$ and an utterance $S = \{s_1, \ldots, s_M\}$, a similarity matrix $M \in \Re^{N \times M}$ is computed, whose rows and columns correspond to the frames of the query and the utterance, respectively.
- Coarse search. Once the similarity matrix is obtained, a coarse search is carried out to obtain a set of candidate matches. To do so, a sliding window of the size of the query (i.e., N frames) is used with 50 % overlap, and the estimated DTW value of the current window defined by features $(q_{n_i}, \ldots, q_{n_{i+N}})$ is computed as follows:

$$\text{Estimated DTW}(n_i) = \sum_{n=n_i}^{n_{i+N}} \min M_{*,n} \tag{3}$$

where $M_{*,n}$ represents the nth column of the similarity matrix M.
A set of candidates is obtained from this step; specifically, as in [29], the number of candidate detections was selected as the maximum between 100 and the duration of the utterance in seconds. Therefore, those candidate detections that obtain the lowest estimated DTW values are selected.
- Fine search. After selecting the candidate detections in the previous step, an additional DTW search of the query on these candidate positions is conducted. A window of the size of the query is employed for this search, and the actual DTW distance calculated during the DTW search is output.
- Selection of best detections. Once the DTW of the fine search has been carried out, those detections that are separated by at least 0.5 s are kept. Next, the detections whose DTW value is less than a threshold are output as final detections, while the rest are discarded.

Train/dev data were used to train the decision threshold. Next, this threshold was used to hypothesize query detections of the test data.

Phoneme posteriorgram-based QbE STD system

This system is the same as the MFCC fingerprint-based system except that, instead of using MFCC fingerprints for query and utterance representation, phoneme posteriorgrams were employed to build the feature vectors. Specifically, phoneme posteriorgrams [72] were computed for the queries and the utterances by means of a long temporal context-based phoneme recognizer [73]. From all the available phoneme recognizers in [73] (English, Czech, Hungarian, and Russian), the English phoneme recognizer was employed, as it maximized ATWV performance on train/dev data. This phoneme recognizer bases on a TANDEM feature extraction architecture, which merges state phoneme posterior vectors computed from a neural network and PLP coefficients to build the feature vectors. These feature vectors are next fed within a standard GMM/HMM system. Vocal Tract Length Normalization, and speaker based mean and variance normalization are applied in the TANDEM architecture, along with gaussianization and Heteroscedastic Linear Discriminant Analysis for decorrelation and dimensionality reduction. Finally, the same DTW search used in the previous system was employed to hypothesize query detections.

Train/dev data were used to train the decision threshold. Next, this threshold was used to hypothesize query detections of the test data.

System fusion

System fusion combines the output of the three systems described above to produce a more discriminative and better-calibrated score for each detection, aiming at taking advantage of the strengths of the individual approaches [30]. First, a per-query zero-mean and unit-variance normalization (q-norm) was applied in order to prevent the scores of the individual systems to be in different ranges and to obtain query-independent scores. At this point, fusion is not straightforward, as not all the systems to be fused output a score for every possible trial (i.e., detection); hence, before fusion, the detection problem is transformed into a verification problem. To do so, all the time instants of the detections found by the different systems are first merged and next aligned with the ground-truth to obtain a set of client and impostor trials (i.e., hits and false alarms). As not all the systems may have produced a score for each of these trials, missing scores are hypothesized by computing the average of the other scores for that detection [74]. Once every trial (detection) has a score for each system, fusion is carried out using the Bosaris toolkit [75]; specifically, a logistic regression fusion scheme was trained using the training/development data and then applied to the test data. This procedure results in a new score for each detection, which is then used to output the final detections of the Fusion (SGMM)+Posteriorgram system. The overlapped detections, i.e., detections of different queries at the same time interval, were removed by keeping the query detection with the highest score.

ALBAYZIN QbE STD 2014 evaluation: Fusion (SGMM) system

This system, whose architecture is shown in Fig. 3, consists of the fusion of two different subsystems: the Kaldi LVCSR-based system and the DTW search-based system with an MFCC fingerprint representation of the queries and utterances, both described previously. These two systems are again fused with the fusion strategy described in the previous section.

ALBAYZIN QbE STD 2014 evaluation: Fusion (DNN)+ Posteriorgram system

This system is the same as the Fusion (SGMM)+ Posteriorgram system with the only difference of the type of acoustic models used in the Kaldi LVCSR-based system. In this case, DNN-based acoustic models have been employed, instead of SGMMs. Specifically, a DNN-based context-dependent speech recognizer was trained using the Kaldi toolkit [60] following Karel Vesely's DNN training approach [76]. The network has 6 hidden layers, with 2048 units each. The features employed for DNN training and recognition are computed as follows: 9 frames of 13-dimensional MFCCs are projected down to 40 dimensions using linear discriminant analysis (LDA) and the resulting features are further de-correlated using maximum likelihood linear transform (MLLT); this is followed by speaker normalization using feature-space maximum likelihood linear regression (fMLLR).

ALBAYZIN QbE STD 2014 evaluation: Fusion (DNN) system

This system is the same as the Fusion (SGMM) system with the only difference of the type of acoustic models used in the Kaldi LVCSR-based system. In this case, the same DNN-based LVCSR system employed for the Fusion (DNN)+Posteriorgram system has been used.

ALBAYZIN QbE STD 2014 evaluation: text-based SGMM spoken term detection system (text-based SGMM STD)

A text-based STD system has been used to establish a comparison with the results achieved by the QbE STD systems. This text-based STD system, whose architecture is shown in Fig. 4, follows the same approach as the Kaldi LVCSR-based system described previously in the Fusion (SGMM)+Posteriorgram system, with the only difference being that, in this case, the actual transcription of the query term is given to the system, so the query decoding step is not necessary.

ALBAYZIN QbE STD 2014 evaluation: text-based DNN spoken term detection system (text-based DNN STD)

An additional text-based STD system was built from DNNs. This system, whose architecture is the same as the Text-based SGMM STD system, replaces the SGMM-based acoustic models of the Text-based SGMM STD system by the DNN-based acoustic models described in the Fusion (DNN)+Posteriorgram system.

ALBAYZIN QbE STD 2012 evaluation systems

Four different systems were submitted to the evaluation held in 2012, which were already described in the Iber-SPEECH 2012 proceedings [77]. Here, a brief description of their main characteristics is provided:

System DTW-Zero is based on a DTW zero-resource matching approach. First, Gaussian posteriorgram features are used as signal representation, which are obtained

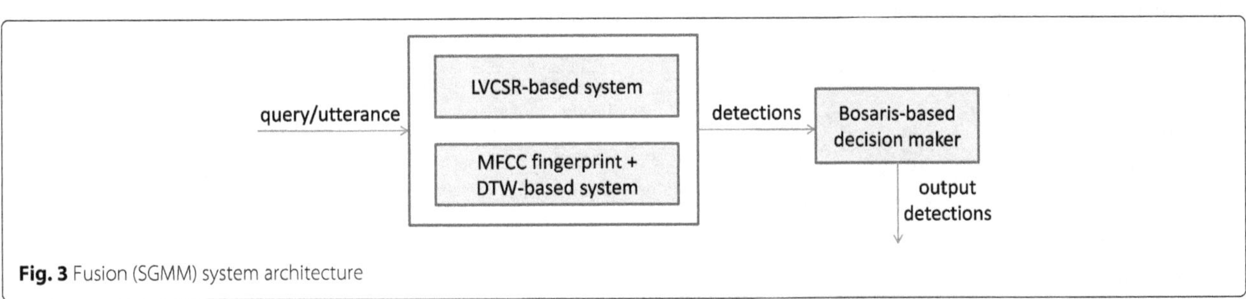

Fig. 3 Fusion (SGMM) system architecture

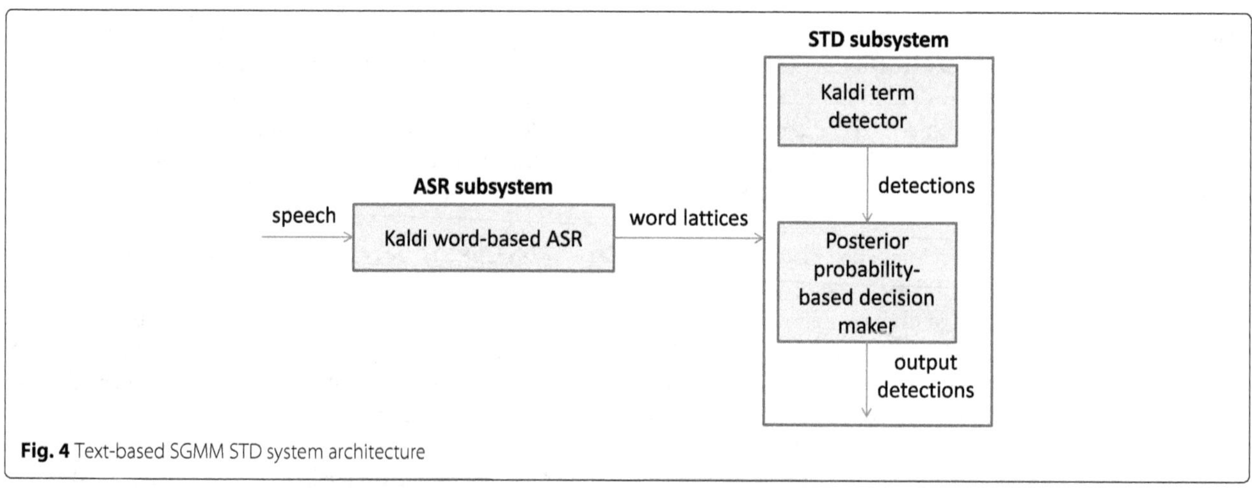

Fig. 4 Text-based SGMM STD system architecture

from a GMM trained from MFCCs. These features are next sent to the subsequence-DTW matching algorithm [19] that employs the cosine distance as similarity measure between query and utterance frames to hypothesize query detections within the utterances.

System P1B-STD is based on an exact match of the phone sequence output by a speech recognizer given the spoken query within the phone lattices corresponding to the utterances. Phone decoders for Czech, Hungarian, and Russian have been employed to produce the phone sequence of each query and the phone lattices. The *Lattice2Multigram* tool [78][9] is used to conduct query search. System P1B-STDa combines the query detections of the Hungarian and Russian phone decoders, and System P1B-STDb combines the query detections of all decoders.

System P1L-STD is based on a search on phone lattices generated from a posteriori phone probabilities. These phone probabilities are obtained by combining the acoustic class probabilities estimated from a GMM-based clustering procedure on the acoustic space and the conditional probabilities of each acoustic class with respect to each phonetic unit [79]. These acoustic classes comprise the set of Spanish phones. Next, the phone probabilities are used in an ASR process to output the phone lattices that represent both the query and the utterance. These query and utterance phone lattices are used for query search, where a search of every path in the query lattice on every path in the utterance lattice is conducted to hypothesize detections. Substitution, insertion, and deletion errors in the lattice matching are allowed. A query-dependent confidence score is assigned to each detection.

System DTW-Spanish employs the same a posteriori phone probabilities as System P1L-STD for query/utterance representation, and these are subsequently used for query search. The query search is based on segmental DTW search [80] with the Kullback-Leibler divergence as similarity measure between query and utterance frames. The same query-dependent confidence score approach used in the P1L-STD system is employed in this system to score each detection.

Results and discussion

ALBAYZIN QbE STD 2014 evaluation

The results of the ALBAYZIN QbE STD 2014 evaluation for train/dev and test data are presented in Tables 9 and 10, respectively. Results show, in general, similar performance for Fusion+Posteriorgram and Fusion systems both for train/dev and test data for SGMM- and DNN-based acoustic models. Paired t tests show that the slight improvement, if any, of the Fusion+Posteriorgram system over the Fusion system is not statistically significant for train/dev and test data ($p \approx 0.3$) for both types of acoustic models. This shows that the phoneme posteriorgram-based system does not pose a complementary behavior

Table 9 Results of the ALBAYZIN QbE STD 2014 evaluation on train/dev data

System ID	MTWV	ATWV	p(FA)	p(Miss)
Fusion (SGMM)+Posteriorgram	0.3023	0.3023	0.00009	0.607
Fusion (SGMM)	0.2957	0.2957	0.00009	0.616
Fusion (DNN)+Posteriorgram	0.3411	0.3388	0.00007	0.590
Fusion (DNN)	0.3394	0.3375	0.00007	0.593
Text-based SGMM STD	0.5639	0.5639	0.00008	0.358
Text-based DNN STD	0.6112	0.6062	0.00006	0.327

Table 10 Results of the ALBAYZIN QbE STD 2014 evaluation on test data

System ID	MTWV	ATWV	p(FA)	p(Miss)
Fusion (SGMM)+Posteriorgram	0.2708	0.2708	0.00006	0.672
Fusion (SGMM)	0.2671	0.2657	0.00005	0.679
Fusion (DNN)+Posteriorgram	0.2894	0.2881	0.00006	0.652
Fusion (DNN)	0.2894	0.2881	0.00006	0.652
Text-based SGMM STD	0.6157	0.6099	0.00006	0.323
Text-based DNN STD	0.6583	0.6469	0.00006	0.282

compared to the MFCC fingerprint-based system, and hence similar results are obtained when one or both are fed to the fusion. In addition, using DNN as acoustic models in the Kaldi LVCSR-based system employed in the fusion outperform the SGMM acoustic models, though the performance gaps are not statistically significant for train/dev data and test data ($p \approx 0.3$). This means that the gains obtained by the DTW-based approach in the fusion compensates the lower performance of the SGMM acoustic models.

Results also show that all the QbE STD systems perform worse than the text-based (SGMM and DNN) STD systems. The improvement of these text-based STD systems over those systems is statistically significant for train/dev data ($p < 10^{-4}$) and for test data ($p < 10^{-6}$). This is mainly due to the use of the correct word transcription of the search query in the text-based STD systems, and indicates that obtaining the word transcription of the query using an ASR system is still problematic. In addition, the improvement obtained by the text-based DNN STD system over the text-based SGMM STD system is statistically significant for train/dev data ($p < 10^{-2}$) and for test data ($p < 0.03$), which shows that the better acoustic modeling plays an important role in the final system performance.

DET curves for train/dev and test data are shown in Figs. 5 and 6, respectively. For train/dev data, it is clear that the Fusion (DNN)+Posteriorgram system performs similar to the Fusion (DNN) system for most of the range, except when false alarm rate is low, where the Fusion (DNN)+Posteriorgram system performs better, and that the DNN-based fused systems outperform the SGMM-based systems counterpart in general. For test data, the Fusion (SGMM)+Posteriorgram and Fusion (DNN)+Posteriorgram systems obtain a slight improvement over the Fusion (SGMM) and Fusion (DNN) systems respectively for most of the range, except for low miss rates, where all systems perform similar. In addition, the Fusion (DNN)+Posteriorgram system performance is worse than that of the Fusion (SGMM)+Posteriorgram when false alarm rate is low, and the same is observed with the Fusion (DNN) and Fusion (SGMM) systems. This confirms our conjecture that the DTW-based approach is able to compensate the lower performance of the SGMM acoustic models.

Foreign query analysis

An analysis on the foreign queries of the test data of the individual and fused systems submitted to the ALBAYZIN QbE STD 2014 evaluation has been conducted, and results are presented in Table 11. There are 5 foreign queries in the test data, with 16 occurrences in total. Due to these low figures, differences in the system results are not, in general, statistically significant for paired t tests. However, we can still shed some light about the performance of the different systems for foreign queries. In general, for QbE STD systems, we observe performance degradation with the LVCSR-based systems compared with the template matching (i.e., DTW)-based systems. MFCC-Fingerprint and Phoneme posteriorgram systems outperform LVCSR SGMM- and LVCSR DNN-based systems. We consider this is due to LVCSR-based systems typically degrade their performance with foreign terms. This is confirmed by the text-based SGMM STD system performance, which is lower than that obtained with the DTW-based systems. However, with the better acoustic modeling of the text-based DNN STD system, the performance is better than that of the DTW-based systems, which confirms the potential of DNNs in STD, especially when the correct term transcription is employed in the search. As expected from the full evaluation results, the fused systems also perform the best for foreign query detection.

ALBAYZIN QbE STD 2012 and 2014 evaluation comparison

Evaluating the systems submitted to the ALBAYZIN QbE STD 2014 evaluation only on the queries of the evaluation held in 2012 produces the results shown in Tables 12 and 13 for train/dev and test data respectively, where these results are compared to those of the systems submitted to the 2012 evaluation.

For train/dev data, all the fusion-based systems improve the rest of the QbE STD systems. Paired t tests show that this improvement is statistically significant ($p < 10^{-3}$). For test data, the same behavior is observed. These results constitute a relevant progress for QbE STD in Spanish from the evaluation held in 2012. The best performance of these fusion-based systems is mainly due to two reasons: (1) The use of a robust LVCSR system for Spanish language in terms of acoustic model (SGMM and DNN) and language model (trained from a large variety of text sources), and (2) all the systems employ a fusion of different kinds of systems. Since word transcription-based systems and template matching-based systems present a complementary behavior for QbE STD, the combination of both yields improvements for QbE STD. When comparing Fusion (DNN)+Posteriorgram, Fusion (DNN), Fusion (SGMM)+Posteriorgram, and Fusion (SGMM) systems,

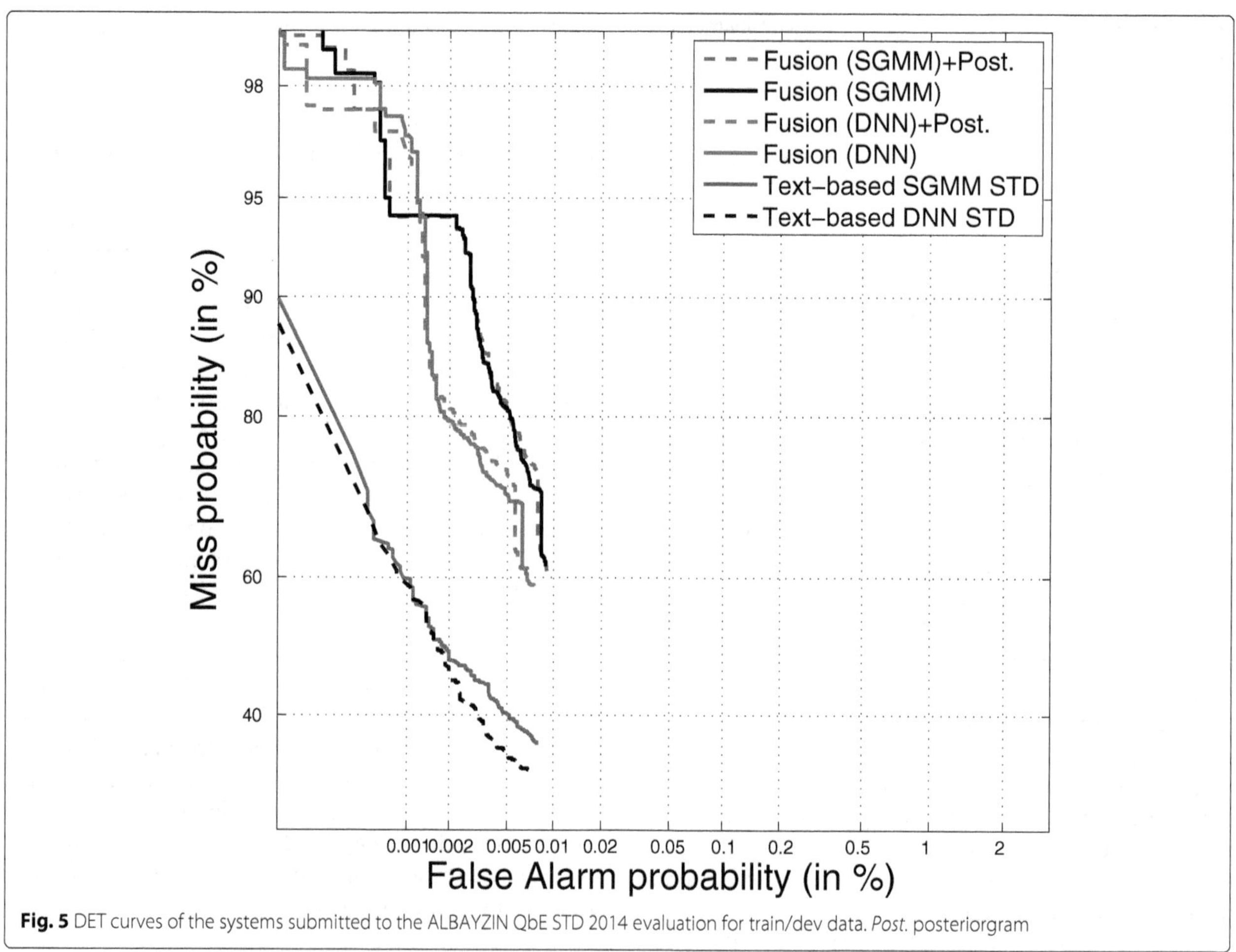

Fig. 5 DET curves of the systems submitted to the ALBAYZIN QbE STD 2014 evaluation for train/dev data. *Post.* posteriorgram

we observe that the DNN-based systems outperform the SGMM-based systems, though there is no significant difference between them ($p \approx 0.4$ for train/dev data and $p \approx 0.7$ for test data), which is consistent with the results presented in the 2014 evaluation.

Among the systems that do not employ system fusion (all the fused systems integrate an LVCSR system), the DTW-Spanish system obtains the best overall performance for train/dev and test data. The query-dependent score normalization produces the smallest difference between MTWV and ATWV. This indicates that the threshold is well set. Paired t tests show that this best performance of the DTW-Spanish system is statistically significant for train/dev data ($p < 10^{-6}$) and test data ($p < 10^{-3}$) compared to the other single systems. This system is language-dependent, and this is probably one of the reasons for the best performance of the DTW-Spanish system. In addition, the similarity measure used to conduct the segmental DTW search (Kullback-Leibler divergence) fits very well the posterior probabilities computed in the feature extraction stage. Aiming at building a language-independent QbE STD system, the DTW-Zero system deserves special mention, since it obtains the best performance in terms of MTWV on test data. In this case, a better threshold setting is needed to get nearer ATWV to MTWV.

The corresponding DET curves for train/dev and test data for the systems submitted to the evaluations held in 2012 and 2014 are shown in Figs. 7 and 8, respectively. We observe similar trends for both sets of data. Fusion (DNN)+Posteriorgram, Fusion (SGMM)+Posteriorgram, Fusion (DNN), and Fusion (SGMM) systems perform the best for almost all the range, except when the false alarm rate is low, where DTW-Zero system performs the best. For test data, the Fusion (SGMM)+Posteriorgram system performs better than the Fusion (SGMM) system for almost all the range, and the Fusion (DNN)+Posteriorgram system outperforms the Fusion (DNN) system, which is consistent with the results obtained in the 2014 evaluation.

Analyzing the DET curves for the single systems, it is clear that the DTW-Zero system, which employs a language-independent approach for QbE STD, outperforms the rest of the systems for train/dev and test

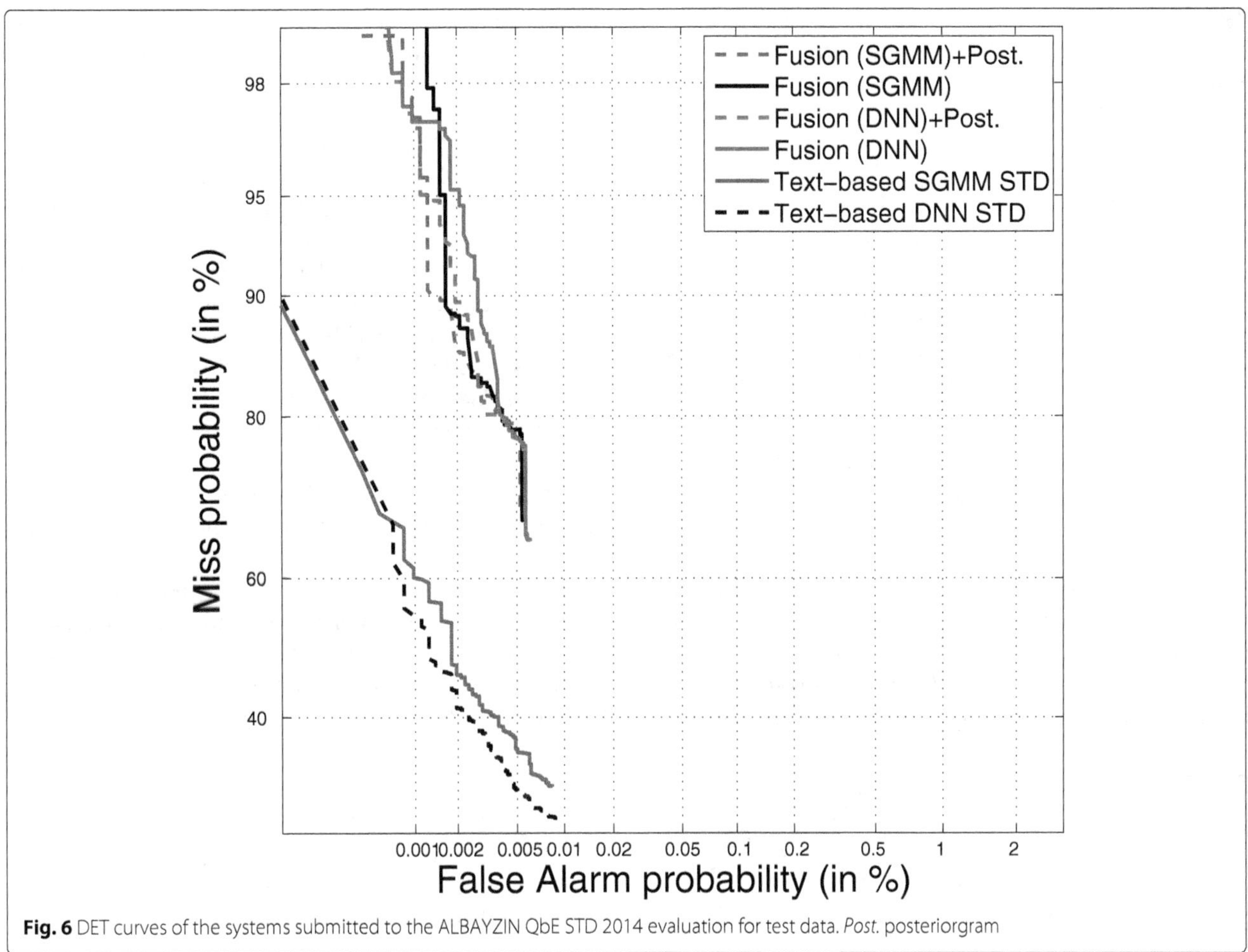

Fig. 6 DET curves of the systems submitted to the ALBAYZIN QbE STD 2014 evaluation for test data. *Post.* posteriorgram

data, except at the best operating point of the DTW-Spanish system, where this performs better than the DTW-Zero system. This makes the DTW-Zero system interesting to face the language independency issue in QbE STD.

Table 11 Results of the ALBAYZIN QbE STD 2014 evaluation on foreign queries of the test data

System ID	MTWV	ATWV	*p*(FA)	*p*(Miss)
Fusion (SGMM)+Posteriorgram	0.5167	0.5167	0	0.483
Fusion (SGMM)	0.5167	0.5167	0	0.483
Fusion (DNN)+Posteriorgram	0.5500	0.5500	0	0.450
Fusion (DNN)	0.5500	0.5500	0	0.450
LVCSR-SGMM	0.2667	0.2667	0	0.733
LVCSR-DNN	0.3000	0.3000	0	0.700
MFCC-Fingerprint	0.3833	0.3833	0	0.617
Phoneme Posteriorgram	0.3833	0.3833	0	0.617
Text-based SGMM STD	0.3675	0.3400	0.00008	0.550
Text-based DNN STD	0.4225	0.4225	0.00003	0.500

Toward a language-independent STD system

From the systems submitted to the ALBAYZIN QbE STD 2012 and 2014 evaluations, insights about the feasibility of building a language-independent STD system can be gained. By comparing the best language-independent

Table 12 Results of the ALBAYZIN QbE STD 2012 and 2014 evaluations on the ALBAYZIN QbE STD 2012 train/dev data

System ID	MTWV	ATWV	*p*(FA)	*p*(Miss)
Fusion (SGMM)+Posteriorgram	0.2850	0.2850	0.00011	0.610
Fusion (SGMM)	0.2824	0.2803	0.00010	0.619
Fusion (DNN)+Posteriorgram	0.3572	0.3514	0.00007	0.578
Fusion (DNN)	0.3579	0.3571	0.00006	0.580
DTW-Zero	0.0455	0.0455	0.00002	0.930
P1B-STDa	0.0128	0.0128	0.00000	0.986
P1B-STDb	0.0092	0.0092	0.00000	0.990
P1L-STD	0.0000	0.0000	0.00000	1.000
DTW-Spanish	0.0612	0.0612	0.00005	0.893
Text-based SGMM STD	0.6866	0.6866	0.00009	0.226
Text-based DNN STD	0.7440	0.7398	0.00006	0.192

Table 13 Results of the ALBAYZIN QbE STD 2012 and 2014 evaluations on the ALBAYZIN QbE STD 2012 test data

System ID	MTWV	ATWV	*p*(FA)	*p*(Miss)
Fusion (SGMM)+Posteriorgram	0.2691	0.2691	0.00006	0.676
Fusion (SGMM)	0.2691	0.2691	0.00006	0.676
Fusion (DNN)+Posteriorgram	0.2815	0.2815	0.00007	0.647
Fusion (DNN)	0.2815	0.2815	0.00007	0.647
DTW-Zero	0.0436	0.0122	0.00000	0.952
P1B-STDa	0.0055	0.0031	0.00001	0.983
P1B-STDb	0.0075	0.0047	0.00000	0.990
P1L-STD	0.0000	-0.0678	0.00000	1.000
DTW-Spanish	0.0238	0.0217	0.00009	0.884
Text-based SGMM STD	0.6795	0.6627	0.00006	0.256
Text-based DNN STD	0.7299	0.7148	0.00007	0.199

QbE STD system (DTW-Zero) with the text-based DNN STD system, we can claim that building a language-independent STD system with a performance similar to that of a language-dependent STD system is still far from being achieved. This means that more research is needed in QbE STD to approximate language-independent to language-dependent STD systems in highly difficult speech domains such as spontaneous speech.

Challenge of the QbE STD task

From the results obtained by all the systems submitted to the ALBAYZIN QbE STD 2012 and 2014 evaluations, we can claim that building a QbE STD system with a performance near to that of a text-based STD system for Spanish is still difficult. The ATWV performance obtained by the best QbE STD system (ATWV= 0.2894) compared to that of the best text-based STD system (ATWV= 0.6583) confirms this. There are still many issues that must be solved in the future, depending on the type of the system.

On the one hand, for systems based on LVCSR, an accurate word transcription of the query must be obtained. Otherwise, the QbE STD system performance dramatically drops, as we have seen in the 2014 evaluation results.

On the other hand, for systems that rely on template matching, a robust template that efficiently represents the queries and the utterances is necessary. In addition, a reliable search algorithm that hypothesizes query detections is also necessary to output as many hits as

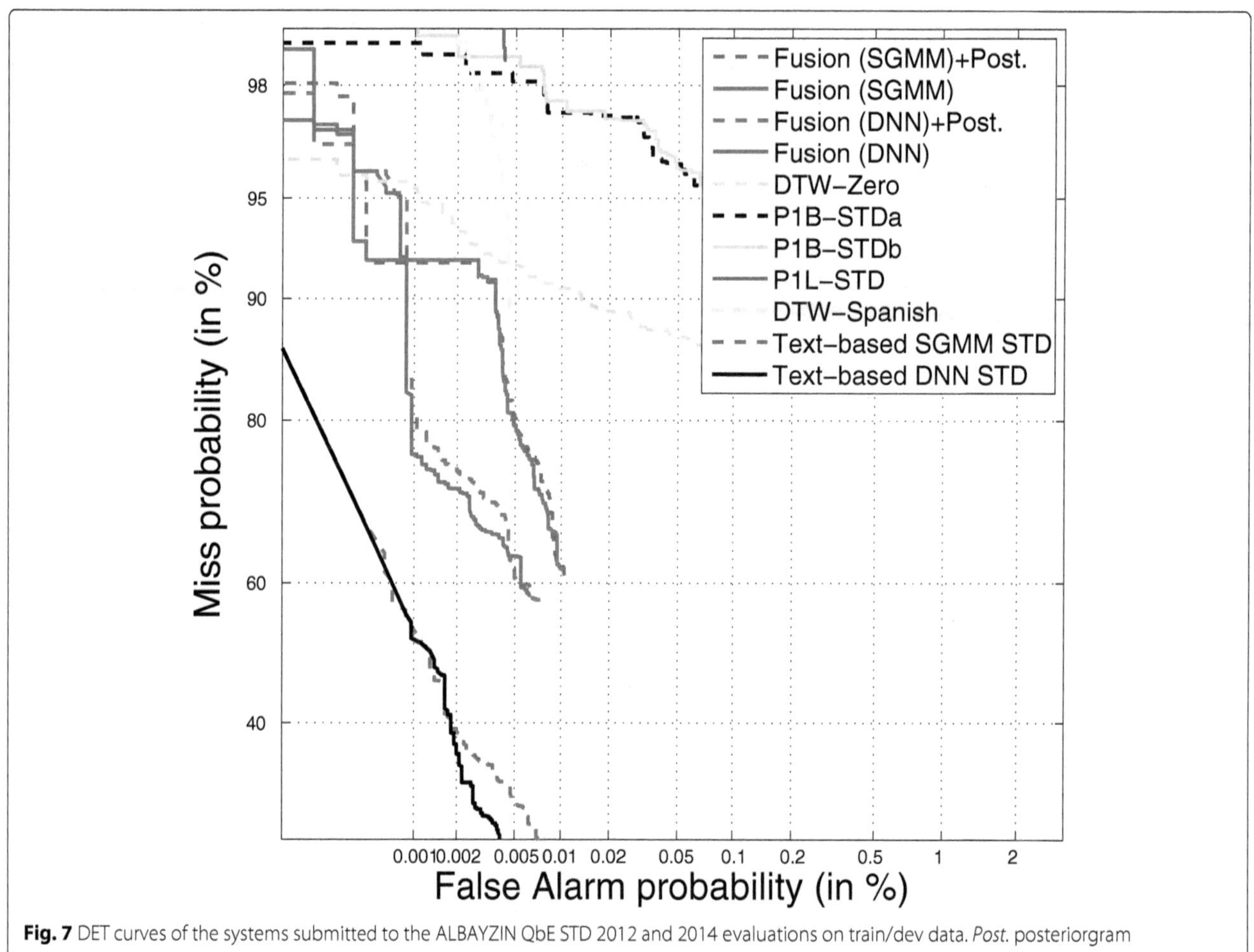

Fig. 7 DET curves of the systems submitted to the ALBAYZIN QbE STD 2012 and 2014 evaluations on train/dev data. *Post.* posteriorgram

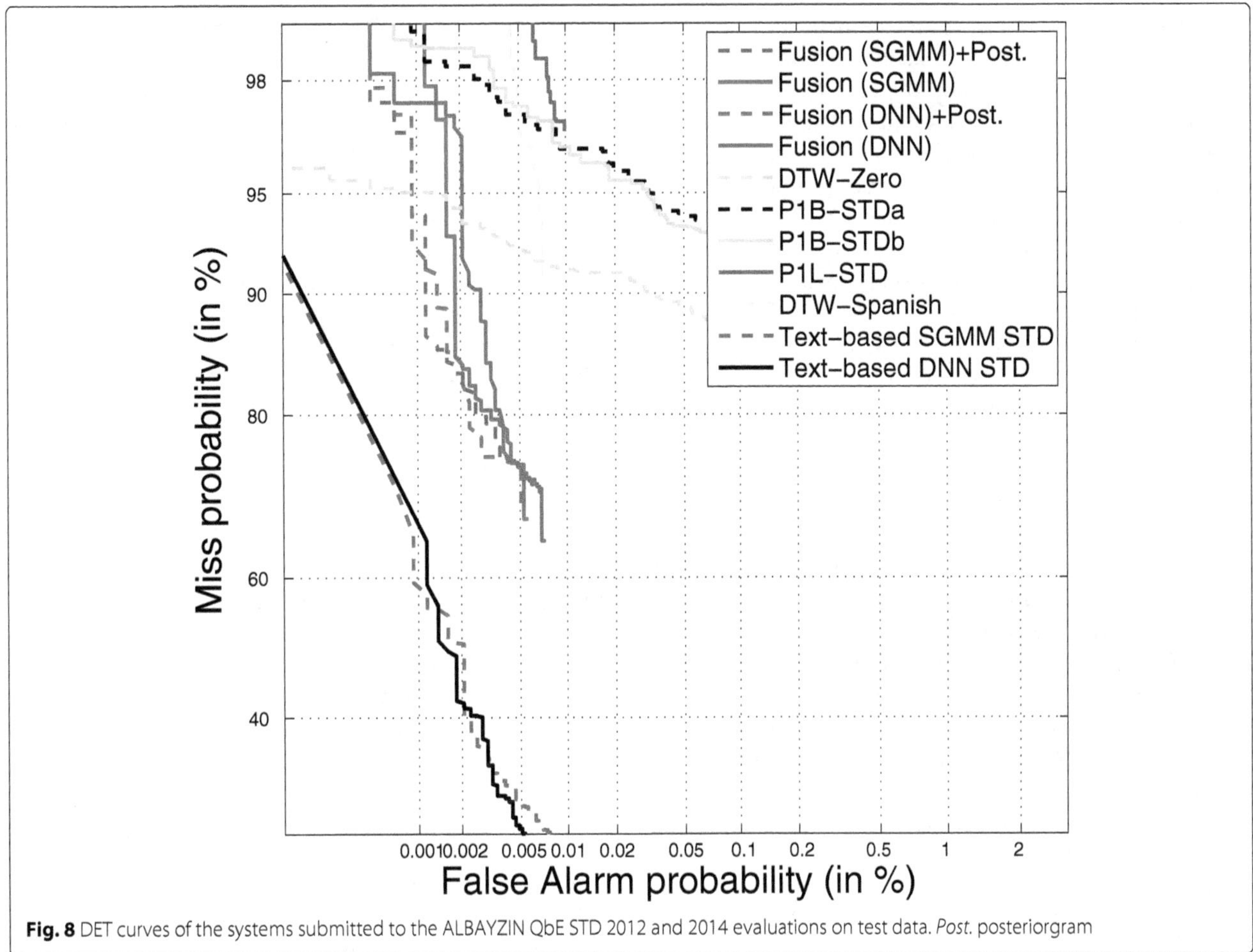

Fig. 8 DET curves of the systems submitted to the ALBAYZIN QbE STD 2012 and 2014 evaluations on test data. *Post.* posteriorgram

possible while maintaining a reasonably low number of FAs.

All the systems must also deal with the type of speech of the evaluation, mainly spontaneous speech, which represents an important challenge for the QbE STD task. In this way, as in standard ASR systems, special attention must be paid to phenomena such as disfluences, hesitations, and noises. The QbE STD system performance can possibly be enhanced by including some pre-processing steps that deal with these phenomena.

Lessons learned

The different types of systems submitted to the first and second QbE STD evaluations in Spanish provide significant lessons that should be taken into account for forthcoming evaluation editions.

The first ALBAYZIN QbE STD evaluation held in 2012 received systems that are mainly language-independent, whereas the evaluation held in 2014 focused on language-dependent systems. Results presented in this paper have shown performance differences between language-dependent and language-independent QbE STD systems. Moreover, for the language-dependent systems, results have also shown performance differences between systems that employ fusion of word transcription- and template matching-based systems, and systems that do not. Therefore, organizers will thoroughly think about dividing the ALBAYZIN QbE STD evaluation for future editions into two different subtasks. The first subtask is suitable for systems that are language-dependent, whereas the second one will be for systems that are language-independent, aiming at building a language-independent STD system and evaluating it in Spanish.

The second ALBAYZIN QbE STD evaluation incorporates multi-word and foreign queries to the evaluation. However, there were just a few multi-word and foreign queries in this second evaluation. Future evaluations should include more multi-word and foreign queries.

To compare the system results of the ALBAYZIN QbE STD evaluations held in 2012 and 2014, the same evaluation metric (ATWV) has been employed. However, in the recent MediaEval 2014 QbE Search on Speech evaluation [39], a different metric called normalized cross entropy cost (C_{nxe}) has been employed. This metric requires calibrated likelihood ratios, and hence, participants will be allowed to submit calibrated likelihood ratios for future evaluation editions. Moreover, the NTCIR-11 QbE STD evaluation also employed a different evaluation metric (F-measure). This metric, contrary to ATWV, assigns the same cost to precision and recall values and hence allows comparing the systems from another perspective. Therefore, organizers will probably propose to use both evaluation metrics (C_{nxe} and F-measure) as secondary metrics for next evaluation editions.

The first and second ALBAYZIN QbE STD evaluations focused on searching a train/dev query list in train/dev speech data and searching a test query list in test speech data. In future evaluations, the cross-data query search should also be considered. The purpose of this cross-data search is to see how critical tuning is for the different systems. For example, searching test queries in train/dev speech data could be enhanced by unsupervised adaptation, whereas searching train/dev queries in test speech data can measure the generalization capability of the systems on unseen data with the same query list for which good classifiers could have been developed.

Regarding the data preparation, organizers used the database of MAVIR project consisting of recordings of seminars and round tables organized at the general meetings of the project. This database has resulted very challenging with many interesting properties (i.e., different noise levels, different speakers, foreign words, etc.). For instance, in the first ALBAYZIN QbE STD evaluation held in 2012, organizers focused on single-word queries in Spanish, but in the second edition, organizers added multi-word and foreign queries in order to analyze the influence of these in system performance. The database was transcribed and aligned at the utterance level. This was very helpful to produce the manual query alignments, but even using this information it took a considerable amount of time to produce the manual alignments. Although MAVIR data have been very useful, we consider that it will be necessary to use additional data (for instance from broadcast news or perhaps more challenging TV programs) to make the evaluations evolve and not become repetitive. Organizers are currently preparing more data in order to perform a new and more challenging evaluation in 2016. Besides using new data, organizers will probably reuse the same MAVIR data to assess technology improvements on a comparable basis.

Conclusions

This paper presented the systems submitted to the ALBAYZIN QbE STD 2014 evaluation along with two systems that conduct text-based STD and compared these with the systems submitted to the ALBAYZIN QbE STD 2012 evaluation. Four different Spanish research groups (TID, GTTS, ELiRF, and GTM) took part in the evaluations. Different kinds of systems were submitted to the evaluations: fusion-based systems (Fusion (SGMM)+Posteriorgram, Fusion (DNM)+Posteriorgram, Fusion (SGMM), and Fusion (DNN)), template matching-based systems (DTW-Zero and DTW-Spanish), and subword transcription-based systems (P1B-STD and P1L-STD).

Results show that the best performance is obtained from fused systems that combine word transcription- and template matching-based systems. These fused systems employ the target language (i.e., Spanish) information.

From the results presented in the ALBAYZIN QbE STD 2014 evaluation, we can claim that significant performance gains have been obtained compared to the ALBAYZIN QbE STD 2012 evaluation. This confirms the progress on QbE STD technology for Spanish language. However, when comparing the QbE STD results with the results of the text-based DNN STD system presented in this paper, it is clear that there is still ample room for improvement to approximate the performance of QbE STD to that of text-based STD. This encourages organizers to maintain this evaluation in the next ALBAYZIN evaluation campaign.

Endnotes

[1]http://www.rthabla.es/.

[2]http://www.isca-speech.org/iscaweb/index.php/sigs?layout=edit&id=132.

[3]http://iberspeech2012.ii.uam.es/.

[4]http://iberspeech2014.ulpgc.es/.

[5] MAVIR was a project funded by the Madrid region that coordinated several research groups and companies working on information retrieval (http://www.mavir.net)

[6]http://sox.sourceforge.net/.

[7]http://www.tc-star.org.

[8]http://cartago.lllf.uam.es/mavir/index.pl?m=descargas.

[9]http://homepages.inf.ed.ac.uk/v1dwang2/public/tools/index.html.

Competing interests

The authors declare that they have no competing interests.

Authors' contributions

Systems submitted to the second QbE STD evaluation for Spanish language are presented. They are hybrid, language-dependent systems. Systems are compared with the systems submitted to the first QbE STD evaluation for Spanish language (which were mainly single, language-independent systems), and show significant improvements. Lessons learned from both evaluations are also presented, which aim to highlight interesting clues for evaluation organizers.

Acknowledgements

This research was funded by the Spanish Government ('SpeechTech4All Project' TEC2012-38939-C03-01 and 'CMC-V2 Project' TEC2012-37585-C02-01), the Galician Government through the research contract GRC2014/024 (Modalidade: Grupos de Referencia Competitiva 2014) and 'AtlantTIC Project' CN2012/160, and also by the Spanish Government and the European Regional Development Fund (ERDF) under project TACTICA.

Author details
[1]GEINTRA, Universidad de Alcalá, Campus Universitario. Ctra. Madrid-Barcelona, km.33,600, Alcalá de Henares, Madrid, Spain. [2]Biometric Recognition Group - ATVS, Universidad Autónoma de Madrid, Av. Francisco Tomás y Valiente, 11. Escuela Politécnica Superior, Madrid, Spain. [3]Multimedia Technologies Group (GTM), AtlantTIC Research Center, E. E. Telecomunicación, Campus Universitario de Vigo, s/n, Vigo, Spain.

References

1. T Zhang, C-CJ Kuo, in *Hierarchical classification of audio data for archiving and retrieving*. Proc. of ICASSP (IEEE, Washington DC, USA, 1999), pp. 3001–3004
2. M Helén, T Virtanen, in *Query by example of audio signals using Euclidean distance between Gaussian Mixture Models*. Proc. of ICASSP (IEEE, Washington DC, USA, 2007), pp. 225–228
3. M Helén, T Virtanen, Audio query by example using similarity measures between probability density functions of features. EURASIP, Journal on Audio, Speech, and Music Processing. **2010**, 2–1212 (2010)
4. G Tzanetakis, A Ermolinskyi, P Cook, in *Pitch histograms in audio and symbolic music information retrieval*. Proc. of ISMIR (ISMIR, Paris, France, 2002), pp. 31–38
5. W-H Tsai, H-M Wang, in *A query-by-example framework to retrieve music documents by singer*. Proc. of ICME (IEEE, Washington DC, USA, 2004), pp. 1863–1866
6. TK Chia, KC Sim, H Li, HT Ng, in *A lattice-based approach to query-by-example spoken document retrieval*. Proc. of ACM SIGIR (ACM, New York, USA, 2008), pp. 363–370
7. A Muscariello, G Gravier, F Bimbot, in *Zero-resource audio-only spoken term detection based on a combination of template matching techniques*. Proc. of Interspeech (ISCA, Baixas, France, 2011), pp. 921–924
8. J Tejedor, M Fapšo, I Szöke, Černocký, F Grézl, Comparison of methods for language-dependent and language-independent query-by-example spoken term detection. ACM Trans. Inf. Syst. **30**(3), 18–11834 (2012)
9. G Mantena, X Anguera, in *Speed improvements to information retrieval-based dynamic time warping using hierarchical k-means clustering*. Proc. of ICASSP (IEEE, Washington DC, USA, 2013), pp. 8515–8519
10. G Mantena, S Achanta, K Prahallad, Query-by-example spoken term detection using frequency domain linear prediction and non-segmental dynamic time warping. IEEE/ACM Trans. Audio Speech Lang. Process. **22**(5), 946–955 (2014)
11. J Vavrek, M Pleva, M Lojka, P Viszlay, Kiktová, D Hládek, J Juhár, in *TUKE at MediaEval 2013 spoken web search task*. Proc. of MediaEval (CEUR, Aachen, Germany, 2013), pp. 73–1732
12. R Jarina, M Kuba, R Gubka, M Chmulik, M Paralic, in *UNIZA system for the spoken web search task at MediaEval 2013*. Proc. of MediaEval (CEUR, Aachen, Germany, 2013), pp. 79–1792
13. A Ali, MA Clements, in *Spoken web search using and ergodic hidden Markov model of speech*. Proc. of MediaEval (CEUR, Aachen, Germany, 2013), pp. 86–1862
14. A Buzo, H Cucu, C Burileanu, in *SpeeD@MediaEval 2014: Spoken term detection with robust multilingual phone recognition*. Proc. of MediaEval (CEUR, Aachen, Germany, 2014), pp. 72–1722
15. S Kesiraju, G Mantena, K Prahallad, in *IIIT-H system for MediaEval 2014 QUESST*. Proc. of MediaEval (CEUR, Aachen, Germany, 2014). pp. 76–1762
16. J Takahashi, T Hashimoto, R Konno, S Sugawara, K Ouchi, S Oshima, T Akyu, Y Itoh, in *An IWAPU STD system for OOV query terms and spoken queries*. Proc. of NTCIR-11 (National Institute of Informatics, Tokyo, Japan, 2014), pp. 384–389
17. M Makino, A Kai, in *Combining subword and state-level dissimilarity measures for improved spoken term detection in NTCIR-11 SpokenQuery&Doc task*. Proc. of NTCIR-11 (National Institute of Informatics, Tokyo, Japan, 2014), pp. 413–418
18. M Gubian, L Boves, M Versteegh, in *Calibration of distance measures for unsupervised query-by-example*. Proc. of Interspeech (ISCA, Baixas, France, 2013), pp. 2639–2643
19. X Anguera, M Ferrarons, in *Memory efficient subsequence DTW for query-by-example spoken term detection*. Proc. of ICME (IEEE, Washington DC, USA, 2013)
20. H Wang, T Lee, in *The CUHK spoken web search system for MediaEval 2013*. Proc. of MediaEval (CEUR, Aachen, Germany, 2013), pp. 68–1682
21. M Bouallegue, G Senay, M Morchid, D Matrouf, G Linares, R Dufour, in *LIA@MediaEval 2013 spoken web search task: An I-Vector based approach*. Proc. of MediaEval (CEUR, Aachen, Germany, 2013), pp. 77–1772
22. LJ Rodriguez-Fuentes, A Varona, M Penagarikano, G Bordel, M Diez, in *GTTS systems for the SWS task at MediaEval 2013*. Proc. of MediaEval (CEUR, Aachen, Germany, 2013), pp. 83–1832
23. H Wang, T Lee, C-C Leung, B Ma, H Li, in *Using parallel tokenizers with DTW matrix combination for low-resource spoken term detection*. Proc. of ICASSP (IEEE, Washington DC, USA, 2013), pp. 8545–8549
24. H Wang, T Lee, in *CUHK system for QUESST task of MediaEval 2014*. Proc. of MediaEval (CEUR, Aachen, Germany, 2014), pp. 73–1732
25. J Proenca, A Veiga, F Perdigão, in *The SPL-IT query by example search on speech system for MediaEval 2014*. Proc. of MediaEval (CEUR, Aachen, Germany, 2014), pp. 74–1742
26. P Yang, C-C Leung, L Xie, B Ma, H Li, in *Intrinsic spectral analysis based on temporal context features for query-by-example spoken term detection*. Proc. of Interspeech (ISCA, Baixas, France, 2014), pp. 1722–1726
27. B George, A Saxena, G Mantena, K Prahallad, B Yegnanarayana, in *Unsupervised query-by-example spoken term detection using bag of acoustic words and non-segmental dynamic time warping*. Proc. of Interspeech (ISCA, Baixas, France, 2014), pp. 1742–1746
28. TJ Hazen, W Shen, CM White, in *Query-by-example spoken term detection using phonetic posteriorgram templates*. Proc. of ASRU (IEEE, Washington DC, USA, 2009), pp. 421–426
29. A Abad, RF Astudillo, I Trancoso, in *The L2F spoken web search system for MediaEval 2013*. Proc. of MediaEval (CEUR, Aachen, Germany, 2013), pp. 85–1852
30. A Abad, LJ Rodríguez-Fuentes, M Penagarikano, A Varona, G Bordel, in *On the calibration and fusion of heterogeneous spoken term detection systems*. Proc. of Interspeech (ISCA, Baixas, France, 2013), pp. 20–24
31. I Szöke, M Skácel, L Burget, in *BUT QUESST 2014 system description*. Proc. of MediaEval (CEUR, Aachen, Germany, 2014), pp. 62–1622
32. P Yang, H Xu, X Xiao, L Xie, C-C Leung, H Chen, J Yu, H Lv, L Wang, SJ Leow, B Ma, ES Chng, H Li, in *The NNI query-by-example system for MediaEval 2014*. Proc. of MediaEval (CEUR, Aachen, Germany, 2014), pp. 69–1692
33. I Szöke, L Burget, F Grézl, JH Černocký, L Ondel, in *Calibration and fusion of query-by-example systems - BUT SWS 2013*. Proc. of ICASSP (IEEE, Washington DC, USA, 2014), pp. 7849–7853
34. H Wang, T Lee, C-C Leung, B Ma, H Li, Acoustic segment modeling with spectral clustering methods. IEEE/ACM Trans. Audio Speech Lang. Process. **23**(2), 264–277 (2015)
35. C-T Chung, W-N Hsu, C-Y Lee, L-S Lee, in *Enhancing automatically discovered multi-level acoustic patterns considering context consistency with applications in spoken term detection*. Proc. of ICASSP (IEEE, Washington DC, USA, 2015), pp. 5231–5235
36. NIST, The Ninth Text REtrieval Conference (TREC 9) (2000). http://trec.nist.gov. Accessed 8 January 2016
37. H Joho, K Kishida, in *Overview of the NTCIR-11, SpokenQuery&Doc Task*. Proc. of NTCIR-11 (National Institute of Informatics, Tokyo, Japan, 2014), pp. 1–7
38. X Anguera, F Metze, A Buzo, I Szöke, LJ Rodriguez-Fuentes, in *The spoken web search task*. Proc. of MediaEval (CEUR, Aachen, Germany, 2013), pp. 1–2
39. X Anguera, LJ Rodriguez-Fuentes, I Szöke, A Buzo, F Metze, in *Query by example search on speech at Mediaeval 2014*. Proc. of MediaEval (CEUR, Aachen, Germany, 2014), pp. 1–2
40. NIST, *Draft KWS14 Keyword Search Evaluation Plan*. (National Institute of Standards and Technology (NIST), Gaithersburg, MD, USA, 2013). National Institute of Standards and Technology (NIST). http://www.nist.gov/itl/iad/mig/upload/KWS14-evalplan-v11.pdf. Accessed 8 January 2016
41. B Taras, C Nadeu, Audio segmentation of broadcast news in the Albayzin-2010 evaluation: overview, results, and discussion. EURASIP Journal on Audio, Speech, and Music Processing. **2011**, 1–1110 (2011)
42. M Zelenák, H Schulz, J Hernando, Speaker diarization of broadcast news in Albayzin 2010 evaluation campaign. EURASIP Journal on Audio, Speech, and Music Processing. **2012**, 19–1199 (2012)

43. LJ Rodríguez-Fuentes, M Penagarikano, A Varona, M Díez, G Bordel, in *The Albayzin 2010 Language Recognition Evaluation*. Proc. of Interspeech (ISCA, Baixas, France, 2011), pp. 1529–1532
44. J Tejedor, DT Toledano, X Anguera, A Varona, LF Hurtado, A Miguel, J Colás, Query-by-example spoken term detection ALBAYZIN 2012 evaluation: overview, systems, results, and discussion. EURASIP, Journal on Audio, Speech, and Music Processing. **2013**, 23–12317 (2013)
45. F Méndez, L Docío, M Arza, F Campillo, in *The Albayzin 2010 text-to-speech evaluation*. Proc. of FALA (Spanish Thematic Network on Speech Technology, Madrid, Spain, 2010), pp. 317–340
46. J Billa, KW Ma, JW McDonough, G Zavaliagkos, DR Miller, KN Ross, A El-Jaroudi, in *Multilingual speech recognition: the 1996 Byblos callhome system*. Proc. of Eurospeech (ISCA, Baixas, France, 1997)
47. H Cuayahuitl, B Serridge, in *Out-of-vocabulary word modeling and rejection for spanish keyword spotting systems*. Proc. of MICAI (Springer, London, United Kingdom, 2002), pp. 156–165
48. M Killer, S Stuker, T Schultz, in *Grapheme based speech recognition*. Proc. of Eurospeech (ISCA, Baixas, France, 2003), pp. 3141–3144
49. J Tejedor, *Contributions to Keyword Spotting and Spoken Term Detection For Information Retrieval in Audio Mining. PhD thesis, Universidad Autónoma de Madrid, Madrid, Spain*. (Universidad AutÃ³noma de Madrid, Madrid, Spain, 2009)
50. L Burget, P Schwarz, M Agarwal, P Akyazi, K Feng, A Ghoshal, O Glembek, N Goel, M Karafiat, D Povey, A Rastrow, RC Rose, S Thomas, in *Multilingual acoustic modeling for speech recognition based on subspace gaussian mixture models*. Proc. of ICASSP (IEEE, Washington DC, USA, 2010), pp. 4334–4337
51. J Tejedor, DT Toledano, D Wang, S King, J Colás, Feature analysis for discriminative confidence estimation in spoken term detection. Comput. Speech Lang. **28**(5), 1083–1114 (2014)
52. J Li, X Wang, B Xu, in *An empirical study of multilingual and low-resource spoken term detection using deep neural networks*. Proc. of Interspeech (ISCA, Baixas, France, 2014), pp. 1747–1751
53. NIST, *The Spoken Term Detection (STD) 2006 Evaluation Plan*, 10th edn. (National Institute of Standards and Technology (NIST), Gaithersburg, MD, USA, 2006). National Institute of Standards and Technology (NIST). http://www.nist.gov/speech/tests/std. Accessed 8 January 2016
54. JG Fiscus, J Ajot, JS Garofolo, G Doddingtion, in *Results of the 2006 spoken term detection evaluation*. Proc. of SSCS (ACM, New York, USA, 2007), pp. 45–50
55. A Martin, G Doddington, T Kamm, M Ordowski, M Przybocki, in *The DET curve in assessment of detection task performance*. Proc. of Eurospeech (ISCA, Baixas, France, 1997), pp. 1895–1898
56. NIST, *NIST Speech Tools and APIs: 2006*. (National Institute of Standards and Technology (NIST), Gaithersburg, MD, USA, 1996). National Institute of Standards and Technology (NIST). http://www.nist.gov/speech/tools/index.htm. Accessed 8 January 2016
57. N Rajput, F Metze, in *Spoken web search*. Proc. of MediaEval (CEUR, Aachen, Germany, 2011), pp. 1–2
58. F Metze, E Barnard, M Davel, Heerden C van, X Anguera, G Gravier, N Rajput, in *The spoken web search task*. Proc. of MediaEval (CEUR, Aachen, Germany, 2012), pp. 1–2
59. NTCIR-11 Spoken Query and Spoken Document Retrieval Task Organizers, Definition of SQ-STD Task at NTCIR-11 SpokenQuery&Doc (2014). http://www.nlp.cs.tut.ac.jp/~sdpwg/ntcir11/SQ-STD.pdf. Accessed 8 January 2016
60. D Povey, A Ghoshal, G Boulianne, L Burget, O Glembek, N Goel, M Hannemann, P Motlicek, Y Qian, P Schwarz, J Silovsky, G Stemmer, K Vesely, in *The Kaldi speech recognition toolkit*. Proc. of ASRU (IEEE, Washington DC, USA, 2011)
61. L Docío-Fernández, A Cardenal-López, C García-Mateo, in *TC-STAR 2006 automatic speech recognition evaluation: The uvigo system*. Proc. of TC-STAR Workshop on Speech-to-Speech Translation (META-NET, Berlin, Germany, 2006)
62. A Stolcke, in *SRILM - an extensible language modeling toolkit*. Proc. of ICSLP (ISCA, Baixas, France, 2002), pp. 901–904
63. D Povey, M Hannemann, G Boulianne, L Burget, A Ghoshal, M Janda, M Karafiat, S Kombrink, P Motlicek, Y Qian, K Riedhammer, K Vesely, NT Vu, in *Proc. of ICASSP*. Generating exact lattices in the WFST framework (IEEE, Washington DC, USA, 2012), pp. 4213–4216
64. G Chen, S Khudanpur, D Povey, J Trmal, D Yarowsky, O Yilmaz, in *Quantifying the value of pronunciation lexicons for keyword search in low resource languages*. Proc. of ICASSP (IEEE, Washington DC, USA, 2013), pp. 8560–8564
65. VT Pham, NF Chen, S Sivadas, H Xu, I-F Chen, C Ni, ES Chng, H Li, in *System and keyword dependent fusion for spoken term detection*. Proc. of SLT (IEEE, Washington DC, USA, 2014), pp. 430–435
66. D Can, M Saraclar, Lattice indexing for spoken term detection. IEEE Trans. Audio Speech Lang. Process. **19**(8), 2338–2347 (2011)
67. P Lopez-Otero, L Docio-Fernandez, C Garcia-Mateo, in *Introducing a framework for the evaluation of music detection tools*. Proc. of LREC (European Language Resources Association, Paris, France, 2014), pp. 568–572
68. C Neves, A Veiga, Sá, F Perdigão, in *Audio fingerprinting system for broadcast streams*. Proc. of ConfTele, vol. 1, (Santa Maria da Feira, Instituto de Telecomunicações, Campus Universitãrio de Santiago, Aveiro, Portugal, 2009), pp. 481–484
69. K Seyerlehner, G Widmer, T Pohle, M Sched, in *Proc. of the 10th Conference on Digital Audio Effects*. Automatic music detection in television productions (LaBRI, UniversitÃ© Bordeaux, Bordeaux, France, 2007)
70. S Kim, E Unal, S Narayanan, in *Music fingerprint extraction for classical music cover song identification*. Proc. of ICME (IEEE, Washington DC, USA, 2008), pp. 1261–1264
71. J Haitsma, T Kalker, in *A highly robust audio fingerprinting system*. Proc. of ISMIR (ISMIR, Paris, France, 2002), pp. 107–115
72. TJ Hazen, W Shen, CM White, in *Query-by-example spoken term detection using phonetic posteriorgram templates*. Proc. of ASRU (IEEE, Washington DC, USA, 2009), pp. 421–426
73. P Schwarz, *Phoneme recognition based on long temporal context. PhD thesis, Brno University of Technology*. (Brno University of Technology, Brno, Czech Republic, 2009)
74. A Abad, RF Astudillo, in *The L2F spoken web search system for mediaeval 2012*. Proc. of MediaEval (CEUR, Aachen, Germany, 2012), pp. 9–10
75. N Brümmer, E de Villiers, The BOSARIS toolkit user guide: Theory, algorithms and code for binary classifier score processing. Technical report (2011). https://sites.google.com/site/nikobrummer. Accessed 8 January 2016
76. Veselý, A Ghoshal, L Burget, D Povey, in *Sequence-discriminative training of deep neural networks*. Proc. of Interspeech (ISCA, Baixas, France, 2013), pp. 2345–2349
77. IberSPEECH 2012, "VII Jornadas en Tecnología del Habla" and "III Iberian SLTech Workshop" (2012). http://iberspeech2012.ii.uam.es. Accessed 8 January 2016
78. D Wang, S King, J Frankel, Stochastic pronunciation modelling for out-of-vocabulary spoken term detection. IEEE Trans. Audio Speech Lang. Process. **19**(4), 688–698 (2011)
79. JA Gómez, E Sanchis, MJ Castro-Bleda, in *Automatic speech segmentation based on acoustical clustering*. Proc. of the Joint IAPR International Conference on Structural, Syntactic, and Statistical Pattern Recognition (Springer, London, United Kingdom, 2010), pp. 540–548
80. A Park, JR Glass, in *Towards unsupervised pattern discovery in speech*. Proc. of ASRU (IEEE, Washington DC, USA, 2005), pp. 53–58

Permissions

All chapters in this book were first published by Springer; hereby published with permission under the Creative Commons Attribution License or equivalent. Every chapter published in this book has been scrutinized by our experts. Their significance has been extensively debated. The topics covered herein carry significant findings which will fuel the growth of the discipline. They may even be implemented as practical applications or may be referred to as a beginning point for another development.

The contributors of this book come from diverse backgrounds, making this book a truly international effort. This book will bring forth new frontiers with its revolutionizing research information and detailed analysis of the nascent developments around the world.

We would like to thank all the contributing authors for lending their expertise to make the book truly unique. They have played a crucial role in the development of this book. Without their invaluable contributions this book wouldn't have been possible. They have made vital efforts to compile up to date information on the varied aspects of this subject to make this book a valuable addition to the collection of many professionals and students.

This book was conceptualized with the vision of imparting up-to-date information and advanced data in this field. To ensure the same, a matchless editorial board was set up. Every individual on the board went through rigorous rounds of assessment to prove their worth. After which they invested a large part of their time researching and compiling the most relevant data for our readers.

The editorial board has been involved in producing this book since its inception. They have spent rigorous hours researching and exploring the diverse topics which have resulted in the successful publishing of this book. They have passed on their knowledge of decades through this book. To expedite this challenging task, the publisher supported the team at every step. A small team of assistant editors was also appointed to further simplify the editing procedure and attain best results for the readers.

Apart from the editorial board, the designing team has also invested a significant amount of their time in understanding the subject and creating the most relevant covers. They scrutinized every image to scout for the most suitable representation of the subject and create an appropriate cover for the book.

The publishing team has been an ardent support to the editorial, designing and production team. Their endless efforts to recruit the best for this project, has resulted in the accomplishment of this book. They are a veteran in the field of academics and their pool of knowledge is as vast as their experience in printing. Their expertise and guidance has proved useful at every step. Their uncompromising quality standards have made this book an exceptional effort. Their encouragement from time to time has been an inspiration for everyone.

The publisher and the editorial board hope that this book will prove to be a valuable piece of knowledge for researchers, students, practitioners and scholars across the globe.

List of Contributors

Yoonchang Han
Music and Audio Research Group (MARG), Graduate School of Convergence Science and Technology, Seoul National University, 599 Gwanak-ro, Seoul, Republic of Korea

Kyogu Lee
Music and Audio Research Group (MARG), Graduate School of Convergence Science and Technology, Seoul National University, 599 Gwanak-ro, Seoul, Republic of Korea
Advanced Institutes of Convergence Technology (AICT), Suwon, Republic of Korea

Stanly Mammen and Ilango Krishnamurthi
Department of Computer Science and Engineering, Sri Krishna College of Engineering and Technology, Coimbatore, Tamil Nadu, India

A. Jalaja Varma
School of Drama and Fine Arts, Dr. John Matthai Centre, University of Calicut, Thrissur, Kerala, India

G. Sujatha
Department of Music, Government Women's college, Trivandrum, Kerala, India

Miguel Melo, Maximino Bessa and Luís Barbosa
Universidade de Trás-os-Montes e Alto Douro, Quinta de Prados, 5000-801 Vila Real, Portugal
INESC-TEC, Rua Dr. Roberto Frias, 4200-465 Porto, Portugal

Kurt Debattista
WMG, University of Warwick, Gibbet Hill Road, CV4 7AL Coventry, UK

Alan Chalmers
WMG, University of Warwick, Gibbet Hill Road, CV4 7AL Coventry, UK
goHDR Ltd., Gibbet Hill Road, CV4 7AL Coventry, UK

Hadas Benisty, David Malah and Koby Crammer
Electrical Engineering Department, Technion–Israel Institute of Technology, Technion City, Haifa, Israel

Nawal Alioua
Ibn Zohr University, Morocco, BP 32/S, Agadir, Morocco
LRIT-CNRST 29, Faculty of Sciences, Mohammed V University, Ibn Battouta Avenue, Rabat, Morocco

Mohammed Rziza
LRIT-CNRST 29, Faculty of Sciences, Mohammed V University, Ibn Battouta Avenue, Rabat, Morocco

Alexandrina Rogozan and Abdelaziz Bensrhair
LITIS, INSA-Rouen, Avenue de l'Université, Saint-Etienne-du-Rouvray, France

Aouatif Amine
LGS, ENSA-Kenitra, Ibn Tofail University, Avenue de l'Université, Kenitra, Morocco

Mingjing Zhang and Mark S. Drew
School of Computing Science, Simon Fraser University, Vancouver, BC V5A 1S6, Canada

Somaya Al-Maadeed, Fethi Ferjani, Samir Elloumi and Ali Jaoua
Department of Computer Science and Engineering, Qatar University, Doha, Qatar

Saghi Hajisharif, Joel Kronander and Jonas Unger
Linköping University, Norrköping, Sweden

Tariq M. Khan and Yinan Kong
Department of Engineering, Macquarie University, Balaclava Rd, 2109 Sydney, Australia

Omar Kittaneh and Mohammad A. U. Khan
Department of Electrical and Computer Engineering, Biometric and Sensor Lab, Effat University, Jeddah, Saudi Arabia

Chengchao Qu
Vision and Fusion Laboratory (IES), Karlsruhe Institute of Technology (KIT), Adenauerring 4, 76131 Karlsruhe, Germany
Fraunhofer Institute of Optronics, System Technologies and Image Exploitation (Fraunhofer IOSB), Fraunhoferstr. 1, 76131 Karlsruhe, Germany

Hua Gao
Signal Processing Laboratory (LTS5), École Polytechnique Fédérale de Lausanne (EPFL), EPFL STI IEL LTS5, ELD 241 (Bâtiment ELD), Station 11, CH–1015 Lausanne, Switzerland

Hazım Kemal Ekenel
Faculty of Computer and Informatics, Istanbul Technical University (ITU), 34469, Maslak, Istanbul, Turkey

Tejas Godambe, Sai Krishna Rallabandi and Suryakanth V. Gangashetty
International Institute of Information Technology Hyderabad, Hyderabad, India

Ashraf Alkhairy
King Abdulaziz City for Science and Technology, Riyadh, Saudi Arabia

Afshan Jafri
King Saud University, Riyadh, Saudi Arabia

Omurcan Kumtepe and Gozde Bozdagi Akar
Department of Electrical and Electronics Engineering, Middle East Technical University, Dumlupınar Blv. No:1, 06800 Ankara, Turkey

Enes Yuncu
ISSD Electronics, METU Technopolis, Dumlupınar Blv. No:1, 06800 Ankara, Turkey

Gustaf Kylberg
Vironova AB, Gävlegatan 22, SE-11330 Stockholm, Sweden

Ida-Maria Sintorn
Vironova AB, Gävlegatan 22, SE-11330 Stockholm, Sweden
Uppsala University, Department of Information Technology, Lägerhyddsvägen 20, SE-75105 Uppsala, Sweden

Kazunori Uruma, Tomohiro Takahashi and Toshihiro Furukawa
Tokyo University of Science, 162-8677 Tokyo, Japan

Katsumi Konishi
Kogakuin University, Tokyo, Japan

Javier Tejedor
GEINTRA, Universidad de Alcalá, Campus Universitario. Ctra. Madrid-Barcelona, km.33,600, Alcalá de Henares, Madrid, Spain

Doroteo T. Toledano
Biometric Recognition Group - ATVS, Universidad Autónoma de Madrid, Av. Francisco
Tomás y Valiente, 11. Escuela Politécnica Superior, Madrid, Spain

Carmen Garcia-Mateo, Paula Lopez-Otero and Laura Docio-Fernandez
Multimedia Technologies Group (GTM), AtlantTIC Research Center, E. E. Telecomunicación, Campus Universitario de Vigo, s/n, Vigo, Spain

Index

Printed in the USA
CPSIA information can be obtained
at www.ICGtesting.com
JSHW052024301024
72690JS00004B/156

9 781632 385291